PRAISE FOR

# Rick Atkinson's *Crusade*

"If you are going to read only one book about the Persian Gulf War, *Crusade,* because of its rich human detail and skillful writing, should be it . . . Atkinson has taken the story of the war to the next level, portraying the passion — anger, fear, pride, ambition, egotism, sorrow — that are equally inevitable in the horrible crucible of war . . . A marvelous book for the ordinary reader and the military buff alike." — *San Diego Union-Tribune,* front page review

"Transfixes us with a crackling-fast, richly reported, absorbing, you-are-there techno-thriller of a military history of this war . . . *Crusade* is paced so as to hurl the reader from one scene to the next . . . you will not be tempted to put it down."
— *Los Angeles Times Book Review,* front page review

"Atkinson pierces the Pentagon's veil of secrecy and gives us a first-rate book about how the war was fought . . . An important book about the American military's forced march from the shame and ashes of Vietnam to the pride and parades of Desert Storm."
— *Chicago Tribune,* front page review

"No one could have been better prepared to write a book on Desert Storm, and Atkinson's *Crusade* does full justice to the opportunity."
— *Wall Street Journal*

"Brilliantly written and rich in important detail and incisive criticism, Atkinson's *Crusade* is an indispensable work that not only shows us how a war was won, but how the U.S. military finally recovered from the debacle in Vietnam." — *San Francisco Chronicle*

"Rick Atkinson . . . has written a superlative account of [the Gulf War]. Singular for its exhaustive, balanced, in-depth portrayal of the multi-

tudinous facets of the conflict, *Crusade* reigns supreme as the best single-volume account to date . . . Certainly *Crusade* sets the standard against which any other military history will have to be measured."
— *Airpower Journal*

"With *Crusade,* [Atkinson] has moved into the forefront of America's military historians. *Crusade* is vital, sweeping, and powerful. [It is] a frank, penetrating and ultimately compelling account of what really happens when the only superpower on Earth sends the best-trained, best-motivated and best-equipped warriors in the world out into the desert — to do our fighting and dying for us."
— *Tampa Tribune-Times*

"Rick Atkinson's *Crusade* is a skillfully crafted, often lyrical piece of work . . . [It] seeks to be many things. It seeks to be news. It seeks to be literature. It seeks to be history. It seeks to be drama. It seeks to be analysis. Surprisingly, it succeeds." — *Kansas City Star*

"Consistently absorbing . . . Military history of a very high order."
— *Kirkus Reviews*

"Atkinson is a compelling writer with a surgeon's attention to detail and a historian's insistence on facts."
— *Virginian-Pilot and Ledger-Star*

"Atkinson . . . succeeds marvelously in recounting a host of untold stories in *Crusade.* It is a narrative steeped in detail, studded with military acronyms, yet altogether absorbing."
— *Hartford Courant*

"First and foremost Rick Atkinson is a journalist with a reporter's eye for detail and nuance. Add to that a large dose of patience and a writing style akin to the best of the techno-thriller authors and you get a book of impressive reportage engagingly told." — *Denver Post*

"Rick Atkinson . . . has written the book that will be the *Guns of August* for the U.S. operations in Saudi Arabia, Kuwait and Iraq. It commands the field and likely will continue to do so for some time."
— *Military Review*

# CRUSADE

Books by Rick Atkinson

*The Long Gray Line*

*Crusade: The Untold Story of the Persian Gulf War*

RICK ATKINSON

# CRUSADE

## The Untold Story of the Persian Gulf War

HOUGHTON MIFFLIN COMPANY
Boston • New York

*Library of Congress Cataloging-in-Publication Data*

Atkinson, Rick.
Crusade : the untold story of the Persian Gulf War / Rick Atkinson.
p. cm.
Includes bibliographical references and index.
ISBN 0-395-60290-4 ISBN 0-395-71083-9 (PBK.)
1. Persian Gulf War, 1991 — United States. 2. United States — History, Military — 20th century. I. Title.
DS79.724.U6A887 1993 93-14388
956.7044'2373 — dc20 CIP

Printed in the United States of America

The maps on pages 513–522 were prepared by Brad Wye.

Book design by Robert Overholtzer

MP 10 9 8 7 6 5 4 3 2 1



Printed on recycled paper

*For Jane and Rush and Sarah, yet again*

# Contents

# Illustrations

"Nothing except a battle lost can be half so melancholy as a battle won."

— *Arthur Wellesley, Duke of Wellington, 1815*

"Every war is its own excuse. That's why they're all surrounded with ideals. That's why they're all crusades."

— *Karl Shapiro, American poet, 1964*

# Prologue

## Safwan, Iraq

The first sound they heard, fittingly, was that familiar din from the last, lost war — the deep *whomp whomp whomp* of helicopter blades beating northward. Half a dozen armed Apaches, darting around a larger Blackhawk like courtiers around their monarch, emerged from the low surf of oil smoke. Veering wide of the armistice tents on the northern stretch of runway, the procession touched down in a furious boil of sand. Soldiers ringing the airfield drew themselves up and squared their shoulders. A murmur ran among them: *Schwarzkopf's here.*

He peeled the radio headset from his ears and heaved himself out of the Blackhawk's bay. The dying rotors churned tiny dust devils around his desert boots. From their distance, the troops took his measure: a great slab of a man, grinning, tented in mottled fatigues, four black stars stitched on his collar. The bill of a round campaign cap shaded his face from the morning sun. A canteen sloshed on his right hip, counterweighted by a holstered pistol on his left. All in all, he cut the very image of an American Mars. In an earlier age, the men might have huzzahed themselves hoarse at the appearance of their victorious commander. Instead, modern decorum kept them mute, and they simply grinned back.

H. Norman Schwarzkopf was the most theatrical American in uniform since Douglas MacArthur, and he strode across the runway like an actor pressing toward the footlights. Delta Force bodyguards, strapping men with automatic weapons, scrambled to form his personal picket line. Possessed of a deep love for ceremony and ritual, he surveyed the set of the spectacle — designed largely by him — that was now about to unfold, a scene drawn in spirit if not in particulars from the parlor at Appomattox, the rail car at Compiègne, and the polished deck of the U.S.S. *Missouri* in Tokyo Bay.

Except for a low bluff to the southwest — known to the Americans as Scud Mountain — the desert pan was drear and barren, a thousand shades of buff. At center stage stood the tents, including the three general-purpose mediums hastily pieced together near the runway to form a single chamber for the talks. Dozens of tanks and armored personnel carriers, their tracks spattered with the mud of occupied Iraq, lined the runway and the narrow road running east to the Basrah highway. A quartet of A-10 warplanes cut lazy circles at four thousand feet. The canisters of a Patriot missile battery stood cocked and waiting south of the airstrip. Artillerymen once again checked the azimuth and elevation of their tubes. The enemy delegates would have no doubt about who had won.

Schwarzkopf's generals gathered round him, swapping salutes and hail-fellow handshakes: Gus Pagonis, the animated Greek logistician; the firebrand Tom Rhame, commander of the 1st Infantry Division; soft-spoken Fred Franks, three-star commander of VII Corps, whose rolling gait bespoke the leg lost in Cambodia two decades before. All had been junior officers in Southeast Asia, forever seared by the war and the hard peace that followed. They had stayed the course after Vietnam, vowing to restore honor and competence to the American profession of arms and, most important, to renew the bond between the Republic and its soldiery. This — Safwan, March 3, 1991 — was their vindication. For Norman Schwarzkopf and his lieutenants, this war had lasted not six weeks, but twenty years.

Schwarzkopf ambled around the armistice tent, nodding his approval. Ducking through the flaps, he eyed the wooden table on which a young major stood, fiddling with the fluorescent lights overhead. "I'm not here to give them anything," he boomed, gesturing toward the chairs where the Iraqis soon would sit. "I'm here to tell them exactly what they have to do." Re-emerging into the wan sunlight, he tromped about the bivouac, happy warrior, stern proconsul.

His satisfaction was fairly won. The last time a Western army had invaded Iraq, the British marched up the Tigris in 1914 and died by the thousands from heat stroke and sickness and Turkish bullets. During the air campaign and land invasion commanded by Schwarzkopf, fewer than three hundred allied soldiers had perished. Among lopsided routs, the victory ranked with Omdurman, where the British and Egyptians in 1898 slew or wounded twenty thousand Dervishes on the banks of the Nile, or Jena, where Napoleon in 1806 won two simultaneous battles, pursued his foes to the shores of the Baltic, and captured 140,000 prisoners.

In Schwarzkopf, the Persian Gulf War would provide America with its first battlefield hero in decades. He had crushed the army of Saddam Hussein at minimal cost, committing no significant error of strategy or tactic. He showed tenacity and fixed purpose, as a good commander must. He also possessed the cardinal virtue of detesting war. Flying north this morning from Kuwait City, he had seethed at the sight of the shattered city below, darkened by the smoke from hundreds of sabotaged oil wells. For the first time, Schwarzkopf had personally witnessed the havoc wreaked by his forces: the endless miles of blackened tanks and trucks, the demolished revetments, the cruciform smudges that had once been Iraqi airplanes. Later he would liken the trip to a flight into hell.

But war was a hell he knew intimately: during a thirty-five-year Army career, including two tours in Vietnam, he had been wounded twice. Retaining a junior officer's feel for the battlefield consequences of his decisions in the gulf campaign, he had adroitly banked "the roaring flux of forces aroused by war" to prevent killing from becoming witless slaughter. The troops revered him, shortening his formal title — commander-in-chief, Central Command — to simply "the CINC." Feared by his enemy, lionized by his nation, Schwarzkopf stood, in the admiring assessment of the British commander Sir Peter de la Billière, as "the man of the match."

And yet what anguish he had caused! He was, as George C. Marshall wryly said of MacArthur, "conspicuous in the matter of temperament." In the cloister of his Riyadh war room, the avuncular public mien disappeared, revealing a man of volcanic outbursts. "That is a stupid idea! You're trying to get my soldiers killed!" he would bellow at some cringing subordinate. During the previous six months, obliquely or directly, he had threatened to relieve or court-martial his senior ground commander, his naval commander, his air commanders, and both Army corps commanders. Secretary of Defense Richard B. Cheney had worried sufficiently about Schwarzkopf's temper and his yen for imperial trappings to consider the possibility of replacing him.

His rage for order — he had mailed home Christmas presents with color-coded wrapping and explicit instructions for their sequence of opening — at times yielded to fury at perceived idiocies and malfeasances, small and large. He raved at the inadequacy of desert boots, at Central Intelligence Agency impertinence, at the ponderous pace of the Army's attack, at the Navy's bellicosity, at Pentagon intrusion. He railed at bad weather.

His headquarters, swept with his verbal grapeshot month after month, became a dispirited bunker, where initiative withered and even senior generals hesitated to bring him unpleasant tidings. Instead, when the tirades began they sat with eyes glassy and averted in what came to be called the "stunned mullet look," until his fury spent itself. In a war with little bad news the sin was forgivable; but even for men who had seen horrific bloodletting in Vietnam, no Asian jungle was more stressful than the endless weeks they spent in Norman Schwarzkopf's Riyadh basement.

For now, however, just three days after the cease-fire, the CINC and his legions could revel in their triumph. Only eight months earlier, on August 2, 1990, Kuwait had been overrun by an Iraqi juggernaut seemingly poised to conquer all of the Arabian peninsula. From a token counterforce of a few thousand soldiers and a few dozen warplanes, the allied host had grown to nearly three quarters of a million troops in an expeditionary masterpiece rivaling the invasion of Normandy. Through months of anguished debate and frantic diplomacy, the world had coalesced around President George Bush's pronouncement that the Iraqi aggression "will not stand." Americans sloughed off their uncertainty, steeled themselves for a war few wanted, and then watched in approval as allied troops fell on the enemy with bloody vengeance.

The consequences seemed as evident as the wreckage strewn from Baghdad to the Saudi border. Kuwait was liberated, Saudi sovereignty assured, Persian Gulf oil secure. An army touted as the world's fourth largest had been smashed. Saddam had been stripped of his conquest, his pretensions, and much of his arsenal. Euphoria swept the United States, renewing the national spirit and reviving an indomitable zeal. As Schwarzkopf stood astride the battlefield, so America seemed to stand astride the world.

This shining moment would quickly pass. All too soon a certain hollowness would set in at home: disdain for a feeble foe; disaffection at the terms of peace; a preoccupation with domestic ills that would disprove Theodore Roosevelt's bromide that "a just war is in the long run far better for a man's soul than the most prosperous peace." A nation accustomed since 1865 to victories of annihilation, with the enemy left prostrate and humble, would chafe at the inconclusive flutter of a limited triumph. A president who had insisted on inflating a military campaign on behalf of national interests into a moral crusade would be trapped by his own rhetoric, unable to explain how he could leave in power a man he had branded the moral equivalent of Adolf Hitler.

Within a year the war would be widely regarded as inconsequential, even slightly ridiculous. An event of the greatest "moral importance since World War II," in Bush's overheated assessment, would dwindle in public consciousness to an expedition of flimsy achievements launched under dubious pretexts — a footnote, a conflict as distant as the Boxer Rebellion of 1900. For most Americans, the Persian Gulf War became irrelevant. A sequence of raids and skirmishes in January 1993, during the waning hours of Bush's presidency, seemed to underscore the futility of the campaign that had ended two years earlier.

Only through the lens of history could the significance of the war be seen as clearly as its shortcomings. Iraq remained unrepentant and Saddam remained in power, depriving the victors of laurels like those won in the conquest of Japan and Germany nearly half a century before. Yet this was a different sort of war and a different sort of victory. The gulf conflict foreshadowed both the promise and limitations of a new collective security order. It also foretold, in the display of smart bombs and stealthy airplanes, how men would kill each other in the twenty-first century. The conflict reaffirmed war as a means of achieving national objectives, and revealed the resurgent vigor of the American military after a generation of convalescence.

Certainly not least, the war neutered a despot, thwarted his ambitions to control the world's pre-eminent oil patch, and denied him a nuclear arsenal. An outlaw nation had been confronted and then crushed by a world unified as never before by enlightened self-interest. If Saddam was no Hitler, the war ensured that he would not become one.

Although flawed, the Persian Gulf campaign boasted accomplishments of which the nation and its allies could be proud. Yet these achievements would be all but forgotten in the discontent that soon besmirched the victory.

Shortly after 11 A.M., a stirring among the soldiers and waiting journalists signaled the arrival of the Iraqis at Safwan. A brigade commander from the 1st Division had intercepted the delegation at a highway cloverleaf six miles to the north. As ordered, the enemy generals abandoned their tanks and trucks and clambered into four American Humvees, the latter-day jeeps. Led by two Bradley armored personnel carriers, trailed by two Abrams tanks, the convoy raced south at thirty-five miles per hour, then west, then onto the runway. A pair of Apaches, skittering overhead, flanked the procession as it rolled up the tarmac between the parallel ranks of American armored vehicles, their flags snapping in the breeze, their crews ordered not to smile.

Schwarzkopf waited outside the main tent. "I don't want them embarrassed. I don't want them humiliated. I don't want them photographed," he told his officers. The Iraqis, clad in black berets and heavy green jackets, climbed from the Humvees and marched through a gap in the concertina wire girdling the tents. Two wore the shoulder epaulets of lieutenant generals: embroidered crossed scimitars with a five-pointed star and a falcon in silhouette. Baghdad had initially proposed sending several lower-ranking officials to Safwan. But under pressure from the Saudis and Americans, the Iraqis had upgraded the delegation to include Lieutenant General Sultan Hashim Ahmad, the thickset, mustachioed deputy chief of staff from Iraq's Ministry of Defense, and Lieutenant General Salah Abud Mahmud, commander of the recently demolished Iraqi 3rd Corps.

Schwarzkopf announced through an interpreter that all delegates and observers would be searched. General Ahmad balked, protesting that the Iraqis had left their weapons behind. Only when Schwarzkopf agreed to be frisked himself did Ahmad yield.

"Follow me," Schwarzkopf ordered, leading the Iraqis into the search tent. A military policeman ran a metal-detecting wand lightly up the CINC's legs and torso, then repeated the procedure with each Iraqi.

Behind this orderly scene lay one of Schwarzkopf's more operatic rages. As the war entered its final hours, he had ordered that the road junction near Safwan airfield be captured and had been assured that it was by the senior Army commander, Lieutenant General John Yeosock.

But when the cease-fire took effect, at 8 A.M. on February 28, the 1st Infantry Division, which was responsible for that sector of the battlefield, remained ten miles from the intersection. In the confusion of the final attack — there were misunderstandings about the cease-fire deadline and a bogus report of friendly artillery fire that temporarily brought VII Corps, including the 1st Infantry, to a halt — the Big Red One had come up short. The division's attack helicopters controlled the sector from above, but the only troops physically occupying the airfield and road junction wore Iraqi green.

After the cease-fire, Schwarzkopf, unaware of the gap, formally proposed Safwan as a site for the armistice. General Colin L. Powell, chairman of the Joint Chiefs of Staff, agreed. He subsequently informed first Cheney and then the White House of the choice (carefully noting that "it's 'Safwan,' not 'saffron' "). Shortly before dawn on March 1, Yeosock learned of the true disposition of his forces from General Franks, the VII Corps commander whose units had been responsible for the main attack against the Iraqi army. Yeosock called Schwarzkopf. "The decision we made about Safwan is probably not a good one," he began. "We don't own Safwan."

A deep flush rose from Schwarzkopf's neck, purpling his face. "What do you mean, 'we don't own Safwan'? You assured me that you had that road junction." The map in Central Command's war room clearly showed the area under U.S. Army occupation.

"We have it under observation," Yeosock replied, "but we don't control it. There are some enemy forces down there."

"Goddam it, that's false reporting!" Schwarzkopf bellowed, each syllable exploding into the telephone. "That's a direct violation of my orders."

Yeosock fumbled for an explanation, but Schwarzkopf cut him off. He had been angry at the Army almost from the onset of the ground attack a week before, and this mishap triggered new paroxysms of wrath. Staff officers in the war room prudently edged away.

After threatening to replace the Army with U.S. Marines and further railing at Yeosock, Schwarzkopf called Powell in Washington.

"Goddam it," he told the chairman, "we don't own the goddam place!"

"Relax, Norm," Powell said. "What the fuck's the matter?"

Calming his commander enough to get a coherent account of the mix-up, Powell agreed that the solution was to eject the enemy from Safwan, cease-fire notwithstanding. Soon, an armored brigade from the 1st Division rumbled forward toward the road junction, encircling the baffled Iraqis, who refused to retreat. For several hours the deadlock persisted. In Riyadh, Schwarzkopf's anger flared with each report. He ordered Franks and Yeosock to write detailed explanations of "why my orders were countermanded." At 1:30 P.M., Yeosock's operations officer, Brigadier General Steve Arnold, scribbled in his log: "CINC has repeatedly threatened to relieve Franks, Yeosock, et al."

Division commander Rhame met with his brigade commander to relay an ultimatum from Schwarzkopf. Tell the Iraqi general, Rhame warned, "that if he doesn't leave by 1600 [4 P.M.] you're going to kill him. You're going to kill all his forces and attack right through him." Shortly before dusk, the Iraqis pulled back. Safwan fell to the Americans.

However justified Schwarzkopf's anger, the episode cast a small shadow over his commanders' exhilaration. As a fleet of Chinook helicopters began to ferry tents, tables, and generators to the captured airfield, Franks retired to the privacy of his command vehicle southwest of Safwan to write Schwarzkopf the required apology.

"Any failure of VII Corps units to seize the road junction with ground forces is mine," Franks wrote. "There was never any intention to disobey orders."

Two days earlier, the corps commander had orchestrated five divisions against the Iraqi Republican Guard with a precision that many believed among the most extraordinary maneuvers in the history of armored warfare. Now he was just another target for one of Schwarzkopf's rages. The VII Corps operations officer, Colonel Stan Cherrie, thought Franks "looked like a whipped dog."

Normally a man of impermeable reserve, Franks permitted himself a small outburst. "My God," he told Cherrie, "how could he not know we were trying to do what was right?"

Even before the talks started, the crowded tent became stuffy. Against one canvas wall, behind Schwarzkopf and Lieutenant General Khalid bin Sultan, the Saudi commander, sat two dozen American and allied officers. Across the table — on which were bottled water, white coffee cups, and notepads — sat the two Iraqi generals and their interpreter, looking like defendants in the dock. Another half-dozen Iraqi officers sat behind Ahmad and Mahmud. Schwarzkopf briefly admitted the press for a few photographs, then opened the talks at 11:34 A.M. He reminded General Ahmad that the agenda had already been set by Washington and sent to Baghdad.

"The purpose of the meeting," the CINC said, "is to discuss and resolve conditions that we feel are necessary to ensure that we continue the suspension of offensive operations on the part of the coalition." Only military matters would be discussed at Safwan, he added. A formal cease-fire, covering such issues as war reparations and the inspection of Iraqi weapons plants, would be negotiated later in the month through the United Nations.

Ahmad nodded, smiling slightly. "We are authorized to make this meeting a very successful one in an atmosphere of cooperation," he said. "We can start."

Quickly the delegates ticked through the agenda: repatriation of prisoners, the return of bodies, the marking of minefields. Schwarzkopf, who could swagger sitting down, was polite, even deferential toward the Iraqi generals. After deflecting complaints about a bloody postwar attack by the 24th Infantry Division against fleeing Iraqis trying to shoot their way north through the Euphrates Valley, he unveiled a map marked with a proposed cease-fire boundary. The Iraqis appeared stunned at the extent of the allied occupation, which stretched from the Persian Gulf below Basrah halfway to Jordan. Nearly a fifth of Iraq lay in Schwarzkopf's possession.

"We have agreed that this is not a permanent line?" asked Ahmad, his eyes narrowing with suspicion.

"It is absolutely not a permanent line," Schwarzkopf replied.

"And this has nothing to do with any borders?"

"It has nothing to do with borders. I also assure you that we have no intention of leaving our forces permanently in Iraqi territory once the cease-fire is signed. But until that time we intend to remain where we are."

Ahmad returned to the issue of war prisoners. How many Iraqis were in allied custody?

Schwarzkopf pronounced the number distinctly so that it could not be misunderstood. "We have, as of last night, sixty thousand. Sixty thousand plus." (In fact, the figure was closer to seventy thousand.)

Again the Iraqi general winced in disbelief. Is this possible? he asked, turning to General Mahmud. The corps commander shrugged. "It's possible," he answered. "I don't know."

"Are there any other matters that the general would like to discuss?" Schwarzkopf asked.

Ahmad nodded. "We have a point, one point. You might very well know the situation of our roads and bridges and communications. We would like to agree that helicopter flights sometimes are needed to carry officials from one place to another because the roads and bridges are out."

Schwarzkopf hesitated for an instant. That morning, flying north from Kuwait City, he had growled, "They better not ask for much because I'm not in a very charitable mood." But the talks were going well. The Iraqis had agreed to abide by the cease-fire line, to disclose the location of all mines, and to release immediately all prisoners. The CINC could risk a little benevolence.

"As long as the flights are not over the part we are in, that is absolutely no problem," Schwarzkopf agreed. "I want to make sure that's recorded: that military helicopters can fly over Iraq. Not fighters, not bombers."

Ahmad looked skeptical. "So you mean even armed helicopters can fly in Iraqi skies?"

"I will instruct our Air Force not to shoot at any helicopters that are flying over the territory of Iraq," Schwarzkopf assured him.

Ahmad agreed that no Iraqi helicopters would cross over allied-occupied territory. "Good," Schwarzkopf replied.

The talks drew to a close, but not before the Iraqis again pressed the American on the cease-fire boundary. "You have my word, I swear to God," Schwarzkopf insisted. "There is no intention of that being a permanent line at all."

The delegates stood and filed somberly from the tent into the crisp midday air. The CINC escorted Ahmad to the Humvee that would carry him back to the Iraqi lines. In a conversation the night before with General Khalid, Schwarzkopf had vowed to the Saudi commander that he would not shake hands with the Iraqi delegates. But when Ahmad saluted and extended his open palm, the CINC returned the salute and shook the proffered hand, a conciliatory gesture that later infuriated senior Pentagon officials. "As an Arab," Ahmad said, climbing into the vehicle, "I hold no hate in my heart."

Schwarzkopf watched the convoy pull away. It was over now, truly over. Whatever tasks remained belonged largely to the diplomats and logisticians. In most respects the conflict had been "a splendid little war," as John Hay, the U.S. ambassador to Great Britain, once described the American pummeling of Spain nearly a century before. Yet like all wars, it had been a pageant of cunning and miscalculation, of terror and exhilaration, of feuding and enmity and the clutching love between comrades that blooms only on battlefields.

At 3:30 P.M., Schwarzkopf marched with his retinue across the runway and hoisted himself into the left rear seat of the helicopter. The engine whine built to a scream. He clipped the shoulder harness and slipped on a headset. Then the tailboom canted up and the Blackhawk lumbered into the air, again escorted by Apache outriders. On the ground, soldiers preparing to strike the tents listened as the spanking blades grew fainter. The helicopter plunged southward. In minutes the dark oil cloud on the horizon had drawn around the speck like a closing curtain.

PART I

# FIRST WEEK

# 1

# First Night

## U.S.S. *Wisconsin,* Persian Gulf

Dark tubes of water peeled back from the battleship's prow, curling along her hull before fanning symmetrically east and west toward the horizons. Watchstanders on the bridge peered fore and aft, checking the navigation lights of the other warships sailing with her at six-mile separations: red to port, green to starboard. Overhead, stars jammed the moonless sky with such intensity that they seemed to hang just beyond the upper poke of *Wisconsin*'s superstructure.

The crew stood at general quarters. Earlier in the day — Wednesday, January 16, 1991 — they had scrubbed the teak deck, scoured the gutters, polished the brass fittings, and swept the corridors. Under Condition Zebra, all watertight doors were latched shut. In the officers' mess, seamen had lifted the ship's silver punch bowl from a glass display case and stowed it in a wooden crate. Even the trash was collected, the bags then punched with holes — so that they would sink and not be mistaken for mines — and heaved overboard. Over the public address system, the Roman Catholic chaplain absolved the crew of sin, then hurried to his office for a box of plastic rosaries and a flask of oil to use in anointing the dying. Now *Wisconsin* waited for war with dreadnought forbearance, silent except for the throb of her four great screws turning beneath the fantail.

Below decks, in the soft blue light of the ship's Strike Warfare Center, thirty men prepared the battleship for combat. In contrast to the tranquillity above, tension filled the crowded room. Electronic warfare specialists listened on their headsets for the telltale emissions of attacking enemy aircraft or missiles. Other sailors manned the radios, the computer consoles controlling *Wisconsin*'s Harpoon antiship missiles, and a dozen other battle stations. A large video screen overhead displayed the radar blips of vessels crossing the central gulf; a smaller screen showed the charted positions of *Wisconsin* and her sister ships, plotted and replotted by a navigation team.

In the center of Strike the battleship's skipper, Captain David S. Bill, perched in his high-backed padded chair. Although he occasionally glanced at the screens above, the captain's attention was largely fixed on the men clustered around four computers lining the far bulkhead. Something had gone awry with the ship's Tomahawk missile system.

Lieutenant Guy Zanti, *Wisconsin*'s missile officer, leaned over a crewman sitting at one of the consoles. "The launch side still won't accept the data," the sailor said glumly. He tapped his keyboard and pointed to the green message that popped onto the monitor. "See, it says 'inventory error.' " Zanti nodded, his forehead furrowed in concentration. Now not only Captain Bill but everyone else in Strike turned to watch the lieutenant and his missile crew.

For nearly six months, *Wisconsin* had prepared for this moment. Five days after the invasion of Kuwait, she had weighed anchor from Norfolk, Virginia, quickly steaming through the Straits of Gibraltar and the Suez Canal to arrive on station off the Saudi coast on August 24. As Gulf Papa, the coordinator of Tomahawk launches from the Persian Gulf, *Wisconsin* was responsible for the seven warships that would shoot an initial salvo of four dozen missiles from the gulf. The targets and their ten-digit authorization codes had arrived with a tinkling of teletype bells just after sunset on January 16. A half-dozen officers and crewmen spent the evening drafting instructions for the other shooters, carefully choreographing their movements so that each ship would steam into the proper launch basket at the correct time. *Wisconsin* would fire first in half an hour; her initial Tomahawk was scheduled to rocket from the gray launcher box at 1:37 Thursday morning for the ninety-minute flight to Baghdad.

But now Gulf Papa faced imminent failure. For reasons no one in Strike could fathom, the Tomahawk computers seemed confused, refusing to transfer the necessary commands from the engagement-planning console to the launch console. The resulting impasse — "casualty," in Navy jargon — meant the missiles could not be fired.

Again the missile crew ran through the launch procedures. All switches were properly flattened, all electrical connections secure. The console operator reloaded the software program and tried once more. Again the infuriating message popped onto the screen: "Inventory error." Still the captain seemed unfazed, as though this was just another repeat of Nemean Lion, the Tomahawk launch exercises — named for the mythical beast slain by Hercules as the first of his twelve labors — that the Navy had practiced before the war. But disquiet spread through the crowded room; one officer's jerky motions and rising voice grew agitated. "Keep your head together," Lieutenant Zanti snapped. "Let's think the problem through."

Failure here, they all knew, would be very bad, not only for the war plan but for the Navy. Skepticism about the Tomahawk Land Attack Missile, or TLAM, was rampant in the military, even among some naval officers. Although more than a hundred missiles had been fired in exercises — including one recently shot from the Pacific at a target in Nevada — none had flown in combat. The closest a Tomahawk had come to being fired in anger was in August 1989, when the United States edged to within hours of attacking Hezbollah camps in Lebanon after the kidnapers of Joseph Cicippio threatened to execute the hostage.

Perhaps the greatest — certainly the ranking — skeptic was the chairman of the Joint Chiefs. Alternately fascinated by and distrustful of the weapon, General Colin Powell in October had warned Norman Schwarzkopf's chief targeteer, "I don't give a damn if you shoot every TLAM the Navy's got, they're still not worth a shit. Any target you intend to destroy with the TLAM, put a fighter on it to make sure the target's destroyed." Tomahawk's role in the attack planning had grown and diminished along with prevailing military confidence in the weapon. The Navy had finally pulled together eight years of test data, sketched a diagram of a baseball diamond, and vowed that if the target was the pitching rubber, the overwhelming majority of warheads would detonate within the perimeter of the base paths, even after a five-hundred-mile journey.

Yet other complications persisted. The gray steel boxes housing the missiles topside contained secrets to which few men were privy. One secret — which would remain classified even after the war — was the route the Tomahawks would fly to Baghdad. The missile's navigation over land was determined by terrain-contour matching, a technique by which readings from a radar altimeter were continuously compared with land elevations on a digitized map drawn from satellite images and stored in the missile's computer. Broken country — mountains, valleys, bluffs — was required for the missile to read its position and avoid "clobbering," plowing into the ground.

For shooters from the Red Sea, the high desert of western Iraq was sufficiently rugged. But for *Wisconsin* and other ships firing from the Persian Gulf, most of southeastern Iraq and Kuwait was hopelessly flat. After weeks of study, only one suitable route was found for Tomahawks from the gulf: up the rugged mountains of western Iran, followed by a left turn across the border and into the Iraqi capital. Navy missile planners in Hawaii and Virginia mapped the routes and programmed the weapons. They also seeded the missiles' software with a "friendly virus" that scrambled much of the sensitive computer coding during flight in case a clobbered Tomahawk fell into unfriendly hands. A third set of Tomahawks, carried aboard ships in the Mediterranean, were assigned routes across the mountains of Turkey and eastern Syria.

Not until a few days before the war was to begin, however, had the White House and National Security Council suddenly realized that war plans called for dozens and perhaps hundreds of missiles to fly over Turkey, Syria, and Iran, the last a nation chronically hostile to the United States. President Bush's advisers had been flabbergasted. ("Look," Powell declared during one White House meeting, "I've been showing you the flight lines for weeks. We didn't have them going over white paper!") After contemplating the alternative — scrubbing the Tomahawks and attacking their well-guarded targets with piloted aircraft — Bush assented to the Iranian overflight. Tehran would not be told of the intrusion. But on Sunday night, January 13, Bush prohibited Tomahawk launches from the eastern Mediterranean; neither the Turks nor the Syrians had agreed to American overflights, and the president considered Turkey in particular too vital an ally to risk offending.

Now it was the Navy's turn to be surprised. Again communications broke down: planners on the Navy flagship U.S.S. *Blue Ridge* learned of the White House prohibition less than four hours before the first launch was to take place. With frantic haste the *Blue Ridge* planners cut new orders, redistributing the Mediterranean shooters' targets to ships in the Persian Gulf and Red Sea, thus increasing the workload of each task force by a third.

On *Wisconsin,* where the scheduled launch time was now just moments away, the men in Strike were running out of solutions. "All right," Lieutenant Zanti announced, "we'll start from the beginning." The data for the eight *Wisconsin* shots — three pages of detailed coding for each missile — would be retyped into the computer. The task was tedious and time-consuming. He turned to Captain Bill and the ship's weapons officer.

"Sir, we need to ask for more time," Zanti told Bill. "If we don't get an extension, we can't shoot."

The captain agreed. As the request flashed up the chain of command to *Blue Ridge,* an excited voice from one of Gulf Papa's nearby shooters crackled through Strike over the radio intercom: "Alpha, alpha. This is the *Paul F. Foster.* Happy trails."

*Happy trails:* the code phrase for missile away. The war had begun without *Wisconsin.* Deep within the battleship the missilemen labored over their keyboards, clicking furiously.

## Ar Ar, Saudi Arabia

Barely seventy-five feet above the dark Nafud, one of Saudi Arabia's three great deserts, the helicopters pushed toward the border in a line as straight as monks filing to vespers. A gap precisely five rotor discs' wide separated each aircraft from the next. Two Air Force Pave Lows stuffed with sophisticated navigation gear led as pathfinders, followed by four Army Apaches, laden with rockets and missiles and extra fuel tanks.

Frigid night air gushed into the lead Apache. The pilot, Warrant Officer Thomas R. (Tip) O'Neal, fumbled with the heating controls. The flapper valves on the helicopter's filtration system seemed to be wedged open, apparently jammed with sand. As O'Neal pressed a gloved hand against the vent, his co-pilot, Warrant Officer David A. Jones, came on the intercom from the back seat. "Tip, you see that glow off to the north? That might be it."

O'Neal scanned the horizon through his night-vision goggles. The headset had two protruding lenses that amplified the ambient starlight to give even the darkest landscape a crepuscular definition. He saw it now, a hazy splotch far ahead. But they were still twelve miles south of the border — they'd just skirted the Saudi town of Ar Ar — and the target lay another thirty miles into Iraq.

Abruptly O'Neal's goggles flushed with light, like small starbursts blooming around the helicopter. "What the fuck is that?" he called.

Jones, now concentrating on the Apache's infrared scope, which registered heat emanations rather than visible light, saw nothing. "What? What?"

"That!" O'Neal insisted. "Down there! God!" He pushed the goggles up, his naked eyes straining through the darkness. Machine gun fire poured from the Pave Low just in front of them. Two streams of bullets slanted down and beneath the Apache. "Don't worry about it, Dave," he said with relief. "It's just the Pave Low clearing its guns."

Three helicopters back, in the trail Apache dubbed *Rigor Mortis,* Lieutenant Colonel Dick Cody knew better. He had clearly seen the first burst of fire from below, followed by a missile streaking just abeam of the line of aircraft. "Jeez, Brian," Cody called to his co-pilot, "did you see that?" The gunfire had come either from nervous Saudis or, more likely, an Iraqi commando patrol aiming at the rotor noise. After the brief retaliatory burst from the Pave Low, the shooting stopped. The helicopters pressed on at 120 knots.

Cody was not by nature a reflective man. Commander of the 101st Airborne Division's attack helicopter battalion, he was a creature of action and instinct, an aggressive pilot with fifteen years' flying experience. But he had occasionally wondered in the past four months whether he had oversold himself for this operation.

The mission planning had begun in late September, when Cody first met with Schwarzkopf's special operations commander, a grizzled Army colonel named Jesse Johnson. Here's the problem, Johnson had explained with conspiratorial zest: in the first minutes of the war, the Air Force intended to destroy Scud missile launchers threatening Israel from western Iraq. But Iraqi air defenses wove such a tight belt around the country's perimeter that bombers would be detected by radar as they crossed the border, providing as much as twenty minutes' warning before the American attack.

If two or three radar sites were destroyed and a keyhole cut through the warning net, the Air Force could strike before the Iraqis had time to defend themselves or launch their Scuds. (A smaller wave of F-117 stealth fighters, essentially invisible to radar, would angle toward Baghdad without the need for such assistance.) The options, Johnson added, included hitting the radars with Air Force bombers, striking with Special Forces troops on the ground, or attacking with Apaches.

Jesse Johnson believed helicopters were best suited to the task, but he wanted guarantees. In April 1980 he had commanded the security force for Operation Eagle Claw, the catastrophic attempt to rescue American hostages in Tehran. Nearly half the helicopters had malfunctioned on that mission, and eight men died when a helicopter collided with a fuel-laden C-130. "In any career you can only afford one time to be waiting in the desert for helicopters," Johnson told Cody. "I've already checked that block. We have to have one hundred percent success."

*One hundred percent success.* The phrase had haunted Cody for months. Schwarzkopf had shown keen personal interest in the mission, at one point — his alarmed subordinates reported — even flinging his glasses in anger after a misunderstanding about precisely who would be crossing into Iraq and when. Clearly, having the raiding party shot down on the Saudi border was not what Johnson or the CINC had in mind.

Cody had tried to anticipate every possible flaw in the plan, which now involved destroying two linked radar outposts twenty miles apart. After picking eight Apache crews for his strike force, Cody divided them into two teams, Red and White, and matched them to the targets. To double the flying time of his helicopters, he halved the number of rockets they would carry and installed an external fuel tank — to the chagrin of pilots leery of the 1500-pound gasoline bomb now bolted to each aircraft's belly. To gauge the reliability of Hellfire missiles, he test-fired a dozen at a fleet of old Saudi school buses, reducing the vehicles to a heap of springs and blackened chassis. To rehearse the plan, he scrutinized satellite photos, built a sand-table model of the targets, and made his pilots explain each and every detail of the attack.

On January 13, Cody and his men had flown far to the west, to their staging base at Al Jouf. This morning at 12:56, Team Red lifted off and headed northwest toward its target, code-named Nebraska. Seven minutes later Team White lifted off with Cody in trail, angling northeast toward target Oklahoma. Cody's orders were to destroy both Nebraska and Oklahoma at precisely 2:38 A.M. That meant having at least eight missiles in the air simultaneously at two locations separated by twenty miles of desert.

Cody checked his watch: just after 2 A.M. The border now lay well behind them. The attack plan called for the Pave Lows to lead the Team White Apaches northeast of the target, as if — should the raiders be spotted — they'd seem to be angling toward an Iraqi armored unit in that direction.

With languid deliberation the Pave Lows swung left in a lazy arc, pointing directly toward the target and dropping almost to the desert floor to mask their movement in the ground clutter. The bright haze from the Iraqi radar compound grew more vivid now, looming just over the horizon.

Two green bundles of chemical sticks tumbled from the Pave Lows, shimmering below like small piles of radium. The signal meant they were 12.5 miles from the target, close enough for the Apaches to find their own way; the Pave Lows hovered for a moment as if in silent salute, then peeled south toward Saudi Arabia.

The Apache pilots updated their navigation systems and began their final stalk. At nine kilometers out, Cody and his wingman moved up on the right. Jones and O'Neal shifted left in battle-spread formation. Four abreast, the helicopters flew toward the waiting glow in the west.

## Riyadh

The city was braced for war. Masking tape cross-hatched the smoked windows of apartment houses and government offices. Saudi soldiers on street corners crouched inside their sandbag bunkers, alternately fingering worry beads and the triggers of their machine guns. In the royal palace, King Fahd ibn Abdul Aziz waited with the appropriate anxiety of a monarch whose throne has just been tossed on the gaming table. He had learned of the imminent allied attack from his nephew and ambassador to the United States, Prince Bandar bin Sultan, who called from Washington with a coded message: "Our old friend Suleiman is coming at nine o'clock this evening. He's sick and I'll ship him out, and he'll get there at nine." As Fahd and Bandar had arranged, the king added six to the hour to calculate H-hour: 3 A.M. At midnight Fahd mustered his closest ministers — not an unusual summons for the nocturnal Saudis — and commanded that no one leave his sight until the first bombs fell.

Yet the center of the kingdom this night was not the palace but a cramped bunker, barely bigger than a tennis court, forty feet beneath the concrete fortress of the Ministry of Defense and Aviation. The select few authorized to enter rode an elevator down three flights, passed through two checkpoints and several steel vault doors, then walked down a fourth flight to the fluorescent netherworld of Norman Schwarzkopf's war room.

It was sparely furnished, stripped to fighting trim. Maps papered the four walls. A pair of television monitors, linked by closed circuit to the operations center next door, listed significant intelligence or combat actions under way. Three clocks told the hour in Riyadh, Washington, and Greenwich (known as Zulu time). Staff officers manned a horseshoe arrangement of desks behind a rectangular table where the senior American commanders sat.

Lieutenant General Calvin A.H. Waller leaned back in his brown leather chair, slightly smaller than Schwarzkopf's black leather throne — now vacant — on his immediate left. The CINC was never one to bungle an entrance, Waller knew. He would be here in due time, probably with a flourish befitting the occasion. Other officers began to enter the room, each bringing a fresh charge of tension. Waller had no doubt that the allies would win, but at what cost? He considered losses of ten to twenty allied aircraft likely tonight; others feared fifty.

Waller again scanned operations order 91–001, dated 17 January 1991. Allied objectives in the coming war had been boiled down to a single sentence: "Attack Iraqi political-military leadership and command-and-control; gain and maintain air superiority; sever Iraqi supply lines; destroy chemical, biological, and nuclear capability; destroy Republican Guard forces in the Kuwaiti Theater; liberate Kuwait." There it was, in thirty-four words. Unlike Vietnam, the mission was succinct, tangible, and limited — precisely the qualities demanded by the military of their civilian masters for twenty years. If things went awry this time, Waller knew, no one in uniform would be able to blame the country's political leaders for ambiguous guidance.

Lieutenant General Waller was a handsome, broad-shouldered man, born in Baton Rouge, much admired for his common touch and geniality. His Army commands had included an infantry brigade, the 8th Infantry Division, and, most recently, I Corps at Fort Lewis, Washington. Now, as deputy commander-in-chief of Central Command, he was Norman Schwarzkopf's principal lieutenant. In more than three decades of military service, this was Waller's fourth occasion to serve with Schwarzkopf, an honor he had stoutly resisted until Colin Powell himself called in early November and barked, "What's this about you not wanting to go to Saudi Arabia? Get your stuff packed." Yet nothing Cal Waller had seen since leaving his corps in the Pacific Northwest and arriving in Riyadh as DCINC had allayed his initial qualms.

Waller knew as well as any soldier that there was much to admire about Norman Schwarzkopf: intelligence, combat prowess, loyalty to his troops. But after two months in country, Waller was weary of the tirades, the histrionics, the regal trappings.

Small, imperial rituals offended Waller's native simplicity. Each of the affectations seemed inoffensive; collectively, he thought, they signified a man infatuated with his position and himself. "He's the CINC," Waller had complained in exasperation; "he's not Your CINC-ness." Before Schwarzkopf entered a briefing room, for example, an enlisted aide would precede him and, with the care of a grandmaster setting chess pieces, place on the table the CINC's polished glasses, a tumbler of water, a glass of orange juice, a cup of coffee, and a glass of chocolate mocha.

Or when Schwarzkopf left his headquarters, he seemed to require a motorcade larger than Fahd's, and he sat in the back seat of a staff car armed with a handgun, surrounded by bodyguards. "That's the last time I ride with you," Waller had declared a few weeks earlier after one foray with Schwarzkopf. "Shit, man, I'm more afraid of you with that damned weapon and all these stupid guards trying to force people off the road than I am of any terrorist attack."

The latest move that irritated Waller was the CINC's relocation of his bedroom into the headquarters' basement. Commandeering an office down the corridor from the war room and one of the two latrines on the floor, Schwarzkopf had converted them into his own quarters. Though the CINC's reasons were sound — he wanted to be close to the war room at all times — the change threw the crowded basement into an upheaval and forced more than a hundred people to share the single remaining toilet.

Waller swiveled in his chair. Such grievances were certainly petty in the grand sweep of the war about to begin, but they galled him. Flamboyance could be a useful element in the general's art: a George Patton packing his ivory-handled revolvers, or a Lucian Truscott in his bright leather jacket, visible to all the troops at the front. But Waller wondered whether Schwarzkopf's showmanship was becoming an end in itself.

The war room door swung open. "Gentlemen," someone announced, "the commander-in-chief." Schwarzkopf entered, jaw set, eyes bright with emotion. Waller and the other officers came to their feet. As the men fell silent, the CINC moved to the front of the room and nodded to the chaplain, Colonel David Peterson. Heads were bowed, eyes pinched shut.

"Our Father," Peterson began, "on this awesome and humbling occasion we are grateful for the privilege of turning to you, our sovereign and almighty God. We believe that, in accordance with the teaching of your word and revelation, we are on a just and righteous mission."

Drawing from the psalmists, the chaplain prayed for the souls of those about to die, then asked "for a quick and decisive victory. Your word informs us that men prepare for battle, and we have. But victory rests with the Lord. Therefore, we commit our ways to you and wait upon the Lord. In the name of the Prince of Peace we pray. Amen."

Amen, the officers echoed, many with a catch in their voices.

Then Schwarzkopf read his message to his troops, nine sentences addressed to the "soldiers, sailors, airmen, and Marines of the United States Central Command," and ending: "Our cause is just. Now you must be the thunder and lightning of Desert Storm. May God be with you, your loved ones at home, and our country."

Finally, Schwarzkopf's aide produced a portable cassette recorder and punched the play button. The room filled with the cloying warble of Lee Greenwood's "God Bless the U.S.A.," which already had become the unofficial anthem of the American expeditionary force. As the last strains died away, Schwarzkopf looked around. "Okay, gentlemen," he said brusquely, "let's go to work."

## Washington, D.C.

The Washington counterpart to Schwarzkopf's war room lay deep inside the Pentagon in the National Military Command Center, where the eight-hour time difference with Riyadh meant it was only 6 P.M. on January 16. Contrary to the popular image of the Pentagon's nerve center as an immense, Strangelovian auditorium with oversized video screens, most of the work in the NMCC was done in a warren of offices with subdued lighting, wall-to-wall carpeting, and Wang computers atop the desks. Soft music droned from aluminum loudspeakers to provide "cover sound" against electronic eavesdropping. The yellow corridor walls were bare except for several aphorisms spelled out in block letters: "The Best Strategy Is Always to Be Strong — Clausewitz," "Forewarned, Forearmed — Benjamin Franklin," and "A Wise Man in Time of Peace Prepares for War — Horace."

Although war was less than an hour away, the atmosphere in the NMCC remained calm, almost sedate. Twenty-five thousand people worked in the Pentagon, but fewer than a dozen had been told in advance that the attack would begin at 7 P.M. Eastern Standard Time. Even in the NMCC, elaborate measures had been taken to ensure security. The Crisis Action Team — a group of 135 military and civilian specialists formed in August after Iraq's invasion of Kuwait — changed shifts, as usual, at 6 P.M. without any disclosure that war was imminent. Several senior officers, including the Joint Chiefs' operations officer, Lieutenant General Tom Kelly, made an ostentatious display of going home for supper, strolling casually out of the NMCC and past vigilant reporters staking out the Pentagon's adjacent E-Ring. (Kelly would return through a back door precisely at 7 P.M., H-Hour.) Only a handful of officers standing watch in the cramped Current Situation Room, the sanctum sanctorum of the NMCC, tracked events from the war zone as they were reported from Riyadh.

To further preserve the illusion of normality, Defense Secretary Richard Cheney and General Powell, chairman of the Joint Chiefs, stayed in their E-Ring offices away from the command center. Cheney's day had begun at dawn, with a meeting at 7:15 in the White House West Wing. There, over breakfast, he, Secretary of State James A. Baker III, and National Security Adviser Brent Scowcroft had reviewed a detailed notification chart that listed individuals to be called before the allied attack began, the time they were to be alerted, and the American official who would make each call. The timetable contained several dozen names, including leaders of every nation in the coalition, the

Soviets, the four living American ex-presidents, congressional leaders, and the United Nations secretary general. President Bush would make many of the calls, particularly to foreign heads of state. Baker would call, among others, Prince Bandar, the Saudi ambassador, and Alexander Bessmertnykh, the Soviet foreign minister. (Mikhail Gorbachev's subsequent efforts to contact Saddam Hussein, detected by American eavesdroppers minutes before the war began, would infuriate Pentagon officials, who suspected the Soviet president of trying to alert Baghdad to the impending attack.) Scowcroft's deputy, Robert Gates, was to inform House and Senate leaders in a clandestine trip to Capitol Hill at dusk. The public would be informed of the attack in a three-sentence announcement from the White House after the first bombs fell, followed two hours later with a televised speech by Bush.

Cheney had returned to the Pentagon after breakfast. At nine o'clock he made one of the first notification calls, using a newly installed secure satellite telephone, code-named Hammer Rick, to phone Moshe Arens, the Israeli defense minister. The Americans had promised the Israelis ample advance notice on the assumption that Saddam Hussein would make good on his threat to attack Israel should war erupt. Cheney told Arens the H-Hour and promised to keep the Israelis informed as the campaign unfolded. Arens, who seemed unsurprised by the call, thanked the defense secretary and hung up.

Most of the other preparations for launching the war were already completed. A top secret warning order had been sent to Schwarzkopf on December 29, advising the CINC to be ready to attack by 3 A.M., January 17, Riyadh time, less than twenty-four hours after the expiration of a United States deadline for Iraq to withdraw from Kuwait. In the Oval Office, at 11 A.M. on January 15, Bush had signed the two-page National Security Directive formally authorizing the attack unless a last-minute diplomatic breakthrough resolved the crisis. Six hours later, in the defense secretary's office, Powell and Cheney had signed the execute message necessary to "activate" the warning order of late December. Powell had then personally faxed the document to Schwarzkopf. The operation was code-named Desert Storm.

Now, with little to do and his calendar cleared of routine business, Cheney waited. A Marine guard in dress blues stood at attention outside the anteroom door, a few steps from a framed print, entitled *Passing the Rubicon,* that depicted the American Navy forcing its way into Japan in 1853. The defense secretary knew Schwarzkopf's timetable well enough to realize that another Rubicon had been crossed minutes before with the initial launches of Tomahawk missiles. The TLAMs could not be recalled.

After so many months of tense preparation, Cheney felt a sense of relief unblemished by doubts or second thoughts. He had an image of a huge machine in motion, a machine no longer in his control. Saddam had squandered countless opportunities to avoid war; now he would feel the wrath of the allied coalition. That, Cheney believed, was as it must be. On the television near the secretary's desk, correspondents reporting from Baghdad for the Cable News Network described the placid scene in the Iraqi capital. Cheney listened carefully. It struck him as eerie, even surreal, that they should be unaware of the massive attack about to descend on Baghdad, while six thousand miles away he had such a clear image of the missiles and aircraft pressing toward Iraq.

Suddenly he was hungry. A long night loomed ahead. Cheney's secretary phoned a nearby Chinese restaurant for takeout food, and soon the round table on which the war order had been signed was strewn with tiny paper cartons of rice and eggrolls and steamed vegetables.

One floor below, in the chairman's office, Colin Powell tried to relax. Like Cheney, he listened to the CNN correspondents, straining for any hint that allied security had been breached. So far the grand secret appeared to be intact.

Earlier in the afternoon, Powell had telephoned Schwarzkopf for their final conversation before H-Hour. The chairman knew that his field commander, high-strung in the best of times, would inevitably be more nervous than generals sitting in the comfort of the Pentagon. A thousand tactical details nagged at the CINC, a thousand things that could go wrong. Powell kept the call brief and businesslike, masking his own anxieties behind a bluff optimism. "I won't bother you," he told Schwarzkopf. "You take care of that end, I'll take care of this end. Norm, good luck."

"You're always confident in the plan," Powell once said, "until the shit hits the fan." On paper, the allied war strategy seemed brilliant. But the chairman knew that history was littered with clever war plans gone awry. Sitting alone at his desk, the television droning in the background, Powell reviewed the initial targets: radar sites, electrical plants, communications towers, command posts. Some would be attacked by Tomahawks, others by warplanes. In the next few hours, hundreds of bombs and missiles would detonate across Iraq in a pattern that had been etched with half-second precision. Saddam, the chairman believed, had no conception of the fury of the attack that was about to rip open his country.

But Powell worried that Americans, including many in the government, had come to expect a level of military performance that was impossible to meet, as though an operation of this magnitude were no more complicated than walking out to the parking lot and climbing into a car. Would the TLAMs work as the Navy believed they would? Would pilots find and destroy their targets? Such issues were obviously important, but the overriding question, Powell believed, was how many allied airplanes would return safely. That would not be known for several hours, when the first raiders flew back to their aircraft carriers and air bases. Only then would Powell, Schwarzkopf, and the other commanders be able to gauge the quality of Iraqi defenses; only then would they have a battlefield sense — not one based on computer models or paper projections — of the price to be paid for the liberation of Kuwait.

Colin Powell had spent more than thirty-two years in uniform, thirty-two years preparing for this moment. Now it all boiled down to one question: How many men would come back?

## Baghdad

Baghdad's last day of peace had been preternaturally calm. Warm winter sunshine splashed over the deserted streets, normally crowded with pedestrians and honking motorists. The souks stood empty save for an occasional vendor peddling oranges and dates. Shops, which in recent months had been flush with the booty of ransacked Kuwait, from Rolex watches to Armani dresses, now were shuttered and padlocked. Stolen Kuwaiti Land-Rovers and Mercedes Benzes, so commonplace on Baghdad's boulevards, had been locked in garages or hidden in the countryside. Along the meandering Tigris River, pretty pastel houses wore the forlorn look of homes abandoned in haste.

Not all of Baghdad's five million souls had fled, of course. Republican Guardsmen in berets strolled the sidewalks with studied insouciance. At street corner checkpoints, militia troops wearing red-and-white-checked kaffiyehs cradled their Kalashnikovs, swapping jokes and cigarettes, as peasant women with tattooed faces hurried past. Saddam Hussein was everywhere: a thousand smiling portraits beamed down from billboards and lampposts, reassuring no one. Deep beneath the capital, his eleven-room command vault — complete with swimming pool, wainscoting, and chandeliers — waited for a marked man too wary to use it.

Even on the eve of war Baghdad was a city with pretensions, bent on recapturing a glorious past. Founded in the eighth century, the metropolis had been a hub of Islamic learning and culture for five hundred years at a time when Europe was mired in barbarism. Baghdad's own dark ages began in 1258, after Mongols slaughtered the caliph and several hundred thousand of his subjects. The city was a provincial backwater for more than seven centuries until it revived as the new capital of independent Iraq in 1921. After Saddam and the Ba'athist Party seized power in 1968, Iraqis forged a tacit social compact with their president-leader-marshal: he would provide the trappings of modernity — skyscrapers, universities, hospitals — in exchange for absolute dominion enforced with a huge army and a vast secret police network. The endless convoys of Kuwaiti loot rolling into the capital since August were but latter-day payments on the covenant.

The city in recent weeks had prepared for war, after a fashion. Iraqi officials staged mock evacuations and civil defense drills. Citizens received instructions on how to construct gas masks from kitchen towels. Saddam appeared on television in green Ba'athist fatigues or Bedouin robes to urge pluck and fortitude. After a few American air strikes, he had predicted, the coalition would weary of war and leave Iraq to pursue its destiny.

In the southwestern suburb of Amiriyah, schoolgirls in blue uniforms had attended English class as usual on this final day of peace. "I am happy, you are kind, he is sad," they repeated dutifully. Down the street stood a fenced compound with a sign at the gate in English and Arabic — Public Shelter Number 25 — listed on American target documents as the Al Firdos bunker.

At the nearby Baghdad Horsemanship Club, jockeys in scarlet, gold, and green paraded their ponies around the paddock while several thousand tweedy bettors cheered from the grandstand with the jaunty air, as one journalist observed, of sportsmen "watching a game of deck quoits on the *Titanic.*" In the first race, at 4:10 P.M., the favorite, a long-legged filly named Scheherezade, had galloped down the stretch to a win. She paid five to one.

The American view of Saddam's seizure of Kuwait had not been troubled by shades of gray. The deed soon was equated with the German invasion of Belgium in 1914, the Japanese occupation of Manchuria in 1931, and the German conquest of Poland in 1939. By the standards of modern sovereignty, the invasion was unpardonable; by the standards of human decency, the subsequent pillage and murder were criminally vile. Yet, however tawdry, the attack was not entirely the act of "unprovoked aggression" denounced by George Bush. Saddam nurtured grievances. Some were preposterous, others legitimate.

Only Iraq had been bold enough in the 1980s to thwart Iranian ambitions for hegemony in the Persian Gulf, although certainly Saddam was motivated less by a desire to champion Arab moderates than by his desire to dominate the Arab world. The eight-year war with Tehran had cost Baghdad 375,000 casualties and half a trillion dollars. When the conflict ended, in August 1988, Iraq found itself $80 billion in debt, a condition aggravated by a plunge in oil prices from $20 a barrel to $13 during the first six months of 1990.

Saddam considered his Arab brethren both inadequately grateful for his challenge to Iran and indifferent to plummeting petroleum prices. He had, with some cause, accused Kuwait and the United Arab Emirates of cheating on oil production quotas and flooding the world market. This latest offense compounded other Iraqi complaints against Kuwait, including the "theft" of more than $2 billion by slant-drilling into the vast Rumaylah oil field, the bulk of which lay beneath Iraqi territory.

Iraq also had long disputed the border with Kuwait and even the emirate's sovereignty, claiming that both were imposed by colonial powers earlier in the century. (Here Baghdad ignored several inconvenient facts: Kuwait's historical autonomy within the Ottoman Empire; the two-hundred-year rule of the emirate by the al Sabah family; and Kuwait's status as a British protectorate from 1899 until independence in 1961.) Saddam felt particularly aggrieved by his neighbor's possession of two Persian Gulf islands, Bubiyan and Warbah, which effectively controlled Iraq's access to the sea.

Bankrupt, desperate, and congenitally bellicose, he chose to attack. In seizing Kuwait — a country slightly smaller than New Jersey — he gambled on Arab apathy toward the fate of a nation long resented for its arrogance and wealth. (Per capita income in Kuwait was twenty times that, for example, of Egypt.) He also counted on Western indifference toward a feudal monarchy where only 4 percent of the population was enfranchised, where parliament and the national constitution had been suspended, and where advocates of democracy had been suppressed with police truncheons and mass arrests.

By annexing the emirate as the 19th Province of Iraq, Saddam would reap a windfall of $20 million a day in oil revenues. He would control 20 percent of the world's petroleum reserves. And he would demonstrate the perilous consequences of ignoring Iraqi demands.

But now, five months after the invasion of Kuwait, Iraq would reap only the whirlwind of miscalculation and overweening ambition. Dusk had overtaken Baghdad, velvety and Mesopotamian, rolling across the city from the east like a dome. The aroma of kebob and carp, grilled by a few intrepid restaurateurs on Abu Niwas Street, drifted along the riverbank. Streetlights flickered to life. Old men in seedy cafés drained their teacups, wiped their mustaches, and shuffled home. Midnight passed, yielding to the wee hours of Thursday morning. The city waited, sensing catastrophe.

Baghdad had been here before: sacked by the Mongol Hulegu Khan and again by Tamerlane, overrun by Turks and Persians, Ottomans and Englishmen. The tribal memory of siege and disaster was strong; foreboding seemed familiar. But in this city of a thousand and one nights, few had been longer than the one now fallen.

## U.S.S. *Wisconsin,* Persian Gulf

Shortly before 2 A.M., the missilemen in Strike finished retyping twenty-four pages of codes into the Tomahawk engagement console. Once again they loaded the software program and struck the keys to shift the commands to the launch console. This time it worked. Lieutenant Zanti turned to Captain Bill. "Sir, the casualty has been corrected."

Bill issued a quick order to the bridge. Now they had to uncage the gyros, a procedure necessary to help the Tomahawk guidance systems get a precise fix on their launch location. Almost imperceptibly the battleship began to swing right, then left, on a zigzag course.

Topside, starboard forward, armored box launcher number one canted upward thirty-three degrees with a shrill hydraulic whir. In Strike, at 2:13 A.M., the captain picked up the public address microphone as the watchstander lifted a clear plastic shield protecting the EXECUTE ON button. "Stand by," Bill told the crew. "Five, four, three, two, one." The watchstander mashed the button.

Nothing happened. Three seconds passed, four, five. The missile crewmen exchanged nervous glances. Seven, eight.

With a low roar, white flame licked through three narrow exhaust vents in the rear of the launcher box. An explosive charge blew free the bolts securing the Tomahawk. The blunt nose burst through a yellow membrane covering the launcher and the eighteen-foot missile vaulted across the port beam. For an instant it hung tail down, the molten plume silvering the dark water, before springing toward the stars. A mile from the ship, the rocket booster fell away. Stubby wings popped from the fuselage and the jet engine ignited in a coil of thick smoke. The missile climbed to a thousand feet — high enough to clear any oil platforms in the gulf — and headed northwest, toward Iran.

Crewmen at battle stations on the main deck watched the missile fade from sight. Among them, manning a triage station forward of gun turret two in case of an enemy counterstrike, stood the ship's dentist, Commander R. J. Turner. "War is the remedy our enemies have chosen," Turner murmured, quoting General William Tecumseh Sherman, "and I say let us give them all they want."

Two minutes later, the second missile burst from its launcher. Then another and another. From his high-backed chair in Strike, the captain again addressed his crew. Many people would be killed in the attack now under way, including innocents, Bill warned. "But there's some solace in that the Tomahawk missiles we just fired have gone against military targets."

Yet again there were secrets, known only to the captain and his senior officers. Several hours before *Wisconsin* had sailed from Norfolk in August for the Middle East, she took aboard a mysterious cargo of new Tomahawks. Some crewmen speculated they carried nuclear warheads. In reality, the missiles were known as Kit 2s, and their purpose was to cripple the Iraqi electrical grid.

For at least a decade, the American military had tinkered with devices for sabotaging enemy electrical systems. Some of the results were serendipitous. In the early 1980s, during a Navy exercise code-named Hey Rube, long strands of rope chaff — glass filaments in which metal shards were embedded — had been dropped over the Pacific Ocean as part of a standard tactic to befuddle an opponent's radar. An unexpectedly stiff westerly wind carried some of the chaff ninety miles to the coastline, where it got draped across power lines, shorting out transformers and causing power failures in parts of San Diego. The Navy quietly settled the damages — while carefully noting the effects of its unintended attack.

During the Iranian hostage crisis in 1980, rescue planners contemplated dumping carbon fibers over segments of Tehran's power grid to make the city "go midnight" and sow enough confusion for commandos to free the imprisoned Americans. In the years since, the Navy had continued developing antielectrical weapons, including fibers designed to penetrate the air intakes and reactor equipment on Soviet nuclear-powered ships.

The projected use of Kit 2s against Iraq had triggered a sharp, secret debate in Washington. Enthusiasts argued that if the enemy's electrical systems were disrupted, its air defense network would be crippled and, simultaneously, the discomforts of war visited on the Iraqi citizenry. That might provoke a rebellion against the Ba'athist regime. Kit 2s, they also argued, could "soft kill" a power grid with less raw destruction than would conventional bombs, permitting easier postwar reconstruction.

Opponents countered that revealing the weapon would ultimately jeopardize the most vulnerable electrical network in the world: the American power grid. The State Department also worried about unforeseen consequences for Iraqi society if the lights went out indefinitely. But at the Pentagon Richard Cheney urged, "Let's give them the full load the first night." Anything less smacked of escalation, of the stench of Vietnam. Consequently, on the American attack list were twenty-eight electrical sites, many of them, particularly around Baghdad, targeted with Kit 2s. The operation was code-named Poobah's Party, a name derived from the radio call sign of the Air Force general responsible for electronic warfare against Iraq.

On *Wisconsin,* the eighth and final shot ripped from the deck at 2:39 A.M. A strange silence settled over the battleship, magnifying the ordinary sounds of a ship at sea: wind singing across the fo'c'sle, the creak of an open hatch, water lapping gently against the hull.

## Objective Oklahoma, Iraq

The radar complex, now barely three miles away, lay before them as toylike and precise as Dick Cody's table model. Through their night-vision goggles, the Apache pilots saw a dozen buildings spread across the square-mile compound; three antiaircraft gun pits faced south. Team White had arrived ninety seconds early, exactly as planned. The helicopters hovered four abreast beyond Iraqi earshot.

In *Rigor Mortis* on the far right of the flying line, Lieutenant Colonel Cody began matching the structures below with the satellite images he had burned into his memory: a Spoonrest radar, another Spoonrest, a Flatface radar, the generator pit, a cluster of command vans, the troop barracks, the troposcatter radio antennae. In the 36-power magnification of the Apache's infrared sights, each target glowed like alabaster, radiating the sun's heat absorbed the previous day.

But now the electric lights began to snap off. Cody spotted a man running, then several men, white ghosts leaving the command vans. A faint chirping sounded in his headset and blips flashed on the cockpit video screen, indications that Iraqi radars had begun to paint the intruders.

On the far left, Tip O'Neal watched as more Iraqis scurried across the compound. "They're turning out the lights, Dave," he told Jones over the intercom. "I think they know we're here."

O'Neal tugged the laser trigger with his right index finger and spun the control knob with his thumb to ease the cross hairs onto the generator pit. The range flashed on the scope: 5026 meters. He checked the computer clock — 0237.30 — as the first radio call of the mission came from one of the other Apaches: "Party in ten."

O'Neal counted down. Eight, seven, six. He and Jones had agreed privately before leaving Saudi Arabia that shooting a few seconds early was a negligible risk compared with the glory of firing the first shots of the war in Iraq. When the countdown reached three, O'Neal pulled the weapons release trigger with his left hand. "This one's for you, Saddam," Jones declared from the back seat. "Iraq, you want it, you got it."

With a soft *whoosh* the left outboard Hellfire leaped from its rail in a shower of sparks and darted toward the target at more than Mach one. A nose sensor picked up the thread of O'Neal's laser beam — he had stopped breathing, so intent was he on keeping the cross hairs steady — and the missile climbed slightly to hit the target from above.

Twenty seconds after launch, the Hellfire burrowed into the generator pit. Two internal plates, crushed together by the impact, electronically detonated a seventeen-pound cylinder of explosives, which in turn squirted a jet of molten copper into the target. The generators vanished in smoke and flame.

Now the air thickened with Hellfires. Iraqi soldiers boiled from the barracks, scurrying about with a panic that reminded O'Neal of an anthill stirred with a hot stick. He slaved the cross hairs to the left, then noticed that the command vans were not burning. Two missiles had jammed on his wingman's Apache. As Jones opened the throttle, pushing forward at sixty knots, O'Neal centered the laser and fired again. He saw an Iraqi pivot to run back into the van just as the missile hit; it flung the enemy soldier high into the air in a fractured jackknife.

O'Neal swung onto the other targets, first the Flatface radar, then the Spoonrest. A small building they had presumed was the compound's latrine blew up with the force of an ammunition locker. "Keep her steady! Keep her steady!" he demanded. "I'm trying," Jones replied, wrestling with the helicopter controls.

From his seat in *Rigor Mortis,* Cody drew a bead on the troposcatter, a large box with four protruding antennae, "mouse ears," that Air Force intelligence believed capable of communicating directly with Baghdad. He had vowed, in a fit of vainglory, to destroy the target even if he had to crash his helicopter on it; two Hellfires spared him the sacrifice.

Now, from a range of four thousand meters, the Apaches unleashed a volley of slender Hydra rockets. Each warhead carried scores of tiny finned darts that blew across the radar site like buckshot, ripping through wires, electronics, and men. At two thousand meters, the helicopters opened up with 30mm cannon fire. Hundreds of slugs churned through the compound, kicking up geysers of sand, riddling walls and roofs, rekilling the dead.

And then it stopped. Four minutes after the first missile exploded, the Apaches broke off the attack. Cody swept back for a final look. Every building and radar dish had been destroyed. Bodies lay scattered like bloody rugs. Nothing moved but the flames.

Exhilarated, exhausted, appalled, the raiders turned south. Cody keyed his microphone to radio the waiting Pave Lows with a message that would subsequently be relayed to Riyadh. Under an arranged code, Charlie meant minimal destruction, Bravo meant partial destruction, Alpha meant total destruction. Alpha Alpha meant the target was destroyed, with no friendly casualties. "This is White Six," Cody reported. "Oklahoma. Alpha Alpha."

Twenty miles west, Team Red had found comparable success at Nebraska. Behind the two Apache teams, special forces troops in Chinook helicopters implanted eleven reflector beacons along the border to mark the keyhole that had now been cut for the Air Force. Cody had delivered on his promise: 100 percent success. But the mission was not without a small setback. Moments before the Apaches struck, U.S. intelligence eavesdroppers had intercepted a frantic radio call. "We are under attack," the Iraqi message began, then went dead.

Through their night-vision goggles, Cody's pilots spotted a swarm of lights sixty miles to the south. The swarm grew closer, brighter, and they saw that it was the Air Force strike package: two EF-111 electronic jammers followed by nearly two dozen F-15E bombers and British Tornados heading for the Scud batteries in western Iraq.

As the low-flying jets neared the border, their lights winked out. With a deep growl, the formation streaked directly over the four Apaches. The wolves were in the fold.

## Baghdad

At 2:51 A.M., thirteen minutes after the Apaches began their attack on the radar sites, the first bomb fell on Iraq. Dropped from an F-117 stealth fighter, the laser-guided GBU-27 demolished half of the air defense center at Nukayb, about thirty miles north of Objective Oklahoma. One minute later a second bomb, from another F-117, destroyed the other half.

Like the longbow at Crécy and the machine gun at the Somme, the Persian Gulf War's mission number 3336-S — as it was listed in Air Force targeting documents — would forever alter the way men contemplated killing one another in combat. The marriage of "low observable" technology — stealth — and precision munitions — smart bombs — brought a new lethality to the battlefield by permitting an attacker to slip unseen into the enemy camp and strike with virtual impunity.

Stealth did not obviate more prosaic weapons; tens of thousands of tons of old-fashioned dumb bombs would also fall from nonstealth planes in the next six weeks. Moreover, the accuracy of F-117 pilots in the initial nights of the war would leave much to be desired, contrary to the public pronouncements of the American high command. But an attacker armed with stealth and smart bombs had an immense psychological and military advantage over an opponent without them. The technology also distorted, even perverted, the American concept of combat, which quickly came to be seen as surgical, simple, and bloodless. War was none of these and never would be.

Stealth had been developed in deepest secrecy beginning in the 1970s. The Lockheed Corporation built two prototypes, code-named Have Blue, followed by a full-scale successor called Senior Trend, which became the F-117. Heavily coated with a radar-absorbing skin similar in texture to Styrofoam, the black, swallowtailed aircraft also was intricately faceted with small flat planes intended to deflect or attenuate radar waves. Flat and narrow "platypus" exhaust vents reduced the jet's infrared signature by quickly mixing hot gases from the engines with cold air. For years, a secret F-117 unit flew from a clandestine base at Tonopah, Nevada, practicing nocturnal missions that included "attacking" the state capitol in Carson City, swinging west to the California coast, then returning to base across the Yosemite Valley.

If stealth was a relatively recent addition to the American arsenal, smart munitions — despite a popular misconception — were not. U.S. and German scientists had labored on crude guided bombs in World War II. Roughly 21,000 precision-guided munitions — more than would be dropped against Iraq — fell on North Vietnamese targets, particularly in the final two years of the war. But during the past decade further refinements had given the weapons a new sophistication in pursuit of a "one bomb, one kill" ratio of destruction.

The most common smart weapons used in the gulf war were LGBs, laser-guided bombs, such as the 500-pound GBU-12 and the 2000-pound GBU-24. (The GBU-27 resembled the latter but was specially configured to fit inside an F-117 bomb bay, because large, radar-reflecting weapons hanging under the wing impaired the jet's stealthiness.)

The bomb dropper, another airplane, or an ally on the ground "designated" the target by pointing a laser beam at it. As the bomb glided to earth, a light-sensing "seeker" in the nose detected the reflected laser radiation. Optics in the seeker carved the field of view into quadrants, and sensors measured the relative intensity of laser "sparkle" in each quadrant.

If all four quadrants measured identical amounts of laser radiation, the weapon was on course; if not, a computer made flight adjustments with signals to the bomb's control fins. Given good weather — fog, clouds, and smoke disrupted the laser beam — a deft air crew could put an LGB within ten feet of the desired impact point from a range of up to seven miles.

During Desert Shield about forty stealth fighters and hundreds of GBU-27s had been flown to Khamis Mushait, a Saudi mountain redoubt near the Red Sea, nine hundred miles from Baghdad. Six hours before the war was to begin, the pilots slipped into their G-suits and parachute harnesses. Only four of them had flown in combat before, and they traded nervous banter about the presumed invisibility of their aircraft. In the hangar bays, flight crews, like British sailors chalking DEATH OR GLORY on their guns before Trafalgar, scribbled TO SADDAM WITH LOVE and other graffiti on the bombs, a reminder that banality, no less than fear and bravery, was a timeless feature of warfare.

Shortly after midnight on January 17, ten planes took off in the first wave, including the two that struck the air defense center at Nukayb. The rest pressed north toward greater Baghdad.

Piloting tail number 816 was a tall, dark-haired Air Force Academy graduate whose father had flown helicopters with the Army's 1st Cavalry Division in Vietnam. Major Jerry Leatherman's assigned target was the Baghdad International Telephone Exchange, dubbed the AT&T building because the formal name was too cumbersome to fit the allotted space on Air Force targeting sheets. The twelve-story switching facility was the only building in Baghdad targeted by two aircraft on this first strike, evidence of the importance attributed to it by Schwarzkopf's targeteers. More than half of Iraqi military telecommunications traveled through the civil telephone system; most of those calls were routed through the AT&T building.

Leatherman steered his black jet eastward at 480 knots, deliberately bypassing the city in order to attack out of the north. One aircraft flew a few miles ahead of him; the others had peeled away to strike from different vectors. Baghdad's glow had become visible from a hundred miles away, but now he could pick out wide boulevards and traffic circles and the crimson glint of neon. The black Tigris snaked through the capital in a double loop. On this clear, cold night, from three miles up, the metropolis bore an odd resemblance to Las Vegas or Phoenix or half a dozen other cities in the great American desert.

As supervisor of mission planning at Khamis Mushait, Leatherman knew the geography of Baghdad as well as any pilot in the wing. He also knew that sixty surface-to-air missile (SAM) batteries and three thousand antiaircraft guns protected the capital. What he did not anticipate was that all of those guns and missile sites, apparently warned by the desperate call from Objective Oklahoma, would begin shooting even before he made his final approach into the Iraqi capital.

In vivid fountains of red and orange and gold, the enemy fire boiled up with an intensity that initially mesmerized more than it frightened. Missiles corkscrewed skyward or streaked up on white tubes of flame. Antiaircraft rounds — 57mm, then 100mm — burst into hundreds of black and gray blossoms. Scarlet threads of gunfire stitched the air, woven so thickly as to suggest a solid sheet of fire — all the more alarming when Leatherman remembered that only every fifth round was a visible tracer. Yet for all its volume, the shooting seemed unguided; the Iraqis were flinging up random barrages in hopes of hitting something. Enemy search or tracking radars might occasionally glimpse a stealthy intruder, but not long enough for gunners or SAM crews to draw a bead. In effect, Leatherman realized, he was still invisible.

Tearing his eyes from the spectacle outside the cockpit, he began to search for the target. Almost immediately it loomed into view on the infrared video screen: a tall green rectangle near the river. Stealth pilots talked about "Stevie Wonder targets" — so distinct that even a blind man could spot them — but this seemed too simple. Leatherman swung the infrared sensors across the city to check his bearings. He spotted the distinctive tulip shape of the Iraqi Martyrs Monument, exactly where it was supposed to be. Now he was certain: he had found his target.

At precisely 3 A.M., H-hour, the lead aircraft struck the AT&T building with a GBU-27. Leatherman saw neither explosion nor flames, but the target began to glow on his scope, suddenly hotter than the adjacent office towers. Centering his laser cross hairs on the upper half of the target, Leatherman hit the pickle button at 3:01 to release one bomb, then a second. The first struck slightly low; the second split the cross hairs. The upper four floors of the building exploded in a scorching flash. Leatherman let out a victorious whoop and immediately banked to the west, heading for the safety of the open desert.

Elsewhere in Baghdad, the other stealth pilots had also found their marks. Across the river on Al Rashid Street, Captain Marcel Kerdavid put a GBU-27 through the top of the Al Khark communications tower; the bomb burrowed partway down the 370-foot concrete structure and exploded, snapping the tower in half. Kerdavid swung around to the northern suburb of Taji and dropped his second bomb on Military Command Bunker Number 2, which, American intelligence later concluded, proved impervious even to the 2000-pound penetrator. Other pilots struck the Iraqi air force headquarters, two air defense command posts, the air defense sector center in Taji, and Saddam's retreat outside the capital at Abu Ghurayb. Of seventeen bombs dropped in this first wave, the stealth wing would claim thirteen hits.

Swinging south for the long flight back to Khamis Mushait, Major Leatherman slipped a pair of earphones under his flight helmet, clicked on his Walkman, and turned the volume up as loud as he could bear it. The heavy metal band Def Leppard pounded through his skull. He glanced over his left shoulder for a final glimpse of Baghdad. Tracers still etched the sky, but now brilliant orange mushrooms also began to sprout across the city. The Tomahawks had arrived.

The missiles from *Wisconsin* and her sister ships swept into Baghdad at nearly five hundred knots, heard but not seen. About eight miles from their targets, the Tomahawks' terrain-contour matching yielded to a different navigation system. Tiny television cameras clicked on and began scanning the landscape, which was periodically illuminated by a small strobe light in each missile's belly.

An optical sensor called DSMAC — digital scene matching area correlator — scanned the passing scenery and divided the image into a matrix of black and white squares. Comparing the televised images with those stored in the Tomahawk's memory, a computer relayed commands to the missile's stubby wings and tail fins for final course adjustments.

The first wave struck between 3:06 and 3:11 A.M. Eight missiles plowed into the presidential palace, their thousand-pound Bullpup warheads delaying for a few nanoseconds before erupting in a roar of flame and flying masonry. Six more hit the Ba'ath Party headquarters compound. Thirty struck the vast missile complex at Taji, with a redundancy that reflected both the size of the target and lingering doubts about Tomahawk accuracy. Within twenty-four hours, 116 missiles would be fired from nine ships. Most struck greater Baghdad.

The secret Kit 2s employed guile rather than brute force. Making multiple passes over key power plants, the missiles spewed out thousands of tiny spools less than an inch in diameter, from which long carbon filaments uncoiled, drifted to earth, and, as in the unintended result a decade earlier, fell over transmission lines, shorting them out in a medley of bright flashes and loud pops.

Nearly all of Iraq's electrical power was supplied by twenty generating plants lashed together by 400-kilovolt transmission lines. Five plants were attacked immediately and eight others would be struck — by bombers and Tomahawks — in the coming hours. Kit 2 filaments fell on Beiji, the nation's largest power plant, shorting the complex so effectively that all six generators seized up. Other strands fell on the city's Doura plant along the Tigris and on the gas turbine complex at Taji. Three Tomahawks attacked the electrical transformers at Salman Pak, believed by American intelligence to house Iraq's biological weapons program.

From Muthana in the east to Mansur in the west, the lights in the capital began to vanish, block after block, neighborhood on neighborhood. In twenty minutes the city's defiant glow was gone, and Baghdad's dark night grew darker yet.

## Riyadh

A mile north of Schwarzkopf's command post, beneath the three-story headquarters of the Royal Saudi Air Force on King Abdul Aziz Boulevard, eighty men, crowded into a hot bunker, watched the war unfold on a four-foot television screen. Electronically linked to a quartet of AWACs planes — airborne radar platforms capable of peering deep into Iraq from their orbits in northern Saudi Arabia — the screen displayed hundreds of tiny green circles, each representing an allied aircraft, as well as the occasional red circle of an Iraqi fighter on patrol.

South of the border flew 160 aerial tankers, stacked in cells five deep. Around them, the strike packages massed for attack: bomb droppers and fighters, radar killers and electronic jammers, eavesdroppers and rescuers. From the Red Sea in the west to the Persian Gulf in the east, the great menagerie of Eagles, Ravens, Tomcats, Weasels, Harriers, Warthogs, Aardvarks, and Falcons surged forward. Some, like the aircraft streaking toward Scud targets near Jordan, already had crossed into Iraqi airspace; others, notably the low-flying Apaches flickering dimly on the screen near Ar Ar, were now headed home. Only the stealth fighters remained unseen, invisible even to the AWACs.

At the center of the room, exuding an unmistakable air of proprietorship, sat the man responsible for the thousand or so lives represented by those swarming green circles. Lieutenant General Charles A. Horner, commander of the U.S. Ninth Air Force, was thickset and inelegant, possessed of brawny forearms and an oval face that thrust itself at the world with fighter pilot self-assurance. When displeased or skeptical, he squinted slightly with one eye, like Popeye, and the deep rumble of his voice seemed to drop yet another octave. He could be blunt even to his fellow three-stars. A day earlier, after noticing that few Marine Harrier jets were scheduled to fly in the first attacks, he fired a message to Lieutenant General Walter E. Boomer, the Marine commander: "Walt, tomorrow morning the war starts and you're going to look out across your lines to see Air Force A-10s and F-16s killing the Iraqi army that you're eventually going to have to attack. I hope Air Force pilots aren't the only ones who die out there."

Capable of playing the dull-witted rube from Iowa, Horner was in fact highly intelligent and shrewd unto cunning. No officer in Riyadh was more skillful at handling the tempestuous CINC (as he had adroitly handled the two waspish generals who preceded Schwarzkopf as head of Central Command). When pressing his case in a roomful of officers, Horner would lean close and murmur in tones audible only to Schwarzkopf, much to the distress of the CINC's staff, who decried this "whispering style" while envying Horner's ability to get what he wanted. Melding thirty-two years of flying experience — including 111 missions over North Vietnam — with an occasional burst of aviator double-speak, Horner showed his expertise in air warfare, about which Schwarzkopf knew little. Of subjects terrestrial, Schwarzkopf was king; but in the air, Horner was his regent.

Now, in the stuffy chamber of his command center, the regent could do little but watch and wait. The first two days of the air campaign had been plotted in such painstaking detail that a computer simulation of the initial air strikes took two months to construct. Any changes at this point would have snarled the nexus of refueling assignments, electronic warfare coverage, and strike sequences required to fling two thousand aircraft at Iraq without a rash of midair collisions. "It's so precise," Horner had fretted aloud a week earlier. "I'd be happy if only a portion of it worked. What we're trying to do is too perfect."

Even victorious air campaigns, he knew, often came at a frightful price. In a single raid on the Nazi ballbearing plants at Schweinfurt in October 1943, sixty American B-17s had been shot down. Over Normandy, despite feeble Luftwaffe opposition, the Allies lost 127 planes. Forty Israeli jets had been destroyed on the first day of the Six Day War in 1967. Horner anticipated coalition losses of perhaps one hundred planes in this war, a figure some Air Force analysts considered optimistic.

And yet he felt an inner calm. Every pilot's death wounded him, even those from foolish training accidents, but two events had forged a steely core. The first occurred in 1964, when his parents, sister and brother-in-law, and their children died in an auto crash. The second was his own near-death while flying as a young captain over North Africa. At three thousand feet, he had rashly pulled a split S — a fighter maneuver, typically requiring at least twice as much altitude, in which the pilot loops the plane back under itself. As the F-100 leveled off, Horner could see sand dunes looming *above* him on either side and a dusty plume below where the jet's exhaust flame was scorching the desert floor.

He had never looked at life quite the same after that flight but, rather, with the liberating conviction that mastery of his own fate was largely illusion. *Inshalla* was the Arab term — God's will. In the months leading up to this night, he had done what he could for those green circles inching into Iraq; now the outcome lay with providence.

Horner glanced around the room and spied Major Buck Rogers, a young pilot who had spent weeks interviewing telecommunications contractors so that he could comprehend the intricacies of Iraq's phone system. If the F-117s hit the AT&T building as scheduled, Horner knew, any television broadcasts routed through that network would be severed. "Buck," he ordered, "go upstairs to my office. There's a TV up there. See what's happening."

Ten minutes later Rogers called from Horner's desk on the third floor. "Sir," he announced, "Baghdad just went off the air."

Horner hung up, then stood and repeated the message to the hushed room. "Shit hot!" the men crowed. "Shit hot!" Horner turned back to watch the screen, a smile playing across his face for the first time that night. *Inshalla.*

Allied strategy in the first three hours of war had many objectives: disrupting Iraqi communications and electrical power; suppressing Scud missile launchings; killing Saddam and his lieutenants, or at least impairing their ability to command; paring away Iraq's weapons of mass destruction.

Yet the overriding purpose in the initial attacks was to seize control of the skies. "Anyone who has to fight, even with the most modern weapons, against an enemy in complete control of the air, fights like a savage against a modern European army," Field Marshal Erwin Rommel had observed. Israeli success in the 1967 war resulted primarily from the destruction of the Egyptian and Syrian air forces in a single day. Horner hoped to match that feat.

Iraq's air defense system, modeled after the Soviets' and built largely by the French, was known as KARI (the French name for Iraq, spelled backward). KARI was carved into five sector operation centers, or SOCs, which controlled aircraft interceptors, surface-to-air missile batteries, and antiaircraft guns in their respective regions. SOC-1 was buried within the air defense headquarters in Baghdad (with a supplemental post in Taji). SOC-2 was at the large airfield near Jordan known as H-3. Other SOCs lay in the far north, at Kirkuk, and in the Euphrates Valley, at Talil. A fifth SOC had been established in Kuwait, at the occupied Ali al Salem Airfield. All Iraqi SOCs were built identically, with a reinforced bunker thirty feet underground.

Each SOC controlled three to five interceptor operation centers, or IOCs, typically tucked into two bunkers spaced about fifty yards apart. Three to seven radar sites reported to each IOC. Several of the initial stealth attacks had been against SOCs and IOCs, including the bombs dropped shortly before 3 A.M. on the Nukayb IOC, which reported to SOC-1 in Baghdad and controlled the two radar sites destroyed by the Apaches. Attacks on the power grid — intended in part to inconvenience the Iraqi population — were designed primarily to knock out much of the electricity necessary to run the air defense network.

That, however, was but the first step. Leading the vast armada heading into Iraq once the initial stealth attacks were completed came a wave of twenty F-15C Eagles, flying in a line that extended from the Jordanian border to Kuwait. Their mission was to sweep like a scythe halfway up the length of Iraq, staking further claim to Iraqi airspace by destroying any enemy fighters that challenged them. Similar sweeps had been flown to good effect in World War II and over MiG Alley in Korea.

Much to the Navy's chagrin, this sweep was solely an Air Force production. Horner's staff, citing computer models that predicted six to twelve cases of air-to-air fratricide in the first two weeks of the war, had yanked all F-14 Tomcats from the operation because the Navy planes lacked some of the F-15s' ability to distinguish friend from foe at long range. The contretemps was not the first between the two services, nor would it be the last.

## Al Taqaddum, Iraq

Leading the middle flight of Eagles in the air sweep was one of the Air Force's most experienced F-15 pilots. Although only a captain, Alan Miller — known, inevitably, as Killer Miller — had logged nearly two thousand hours in an Eagle cockpit and was a Fighter Weapons School graduate, a distinction reserved for the best. Miller had worked after college for the McDonnell Douglas Corporation, maker of the F-15, until deciding he would rather fly the plane than build it.

He crossed the border at thirty thousand feet, searching for the six Iraqi Fulcrum fighters spotted by AWACs near Al Taqaddum Air Base west of Baghdad. Never before had he flown "clear BVR," authorized to shoot any aircraft beyond visual range without confirming it as an enemy with his eyes. For Miller, as for many pilots on January 17, this first combat mission was a chance to validate his mastery of airmanship and prove himself worthy of that exclusive fraternity of American fighter aces stretching back to Eddie Rickenbacker.

The honor of leading the initial mission also was the first tangible benefit of the miserable weeks of training flights in the dark. The issue of which pilots would fly during the day and which at night had caused a rift in Miller's squadron, one that deepened every day. In December, after arriving in Saudi Arabia from their base in Germany, the pilots had been evenly sorted into two shifts by the 53rd Tactical Fighter Squadron commander, Lieutenant Colonel Randy Bigum.

Inherently more dangerous, night flying was practiced only fitfully in peacetime; for an extended campaign, it also required the pilots to become nocturnal creatures capable of sleeping through the daytime bustle of a busy air base. Excluded from diurnal decisions and mission planning during the final weeks before the war, the night crew of the 53rd TFS had begun to feel slighted. Many had trouble staying awake on their long night training flights, only to find insomnia a problem during the day. Querulous and short-tempered, they snapped at their daytime comrades or vented their frustrations in the green-covered "doufer book," a squadron log used to register complaints and grievances.

Of greater concern to Lieutenant Colonel Bigum was drug addiction. Issued the potent amphetamine Dexedrine and the sedative Temazepam by flight surgeons to help them adjust to their new schedules, some pilots during the preceding month had become psychologically if not physically dependent on the medications, dubbed "go" and "no-go" pills. Although fighter pilots had used amphetamines since World

War II, particularly for monotonous transoceanic flights, drug use in the Persian Gulf operation was unprecedented. Nearly two thirds of American fighter pilots took amphetamines during Desert Shield and more than half would do so during Desert Storm — a remarkable development for an institution that considered itself "drug free."

When an alarmed Bigum finally realized from the flight surgeon's records how many go and no-go pills were being consumed, he promised to switch the day and night shifts a month into the war. In a shrill, bruising meeting in mid-January, he had confronted the night pilots, challenged them to adjust their inner clocks without chemicals, and banned amphetamines. No-go pills would be permitted in measured doses to help the men sleep after the stress of a night mission. "You don't know how tough this is," one pilot pleaded. "It's not as easy as you say. It's the hardest thing we've ever done." Bigum was sympathetic but adamant; his pilots, angry and frustrated. Some, he suspected, would find ways to obtain bootleg pills.

Captain Miller was not among them, but the schism in the squadron upset him. The war had just begun and already many pilots had purple bags beneath their eyes; creases marred their cheeks from the oxygen masks they wore during endless training flights. Miller still found it impossible to sleep soundly during the day without sedatives.

Scanning his radar scope, listening for AWACs' instructions, he tried to focus on his job. Fighter pilots hoped to begin sorting out the enemy at a range of forty miles, firing their air-to-air missiles at fifteen. Once the opposing aircraft crossed paths, a point known as "the merge," the dogfight often became a mêlée, a "furball" in pilot slang.

Yet as Miller raced closer to Al Taqaddum, he began to suspect there would be no merge, no furball, no enemy planes twisting to escape. The Fulcrums spotted earlier had fled; his radar scope was confoundingly clean. And instead of Iraqi jets, the sky abruptly came alive with missiles and antiaircraft fire. As he crossed the Bahr al Milh, a large lake west of Karbala, the Eagle's cockpit filled with electronic beeps and flashes warning him of Iraqi SA-6s, SA-8s, and other SAMs.

Miller banked sharply to the west. To gain speed, he punched off his auxiliary fuel tanks, despite the risk of having them collide with his turning aircraft. The tanks tumbled clear into the darkness. Three other Eagles trailed behind him, but within seconds the tidy formation collapsed as the pilots frantically jinked to evade the oncoming missiles. Miller dipped one wing in a sharp, slicing dive to avoid a SAM, then pressed the buttons to eject a cloud of chaff, for decoying radar-guided missiles, and four hot flares, designed to lure away heat seekers.

It was a foolish move, prompted by ten years of daytime experience.

The flares ignited with milky brilliance, enveloping the jet in light and momentarily blinding Miller. Unable to see the rising salvo of missiles but aware of the audio warning of a second SAM lock-on, Miller again dipped a wing to break toward the earth. The enemy missile shot past him. A third warning signal sounded and once more he pitched the control stick into a sharp plummet.

After easing the plane out of its dive, Miller checked the altimeter and was alarmed to see that he had slipped below fifteen thousand feet and was well within reach of antiaircraft guns. He instinctively shoved the throttle into afterburner for extra thrust. A yellow nozzle of flame gushed from the jet's tailpipe, drawing the attention of the Iraqi gunners below. As their bursting rounds pursued him, the Eagle struggled for altitude. In sixty seconds Miller had climbed to twenty thousand feet. The gunfire died away.

Pilots had long known that they were most vulnerable during their first eight or ten combat missions. For those who lived through the initial panicky mistakes, the odds of survival increased dramatically. Killer Miller had survived, but he knew his performance was unworthy of his training, of his airmanship skills, of the grand, silk-scarf tradition of American fighter pilots. Angry and embarrassed, he eased the Eagle back to thirty thousand feet. He had been lucky this time, but luck was nowhere more fickle than in the air.

Past the scattered Eagles swept several dozen aircraft headed toward greater Baghdad. To Iraqi gunners, the armada appeared to be a bomber fleet intent on pummeling the Iraqi capital with blows to complement the earlier strikes by F-117s and Tomahawks. Search and tracking radars, many of which had been kept unused lest they betray their positions, now flickered to life. Unlike the stealth fighters, these planes were easily detected. Jubilant Iraqi batteries began reporting the flaming wreckage of many allied aircraft.

In this, as was so often to be the case, the Iraqis displayed fatal misjudgment. The aircraft were not bombers; they were drones. Some, the BQM-74s, had been launched from airfields south of the Saudi border. They flew programmed routes that put them in orbit near Baghdad; only thirteen feet long, each powered with a small jet engine, they bore radar reflectors that lent them the appearance of larger warplanes. Others, known as Tactical Air-Launched Decoys, were gliders flung from Navy aircraft.

Three minutes behind the drones — and forty minutes after the AT&T building had first been struck — a real force of seventy aircraft formed a picket line thirty miles south of Baghdad. Flying in silence —

"zip lip" — the pilots were permitted a single radio call: *magnum*, the code word for launching a radar-killing HARM missile. Again and again the call rang out, and more than two hundred HARM missiles sped north. Designed to home in on enemy radar beams, the HARMs fell like a volley of arrows on dozens of Iraqi radar dishes. British pilots also launched new and even more sophisticated ALARM missiles. After climbing to seventy thousand feet, the ALARM popped open a parachute and floated to earth while searching for Iraqi radar emissions; on finding a suitable target, the missile jettisoned the parachute and dived into the radar site.

The allied attack was clever and sophisticated, relying on a tactic similar to that employed by the Israelis against Syrian troops in the Bekaa Valley in 1982. In less than an hour, the Iraqi air defense network, which had cost billions of dollars and taken years to construct, was crippled. Hundreds of SAMs and tens of thousands of antiaircraft rounds would still be fired and several dozen allied planes lost. But the Iraqis' ability to orchestrate a coherent, unified defense was gone. If they were to fight, it would be, as Rommel had foretold, as savages.

## Salman Pak, Iraq

Two more waves of stealth fighters swept across the border roughly an hour behind the first, oblivious of the earlier success because of a strict radio blackout. A dozen aircraft flew in one batch, eight in the other. Their targets included additional strikes against the Iraqi air force and air defense headquarters, and the presidential retreat at Abu Ghurayb. They also headed for biological weapons bunkers, television towers, Saddam's bunker in Baghdad, ammunition dumps, more air defense complexes, and a radio transmitter. The barrage of missile and antiaircraft fire around the capital had not abated; on the contrary, the earlier attacks by Tomahawks, F-117s, HARMs, and decoys provoked the Iraqi gunners to shoot with abandon.

Fifty miles south of his target, Major Mike Mahar saw what he assumed were the fiery splashes of bombs rippling across the landscape. Yet there were no allied bombers in front of him. Puzzled, he craned his neck for a better look over the stealth fighter's beaklike nose. The flashes, he abruptly realized, came from gunfire and SAM batteries.

Sheets of fire swept up toward him. Great black puffs detonated above and below, buffeting the airplane. The stench of cordite seeped into the cockpit. Looking down through the tracers, Mahar spotted long strings of headlights from Iraqis fleeing their capital. Gooseflesh covered his arms; the hair stood up on the back of his neck. Trembling uncontrollably, he lowered the seat as far as it would go until no more than the top of his helmet was visible from outside the cockpit. With the target area still more than five minutes ahead, Major Mahar had not the slightest doubt that he was about to die.

He considered his options. He could press on and perish, or cut and run. If he turned back, someone else would have to fly the mission; another pilot would die in his place. The shame would be unbearable. But the thought of his children growing up without a father stabbed at him. Before taking off from Khamis Mushait, he had written farewell letters — one to his four-year-old daughter, one to his two-year-old son. He had tried to imagine what he would have recalled if his own father had vanished when he himself was two. He explained in the letters that he had sworn an oath to give his life, if so ordered, to defend the Constitution. Yet the vow taken when he was a seventeen-year-old cadet had a different cast now that he was thirty-six and heading into combat.

Even as he sealed the letters Mahar hadn't expected war. Through the suiting up, the final briefs, the preflight checks, he awaited the order to stand down, the word that Saddam Hussein had begun to withdraw from Kuwait. When he taxied down the runway and lifted into the night sky he had listened for the order to return to base.

The airplane dragged him northward. Now he could hear the fierce boom of explosions outside the cockpit. Mahar had spent eighteen years in uniform preparing for this; before flying stealth fighters he had been a Strategic Air Command pilot, trained to attack the Soviet Union with nuclear weapons. But never had he anticipated the raw terror that had hold of him. How foolish he had been to think of war as a glamorous adventure. When the first stealth squadron was deployed to Saudi Arabia in August, Mahar had been among those ordered to remain in Nevada; refusing to obey, he stowed away on an Air Force tanker to Jiddah, then hid for a week at Khamis until the colonels finally relented and let him stay. Now that seemed a childish stunt.

In his mind's eye he pictured the lethal Iraqi round leaving the gun muzzle below, streaking up, blasting the plane to pieces. He was grateful the F-117 had only a single seat; at least he didn't have another life to worry about. He shifted uncomfortably, feeling a bruise rising beneath the 9mm pistol holstered under his arm. The booming around the plane seemed to grow louder. If he could make the noise stop, drown it out, perhaps his nerves would steady. He thought of Christmas three weeks before, and the caroling at Khamis. He began to sing, softly at first:

Oh, the weather outside is frightful,
But the fire is so delightful . . .

By the time he reached the refrain, Mahar was in full throat:

Let it snow, let it snow, let it snow!

He belted out another verse, and another, while scanning the terrain with his infrared sensors until he found his target just southeast of the capital: a biological weapons bunker at Salman Pak, one of Iraq's best-defended military installations. Centering the cross hairs, Mahar punched loose a bomb. As it swept toward the bunker, guided by the F-117's laser beam, a quick movement above the plane distracted him. He glanced up to see a missile streak past the canopy and detonate in a brilliant burst of yellow. The aircraft shook and seemed to falter.

For ten seconds the flash left him sightless, beset with new spasms of fear. As his vision returned, he saw that the altimeter and other instruments showed the plane was plummeting. But before he could react, the needles stopped whirling and the red warning lights went out. The shock wave from the explosion had temporarily knocked the instruments askew, but the plane appeared to be unscathed.

The bomb, he knew, had missed, fallen far wide of the target. Mahar hardly cared. He was alive. Exhilaration began to creep over him, as intense as terror had been moments before. He would be back to drop more bombs, as many as it took. Like Killer Miller in his F-15, like hundreds of novice pilots across the war zone, he now was blooded.

Banking the aircraft around to the south, Mahar was troubled only by the prospect of being the sole survivor of this terrible mission. No one else, he was certain, could possibly come home alive.

## Riyadh

Major Mahar was wrong. They all came home. All of the stealth fighters, all of the Eagles and Ravens, Falcons and Harriers, Warthogs and Aardvarks, Tomcats and Weasels. Eight search-and-rescue helicopters sent deep into Iraq to snatch up the expected covey of downed pilots also returned, happily empty. Of the thousand sorties flown this first night, one allied pilot died. Lieutenant Commander Scott Speicher, flying from the U.S.S. *Saratoga* in the Red Sea, was shot down by a MiG-25 that apparently slipped through a gap in AWACs coverage, spotted the flash of a HARM fired from Speicher's F-14, and killed him with a missile.

Across a 600-mile front, from the Iraqi naval base at Umm Qasr in the east to six Scud sites in the far west, the marauders bombed, rocketed, mined, and missiled. Lumbering B-52s, flying in combat for the first time since Vietnam, loosed hundreds of bombs on four airfields — As Salman, Glalaysan, Wadi al Khirr, and Mudaysis. Twenty-two British Tornados either dropped runway-cratering munitions or fired radar-killing ALARM missiles at targets that included Al Asad and Al Taqaddum airfields in central Iraq.

Navy pilots from *Saratoga* also struck Al Taqaddum aircraft hangars with 2000-pound laser-guided bombs; they were followed by planes from U.S.S. *Kennedy,* which hit a MiG maintenance depot. Marine Corps A-6s struck a rail yard southeast of Basrah. Thirty-eight Air Force, Navy, and Saudi planes together shattered Shaibah Airfield near Basrah. Fifty-three F-111Fs, flying in attack packages of four to eight jets each, bombed chemical bunkers at H-3 airfield, Salman Pak, and Ad Diwaniyah; others hit airfields and control towers at Jalibah and Balaad Southeast in Iraq, and Ali al Salem and Ahmed al Jaber in Kuwait.

There were, inevitably, mistakes: targets missed or never found; planes aborted before takeoff; pilots' sins of omission or commission; missiles that misfired or avionics that malfunctioned. Half of the attackers of Ali al Salem veered away in the face of intense Iraqi fire. A strike against Saddam's summer home in Tikrit went sour when one bomber turned back and another — aiming for the vast skylight in the palace roof — abruptly broke off the attack to evade three missiles. F-111s returning to their base at Taif in western Saudi Arabia found themselves so low on fuel — the pilots screamed over the radio for aerial tankers — that half the wing ended up scattered at emergency airfields across Saudi Arabia. And of the thirty-two bombs dropped by the second and third waves of F-117s, only fifteen hit their targets.

It mattered not a whit. In the twentieth century, only one sizable war had been decided by a single battle on a single day: the 1967 conflict between Israeli and Arab. Now there were two. Schwarzkopf, Calvin Waller, and Chuck Horner watched with ebullient disbelief from their Riyadh headquarters as squadron after squadron reported targets struck with no losses; before first light they knew that the plan had worked. Iraq was doomed.

In the nine major wars waged heretofore by the United States, the nation had been chronically unprepared to fight. From Bull Run to Kasserine Pass, from the Pacific Fleet at Pearl Harbor to Task Force Smith in Korea, the historical record of initial American sallies into combat was dismal. Rarely striking the first blow, habitually underrating the enemy and overestimating American fighting prowess, the country had forfeited thousands of lives before rousing and steeling itself. This first battle, by contrast, had found the nation ready. Though hardly flawless, the combination of tactical ingenuity and raw power in the initial attacks helped ensure that war would not become a quagmire.

Across the war zone, combatant and spectator sifted through the battle reports and wondered what would happen next. At Khamis Mushait, maintenance crews lined the runways, every man saluting, as the returning F-117s touched down. Like bats, the stealth fighters retired to their caves until covering night fell again. But at twenty other bases across the Arabian subcontinent and on six aircraft carriers, fresh pilots roused themselves to prepare for new attacks, which would be launched with the rising sun.

And in Baghdad, where clapping Iraqis had trooped to their shelters chanting, "Palestine belongs to the Arabs and Kuwait belongs to Iraq," the long night of bombs and booming guns had doused any trace of festivity. Now terrified children clutched their mothers, and their wailing found echo in the muezzins' call to morning prayer, broadcast from minarets at 5:20, as the last allied planes sped away. Then the guns fell silent and a false dawn glowed in the east, and the criers' pious chants carried over the city with the precise pitch of lamentation.

# 2

# First Day

Long after the last bomb had fallen, the brilliance of the allied campaign would overshadow all but the proximate cause of the conflict — Iraqi cupidity. Then, slowly, a larger tale of mutual miscalculation and blurred vision would emerge, staining the image of virtue triumphing over iniquity. Whether war could have been averted is unknowable. Given Saddam Hussein's greed and the clear danger he posed, his defeat would stand as a victory for the United States and its allies. But the prelude to war, not unlike its aftermath, was neither simple nor unambiguous.

American policy toward Iraq parsed neatly into two phases: pre- and post-invasion. The former was distinguished by realpolitik run amok. In the early 1980s, eager to contain Iran's Islamic fundamentalism, Washington had begun tilting toward Baghdad during Saddam Hussein's war with Tehran. In 1982, Iraq fell from the U.S. government's list of nations supporting international terrorism, paving the way for Saddam to start receiving American aid a year later. In 1984, the CIA covertly began sharing intelligence data with Baghdad to forestall an Iraqi battlefield collapse. With Washington's secret approval, several Arab nations also shipped American-made howitzers, bombs, and other military equipment to Saddam.

From this expedient beginning grew a feckless strategy that sought, in President Bush's sad idiom, to draw Iraq "into the family of nations." Weeks after the Bush administration took office, in early 1989, a U.S. State Department report noted that Iraq was sheltering Palestinian terrorists while amassing a vast arsenal and again hectoring Kuwait over the disputed border. Other intelligence briefs recounted Saddam's efforts to purchase nuclear bomb components and his barbarous use of chemical weapons to purge Kurdish villages in northern Iraq. Nevertheless, in October 1989, Bush signed National Security Decision Directive 26, secretly commanding the U.S. government to expand trade and other ties with Baghdad in an effort to moderate Iraqi behavior while helping to rebuild the war-ravaged country. NSDD 26 also decreed that the United States would impose economic and political sanctions if Saddam continued to develop weapons of mass destruction.

This philosophy of reward or punishment, perhaps plausible in theory, proved bootless in practice. Preoccupied with the collapse of the Soviet Union and the reunification of Germany, the administration ignored ample signs that Saddam, after his war with Iran ended in 1988, was not so much rebuilding as rearming. Despite the Pentagon's qualms over his voracious appetite for military hardware, the decision to cultivate better commercial and political ties held sway. More warnings accumulated in Washington — from U.S. Customs, the Defense Intelligence Agency, the CIA, and officials in the Departments of State and Energy — but they were silenced by the weight of NSDD 26.

In the fall of 1989 a federal prosecutor and the CIA both noted evidence of possible illegal transactions by the Atlanta branch of Italy's Banca Nazionale del Lavoro. The bank had funneled more than $4 billion in loans and loan guarantees to Iraq, much of which was diverted for weapons purchases. Even so, Secretary of State James A. Baker III and others successfully pressed for White House approval in November 1989 of another $1 billion in government-backed agricultural credits for Iraq, already the world's major beneficiary of the program. Three months later Baker's own State Department concluded that Iraq's "human rights record [remains] abysmal."

Yet so, unfortunately, was the American record of courting Saddam. Not unlike the periodic efforts by Arab monarchies to control Baghdad with cash payoffs, Washington wagered billions of dollars in its attempt to convert a sow's ear into a silk purse. (When Iraq defaulted on loans guaranteed by the agricultural program, U.S. taxpayers would be stuck with a bill of more than $1 billion.)

In the five years preceding Saddam's attack, the United States also had approved nearly eight hundred export licenses permitting Saddam to buy technology with potential military applications, including products eventually used in missile, artillery, chemical, biological, and nuclear weapons programs. Such purchases were only a small fraction of those made by Iraq in Europe, the Soviet Union, and elsewhere. The British government, for example, permitted illegal sales of arms-manufacturing equipment in order to maintain an intelligence pipeline into Iraq.

Subsequent claims by Bush's political adversaries that the United States substantially rebuilt Saddam's army were exaggerated. But the administration certainly contributed to Iraq's emergence as a Middle

East military power rivaling Israel. Despite the dismay of some U.S. officials over a policy that had produced no perceptible modification in Saddam's behavior, an interagency meeting in early 1990 concluded it was too soon to abandon the effort to "engage" Baghdad. In March, Iraqi agents were caught trying to smuggle nuclear triggers from Britain, yet secret American intelligence sharing continued until early summer. As late as June, the administration resisted congressional proposals to slap Iraq with economic sanctions.

In the final weeks before the August invasion of Kuwait, American efforts to deter Saddam remained inconsistent if not incoherent. On July 17, 1990, the Iraqi leader had publicly threatened to use force to resolve his disputes in the gulf. Later, recalling such warnings, Bush's critics would accuse the administration of unconscionable passivity, but this distorted the record. On July 19 Baker cabled the American ambassador in Baghdad, April C. Glaspie, who subsequently informed the Iraqis that "disputes should be settled by peaceful means, not through intimidation." Five days later Glaspie delivered another message from Baker, which noted that "we remain committed to ensure the free flow of oil from the gulf and to support the sovereignty and integrity of the gulf states."

In a much-debated meeting with Saddam on July 25, Glaspie repeated the American desire for a settlement by "peaceful means," while noting that "we have no opinion on the Arab-Arab conflicts, like your border disagreement with Kuwait." President Bush, she added, hoped to "broaden and deepen our relations with Iraq." Glaspie's casual attitude alarmed some officials in Washington, particularly at the Pentagon. "This stinks," declared Paul Wolfowitz, the defense under secretary, after reading the ambassador's dispatch. "We've got to get a stronger presidential message." On July 28, Bush sent his own three-paragraph dispatch to Saddam, asserting that "differences are best resolved by peaceful means and not by threats . . . We still have fundamental concerns about certain Iraqi policies and activities."

Although not wholly passive, the administration's signals were hardly tantamount to a warning shot across the bow. Washington dithered in a funk of disbelief even as Iraqi forces marshaled for the southern onslaught. Most policymakers assumed that Saddam was bluffing or — as Norman Schwarzkopf believed — that at most he would occupy the Rumaylah oil field and Bubiyan Island. Bush's aides met on August 1 to consider a sterner warning, but the opportunity for an effective démarche had gone begging.

At 1 A.M. on August 2, three Iraqi Republican Guard divisions crossed the frontier with nearly a thousand tanks. The Tawalkana and Hammurabi divisions rolled south to seize the Al Jahra heights overlooking Kuwait City, while the Madinah veered farther west to block approaches from the Saudi border. Thirty minutes later, special forces landed by helicopter in the Kuwaiti capital and began seizing key government buildings. The emir fled to Saudi Arabia, but his brother was killed in a shootout at the Dasman Palace.

By early evening, the capital had fallen. Iraqi tanks advanced south to capture Kuwaiti ports. Within four days portions of eleven divisions occupied the emirate. Kuwait, Saddam proclaimed in his message of annexation, no longer existed except as "the 19th Province, an eternal part of Iraq."

American maneuvering would prove as adroit, resolute, and ingenious after the invasion as it had been clumsy, amoral, and unimaginative before. The United Nations Security Council on August 2 condemned the attack in the first of a dozen resolutions orchestrated by Washington to confront Saddam with a wall of international opposition. Three days later, with the obduracy that would mark his public pronouncements for the next six months, Bush declared that "this will not stand, this aggression against Kuwait." He and Baker began a 166-day blitzkrieg of "coercive diplomacy" that imposed onerous economic sanctions on Iraq and eventually extracted military contributions from thirty-eight nations.

Among the first and most important allies in the coalition were the Soviet Union — Iraq's erstwhile Cold War patron — and Saudi Arabia, where King Fahd recognized that the House of Saud was unlikely to survive in a Persian Gulf dominated by Saddam Hussein. Four days after the invasion, Fahd agreed to a massive buildup of American troops. Within ten days, more than two hundred U.S. combat planes had flown to Saudi bases. In late August Schwarzkopf, who as head of Central Command assumed the role of theater commander-in-chief, arrived in Riyadh, where he began planning for the ejection of Iraqi forces from Kuwait even as he prepared to defend the Saudi kingdom.

Whether Saddam had designs on the even vaster oil fields of eastern Saudi Arabia was furiously debated in Washington and Riyadh. In retrospect, it appears he did not. But in early August, the CIA concluded that Saddam's troops had the wherewithal to reach the Saudi capital in three days. CENTCOM postulated three possible invasion routes, including a drive straight down the coast to Dhahran. By late August, Schwarzkopf had assembled a force of seven Army brigades, three aircraft carrier battle groups, fourteen fighter squadrons, seventeen thousand Marines, and, on the Indian Ocean island of Diego Garcia, a squadron of B-52s. By early October, Schwarzkopf reported that the "window of vulnerability" had sufficiently narrowed to guarantee a successful defense of the kingdom.

Through the fall and into the winter the extraordinary allied buildup continued at a pace far exceeding that in Vietnam a quarter-century before. A transport plane landed at Dhahran Air Base every seven minutes around the clock, hauling war matériel that ultimately reached seven times the cargo tonnage flown in the Berlin airlift of 1948. The call-up of National Guard and reserve units begun in late August eventually grew to 245,000 men and women, nearly half of whom served in Southwest Asia. By November, the United States had deployed sixty Navy ships, more than a thousand airplanes — including 590 "shooters" — and a quarter of a million troops. Bush's decision at that time to double the American commitment, including the unprecedented movement of an entire Army corps from Europe, would raise the total allied force to four thousand tanks, 150 ships, and more than 600,000 troops by the time the war began.

A diplomatic juggernaut accompanied the military buildup. White House telephone logs in the month after the invasion listed sixty-two calls from the president to various foreign leaders. Bush and Baker cajoled, pleaded, and offered sundry compensations to weave together their alliance: Egypt was forgiven its $7 billion debt to the U.S. Treasury; Turkey got textile trade concessions; China was again pardoned for suppressing dissident democrats in Tiananmen Square; Syria received absolution for many of the same sins of which Saddam now stood accused, including state-supported terrorism. The president also avoided sharp condemnation of the Soviet crackdown in the Baltics in order to keep Moscow as an ally in the gulf.

Bush's diplomatic virtuosity was matched by his deft political touch at home. On November 30, a day after the Security Council authorized "all necessary means" to liberate Kuwait, the president offered to exchange envoys with Baghdad in an attempt to resolve the crisis short of war. To further mollify a divided Congress and a wary American public, he subsequently agreed to a congressional vote on whether military force should be used against Saddam. Gradually, the president outmaneuvered those advocating an indefinite reliance on economic sanctions to choke Iraq into submission.

Bush consistently couched his Persian Gulf policy as a crusade "for what's right," but he was not beyond the use of deception. He initially denied considering American military intervention; exaggerated the Iraqi threat to Saudi Arabia; deliberately minimized Arab efforts to effect a diplomatic solution; dissembled regarding the size of the American deployment and his willingness to link Saddam's withdrawal from Kuwait to the settlement of other problems in the Middle East; and concealed his own belief, arrived at soon after the invasion, that war was more likely than not. Whether the president was uncertain of his own mind or simply convinced that the end justified the means is difficult to assess; perhaps a bit of both.

Certainly his greatest ally through the five-month prelude to war was Saddam Hussein. It is no small irony that the Iraqi leader appears to have believed even before August 2 that the United States, in league with the Arab monarchies, was determined to crush his regime, a baffling assessment, given Washington's efforts to appease him. He compounded his paranoia with another strategic miscalculation: that the industrialized West could turn the other cheek when a fifth of the world's trillion-barrel oil reserve fell under Baghdad's dominion.

Saddam's subsequent miscalculations were legion. He had attacked when world oil supplies were plentiful — thus weakening his economic leverage — and when relations between world powers were better than at any time since the Congress of Vienna, in 1815. In confronting the United States, he picked a fight with a nation that had fifteen times the population of Iraq and eighty times its gross national product, on the apparent assumption that Washington would display no more staying power in the Middle East than it had after the U.S. Marine Corps barracks was bombed in Beirut in 1983.

Saddam failed to reassure Fahd of his benign intentions toward Saudi Arabia, thus pushing Riyadh into Washington's arms. He forfeited any hope of world sympathy by seizing thousands of Western hostages in Iraq and Kuwait, then simplified Schwarzkopf's military planning by setting them free in December. He rejected French, Arab, and Soviet mediation efforts that could well have given him at least a partial victory without bloodshed. He calculated, wrongly, that his Arab opponents could not survive the political consequences of war with a brother Arab, particularly if he lured Israel into the fight.

A generation earlier, Omar Bradley had warned against waging "the wrong war, at the wrong place, at the wrong time, and with the wrong enemy." Whatever policy failings had contributed to the crisis in the Persian Gulf, George Bush was now convinced, with a certitude not widely shared by his fellow Americans, that he had the rare opportunity to wage a wholly righteous conflict: the right war, at the right place, at the right time, and against the right enemy.

On January 9, James Baker met for nearly seven hours in Geneva with the Iraqi foreign minister, Tariq Aziz. The secretary presented a three-page letter from Bush to Saddam in which the president warned,

"There can be no reward for aggression. Nor will there be any negotiation. Principle cannot be compromised." Aziz refused to accept the missive and repeated Baghdad's intransigent stand. "Perhaps," he told Baker, "it will just come down to fate."

Throughout the nation, in offices and on factory floors and across millions of supper tables, Americans passionately debated the proper course. On January 12, an emotional congressional debate ended with the House of Representatives authorizing the use of force by a vote of 250 to 183; the Senate concurred, 52 to 47. More in sorrow than in anger, the nation steeled itself for combat. Not since the Cuban missile crisis had the world endured such gut-wrenching brinksmanship. Not since August 1914 had the world witnessed such a ponderous prelude to conflagration: armies trudging forward; reserves and resources mobilized; civilians evacuated; trenches deepened. "For whatever purpose God has," General Schwarzkopf wrote in a letter to his family, "we will soon be at war."

And now, war would serve as its own justification.

## Washington, D.C.

Beneath the concrete portals of the Pentagon's River Entrance, in a drab basement vault numbered BF922, the pummeling of Iraq in the war's initial hours was played out, bomb by bomb, in the imagination of the man most responsible for plotting the aerial destruction of Saddam's war machine. Colonel John A. Warden III saw it all in his mind's eye: the phones gone dead, the lights gone out, the flames licking from a hundred shattered targets. Sitting at a wooden table with the inch-thick timetable of the first night's attack before him, he watched the clock and punched the air in a small gesture of triumph whenever the moment passed for the obliteration of another crucial target. Five months before, he had assured Norman Schwarzkopf and Colin Powell that air power would lay Iraq low; now, he believed, that promise was redeemed.

Less than a week after the invasion of Kuwait, Schwarzkopf had asked the Joint Chiefs for help in designing a counterstrike in the event of war. The job fell to Colonel Warden, who headed an Air Force planning cell called Checkmate. His first blueprint bore a striking resemblance to the war plan executed in January. In war no less than in peace, success has a thousand fathers, and paternity claims in the Persian Gulf War would mount in direct proportion to allied achievements. Yet no claim was stronger than Warden's.

Outside the Air Force he was largely anonymous. Within the service he was a figure of controversy — revered as a visionary, disdained as a zealot. A manner that seemed courtly and pensive to some struck others as patronizing and bookish. He roundly rebuked the Air Force — particularly the so-called fighter mafia at Tactical Air Command — for a narrow and defensive concept of war that dwelt excessively on support for the Army against a Soviet tank assault. The service, Warden insisted, had "lost its focus" by failing to think of combat in bold, strategic strokes, by wanting to hack at an enemy's limbs rather than thrusting at its heart.

Such criticisms — and the pontifical tone in which Warden sometimes delivered them — sat badly with many of his colleagues. Although Warden had flown nearly three hundred combat missions in Vietnam and later commanded an F-15 wing in Germany, his credentials as a warrior-aviator were considered suspect. Within the fraternity of senior Air Force generals, so tightly woven as to be dubbed the Blue Curtain, he was said to lack "good hands." No more damning slur could be cast at any pilot; the dexterity needed to operate the toggles and buttons in an F-15 was such that air crews spoke of "playing the piccolo."

But into those hands was committed the planning for the air campaign against Iraq. Warden had set to work beneath the stern stare of Henry (Hap) Arnold, the World War II air chief whose portrait hung on his office wall. Exactly eighty years before Desert Storm began, in January 1911, the first bomb had fallen from an airplane when a barnstorming pilot in San Francisco dropped a pipe full of black powder as a fairground stunt. Millions of tons of high explosives had fallen since, yet air power still was deemed a developing art fraught with uncertainty.

Pilots also had an unfortunate tradition of overpromising, as exemplified by Claire Chennault's boast to Franklin Roosevelt in 1942 that with just 150 fighters and forty-two bombers he could "accomplish the downfall of Japan." In the same year, the British air marshal, Arthur (Bomber) Harris, had predicted that the razing of German cities would effect the destruction of Nazi industry and morale. Yet the enemy's fighting spirit remained intact, industrial production continued to rise for two more years, and the national telephone system worked even as Red Army troops smashed through the suburbs of Berlin. Strategic bombing in 1945 was discredited and, in some quarters, condemned as an atrocity.

Warden, a tireless partisan of the supremacy of air warfare, was undaunted. He believed every warring state had "centers of gravity," the phrase of Clausewitz that Warden defined in his book, *The Air Campaign,* as "the point where the enemy is most vulnerable and the point where an attack will have the best chance of being decisive." Alexander the Great had destroyed Persian sea power by seizing enemy bases on the Mediterranean littoral, Warden wrote, thus successfully attacking the Persian center of gravity without ever waging a sea battle. The German Luftwaffe in 1940 had misjudged the British center of gravity by prematurely abandoning attacks on the Royal Air Force in favor of peripheral strikes on London and other cities.

Warden also believed that every modern state, regardless of size, had three to five thousand "strategic aim points." Find and strike those points, he decreed, and you paralyze the enemy whether he is Soviet or Iraqi. This was not a new premise. French aviators before World War I spoke of bombing *points sensibles,* sensitive spots whose destruction would block critical supply lines or production chains. Early air power theorists also postulated that a modern, centralized state could not endure substantial damage to its capital; some enthusiasts had predicted that an aerial "knockout blow" could cause enough civil upheaval to result in the overthrow of an enemy government. Giulio Douhet, the dogmatic and imperious Italian artilleryman who became air power's most celebrated theorist in the 1920s, even argued that combat in the air offered a merciful alternative to the carnage of ground fighting — a view still supported by many in the modern United States Air Force.

Yet theory had never quite lived up to expectations. "The potential of the strategic air offensive was greater than its achievement," British Bomber Command historians admitted after World War II. Warden believed, however, that modern munitions offered a precision that would marry theory and practice. A B-17 bomber in World War II had a "circular error probable" of 3300 feet, meaning that half of the bombs dropped would likely fall within that radius; mathematically, that meant it took nine thousand bombs to achieve a 90 percent probability of hitting a target measuring sixty by a hundred feet. (A bomber "can hit a town from ten thousand feet," one military writer observed, "if the town is big enough.") Even in Vietnam, hitting that sixty-by-hundred-foot rectangle required three hundred bombs. But an F-117 carrying laser-guided munitions could theoretically destroy the target with a single bomb.

On August 8, immediately after receiving his orders, Warden gathered a dozen officers from his Checkmate staff and began translating his beliefs into a war plan. On a blackboard, he drew five concentric circles, representing Iraq's centers of gravity. The innermost circle, he said, stood for Iraqi leadership. The next circle — the second most critical center of gravity — represented petroleum and electricity targets, without which a modern military was crippled. Third was Iraqi infrastructure, particularly transportation. Fourth was the Iraqi population. The fifth and outermost circle symbolized Saddam's fielded military force. "Strategic warfare," Warden declared, "is only indirectly concerned with what is happening on some distant battlefield." Air power could leap over the outer rings to devastate the enemy's core.

Each of Warden's five categories was then listed in column form for targeting purposes. Under LEADERSHIP, the Checkmate officers initially scribbled, "Saddam." (Two days later, after reflecting on the American legal prohibitions against assassinating foreign leaders and the importance of achievable objectives, this was amended to "isolate and incapacitate Hussein's regime.")

Warden subsequently propped a large satellite map of Baghdad against the wall. Using data from the CIA, Defense Intelligence Agency, and elsewhere, he began filling the map with target pins, eventually forty-five of them. To illustrate the devastating consequences of hitting just a handful of key sites, he also produced a map of greater Washington with equivalent targets: the White House; the Capitol; the Pentagon; Blair House; Camp David; CIA headquarters in Langley; National Security Agency headquarters at Fort Meade; power plants in northern Virginia and southern Maryland; telecommunications facilities for AT&T, MCI, and Sprint; FBI headquarters; Andrews Air Force Base. A two-inch snowfall paralyzes Washington, Warden told his officers, so imagine what blows to these targets would do to the city.

Checkmate's writ was to draft an attack scheme that could be executed by the end of August. As the plan took shape, Warden called it Instant Thunder in deliberate contrast to the gradually escalated Rolling Thunder campaign in Vietnam. (In 1964, the Pentagon had drafted a remarkably similar proposal with ninety-four strategic targets in North Vietnam; not until the Linebacker I air campaign in May 1972 did the White House authorize the bombing of most of those sites.)

On August 10, Warden and several other Checkmate officers flew to Central Command headquarters in Tampa for a quick conference with Schwarzkopf. The CINC looked weary after more than a week of frenzied preparations. On the flight to Florida, Warden and his deputies had talked about translating the concept of an air campaign into terms that would excite an Army infantry officer like Schwarzkopf. They latched on to two historical analogies, which Warden proposed to the CINC after a brief sketch of Iraqi vulnerabilities. The first was the famous Schlieffen Plan, drafted by the Germans before World War

I as a stratagem for skirting French strongholds; it called for attacking through Belgium and along the Channel coast. With air power, Warden believed, Schwarzkopf could similarly bypass Iraqi strongholds to attack the more critical centers of gravity. The Checkmate theorists discreetly avoided noting that Count Schlieffen's plan, when modified and executed in August 1914, led to four years of horrific trench warfare.

Warden's second analogy was more contemporary and eminently American. "You have a chance," he told the CINC, "to achieve a victory equivalent to or greater than MacArthur's Inchon landing — by executing an air Inchon in Iraq." (As the historian Barbara Tuchman once wrote, "Dead battles, like dead generals, hold the military mind in their dead grip.") Schwarzkopf stirred with renewed animation. He had been leery of Warden as a devotee of "the Curtis LeMay school of Air Force planners," but he liked the plan and the ideas behind it. I want to see more, he said. Come back in a week.

The next day, August 11, Warden outlined Instant Thunder for Colin Powell in the chairman's Pentagon office. "This plan may win the war," Warden said. "You may not need a ground attack . . . I think the Iraqis will withdraw from Kuwait as a result of the strategic campaign."

Powell was intrigued, but he nurtured an infantryman's instinctive wariness of "flyboy promises." He ordered Warden to draw other services into the planning. "I don't want to end the war without certain things being accomplished," the chairman added. "I don't want Saddam Hussein to walk away with his army intact. He could claim a political victory." He leaned forward. "I want to leave those tanks as smoking kilometer signposts all the way back to Baghdad."

For the next week, Checkmate labored round the clock to refine Instant Thunder and to draft a rudimentary operations order laying out an attack sequence. Warden napped on the couch in his office, leaving only long enough to dash to the Pentagon gymnasium for a shower and shave. On August 17, he again flew to Tampa. At 10 A.M. in the CENTCOM amphitheater, Warden flashed the first slide of his presentation, classified Top Secret LIMDIS (limited distribution): "What It Is — A focused, intense air campaign designed to incapacitate Iraqi leadership and destroy key Iraqi military capability in a short period of time. And it is designed to leave basic Iraqi infrastructure intact. What It Is Not — A graduated, long-term campaign plan designed to provide escalation options to counter Iraqi moves."

Warden showed Schwarzkopf eighty-four targets, now classified in ten target categories. Among them were ten air defense targets, eight chemical, five leadership, nineteen telecommunications, ten electrical. A single target might have many strategic aim points. This attack, Warden estimated, would knock out 60 percent of the electrical power in Baghdad, 35 percent in Iraq overall. Seventy percent of the country's oil-refining capacity would be destroyed. Several squadrons of A-10 Warthogs would be held in reserve to deal with any Iraqi tanks that ventured south. When he was listing potential hazards, Warden flashed another slide that read: "Scuds. Long-range missiles. Chemical capable. Present real problem."

With fair weather the campaign should take about six days to complete, he said, using 670 airplanes averaging roughly a thousand sorties a day. Under a slide entitled "Expected Results," Warden predicted, "National leadership and command and control destroyed. Iraq's strategic offense and defense eliminated for extended period. Iraq's economy disrupted. Iraq's capability to export oil not significantly degraded."

Schwarzkopf complimented Warden but stressed the importance of also bombing Iraq's army of occupation, particularly the Republican Guard divisions, which contained Saddam's best troops. As the briefing ended, an Air Force general sitting in on the session asked Schwarzkopf, "Sir, is there anything else we can do for you?" Schwarzkopf pointed to Warden. "Give me him. I want him to take this plan to Horner."

On August 20 at 2 P.M., Warden and three of his lieutenant colonels trooped into a third-floor office at the Royal Saudi Air Force headquarters in Riyadh. The Checkmate officers lugged a crate of razors, lip balm, and sun block brought as a good-will offering to their Air Force comrades in Saudi Arabia. Lieutenant General Horner, who had arrived two weeks earlier to serve as Schwarzkopf's point man during the initial deployment, sat at the corner of a conference table. Glancing disdainfully at Checkmate's crate, he uttered a curt *harrumph.*

From that sour beginning, the session dissolved in acrimony. Horner had never read Warden's book on air campaigns, but he was wary of his brethren in blue who made sweeping claims for Air Force hegemony. "Air power airheads," he called them. The previous April, Horner and his Ninth Air Force staff had worked closely with Schwarzkopf on an exercise code-named Internal Look, which postulated an attack by Iraq against Kuwait and Saudi Arabia. Horner had his own ideas about which targets to hit, particularly oil refineries and bridges. Like many senior officers, he resented meddling from Washington; it reminded him of Lyndon Johnson picking targets in the White House while boasting that American pilots in Vietnam couldn't "even bomb an outhouse without my approval." In Horner's view, Warden was a Pentagon intruder peddling an inflexible scheme, one not open to further discussion.

Warden struggled through his briefing amid muttered gibes about "armchair generals" and "the typical academic crap you'd expect out of Washington." Horner's immediate concern centered on several Iraqi armored divisions poised to strike from southern Kuwait. "What happens if the Iraqis attack?" he asked. "They have a lot of tanks ready to come south."

Warden shook his head. "Sir, I don't think they're going to attack. You're too concerned with the defense."

Warden's deputies winced, but it was too late. An awkward silence filled the room. "Sir, I apologize," Warden said at last. "I may have gone too far." Horner replied gruffly, "Apology accepted," and the briefing sputtered to a close. Horner then turned to each of the Checkmate lieutenant colonels and asked if he would remain in Riyadh temporarily. For Warden, there was no invitation, only a terse and final "Thank you."

That evening Warden reported to Riyadh air base for the return flight to Washington. He sagged with disappointment and exhaustion. His deputy, Lieutenant Colonel David A. Deptula, tried unsuccessfully to brighten his spirits: "Don't worry, sir. We'll try to maintain the integrity of the plan." Warden nodded glumly, turned, and walked to the waiting jet.

John Warden's Instant Thunder underwent many changes in the next five months. The number of target categories grew from ten to twelve; the number of strategic targets swelled from eighty-four to 386 on January 17 (by war's end the figure would surpass seven hundred); and the number of aircraft to attack those targets nearly doubled after Bush's decision in November to expand the American forces deployed to the gulf. At Horner's insistence, even the code name Instant Thunder vanished from all planning documents, supplanted by the more prosaic Offensive Campaign, Phase I.

The vast apparatus of American intelligence agencies dumped tens of thousands of pages of information and photographs on the strike planners. Some of it proved wretchedly inadequate, notably the analyses of Iraq's nuclear weapons program; other intelligence was sterling, including detailed descriptions of the KARI air defense system, first delivered in a secret analysis named Proud Flame by the Joint Electronic Warfare Center, then refined by SPEAR, a Navy intelligence unit outside Washington.

Beyond Horner's pot shots, Instant Thunder also took salvos from the Navy, which condemned the plan for focusing on Baghdad instead of enemy airfields and SAM sites. Warden's vision of an air assault that quickly brought Iraq to her knees was deemed politically naïve, requiring what one naval officer ridiculed as "the Mussolini ending, with the plucky natives throwing off their chains to string up the hated dictator."

Yet despite the sniping, the blueprint survived. Warden's essential ideas — emphasizing centers of gravity, focusing on leadership targets, leaping over Iraq's fielded forces to strike at the country's core — remained the principles behind Desert Storm. For Warden had left in Riyadh not only his plan but his proxy: Dave Deptula, who had promised to "maintain the integrity" of Instant Thunder. He did precisely that.

Deptula had been lent to Checkmate from the office of the Air Force secretary, where he had been a staunch advocate of strategic air power. Good-humored and unassuming, Deptula was an F-15 pilot and Fighter Weapons School alumnus with degrees in astronomy and engineering from the University of Virginia. Soon after arriving in Saudi Arabia, he gained Horner's confidence and became chief planner for Iraqi targeting in the top secret cell later known as the Black Hole. Deptula kept the Warden flame, talking often — though always discreetly — to his former boss by phone. (With similar tact, messages from Pentagon office BF922 were shorn of Warden's signature and the Air Force code designator for Checkmate — XOXW — before being shown to Horner.)

Deptula also was keeper of the master attack plan, a battered looseleaf notebook in which he listed the specific targets to be hit. His imprint could be found throughout the air campaign. One day, for example, he had realized that paralyzing the fortified Iraqi sector operation centers in the initial attack did not require eight 2000-pound bombs, as had been widely assumed. One or two GBU-27s, he calculated, would disorient the Iraqis enough to render them impotent the first night; the SOCs could then be pulverized at leisure on subsequent days. Deptula amended the master attack plan accordingly, freeing more than a dozen F-117s to hit other targets simultaneously. When the war ended, Horner would present him with a photograph inscribed: "Dave — you were the guru of the strategic air campaign that rewrote history. Thank you so much."

One other figure completed the Air Force troika steering the air war from Riyadh. Among the theater's flamboyant fighter pilots, Brigadier

General Buster C. Glosson was perhaps the most baroque. Brusque and profane, tireless and supremely confident, Glosson was anointed by Horner to be both chief targeter and commander of all Air Force wings in the gulf. That investiture, together with his autocratic bearing, lent Glosson's single star the authority of three or four. With Horner's blessing, he wielded his sovereignty like an ax. "If I don't see you here in two minutes," Glosson would bark on the phone to some delinquent subordinate, "I never want to see you again." Obstacles to his will he dismissed as "pimples on the ass of progress"; impertinence he brushed aside with a tart "Bullshit! Do you think I'm brain dead?"

Moon-faced, with a shock of white hair, incipient paunch, and North Carolina drawl, Glosson resembled less the prototypical fighter jock than a Capitol Hill lobbyist, which is what he would become after the war, when he served as the Air Force's legislative chief in Washington. In a quarter-century career he was said to have left behind more enemies than friends; in Riyadh he made more of both. Many disliked him for being a poseur — Navy officers privately called him Bluster — but even detractors conceded his ability to get things done. He spent countless hours on the telephone — coaxing better intelligence from Washington, badgering his wing commanders, cutting deals. Those whom he could neither command nor charm, he often outwitted. Like Horner, he employed the artful touch with Schwarzkopf, blending cajolery, dialectic, and raw resolve to prosecute the air war as he thought best. Never quite disobeying, he could be willfully dense, feigning confusion or choosing, as he put it with a faint smile, "to clarify the CINC's intent at a later time."

Of his airmen's prowess Glosson had no doubt. "It doesn't matter what the ground attack plan is," he assured Schwarzkopf in the fall. "Whatever you come up with will work because the enemy will have been so reduced from the air." Air power alone, he added, could eliminate half of the Republican Guard in just five days and the rest of Iraq's forces in the Kuwaiti theater in twelve days. In late December, Glosson met with the Army's generals outside Riyadh in a session that left many amazed by his brass. "I own every one of those sons of bitches who will be dropping bombs and they're going to do exactly what I tell them to do," Glosson declared. "When I tell them something is going to happen, it's going to happen even if I have to stack airplanes so high that they're the largest obstruction you face going north."

In addition to smashing strategic targets in Baghdad and elsewhere, allied planes would wage a relentless interdiction campaign to sever Iraqi supply lines and then "prepare the battlefield" with thousands of sorties against entrenched enemy forces. The first American soldiers attacking into Iraq, Glosson vowed, would find at least half of the enemy armor, artillery, and troop concentrations destroyed.

"Don't worry about the Iraqi fire trenches. There won't be one drop of oil in them. Don't worry about enemy helicopters. There won't be anything flying overhead but what we own. All of the bridges will be out over the Euphrates and the canal near Basrah. I guarantee it." The session ended and the generals disbanded, joking among themselves that the United States Army might as well pack for home since Buster Glosson planned to win the war alone.

He and Deptula were an effective pair — "symbiotic," the latter called it, aware that his studied patience complemented Glosson's impulsive dash. Deptula liked and admired Glosson. But when Major Buck Rogers arrived from Checkmate to work the graveyard shift in the Black Hole, Deptula told him, with a smile, "Buck, your job at night is to be the BCO — the Buster Control Officer. Make sure he doesn't change anything in the master attack plan."

Glosson's overriding passion — which many thought made up for his less endearing traits — was to preserve the lives of his pilots. In May 1971, when he began flying F-4s from Da Nang, his squadron had comprised twenty-six airplanes; three months later, when the squadron moved to Thailand, twelve were left. He was determined to avoid incurring such losses again. Shortly before the war began, Glosson toured all the wings in his command to stress prudence. "The outcome of this war is not in question," he told the pilots. "The only issue is how many body bags we're going to send back across the Atlantic. The bottom line is that there's not a damn thing worth dying for in Iraq. Nothing."

Though the sentiment was worthy, the last part would come back to haunt him. For some men would die in Iraq — terrible, plummeting deaths — and, inevitably, their surviving comrades could only wonder why.

## Riyadh

Only by the clock on the wall did General Schwarzkopf and his battle staff know that forty feet above their bunker the war's first night had given way to morning. Hour by hour tension drained from the war room as reports of allied success poured into Riyadh. Schwarzkopf spent much of his time on the telephone, receiving accounts from his subordinate commanders. His booming voice periodically lifted above the hubbub: "Goddam it! Good news! That's great!"

On a yellow notepad the CINC kept a log of such events as "0310: phones out Bag." By midmorning on January 17 the pad was covered with glad tidings in his nearly indecipherable left-handed scrawl.

Chuck Horner phoned periodically from his war room across town in the Royal Saudi Air Force headquarters. Nearly seven hundred allied combat planes had entered Iraq the first night, supplemented by hundreds of sorties flown south of the border by tankers, Rivet Joint electronic eavesdroppers, Compass Call communications jammers, and sundry others. Now the first U-2 spy planes had been dispatched from their base at Taif; one mission would crisscross western Iraq to search for mobile Scud missile launchers. Some of the officers sitting in the Central Command war room, including the British commander, General Sir Peter de la Billière, dared to hope that the relentless attacks had destroyed any Scuds capable of hitting Israel.

The first day's sorties were in full swing, Horner reported. Marine bombers at dawn had hit enemy airfields at Shaibah, Al Qurnah, Ar Rumaylah, and Talil; others struck bridges and rail yards outside Basrah and Iraqi armor and artillery positions in southern Kuwait. Marine Harrier jets had silenced enemy gun tubes after a brief artillery barrage fell on the northeastern Saudi town of Ras al Khafji. Air Force and Navy planes were hitting targets across central and southern Iraq. There was a massive strike against the oil refinery at Habbaniyah and the airfield at Al Taqaddum by thirty-two F-16s, sixteen F-15Cs, eight Wild Weasels, and four EF-111 electronic jammers. An allied jet had been lost — a British Tornado shot down near Basrah. By midmorning, coalition losses stood at two aircraft.

Schwarzkopf hardly bothered to conceal his elation, yet he was puzzled. Central Command had declared Defense Condition One across the theater — a formal state of war — but only one side seemed to be fighting. Except for Iraq's air defense batteries, the enemy remained virtually quiescent. Some fifty enemy fighters had taken to the skies; most, however, orbited aimlessly near their bases. In a call to Colin Powell in Washington, the CINC recounted the latest war news, then added, "We keep waiting for the other shoe to drop."

In analyzing the first hours of the war for Schwarzkopf, CENTCOM's intelligence staff noted, "There does not appear to be a coordinated Iraqi plan of action . . . Initial Iraqi air reaction to the coalition attack was minimal, as the Iraqi air force seemed to have been taken by surprise." Another intelligence assessment, entitled "Saddam's Strategy," concluded that the Iraqis were husbanding their forces in hopes "that a protracted defensive campaign will afford the time needed to achieve escalation of the casualty count."

Schwarzkopf agreed. No other explanation made sense, unless cowardice had afflicted the entire Iraqi military. If that was Saddam's game, the CINC was confident it would fail. Allied bombers would hunt down and destroy enemy forces wherever they tried to hide. Another report from the field gave Schwarzkopf further cause for good cheer: the American Army had begun its long migration toward attack positions along the Saudi-Iraqi border. A vast convoy stretched to the west, and C-130 cargo planes hauled men and supplies to the rude bases from which the ground invasion would be launched several weeks later.

The image of tanks and trucks and howitzers rumbling forward pleased Schwarzkopf. High spirits filled the war room. Men smiled and even laughed out loud. For those who had spent five months living in the shadow of the CINC's perpetual choler, jubilation was a most unusual experience.

Like Leo Tolstoy's famous observation about families, all happy military headquarters are alike but every unhappy headquarters is unhappy in its own way. Central Command suggested several cheerless antecedents: the fractious camp of Sir John French, saturnine commander of the British Expeditionary Force in August 1914; or the World War II fleet headquarters of the splenetic Admiral Ernest J. King (who was said by his daughter to be the most even-tempered man in the Navy because "he is always in a rage"); or even the internecine Greeks besieging the walls of Troy. Yet none quite captured life in the Riyadh bunker. In its misery, it was unique.

Lieutenant General Calvin Waller's first inkling of what he would find came in early November 1990, when he phoned Riyadh from Fort Lewis to talk to Schwarzkopf about Waller's new appointment as deputy CINC. After placing the call at 7:30 A.M. Saudi time, he encountered resistance from first one staff officer, then another. Schwarzkopf, they explained, had worked late the previous night and was still asleep. "Well, wake his ass up," Waller suggested. They refused. A few minutes later, Waller took a return call from the CINC's chief of staff, Major General Robert Johnston, who was courteous, solicitous, and firmly disinclined to rouse Schwarzkopf. "Sir," Johnston said, "we don't want to wake him up. That would be terrible."

On November 16, Waller arrived in Saudi Arabia. Colin Powell had worried that ground forces in the gulf lacked a single ground commander to oversee both Army and Marine divisions, as Chuck Horner presided over all air forces in the gulf. "If I'm Marshall and you're Eisenhower," Powell asked Schwarzkopf, "who's Omar Bradley?" Waller was a potential candidate, but after touring the theater for two weeks, Waller recommended to Schwarzkopf that the CINC keep the ground commander's authority to himself — in effect playing both Eisenhower and Bradley — rather than formally transferring it. Setting up another command structure would be difficult and time-consuming, he observed, and would complicate the CINC's relations with the Saudi leadership. Also, the Marines and Air Force — already irritated by having Army officers Schwarzkopf and Waller filling the top two CENTCOM billets — would be further upset if Waller's role were expanded.

Moreover, Waller could see himself performing a better function: shielding the staff from the CINC's fiery temper and injecting some esprit into CENTCOM headquarters, which since August had grown steadily more miserable. That, he believed, was his unspoken writ from Powell, who was well aware of how dispirited life had become in Riyadh.

And dispirited it was. The low morale shocked Waller, despite his experience in working with Schwarzkopf. He was astonished at the extent to which even senior generals were intimidated by the CINC. He came to think of Schwarzkopf as a volcano — at times nearly dormant but for a small hiss of steam, at other times erupting with molten rage. His temper built progressively: the voice climbing to a bellow, the complexion flushing from pink to red to purple. If the CINC was irritated before a staff meeting, he would stalk into the room and throw himself into his chair, glowering at those assembled. "Whoa, shithouse mouse," the deputy operations officer, Brigadier General Richard (Butch) Neal, would murmur, "here we go again." Bob Johnston, the chief of staff, could only give a sad shrug and whisper, "I don't know what's going on."

Sometimes, by acting quickly, Waller could defuse him. "I got it, CINC, I got it," he would interject. "I see what you want. Leave it to me." If the tirade continued, Waller would launch into a corny joke or offer a bit of folklore passed down from his grandmother in Louisiana. Once, while scanning an ambiguous message, Schwarzkopf fumed, "What stupid son of a bitch sent this?" "Give me the damned message," Waller said, plucking it from his hand. "I'll take care of it. Now let's move on to something else." On other occasions, he would simply kick Schwarzkopf's boot under the table to calm him down.

To a staff terrified of the commander's wrath, Waller urged, "If there's a problem, bring it to me. I'll take it to the CINC. Bad news does not improve with age." One afternoon, when Schwarzkopf began to berate an intelligence officer, Colonel Chuck Thomas, Waller wrapped an arm around Thomas and sent him from the room, whispering, "Go get it straightened out. I'll buy you an hour." On another occasion, Schwarzkopf interrupted an explanation of why the Army lacked enough heavy-equipment transporters to reposition both corps simultaneously. "Bullshit!" the CINC began. Waller cut him off with a sharp "Bullshit again! Can't do it and I'll show you why."

The CINC could excoriate younger officers. (One group of majors and lieutenant colonels kept count of how many generals' stars had been in the room at the moment of their greatest humiliation by Schwarzkopf; the "winner" claimed twenty-two.) But Schwarzkopf's hottest fire was saved for the generals themselves, particularly those he deemed insufficiently aggressive or, as he once put it, "those who think their primary mission is to improve their golf scores."

He was openly disdainful of the officer who should have been his closest confederate: his J-3, the operations officer, Major General Burton Moore. Schwarzkopf called Moore "a thorn in the side" and was particularly contemptuous of Moore's frequent practice of excusing himself from the war room promptly at 8 P.M. for a hot supper and a good night's sleep in the hotel across the street. The CINC alternately blistered Moore and ignored him, leaving the Air Force pilot to wonder aloud one day, "How do I get this guy's confidence? How do I speak to him so he listens?" Schwarzkopf instead relied on Richard (Butch) Neal, Moore's deputy, even hinting on his efficiency report after the war that Neal should have been the J-3.

The CINC's scorn was infectious. Lieutenant General Tom Kelly, the Joint Chiefs' operations officer, privately referred to Moore as "a major in a major general's uniform." And Neal, angry at what he took to be Moore's cold shoulder, confronted the senior officer in his office one day. "Look," he snapped after closing the door, "I've never been treated so badly in twenty-six years in the Marine Corps. Either you change your ways or I'm going to see Schwarzkopf right now." Though never convivial, relations between Moore and Neal improved.

Victims of the CINC's ire developed their own argot. When Schwarzkopf lost his temper, he was said to have "gone ballistic." An officer could "have his face ripped off" or be "clawed by the bear." Those claws often were sharpest when raking the Army, Schwarzkopf's own clan. Lieutenant General John Yeosock, the senior Army commander, was so frequently berated that he seemed reluctant to leave his headquarters outside Riyadh for the daily CENTCOM meetings. Again, the public upbraiding of a senior officer — considered bad form — bred contempt among subordinates, who privately and unfairly referred to Yeosock as General Halftrack, the confused, aging commander in the cartoon strip "Beetle Bailey."

*

What are we to make of this man, so conspicuously gifted and yet so plainly beset with demons? "All very successful commanders are prima donnas and must be so treated," the prima donna George Patton once observed. Analyzing Schwarzkopf became a cottage industry in Riyadh, but conventional wisdom remained elusive.

He possessed, indisputably, both a formidable intellect and an adhesive memory. A staff officer presenting him with a statistic — the number of Iraqi SA-6 missile launchers, the tonnage of food rations entering Saudi Arabia — could expect minute cross-examination if he amended the figure even months later. When Schwarzkopf had arrived in Tampa to assume command of CENTCOM two years earlier, his predecessor designed an absurdly complex ceremony for the occasion, using a scale model and clothespins to represent the main participants. Schwarzkopf listened to the graphic explanation — the clothespins jumping around the model like checkers — then repeated the instructions verbatim, syllable by syllable.

His preoccupation with the welfare of his troops knew no bounds, whether the issue was mail delivery or ammunition stocks. Disgruntled at the Army's inability to make a satisfactory desert boot, he organized his own "boot fair." He ordered more than a dozen prototypes from various American footwear manufacturers, lined them up in Waller's office, then tugged them on and off to determine which measured up to his standards. On more substantive combat issues he relied heavily on his field commanders for advice, allowing them wide latitude in drafting their own battle plans within the framework of his strategy. Distractions and sideshows intrigued him not at all unless they furthered his larger purpose: ousting Saddam from Kuwait. And he was oddly loyal, threatening to relieve subordinates or treating them with harsh indignity, yet never sending them home.

His constitution seemed cast of iron. Routinely working eighteen or twenty hours a day, he often napped in the afternoon to gain a second wind for a run to the wee hours, thus carving two working days from one, as Winston Churchill had done during World War II. He suffered neither fools nor shoddy performance. "Quit bashing the goddam helicopters into the ground," he growled after a rash of accidents. The accidents subsided. In a military culture that often bred inexpressive commanders, he wore his emotions like campaign ribbons for all to see. An earnest letter from a soldier's mother could move him to tears. Among troops in the field he was warm, even gregarious; he seemed to like the *idea* of "my soldiers" as much as the creatures themselves. But with his staff in Riyadh he remained aloof, occasionally sharing an article from *Sports Illustrated* or his stack of hunting magazines, but more commonly keeping his distance.

Though personally fearless — three Silver Stars from Vietnam bore testament — the prospect of sending men to their death seemed almost to unhinge him. He often slept badly, padding about in his running suit and fretting over a thousand conjured calamities. Knowing that doubt spreads through an army like cholera, he kept his qualms admirably concealed. Publicly he remained robustly confident; perhaps, some of his generals speculated, anxiety and insecurity required him constantly to assert his dominion. Those who had known him longest thought his ego had swelled with each additional star on his shoulder to the point that once, without blushing, he likened himself to Abraham Lincoln, with "nobody to turn to but God."

Of his temper, theories abounded: that he was tormented by chronic back pain from an old parachuting injury; that the stress of command could be relieved only by periodic tirades; that he was still haunted by the horrors of Vietnam. In his memoirs Schwarzkopf would paint an affecting self-portrait of a nomadic child with an alcoholic mother and an unstable home. In one revealing episode he described pummeling a canvas punching bag with bare knuckles until both fists bled, imagining the bloody bag "to be me."

He was not unaware of his shortcomings, including the temper, which he once described as "without question my major weakness as a commander." At six-foot-four and 250 pounds, Schwarzkopf knew that he intimidated those of lesser stature. Whenever he assumed a new command, he advised his subordinates: "I never get mad at you personally. I wear my heart squarely on my sleeve. When I like something, you'll know it. And when I don't like it, you'll know it . . . I don't get mad at people, I get mad at principles, at things that don't happen."

Those who worked closest with him, however, including Waller and Glosson, thought the rages immature and dysfunctional. Like the French field marshal Joseph J.C. Joffre, he appeared to want generals who were lions in action and dogs in obedience. General de la Billière suspected that some of the CINC's "storms" were "laid on to keep people sharpened up"; if so, the British commander concluded, Schwarzkopf overplayed his hand and badly inhibited "the free thinking of his staff." Waller worried that he was creating a band of yes-men; several times the DCINC urged him to be more gracious and to encourage debate. You're right, Schwarzkopf would concede, only to explode anew at some infraction and plunge the staff once again into their stunned-mullet stupor.

At times, certainly, they gave him cause for anger. Tim Sulivan, a British brigadier and the only non-American officer permitted a seat in the war room (he wore U.S. fatigues to avoid offending excluded allies) considered many a tantrum to be justified. Vice Admiral Stanley Arthur, the Navy's senior commander, thought the CENTCOM staff mediocre and indecisive. In a curious way, Schwarzkopf's temper also helped quell interservice squabbles by unifying natural rivals beneath a common fear.

Moreover, he prudently spared the allies his wrath. Here he showed himself most competent at that for which he was presumed least prepared by training and constitution: the muster and mastery of a huge coalition drawn from three dozen nations. In unveiling the attack plan for the first time to his Saudi counterpart, Lieutenant General Khalid bin Sultan, he was said to have presented the scheme so nimbly that Khalid left the session believing it was his own idea. (To win support from the Egyptians, whose divisions would be the linchpin in any Arab attack, the CINC, before showing Khalid the plan, had secretly flown to Cairo for consultations with President Hosni Mubarak's high command.) Privately, Schwarzkopf despised Khalid as a pompous, arrogant dabbler, "a joke." Yet he sat with the Saudi general for endless hours, drinking coffee and passively listening as Khalid spun his grand theories of combat. "Sir," Bob Johnston told his commander-in-chief, "you must have the patience of Job."

With the French, the Syrians, the Kuwaitis, and other high-strung allies he showed a gift for bluster, flattery, cajolery. In a message to Washington, Schwarzkopf wrote, "I finally have come to understand the meaning of the word 'byzantine.' " He believed that a coalition commander required "a multifaceted personality," an ability to be both "the ultimate persuader and the ultimate bully." Whatever his measure as a field marshal, as a diplomat Schwarzkopf was peerless.

Toward the Pentagon he could be splenetic and supercilious, exhibiting a field commander's common wariness of higher headquarters. "I'll shoot the first son of a bitch who sends a message in hard copy to Washington without me seeing it first," he told the staff. At least one general in Riyadh began using code words when the Pentagon phoned, as a warning that the CINC was eavesdropping in the war room. His contempt for others in Washington could be even sharper, as displayed in his occasional jibes about the CIA and Brent Scowcroft, the national security adviser.

Yet Schwarzkopf often sought counsel from home, particularly from General Carl Vuono, the Army chief of staff, whom he had known since they were West Point cadets in the mid-1950s. Vuono, both patron and close friend, also could play the Dutch uncle. Once asked to list terms that described Schwarzkopf, Vuono ticked them off: "competent, compassionate, egotistical, loyal, opinionated, funny, emotional, sensitive to any slight. At times he can be an overbearing bastard, but not with me."

Nor with the man in Washington who mattered most, Colin Powell, whose genius in managing the temperament of his field commander was one of the war's best kept secrets. Technically, the chairman stood outside the chain of command that extended from the president to the secretary of defense to fighting CINCs like Schwarzkopf; in reality, Powell served as the primary link between Riyadh and the civilian leadership. Connected by a direct phone line that permitted them to confer at the press of a button, Powell and Schwarzkopf talked two to five times a day.

As a historian once wrote of MacArthur, Schwarzkopf was "an essentially thespian general who required constant backstage handling." The chairman tried to give the CINC his head, recognizing how herculean was the task of building a military operation, from scratch, seven thousand miles from home. Powell massaged his ego, bolstered his spirits, addressed his frustrations, and, occasionally, absorbed his impertinence. "How's the CINC doing?" he would ask Waller. "His hair still on fire?" Powell's attitude helped to reassure those in Washington who had doubts about the CINC's fitness for command in Riyadh. Powell and Schwarzkopf needed each other, but the latter's need was greater.

At times, after hanging up the phone, Schwarzkopf seemed to portray himself as the warrior and Powell as the politician. "Colin is damned good at what he does, but he's really concerned about his political future," he once said. "Some of this is for political reasons." Waller considered such comments benign and respectful; others detected a tinge of contempt, even malice.

And thus into war they plunged. Schwarzkopf was hardly the first American general with a temper. George Marshall's ranged from "white fury to cold and quiet contempt." The curmudgeon Admiral King was credited with observing, "When they get in trouble they send for the sonsabitches." They were in trouble and they had sent for Norman Schwarzkopf; he would lead them to victory. It was a mark of the man's complexity that one of his favorite quotes, saved from a lecture at the Army War College years before, suggested both his task and himself:

> For what art can surpass that of the general — an art which deals not with dead matter but with living beings, who are subject to every impression of the moment, such as fear [and] exhaustion — in short, to every human passion and excitement?

## Al Kharj, Saudi Arabia

An hour's drive southeast of Riyadh, on the lip of the vast wasteland known as the Empty Quarter, the bomb loaders worked with controlled frenzy throughout the war's first day. They called themselves "candy men" or "BB stackers." From immense caches in the open desert, they wheeled in sleek, dark tubes of explosive tritonal. Almost quicker than the eye could follow, the crews slapped on booster charges, fore and aft; screwed in fuses, fore and aft; clipped the slip rings; and, with a few practiced turns of a wrench, bolted the tail fins. In ninety seconds, a tube became a bomb.

Most were unguided, dumb, little different from those dropped on Yokohama or Schweinfurt nearly half a century before. (Thousands, in fact, came from Vietnam-era stockpiles and bore manufacturing stamps from the early 1970s.) They included 500-pound Mk-82s and 2000-pound Mk-84s. More sophisticated were the cluster bombs, like the CBU-87 and Mk-20 Rockeye, designed to bloom in flight and scatter 247 submunitions capable of cutting through six inches of armor plate. More sophisticated yet were smart bombs like the laser-guided GBU-10s, each weighing two thousand pounds, destined for bridges and bunkers.

Beyond the bomb loaders lay the flight line, blurry through the exhaust shimmer of a hundred jets. No base would be responsible for more dropped bombs during the war than Al Kharj. Two months before, it had been little more than a concrete runway known locally, with cause, as Snake Hill. Then the Americans arrived. Contractors scraped away the sand, which was too powdery to give tent pegs a purchase. They hauled in 25,000 truckloads of red clay, rolled it flat, erected six hundred tents in neat rows lettered A through X, built four field kitchens, a laundry, a hospital, nineteen latrines, and a shopping arcade known as the Mall, complete with a gazebo and a Baskin Robbins ice cream parlor. After posting a sign at the edge of the desert — WELCOME TO AL KHARJ NATIONAL FOREST — the airmen christened the base Camel Lot and declared themselves ready for war.

Throughout the day the planes took off singly, at twenty-second intervals, with a cacophony that made even the soundest sleeper yearn for a no-go pill. To some extent, the daytime attacks designed in the Black Hole by Brigadier General Buster Glosson and Lieutenant Colonel Dave Deptula were variations on the initial night attacks, although with greater numbers of planes and fewer precision bombs. French Jaguars flew for the first time and came home bloodied. Of a dozen French jets to attack Ahmed al Jaber Air Base in Kuwait, four limped back with antiaircraft damage; one lightly wounded pilot landed with a bullet hole in his flight helmet. Italian airmen, in their first combat foray since World War II, also would pay a price. Of eight Italian Tornados to take off from Abu Dhabi in the United Arab Emirates, one aborted for mechanical problems and six turned back because of refueling difficulties; the one pilot who pressed on toward the target was shot down and captured.

From Al Kharj, a twelve-ship strike force of F-16s, using the radio call sign Tumor, had closed to within a dozen miles of their targets — surface-to-air missile sites in southern Iraq — when heavy fire drove them off to seek a less fortified alternative. "All Tumors," the flight leader called bluntly, "let's get the hell out of here." A flight of four F-15C Eagles, led by squadron commander Randy Bigum, returned to Al Kharj equally frustrated, although for different reasons. After leading twenty attackers to the Iraqi air base at Talil shortly after noon and expecting to fight as many as 150 Iraqi fighters, the Eagles instead found none — and subsequently learned that a sister squadron to the west had shot down four enemy jets.

In the main, however, the day attacks were as successful as those launched the previous night. Seven squadrons of A-10 Warthogs flew more than three hundred sorties on the first day from King Fahd Airport north of Dhahran. Agile but slow — the Warthog pilot was said to carry a calendar rather than a watch — the A-10 was designed primarily for close air support of ground troops; instead, Horner and Glosson sent them against dozens of targets along the Iraqi border, including eight radar sites similar to those destroyed by the Apaches at the start of the war.

Some of the targets missed or only damaged in the first wave were attacked again. Four F-111s, headed north toward Tikrit, Saddam's summer home, left unscathed on the first night. One turned back, one fled from a pursuing MiG, one missed the target, and the fourth put a 2000-pound bomb through the palace skylight.

More Tomahawks from the Red Sea and Persian Gulf struck Baghdad; ten of the missiles progressively battered the Ministry of Defense and Iraqi army headquarters into rubble between 10 A.M. and 3 P.M. Others hit the Al Zubayr petroleum plant near Basrah, and the Scud propellant factory at Latifiyah. TLAMs also attacked many of the electrical targets struck the previous night. After the initial euphoria of seeing the lights go out, Tomahawk planners realized that they could not tell whether the power grids had been destroyed — as later proved to be the case — or whether the Iraqis had simply shut them down. This was the first glimpse of a problem that would trouble the Americans for the rest of the war: battle-damage assessment, the art of gauging how badly the enemy had been hurt.

And from the United States arrived the very symbol of American air power. Eight B-52s, flying from an Air Force base in Michigan's upper peninsula, would drop more than a hundred tons of high explosives on the Republican Guard's Tawalkana Division, a target endorsed by Schwarzkopf as a warning to Iraqi ground forces of what they could expect in the days and weeks to come. The bombers then flew to a new base on the Red Sea coast. The Saudis, wary of basing the B-52s so close to Mecca in peacetime, had secretly agreed in the fall to open up the huge airport at Jiddah once war began.

Seven other B-52s flew from Barksdale Air Force Base in Louisiana on a thirty-five-hour round-trip mission. The bombers carted a secret, less conventional payload: air-launched cruise missiles, or ALCMs. To avoid a number of difficult and potentially embarrassing arms control issues, the United States had never disclosed the development of a nonnuclear ALCM. Now, nearly three dozen of them were about to fly into Iraq. The strike planners, who began drafting the ALCM attack in August, renamed them LRBs, or long-range bombs, to further obfuscate the true nature of the weapon.

Before dawn on January 17, the B-52s crossed Egypt and headed for northern Saudi Arabia. Glosson had delayed their original flight schedule by more than five hours, fearful that Libya would detect the bombers flying past in the Mediterranean and sound the alarm. From a launch point near the Saudi town of Ar Ar, the missiles dropped from their mother ships and sped toward hydroelectric and geothermal plants in Mosul, the telephone exchange in Basrah, and several other targets in Iraq.

Precisely what they hit would remain the subject of contention long after the war. The Air Force claimed at least partial success in destroying the targets. The Navy, ever suspicious of her sister service, disputed the claim and privately ridiculed the raid as a stunt to further the cause of the new B-2 bomber by demonstrating global power projection from American bases.

At Al Kharj, as afternoon faded to dusk, the relentless demands of launching yet another wave of planes left little time for such concerns.

Returning pilots climbed from their cockpits and distributed the metal slip rings from empty bomb racks to their crew chiefs as souvenirs. Then fuel trucks and maintenance teams swarmed around the jets. Hydraulic hoists lifted new bombs beneath the wings. Fresh pilots tugged on their helmets and scrambled up the boarding ladders. Chaplains strolled along the flight line, clasping their hands in prayer or flashing an upturned thumb. Once again the planes taxied toward the runway for take-off. The sharp crack of engines at full throttle washed over Camel Lot. On a squadron room bulletin board, an airman tacked up a truculent sign: IRAQIS WERE BORN TO DIE.

## Dhahran, Saudi Arabia

To parry an expected barrage of Iraqi Scud missiles, the U.S. Army had positioned 132 Patriot missile launchers around Riyadh and other "high-value" targets across the northern Arabian peninsula. Each launcher held four missiles: typically, one to shoot down enemy airplanes, the task for which Patriot was originally built, and three others — more sophisticated in their design — intended to destroy Iraqi tactical ballistic missiles. Patriot was a relatively new system, having first joined Army air defense units in Europe in 1985, and its role as an antimissile weapon was still evolving when Saddam invaded Kuwait. But the dream of shooting down a missile with another missile had been nurtured in the Army since at least 1946, when General Dwight D. Eisenhower likened the technical difficulties involved to "hitting a bullet with another bullet."

In eastern Saudi Arabia the air defense "architecture" was designed for the protection of key ports and airfields, particularly Dhahran Air Base and its immense fleet of allied warplanes, cargo jets, and helicopters. Four Patriot batteries from the 2nd Battalion of the 11th Air Defense Artillery Brigade covered the kingdom's northeast corner: Foxtrot battery at Jubail, Echo at King Fahd Airport, Delta at Dammam port, Bravo and Alpha on opposite sides of the airfield at Dhahran. A fifth battery, Charlie, was entrenched in nearby Bahrain. Each battery controlled five launchers.

For Alpha, as for her sister batteries, the first twenty-four hours of war had been quiet though tense. The missile crewmen — known derisively in the Army as "scope dopes" and "van rats" — watched wave after wave of allied warplanes head north before vanishing from the upper rim of the Patriot's radar screens. Not a single Iraqi jet, much less an enemy missile, had ventured south of the border. Alpha's soldiers began to wonder whether their long months of waiting in the desert had been for naught.

Then, shortly after 3 A.M. on Friday, January 18, the base sirens burst into a fearful howl. From public address speakers, a voice boomed, "Condition red! Condition red! Don your gas mask." Over the battalion radio network an officer warned, "We've got a positive Scud launch. Stand by for trajectories." A pair of Air Force Space Command satellites, each scanning Iraq with a twelve-foot infrared telescope, had spotted the hot plume of an apparent missile firing. From Cheyenne Mountain in Colorado, where the launches were detected, the alert warning had been flashed to the Pentagon and to Saudi Arabia.

Outside, the missile crewmen crouched behind sandbag revetments, their faces hidden by the black snouts of their gas masks. Scraps of paper and plastic bags cartwheeled past, carried by the wind toward the black silhouettes of the Patriot canisters cocked skyward a hundred yards to the north.

Inside Alpha battery's pressurized van, crowded as a cockpit, four men watched their radar scopes with unblinking eyes. In the right-hand chair reserved for the tactical control officer sat a twenty-seven-year-old lieutenant from Montgomery, Alabama, Charles L. McMurtrey. The screen before him resembled a baseball diamond, with a 120-degree field of vision searching north toward the Kuwaiti border. At most, McMurtrey knew, he had three to five minutes from the initial warning until an incoming Scud plummeted into Saudi airspace at several times the speed of sound.

Yet nothing happened. Three minutes passed, then five, then ten. The scope remained clean. Battalion reported that the missile launches evidently had occurred much farther west, hundreds of miles from Dhahran. Tension drained from the van. Lieutenant McMurtrey's eyes felt raw from the strain. He listened to the radio net for an all clear canceling the alert.

Then, at 4:28 A.M., he saw it: a small green triangle on the right edge of the scope, moving faster than anything he had ever seen before. "Oh, shit!" McMurtrey yelled, lurching forward in his chair. The missile appeared headed for the air base. He scanned the launch indicators on the console. The system was set to fire automatically on the assumption that Patriot computers could react faster and with more precision than human operators. But the system had yet to react. The voice on the battalion radio net shrieked over the speaker box, "That's a real one! That's a real one!"

Without moving his eyes from the green triangle, McMurtrey pivoted toward the sergeant in the next chair. "Shoot!" he ordered. The sergeant's hand darted toward the override button to shift the system from automatic to manual, but before he could flip the switch, an explosive crack rocked the van.

A seventeen-foot Patriot leaped from launcher number five and accelerated past Mach one even before the tail fins had cleared the canister. The missile cut an orange slash into the moonless sky, guided by commands from the battery radar below. Modifications made to the missile and its software before the war gave the Patriot a bigger warhead, a quicker fuse, and the ability to look "up" more than "out" in order to better defend against plummeting ballistic missiles rather than enemy bombers. An incoming Scud, after arcing far above the earth in a steep parabola, would dive toward its target at more than four thousand miles an hour. The Patriot was supposed to pass abeam of the incoming warhead and detonate in a quarter of a millisecond, spewing out three hundred lead blocks the size of ice cubes; each fragment carried the kinetic energy of an automobile hitting the ground after falling from a nine-story building. But a split-second miscalculation could let at least part of the enemy missile slip through.

Several seconds after the Patriot launch, a violent boom washed over the engagement van. On McMurtrey's console a dark green # flashed at the bottom of the screen, the so-called tic tac toe symbol standing for "probable kill." "We got — yeah — we got him!" McMurtrey yelled. "We got him!"

Yet it was not to be so simple. Precisely what Alpha battery fired at that night would remain a mystery. The Army credited McMurtrey and his crew with the first Scud kill of the war, then reversed itself a year later. No confirming missile wreckage was ever found; some analysts concluded McMurtrey's battery had been deceived by a computer error. But within hours of the incident Schwarzkopf publicly proclaimed that an Iraqi missile had been "destroyed by a United States Army Patriot." The myth of Patriot invincibility was born.

Iraq soon followed what came to be known as the "phantom Scud" with several incontrovertible missiles, all apparently parried by Patriots or misdirected into the empty desert. Even before the cheering stopped, however, the Patriot crews noted several puzzling phenomena that would perplex American missile experts for the next several days. The incoming missiles traveled about 40 percent faster than Soviet Scuds. They also possessed a smaller radar signature than anticipated and fell with a peculiar corkscrew motion instead of following a smooth ballistic path. Finally, the Scuds seemed to multiply in flight; for each launch detected by the satellites, Patriot crews reported several incoming targets.

Of more immediate concern were television and radio reports from Tel Aviv. The earlier satellite warning on Friday morning had indeed been accurate. Enemy crews had launched missiles from western Iraq at Israeli cities, where there were no Patriots at all.

We keep waiting for the other shoe to drop, Schwarzkopf had told Powell earlier that day. Now it had fallen, at Mach six, on Israel.

# 3

# An Event in Israel

## Washington, D.C.

For twenty-four hours, the White House was swept with euphoria. Reports from the front portrayed a military triumph beyond even the most optimistic hopes of the president and his war councilors. Before dawn on January 17 — it was shortly after noon in Saudi Arabia — Bush, his hair still damp from the shower, pulled a trench coat over his blue knit shirt and padded down to the Situation Room in the West Wing, not unlike Lincoln on his solitary visits to the telegraph office.

In contrast to dispatches from Bull Run and Fredericksburg, however, the news from Riyadh was wholly good: targets destroyed, allied losses almost nil, Iraqi defenses disorganized and ineffective. The president poked his head into the White House briefing room and, with a twinkle, told a handful of groggy reporters, "Things are going pretty well." When the financial markets opened that morning, the Dow Jones soared 117 points. Oil prices dropped eleven dollars a barrel. As the war's first day ended in Washington, officials who had expected to keep another all-night vigil instead began packing their briefcases to go home early for dinner.

Then several Scuds fell on Israel — six in Tel Aviv and two in Haifa, by Israeli count — and the euphoria vanished. A watch officer at Cheyenne Mountain in Colorado flashed the warning of a "missile event" to the Pentagon's National Military Command Center. Public address speakers throughout the NMCC crackled with the command "Teams to station! Teams to station!" While others plotted the missile trajectories, an officer in Room 2B902 activated Hammer Rick, the secure satellite telephone to Israeli Defense Forces headquarters in Tel Aviv. "IDF, this is NMCC," the American officer began, then quickly listed the projected impact zones. Moments after the sirens began to wail in Israel, the first warhead exploded.

Colin Powell and Richard Cheney remained in the Pentagon, and Bush was occupied in the White House at a working supper on environmental issues. But shortly after the gilded pendulum clock in the West Wing anteroom struck seven, the rest of the president's brain trust hurried to the northwest corner office of National Security Adviser Brent Scowcroft. James Baker arrived from the State Department, nervously twirling his key chain as he settled onto the blue damask couch next to his deputy secretary, the droll, corpulent, wheezing Lawrence Eagleburger. Scowcroft, small and elfin, settled into a big wing chair; his deputy, Robert Gates, took Scowcroft's seat behind the desk, swiveling beneath the oil painting of a rainbow above a hay wagon. They were joined by John Sununu, the White House chief of staff, and Richard N. Haass, the National Security Council's director for Near East and South Asian affairs.

Scarcely had the men begun considering their options when an aide came in from the Situation Room with a report from Tel Aviv: nerve gas supposedly had been detected in the Scud debris. Another report cited an unusual number of ambulances racing through the Israeli city. For thirty seconds a heavy silence hung in Scowcroft's office. Eagleburger lurched to his feet and paced the thick green carpet. He had just returned from a weekend mission to Jerusalem and Tel Aviv, where the Israelis — reluctant to abandon their military self-sufficiency — had rejected an American offer of Patriot missiles. "If they've been hit with chemicals, Katie bar the door because they're going to do something," Eagleburger predicted. "I know these people. They're going to retaliate. If it's nerve gas, we'll never stop them."

Baker asked the White House switchboard to place a call to Israel. To his disgust, a recording announced, "All circuits are busy. Please try your call again later." "Holy shit," muttered Scowcroft's press secretary, Roman Popaduik; "the world's burning down and we can't get a line out?" After several more tries, the operator reached the office of Yitzhak Shamir in Jerusalem, but the prime minister was unavailable. "We'll call you back," one of Shamir's aides said. "Where can we reach you?"

"What's our telephone number?" Baker asked Haass, who was listening on an extension line. A few months earlier a frustrated Baker had publicly rebuked Israel for its recalcitrance in balking at a Middle East peace conference. Now Haass indelicately used Baker's line against him: "You should know, Mr. Secretary. 'The White House number is 1-202-456-1414. When you're serious about peace, call us.' "

Scowcroft, Gates, and others cackled with laughter. "Haass," Eagleburger quipped, "you just lost whatever chance you might have had to get an embassy."

Baker smiled thinly and stared at the curtained windows facing Lafayette Park. For five months he had labored to build an international coalition against Iraq that would give substance to the president's inchoate ambitions for a new world order to replace a generation of superpower confrontations. In his wooing of the Soviets he had even gone so far as to reveal the allied war plan as evidence of American resolve. He had battled the British prime minister, Margaret Thatcher, and other skeptics to convince them of the need for a United Nations resolution authorizing the use of force against Baghdad. (Thatcher in turn had braced Bush by urging, "This is no time to go wobbly, George," a line immediately adopted with glee by the president's men.) Baker had browbeaten Congress for moral support and America's rich allies for hard cash. From the leaders of Egypt, Saudi Arabia, Kuwait, and even Syria, he had extracted secret pledges that they would remain true to the coalition if Iraq drew Israel into the war. Now that fragile scaffold of vows and troths was imperiled. The exigencies of war, Baker knew, could scuttle the best-intentioned peacetime promises.

Another message arrived from the Situation Room. The earlier report of a chemical attack was erroneous; an Israeli civil defense team had drawn a false reading from a Scud fuel tank. Baker placed quick calls to Arab leaders in Egypt, Syria, and Saudi Arabia, urging calm.

But now new alarm bells sounded, this time from the Pentagon. More than sixty Israeli warplanes had roared into the night, although it was unclear whether the fighters were bound for Iraq or were simply setting up defensive orbits against a possible Iraqi air attack. Cheney phoned Scowcroft to recount his conversation on Hammer Rick with Israeli Defense Minister Moshe Arens. Arens wanted American and Israeli planners to separate their forces — to "deconflict," in military jargon — so that Israel could attack Scud targets in western Iraq without inadvertently taking on coalition aircraft.

Both Powell and Paul Wolfowitz, the defense under secretary, believed that further American demands for Israeli restraint would be difficult to justify. Shamir had kept his promise to Washington, made in December, not to launch a pre-emptive strike against Iraq; now his country had been pelted with missiles. If necessary, under a contingency plan concocted by Central Command and the Joint Staff, U.S. planes could remain east of the forty-second-degree longitudinal line in western Iraq.

Although Cheney seemed pessimistic about keeping Israel out of the war, he had cautioned Arens that any deconfliction agreement required Bush's approval. "You've got to understand why the Israelis want to get involved," Cheney told Scowcroft. "You can't simply say, 'Please don't.' " Israeli policy called for immediate retaliation against any attacker; more to the point, Israel wanted the missile attacks stopped. The Americans had to demonstrate that the coalition could suppress the Scuds, Cheney added, because "the problem is a very real one."

Cheney's call triggered a sharp debate in the White House West Wing. "I don't think it's inevitable that the Israelis get involved," Haass said heatedly. "It's not militarily necessary or politically desirable. I think we should weigh in hard to discourage them." However difficult it might be for Israel to hold back, he added, restraint was in everyone's best interests. Scowcroft and Baker agreed. An Israeli counterpunch carried incalculable risks. Deconfliction might make tactical sense, they concurred, but strategically it would be a mistake.

At this point, Shamir called from Jerusalem. "We are going after western Iraq full bore, Mr. Prime Minister," Baker assured him. "We've got aircraft over the launch sites." The secretary rolled his eyes and shrugged; it was not clear from Pentagon accounts whether this was precisely accurate, but others in the office nodded vigorously. "There is nothing that your air force can do that we are not doing. If there is, tell us and we'll do it. We appreciate your restraint, but please don't play into Saddam's hands."

Shamir said little. His relations with the Bush administration had been strained to the point of incivility, but in addition to his pledge not to attack Iraq pre-emptively, he had agreed not to retaliate without consulting Washington. "We'll get back to you," the prime minister said before hanging up. "I need to talk to my cabinet. Thank you for your concern."

At 9:30 P.M., Baker called Arens. "We regret very much what happened, Misha," Baker told the defense minister. "This is a formal request to withhold a response. We're outraged by the action that the Iraqis have taken. We're very sad and sorry about the attacks, but we've got aircraft going to the launch sites."

Arens had other ideas. His forces in the Negev Desert secretly had been rehearsing strikes against mock Scud batteries. Israel envisioned flying army troops by helicopter to attack the missile bases in western Iraq. The plan, Arens told Baker, required the overflight of either Saudi Arabia or Jordan. Baker knew that neither country would willingly permit such an intrusion; should Jordan refuse permission and try to block the flights, it likely would lose its small air force to Israel's superior fighters in a matter of minutes. Not only would Arab partnership in the allied coalition be jeopardized, but the entire Middle East could be at war.

"Your actions have been exceptional so far," the secretary told Arens. "I hope that you won't respond or retaliate. Please consider this a formal request for moderation."

Baker cradled the receiver. As he and Scowcroft discussed the conversation, another idea emerged: if Shamir and Arens felt obliged to counterattack, then perhaps a better retaliatory weapon would be Israel's Jericho missile — carrying a conventional rather than a nuclear warhead, of course. By arcing high over Jordanian airspace, Jerichos would avoid the overflight problem and still allow Israel to respond proportionately, even biblically, with a missile for a missile. Scowcroft called Cheney at the Pentagon with new guidance from the White House: we'll try very hard to restrain the Israelis, the national security adviser said, but if they insist on retaliating, the Jericho might be preferable to a full-scale attack with air or ground forces.

Now there was little to do but wait and hope. Haass drafted a cable to Jerusalem offering to send a high-level American envoy for further consultation. From the White House mess a steward brought sandwiches. Most of the men in the crowded office stripped to their shirt sleeves. Baker pulled a pen from his pocket and began doodling on a photograph of Scowcroft. As the evening wore on, the secretary of state labored to find just the right rhymes to complete an obscene limerick about the national security adviser.

The bogus report of a chemical attack on Israel was to be the first of many such false alarms during the war, and it underscored American uncertainty over how to deal with Iraq's purported weapons of mass destruction. Before the war, few debates raged with greater intensity than those over chemical and biological weapons.

That Iraq possessed immense stocks of nerve agents and other chemicals was beyond dispute. The country had built a vast web of thinly disguised production and storage facilities, such as the Muthana State Establishment for Pesticide Production. In its eight-year war with Iran, Baghdad employed both mustard gas and nerve gas; in March 1988, more than four thousand Iraqi Kurds in the town of Halabja had been blistered, burned, and asphyxiated with mustard gas as part of Saddam Hussein's ruthless campaign to suppress Kurdish rebels. In April 1990, Saddam had vowed to "make the fire eat up half of Israel if it tries to do anything against Iraq."

Iraq was a signatory, if not an adherent, to the 1925 Geneva Protocol outlawing use of "asphyxiating, poisonous, and other gases." The treaty did not prohibit production or possession, however. By Central Intelligence Agency estimates, Saddam's chemical stockpile now exceeded a thousand tons, including artillery rounds, chemical bombs, and caches possibly moved into occupied Kuwait. In the fall of 1990, the CIA had issued an intelligence assessment reflecting the government's conviction that Iraq would use those stocks in the event of war. Before the invasion of Kuwait, however, intelligence agencies in Washington had repeatedly assured Schwarzkopf that Iraq lacked the means to launch chemical warheads aboard long-range missiles; the CINC had even been shown missile telemetry data and photographs of the Iraqi test-firing range. But after the invasion, intelligence analysts reversed themselves and concluded that Iraq probably *could* launch chemical Scuds. This shift in appraisal enraged Schwarzkopf, who charged that the agencies "were just covering their asses."

If the likelihood of chemical attack by one means or another was now acknowledged, the means of retaliation was still a subject of debate.

In early August, in a plan subsequently known as the "Punishment Air Tasking Order," Chuck Horner had laid out seventeen targets — including the presidential palace — to be struck by allied bombers if the Iraqis used chemical weapons. Buster Glosson later proposed a more radical scheme that targeted three dams — two on the Euphrates River and one on the Tigris above Baghdad. The destruction of the dams, the proposal calculated, would flood the Iraqi capital with two to four feet of water and destroy much of the country's industrial base.

Schwarzkopf went a step further. "I recommend that we send a démarche to Baghdad," he told Powell in the late fall. "We'll say, 'If you use chemicals, we're going to use nuclear weapons on you.' We may never intend to do that, but you've got to understand the Arab mind. The Arabs and Saddam in particular understand brute force. It would not hurt one bit to send that signal to Baghdad."

Powell was intrigued by the CINC's suggestion, but ultimately neither Glosson's nor Schwarzkopf's proposal found favor in Washington. Much of the policymaking debate centered in the Deputies Committee, particularly within a six-man subcommittee, informally known as the Small Group, that was chaired by Robert Gates and comprised senior officials from State, Defense, the CIA, the National Security Council, and the Joint Chiefs. Failure to respond adequately to an Iraqi chemical attack, the Small Group agreed, could demoralize the allies and encourage other Third World autocrats to pursue the "poor man's nuke" provided by a chemical arsenal. But an overzealous retaliation with American chemicals or even nuclear bombs — turning Iraq into "a sheet of glass," which a few truculent military officers espoused — would morally corrupt the coalition and bring disastrous political consequences in the Middle East.

Most preferred a middle ground that widened the Iraqi target base in the event of chemical attack. Scowcroft, for example, advocated bombing Iraq's oil fields and a broader array of industrial plants. Wolfowitz, a Small Group member as under secretary of defense, favored striking military targets that would otherwise be proscribed because of their proximity to civilian areas. Yet upping the ante proved difficult to translate into practical, punitive terms. With three thousand sorties a day already planned against Iraq, most militarily significant targets would be struck anyway. Also, with Iraq's communications and intelligence networks shattered, it was not clear that Baghdad would even recognize the higher level of pain being inflicted by the allies.

In large measure, the policy discussions were guided by the military's assurances that chemical attacks posed little threat to the war plan. Three generations had passed since the Germans used a cloud of chlorine to kill five thousand Frenchmen at Ypres; no longer were defenders reduced to masking themselves with wool socks or urine-soaked handkerchiefs. America and its NATO allies, long accustomed to preparing for a possible Soviet chemical attack, had developed sophisticated tactics and equipment to parry such a blow. The cumbersome protective suits and gloves that soldiers wore cut their speed and dexterity in half but still permitted them to fight. And to effectively saturate a single square kilometer with nerve agents, the Iraqis would need favorable winds, accurate gunners, and sufficient time to fire about 150 chemical artillery shells. Even then, in many circumstances the open desert would allow the Americans to steer clear of contaminated areas.

Allied threats had successfully deterred German use of chemical warfare in World War II, and a similar tack was taken against Iraq with a blustery campaign of intimidation. The Pentagon deliberately publicized American preparations for chemical combat, while Bush and others warned of the gravest consequences. In his prewar meeting in Geneva with Tariq Aziz, Baker cautioned that a chemical attack could cause the allies to amend their war aims and put the Ba'athist regime's existence at risk. Cheney even hinted publicly that Israel could be expected to retaliate with nuclear weapons. The threats were vague, and the question of a proper response remained unresolved through the end of the war. But ambiguity had its virtues: if the American leadership was undecided on its path, Saddam could feel equally unsettled.

Biological weapons provoked even greater puzzlement. At the time of the August invasion, Iraq's ability to fashion weapons from biological strains, particularly anthrax bacteria and botulinum toxin, was considered improbable. But by late August, the Defense Intelligence Agency reversed itself and concluded that Iraq indeed had an active anthrax program; a few days later, DIA expanded the assessment to include botulinum. By October 1990, American intelligence warned that the enemy's botulinum capability was sophisticated enough to begin causing allied casualties within four hours after the weapons were used — and thus posed a serious battlefield threat. (After the war United Nations inspectors would discover an extensive anthrax and botulinum research program, begun in 1986 at Salman Pak, but no evidence that Iraq had produced biological weapons or a practical delivery system.)

The American change of heart during the fall reflected both revised intelligence and prudence. Botulinum is three million times more potent than the nerve agent Sarin; a single Scud warhead filled with the toxin could contaminate 3700 square kilometers. Still other factors were considered. Saddam had been sufficiently demonized by Bush that stuffing germs into bombs was seen as precisely the sort of behavior one could expect of him. The so-called *Götterdämmerung* theory also gained currency, a suspicion that Saddam would pull Iraq down around him in a spasm of violence and death.

Biological weapons — BW — retained a sense of horror exceeding even nuclear munitions. Except for giving smallpox-infested blankets to Indian tribes or throwing a dead horse down a well to poison an enemy's water supply, biological combat was rare in the annals of warfare — although scientists for years had experimented with such exotic strains as Q-fever, Red Tide poison, Rift Valley fever, and Russian spring-summer encephalitis. American BW stocks had been destroyed in 1969, shortly after an Army experiment gone awry killed six thousand sheep in Utah. The residual expertise was minuscule, few battlefield detection devices existed, and, unlike nuclear warfare doctrine, relatively little thought had been given to deterring BW use, particularly by a Third World adversary.

American intelligence identified eighteen Iraqi BW targets, among them heavily fortified storage bunkers, and more would be discovered late in the war. The consideration of whether and how to destroy those targets raised difficult questions. If a storage bunker was bombed, how hot must the fire be to destroy the contents completely? If anthrax spores escaped, how long would the area remain contaminated? If civilians died as a result, who would be guilty of biological warfare, the Iraqis or the Americans? Could a contaminated Iraqi river end up poisoning the Persian Gulf?

Ancillary questions concerned the macabre issue of how to handle corpses contaminated by BW. The British, for example, reluctant to demoralize the home front by shipping back bodies during the war, had leased a fleet of freezer wagons in which to store its dead soldiers until an armistice. General de la Billière commissioned urgent studies to determine whether bodies tainted with chemical or biological agents could be decontaminated or would require immediate cremation.

For six weeks in the fall the debate over BW targets raged on. Some suggested that detonating a small nuclear warhead might be a legitimate employment of one weapon of mass destruction to negate another. Temperatures reaching at least twenty thousand degrees Fahrenheit in three seconds were believed necessary to ensure that no spores survived an attack. "We both know there's one sure way to get the temperature hot enough," Glosson remarked to Powell, alluding to thermonuclear explosions. "Yeah," the chairman replied, "but we don't want to talk about that." In New Mexico, scientists attacked stockpiles of an anthrax facsimile with fuel-air explosives — a fiery bomb made with a petroleum aerosol — to gauge whether the fire was sufficiently devouring. Some, including John Warden of Checkmate, favored avoiding direct attacks against the BW stockpiles and instead denying the Iraqis access to them by, for example, seeding storage areas with mines.

In December, the issue came to a head during a briefing by Air Force commander Chuck Horner for Cheney and Schwarzkopf in Saudi Arabia. Horner proposed striking the BW targets just before dawn, when winds were calm. Stealth fighters would drop GBU-27 bombs to crack open the bunkers, followed immediately by F-111s dropping CBU-89 cluster bombs to fuel the conflagration. Any escaping spores would be subject to the rising sun's ultraviolet rays, which, according to one estimate, could be expected to kill up to 2 percent of the spores per minute of exposure.

"If there's collateral damage in Iraq," Horner said, pointing to a map of potential contamination areas, "perhaps that's not all bad. There has to be a penalty for building and storing these weapons. If there is some fallout in Iraq and it causes death, that penalty also sends a signal to others." Schwarzkopf added, "CENTCOM's position is that we attack these targets."

Horner and Schwarzkopf prevailed. Uncertain whether the war would last a few hours or a few months, Horner, Glosson, and Dave Deptula put the eighteen BW sites high on the target list. For insurance, related electrical targets also went on the list; allied bombers would sever the power used to run refrigeration units in the storage bunkers.

To further protect the troops against Iraqi biological weapons, the

Pentagon launched a vaccination program. This, too, caused consternation, since anthrax vaccine stockpiles were limited, and even less antibotulism vaccine existed. Virtually the entire antibotulism supply came from two Army horses, Abe and First Flight, which had once drawn funeral caissons at Arlington National Cemetery. Both horses had been put to pasture — the plodding Abe for lameness, the skittish First Flight for allegedly bolting through Arlington with a general's coffin in tow — and eventually ended up in Minnesota producing antitoxins. Immunized with gradually increasing dosages of botulinum, the horses built up a tolerance until they could survive "hot toxin" capable of killing a thousand horses. Plasma was then drawn from the animals' blood and converted into vaccine.

Unfortunately, Abe and First Flight could not produce nearly enough plasma to immunize hundreds of thousands of troops in Saudi Arabia. The issue grew more complicated when de la Billière asked that British soldiers also be vaccinated and General Khalid, the Saudi commander, requested immunization for "at least the royal family." On seeking advice from Washington, Schwarzkopf was told to make the decision himself. The rebuff infuriated the CINC, who accused policymakers of "washing their hands like Pontius Pilate." Schwarzkopf also was displeased when his staff recommended that the precious vaccines be administered to "those in critical positions," including the CENTCOM staff.

In the end, two decisions were taken. The first was to begin inoculating those soldiers deemed most likely to be exposed to an Iraqi BW attack. Eight thousand soldiers ultimately received botulinum vaccinations, and 150,000 were inoculated against anthrax.

The second decision was to buy more horses. The Joint Chiefs, now gentlemen ranchers as well as warlords, closely monitored the growing herd. By the beginning of the war nearly a hundred steeds had been pastured at Fort Detrick, Maryland, all bleeding for the cause.

## Tel Aviv

With a luck that could pardonably be interpreted as divine preservation, Israel endured the first salvo of Scuds with moderate damage and no serious injuries. Civil defense analysts, studying the havoc wreaked on Tehran during Iran's War of the Cities with Iraq in the 1980s, had predicted five to ten deaths for every Scud launched. Instead there were none. Hundreds of apartments in Tel Aviv and Haifa suffered broken windows and cracked plaster, and a few dozen were demolished. About sixty people suffered cuts and bruises.

The psychic damage was greater. Except for border skirmishes and terrorist attacks, Israeli civilians had not been directly assaulted since the war of independence in 1948. Hundreds now sought treatment for panic and anxiety, including many who needlessly injected themselves with atropine, the antidote to nerve gas. A group of Hassidic Jews, convinced they were contaminated with mustard gas, leaped into the freezing water of a makeshift outdoor shower wearing only their hats. Mothers and fathers took refuge in *heder atum,* sealed rooms, taping the windows, stuffing soaked rags beneath the door, and counting their children, who all looked alike beneath their gas masks. More than a thousand souls from the poor neighborhoods of south Tel Aviv sheltered in the concrete stalls of the city's new, unfinished bus station, eighty feet underground. Phone calls from the United States to Israel, as the White House had discovered, jumped from the usual three thousand an hour to 750,000. Doomsday prophets tramped among the fearful, reciting Isaiah 26:20: "Come my people, enter thou into thy chambers, and shut thy doors about thee. Hide thyself as it were for a little moment, until the indignation is past."

In a stout command bunker beneath the walled compound of the Keriya, the Israeli defense headquarters, Moshe Arens also awaited the all clear, but his indignation had just begun. The defense minister had driven himself and a bodyguard from his home near Ben Gurion Airport, east of Tel Aviv. As he raced down the Jerusalem highway and through the deserted city streets, the Scuds seemed to detonate all around him. He pulled through the Keriya gates, parked the car, and hurried inside to prepare his forces for all-out war.

Dozens of Israeli warplanes circled to the east, ready to thwart an Iraqi air attack. Nuclear missile crews stood at full alert for only the third time in the country's history. Flares lit the northern border, where soldiers watched for infiltrators slipping across from Lebanon. The Israel Defense Forces waited for orders; the air force particularly strained at the leash. Even the Iceman, as the impassive Arens was dubbed by an Israeli magazine, felt his stomach knot with tension.

For two years the Israelis had badgered the Americans about Iraq's imperialist ambitions, although the warnings reflected more anxiety than clairvoyance regarding Saddam Hussein's intent. In June 1990, they repeated the caution to Lieutenant General Harry E. Soyster, director of the Defense Intelligence Agency, who subsequently reported to Colin Powell, "The Israelis want us to be aware that Iraq has a huge military force looking for something to do."

A month later, shortly before the invasion of Kuwait, Arens and the head of the Mossad flew to Washington to offer Cheney detailed intelligence on Iraqi efforts in Europe to procure centrifuges used to enrich uranium for atomic weapons. On August 28, another delegation, headed by David Ivri, director general of the defense ministry, flew secretly through Zurich to Washington, where they registered incognito at a Holiday Inn. During two days of meetings with Scowcroft, Cheney, and other American officials, the Israelis hammered home their conviction that Saddam was certain to attack Israel if war erupted over Kuwait.

Now that prophecy was realized. In the balance hung not only Israel's role in this war but its standing in the discordant neighborhood of the Middle East. For four decades Israel had swiftly retaliated against any attack; Arens and others feared that failure to strike back would efface forty years of deterrence, embolden Israel's enemies, and batter the country psychologically.

Yet also at stake was Israel's "special relationship" with the United States. Ties between the two had frayed badly in the past year; with the ending of the Cold War, the Americans had begun reassessing their need for a proxy policeman in the Middle East. The Israelis particularly distrusted Brent Scowcroft, who they believed had misjudged Iraqi intentions and was personally hostile to Israel. After the invasion of Kuwait, the United States, reluctant to offend Saudi Arabia and other Arabs, stalled for weeks before agreeing to meet David Ivri's delegation. "I feel like a small child who has been told to sit in a corner and behave," Ivri fumed. When Lawrence Eagleburger and Paul Wolfowitz visited Jerusalem the weekend before the war, Shamir used a different metaphor: "You treat us like a relative who has a social disease. You want to have nothing to do with us."

But Arens knew that if an Israeli counterstrike against Iraq split the allied coalition or harmed the American cause in any way, there would be no hope of reconciliation. The pipe dream some Israelis harbored of building a strategic alliance with the United States, similar to America's links to Great Britain, would vanish forever. At a minimum Israel could lose its $3 billion annual aid stipend, plus the billions more needed to settle thousands of Soviet immigrants arriving every week.

For months the pros and cons had been debated in the cabinet, in the newspapers, in Jerusalem coffee houses and Tel Aviv bistros. In a country said to have five million prime ministers, palaver was the national pastime. But on Arens's shoulders fell much of the responsibility for steering a proper course. The military link between the United States and Israel remained strong even as the political affiliation wavered. Arens, moreover, retained a deep emotional attachment to America. Lithuanian-born, he and his family had fled the Nazis in 1939 to settle in Manhattan. He had served in the U.S. Army, married a New Yorker, attended MIT and Cal Tech, and worked as an engineer for an American aerospace company. He possessed a boyhood affection for Franklin Roosevelt and the New York Yankees. Except for the former prime minister Golda Meir, no American émigré had gone further in Israeli politics than Iceman Arens.

In his conversations with Baker and Cheney after the first Scuds fell, Arens tried to keep his options open. To Cheney, in one of the two dozen conversations they would have during the war on the Hammer Rick line, Arens reversed his earlier refusal of American-manned Patriot missiles. He asked that the batteries be dispatched to Israel as quickly as possible. Later that night, in another conversation with Cheney, the defense minister sharply rejected the idea of relying on Jericho missiles instead of air strikes. "That's crazy," Arens said. "It would accomplish nothing, and it might even make matters worse" by provoking the Iraqis to resort to chemical weapons. As dawn broke in Israel on Friday morning, Arens settled in to await Shamir's authorization for air strikes and Bush's agreement on a deconfliction plan.

But the attack order never came. On Friday morning in Jerusalem, Shamir talked to Bush by phone for the first time in many months. The president repeated the plea for restraint made earlier by his secretaries of state and defense; he pledged a relentless American effort to destroy the Scud sites. Shamir, constitutionally enigmatic, offered no assurances, but he told the president that any decision would be deferred until the Israeli war cabinet met later in the day.

The arrangement satisfied Arens. The extra time would give the allies a chance to unfurl their campaign against the Iraqi missiles; the Americans had assured him that they could suppress the Scuds in twenty-four to forty-eight hours. If the coalition failed, Arens believed, Israel would have another opportunity for vengeance. But for now, it would turn the other cheek.

## Washington, D.C.

Moshe Arens had a counterpart in the Pentagon worthy of his stoicism. Richard Bruce Cheney had elevated inscrutability to an art. An erstwhile *Wunderkind* — at thirty-four he had been President Gerald Ford's chief of staff — Cheney had survived not only the hurly-burly of Washington politics but also three heart attacks and quadruple coronary bypass surgery. Now forty-nine, an avid reader of history and biography, given neither to belly laughs nor tantrums, he rarely permitted his countenance any expression beyond a lopsided smile or a quizzical frown. On Capitol Hill, where he had served more than a decade as a Republican representative from Wyoming, he was known as the Sphinx; his military assistant, Rear Admiral Joe Lopez, occasionally urged him to convert "the best poker face in Washington" into hard cash by playing cards for a living.

Ambitious, intelligent, unaffected, Cheney also could be ruthless. In March 1989, eight days into his tenure as defense secretary, he had publicly rebuked the Air Force chief of staff, General Larry Welch, for usurping presidential prerogative on the issue of strategic missiles. A few months later, before the American invasion of Panama, he dismissed General Frederick F. Woerner, Jr., the U.S. commander in Central America. And six weeks after the invasion of Kuwait, he fired Welch's successor as Air Force chief, General Michael J. Dugan, for disclosing to the *Washington Post* and the *Los Angeles Times* sensitive details about the strategic air campaign after Dugan had visited Horner and Glosson in Saudi Arabia.

Picking able commanders, Cheney believed, was his single most important duty. He had elevated Colin Powell to the chairmanship over fourteen more senior four-stars. They formed an unlikely partnership: the balding, unemotional, fly-fishing politician from Wyoming and the animated, charismatic infantryman from the Bronx. Yet despite occasional friction — Cheney at times wanted more military options and less political advice from Powell — in the main they worked hand in glove.

On no issue did the secretary rely more on his chairman than in the handling of Norman Schwarzkopf. In the months preceding the war Cheney had often asked himself — and Powell — whether Schwarzkopf was the proper general for the job. Cheney carefully cultivated back-channel sources of information, both in the Pentagon and in Riyadh; he called it "pulsing the system." He was well aware of the CINC's reputation as a volatile man who berated his subordinates and was, as the secretary dryly observed, "something of a screamer."

One incident in particular nagged at Cheney. In early August, at Bush's behest, he had flown to Saudi Arabia to secure King Fahd's approval for the deployment of American forces. Schwarzkopf was aboard for the fifteen-hour flight. When the dozing passengers awoke at dawn, a line formed to use the bathroom. The queue inched forward and a major who finally worked his way to the front turned and said, "General?" The officer had been Schwarzkopf's place holder. About the same time, Cheney glanced down the aisle and saw a colonel on hands and knees ironing the CINC's uniform blouse.

The secretary of defense prided himself on his plebeian touch; pretensions offended and irritated him. Powell had assured him that Schwarzkopf's strengths far outweighed his foibles. But Cheney wasn't certain. Given cause, he let it be known, he would find another general without hesitation, as he had found replacements for Woerner and Dugan.

Yet in the succeeding weeks and months, watching Schwarzkopf's performance, the secretary had come to agree with Powell. The CINC's briefings, including presentations for the president, had been crisp and professional. He clearly knew the Middle East well, as he knew the profession of arms. And Schwarzkopf had panache, which played well on television, no small consideration for an American public still wary of generals after Vietnam. Cheney concluded that other capable officers, notably Cal Waller, could be sent to Riyadh to help sweep up the CINC's broken crockery. "Managing the Schwarzkopf account," as Cheney called it, would put a great burden on Powell, but one the chairman seemed ready to shoulder.

Disinclined to accept the status quo in personnel matters, Cheney also took an active role in operations. He had never served in uniform; as a student and young father, he was exempt from conscription in the mid-1960s. But the world of the soldier intrigued him, with its argot, its ethic, and its romance. He sat for hours in the National Military Command Center, listening, watching, questioning. In his third-floor office, Cheney kept what the operations officer Tom Kelly called "the largest goddam collection of maps I've ever seen." Eager to master the nitty-gritty of the coming war, he subjected himself to fifteen tutorials on such arcane subjects as "Building an Air Attack Plan" and "Breaching Iraqi Forward Defenses." Like a field marshal searching for chinks in an opponent's character, he studied intelligence reports and academic analyses of Saddam. (The Iraqi leader, Cheney concluded, misunderstood modern combat and had misread the lessons of American military history by assuming he could destroy U.S. public support for the war through World War I–style trench warfare that created a "great engine of casualties.")

If conventional wisdom seemed misguided or specious, the secretary poked, probed, and wheedled. He encouraged the Navy, for example, to reconsider its reluctance to move aircraft carriers into the confines of the Persian Gulf, where they were more vulnerable to missiles and attack boats than in the open Gulf of Oman; by the middle of the war, four carriers would steam through the Persian Gulf, thus halving the flight time to targets in Iraq and Kuwait. To his commanders he offered an unwritten contract: he would provide whatever military resources they wanted in exchange for guaranteed success. "You want two thirds of the Marine Corps? Done," he declared at one point. "Six carriers? Done. Two Army corps? Done. Tell me what you need. But then you've got to deliver." Within a few weeks of the invasion of Kuwait, the secretary had concluded that force of arms would likely be required to expel Saddam's troops.

As Powell managed Schwarzkopf, Cheney managed the Israelis. He believed that, with the notable exception of Lawrence Eagleburger, the powers at the State Department and White House chose to keep Israel at arm's length and to demand Israeli passivity in the war. Cheney demurred, urging instead that the United States wrap the Israelis in a warm embrace. "We want them to be comfortable," he insisted. "Cooperating with them is going to be a hell of a lot more productive than stiffing them."

Cheney also suspected that CENTCOM failed to appreciate the importance of destroying the Scud sites in western Iraq as a condition for keeping Israel out of the war. In October, he had taken a direct hand in the war planning by proposing an American attack that swept far to the west, seizing the Iraqi strongholds H-2 and H-3. (The curious nomenclature derived from pumping stations along a now-defunct oil pipeline terminating in Haifa.) The operation, he believed, would eliminate the Scud threat to Israel, sever the Amman-Baghdad highway used by smugglers, and provide the allies with a potential bargaining chip — a chance, in effect, to swap western Iraq for Kuwait. Drawn from Operation Scorpion, a top secret plan drafted by his staff and deliberately kept from the Joint Staff, the proposal also was seen as a means to threaten Baghdad and expose counterattacking Iraqi tanks to allied air power.

In Riyadh this scheme was viewed as twaddle, far removed from the strategic objective of liberating Kuwait and a potential cause of humiliation if Saddam decided to pocket the emirate rather than redeem his occupied western wasteland. Even so, suggestions from the secretary of defense could not be ignored. The Army duly analyzed the idea and concluded that to conquer an area roughly sixty by a hundred miles would require up to two divisions, either the 82nd Airborne and the 101st Airborne or the 101st and the 3rd Armored Cavalry Regiment. Logistically the operation would be a nightmare. By Christmas even Cheney had abandoned the concept.

But his focus on the Scuds remained fixed. As the missiles rained down on Israel, Cheney spent the night in his office behind the massive desk once used by General John Pershing. A map of the world covered one wall and mementos littered the room like war trophies: a bust of MacArthur, a grandfather clock, a model ship in full sail. From the windows behind the desk he could see the dark blotch of the Tidal Basin and, in the distance, the lighted dome of the Capitol.

After his conversations with Arens on Hammer Rick, Cheney had sifted through the Israeli's words looking for clues. Arens seemed ready, even eager for war. But Cheney knew a crafty politician when he heard one: Arens was shrewd enough to act bellicose both for the benefit of archconservatives in the Israeli cabinet and for the Americans. The more pugnacious he seemed, the more grateful Washington would be for his restraint. That could pay dividends in Israel's long-term strategic links to the United States and in its requests for greater financial aid.

Yet no Israeli politician could withstand the pressure to retaliate if eight Scuds continued to explode every night. CENTCOM and the rest of the U.S. military, Cheney believed, had to understand that. When he talked to Powell later that evening, the secretary made the point as clearly and concisely as he knew how. "Colin," he declared, "we have got to get those things."

## U.S.S. *Saratoga,* Red Sea

Beneath the harsh orange glare of the flight deck lamps, the pilots and crew of the *Saratoga* were preparing to do precisely that with an attack on the Iraqi base at H-3. The mission would prove to be one of the most difficult and controversial of the war, with consequences that directly affected the frantic hunt for Scud missiles.

An aircraft carrier in combat was a unique amalgam of grace, power, and orchestrated confusion. Some strategists believed that the carrier was steaming into oblivion, a glorious anachronism doomed by the lethality of guided missiles; for now, however, she remained queen of the sea. Five thousand men moved through their paces, each contributing to the larger purpose of flinging a jet from a narrow, floating airstrip and eventually reeling it back again.

Scores of sailors scrambled across the flight deck, their eyes tearing from the acrid fumes, their duties coded by the color of their jerseys: fuelers in purple, medics and safety teams in white, plane captains in brown, crash crews in red, flight directors in yellow, utility men in blue. Like bats wrapped in the folds of their own wings, the jets rode up from the hangar deck on stout elevators to be fueled, armed, and manned. Then one by one they inched toward the catapult on the bow.

A petty officer aligned the nose wheel and hooked the toe links into the steel shuttle. The pilot shoved his throttle to full power, released the brakes, and saluted with a flash of lights. Slicing the air with a quick sweep of his yellow wand, a catapult officer touched the deck to trip the cat. With a hiss and a clatter the great steam slingshot dragged twenty tons of screaming aircraft from dead stop to 160 knots in three seconds, and another sortie flew into the night. A carrier, as a naval officer once observed, was nothing more than a giant aircraft hatchery.

No hatchery in the United States Navy claimed a grander lineage than the *Sara.* The first *Saratoga* served as flagship for Thomas MacDonough's Lake Champlain squadron, which in 1814 repulsed the largest invasion of the United States ever attempted and sent the British reeling back into Canada. The first aircraft carrier to bear the name had, in 1932, simulated a predawn raid on Pearl Harbor, catching the Pacific Fleet by surprise in a lesson regrettably forgotten nine years later. During the war she saw action throughout the South Pacific, only to be sunk by a forty-foot wave in an atomic bomb test at Bikini Island in 1946. The current *Sara* was commanded by Captain Joseph S. Mobley, who had been a North Vietnamese prisoner of war for five years. She had been on station in the Middle East since late August, and had practiced so many launches and traps that the flight deck resin was worn through to bare steel in patches. Even in the placid Red Sea, night operations never became perfunctory: carrier lore held that a pilot's heart beat faster during a routine night landing than in the fiercest combat.

The cruise already had been marred by tragedy. Although the first — and, ultimately, the only — two Iraqi MiGs downed by Navy jets were claimed by *Saratoga* pilots, the first American killed in the air war — F-14 pilot Scott Speicher — also flew from *Sara.* Moreover, four weeks earlier, twenty-one *Saratoga* sailors drowned when a rickety packet boat ferrying the men from a liberty call in Haifa capsized and sank in fifteen seconds. The accident cast a pall over the ship and provoked whispers elsewhere in the Navy that *Sara* was unlucky, an epithet subsequently reinforced by the carrier's loss of three airplanes in the opening days of the war.

At the heart of the raid on H-3 was the carrier's A-6 bomber squadron, known as VA-35. The Navy's oldest attack squadron, formed in 1934, VA-35 had shipped aboard U.S.S. *Hornet* during the Doolittle raid on Tokyo, and later saw action at Guadalcanal, Midway — where her mother ship, *Yorktown,* was sunk — the Coral Sea, and during four deployments in Vietnam.

Leading the attack on the night of January 17–18 was the pilot who had planned it, VA-35's executive officer, Commander Michael J. Menth, an ebullient Minnesotan with more than 3500 flying hours in an A-6 Intruder and a thousand carrier landings to his credit. From the moment his plane left *Sara*'s bow and headed north, Menth saw that the mission would be even more difficult and dangerous than he had anticipated. A storm front had blown across the Middle East earlier in the day, bringing turbulent winds and a solid cloud bank at twelve thousand feet. Ice coated the bombs, and sheets of St. Elmo's fire danced on the wings as the raiders hunted over northern Saudi Arabia for the tankers that would refuel them.

Typically the tankers preferred to fly at twenty thousand feet; tonight they pressed below ten thousand to find a clearing under the clouds. Buffeted by ground turbulence, the fuel baskets smacked against the blunt snouts of the Navy bombers as the pilots struggled to steady their heaving planes. Laden with ten thousand-pound Mark 83s each, the bombers felt slow and sluggish, a reminder of why A-6 pilots referred to their aircraft as the Cement Truck.

Mike Menth, like all young pilots, had heard countless horror stories about how higher authority in Vietnam had dictated battle tactics in minute detail. To his surprise, he had been permitted to plan this attack without interference. For months he had studied the targets. Originally the squadron was to attack both H-3 and H-2. The former, only thirty miles from the Jordanian border, was guarded by six SA-6 and two Roland SAM sites, as well as innumerable antiaircraft (AAA) guns and fighters. H-2, lying farther northeast, had six Rolands and almost as many guns. When the carrier *John F. Kennedy* joined *Saratoga* in the Red Sea, the plan was modified to exploit the additional firepower. Menth kept the H-3 targets, which he considered the more treacherous, and gave H-2 to *Kennedy*. Scud batteries near both sites had been struck the previous night by Air Force jets. Now the Navy bombers were to hit fuel dumps, control towers, and hangars used to support Iraqi missile operations.

Menth planned to attack at low level, between three and five hundred feet, and therein lay controversy. For half a century, aviators had pondered the proper altitude for successful — and survivable — bombing missions. In the summer of 1942, Army Air Corps planners agonized for weeks over whether to attack the critical Nazi oil refinery at Ploesti, near Bucharest, from on high or down low. They chose the latter — one B-24 grazed the tip of a church steeple, another clipped a field of sunflowers with its belly antennae — and forty-one planes were lost. In Vietnam, 70 percent of 2300 downed American aircraft fell to relatively low-level AAA fire. Even so, after Vietnam many pilots concluded that the only way to penetrate the thick SAM belts ringing the Soviet Union was to fly "nap of the earth," just above the ground.

At higher altitudes, the Iraqis offered SAMs and fighters; closer to the ground, antiaircraft batteries and small arms fire. (Machine gun rounds generally petered out above a couple of thousand feet; some large AAA guns, like the Iraqi 57mm, could reach to fourteen thousand feet or higher.) Bombing accuracy diminished at higher altitudes — particularly with the dumb bombs that accounted for more than 90 percent of those dropped on Iraq — and pilots on high often had trouble finding small, elusive, or camouflaged targets, such as mobile Scuds.

The debate had intensified as war with Iraq drew closer. Nearly all senior Air Force and Navy aviators, remembering the harrowing sheets of ground fire in Vietnam, concluded that pilots would be safer at high altitude, where they could maneuver against SAMs or thwart radar-guided missiles with electronic jamming. Chuck Horner and Buster Glosson recommended that their pilots — except those in F-15Es, and F-111s — remain above fifteen thousand feet. Before the war, the Navy commander in the Red Sea, Rear Admiral Riley Mixson, had directed *Saratoga* to practice high-altitude tactics. He had even intervened before *Sara*'s attack on the first night against Al Taqaddum Air Base to ensure that the bombs fell from fifteen thousand feet. He had not, however, studied the tactics to be used against H-3. Commander Menth's attack plan was sent to him aboard his flagship, *Kennedy,* but Mixson was occupied with other duties.

The *Kennedy* A-6s planned to hit H-2 from high altitude. On *Saratoga,* however, Menth reasoned that a low-level attack at H-3 would catch the enemy by surprise and negate the SA-6s, the most feared missile in Iraq's arsenal. As for the guns, eventually the Iraqis would either exhaust their ammunition or melt their barrels. Menth's superiors on *Sara* supported his attack plan, although the carrier air group commander (known as the CAG), Captain Dean M. Hendrickson, wrestled with his own ambivalence before assenting. Among other voices arguing against low attacks, Hendrickson had heard that of Captain Mike (Carlos) Johnson, chief of the Navy intelligence unit SPEAR and a former CAG on *Kennedy.* In the last year of the Vietnam War, Carlos Johnson had flown from a carrier that lost seventeen airplanes. The ship was the *Saratoga.*

Shortly before 11 P.M. on January 17, Menth's jet and three other refueled A-6s roared across the outer defenses at H-3. Had western Iraq boasted steeples or sunflowers, the jets would have clipped them. As the pilots entered "government time" — aviator slang for the most intense portion of a bombing run — SAMs and antiaircraft tracers began to boil up even though the attackers were still fifteen miles from the target. The H-2 attack, Menth learned, had been aborted after *Kennedy's* aircraft found refueling impossible in the stormy weather, but he was determined to complete his half of the mission.

As planned, a volley of TALD drones and HARM missiles preceded the A-6s, which split into two flights to attack from different vectors. To Menth's surprise, the Iraqis responded with a barrage of flares that abruptly lit the airfield at H-3 with midday brilliance and stripped the Navy bombers of their last stitch of surprise. Each A-6 broadcast a signal electronically simulating a flock of airplanes, but the deception was useless when defender could plainly see attacker. Menth and his bombardier-navigator, flying under call sign Dash 1, swerved through the thick hoses of antiaircraft fire, flung their bombs, and fled toward the open desert. Twenty miles out, Menth checked his air-to-air radar and saw no trace of his wingman, Dash 2. Turning to the bombardier, he said, "I'll bet they got bagged."

Bagged they were. A mile from bomb release, the pilot, Lieutenant Bob Wetzel, and his bombardier, Lieutenant Jeffrey Zaun, spotted a SAM streaking toward the right side of the plane. Wetzel sliced beneath the missile, but it detonated with a force that shook the airplane violently from side to side. The tail burst into flame and Wetzel heard one of the engines grind to a halt. "Eject!" he yelled, yanking the black-and-yellow loop above his head. Explosive bolts beneath the seat blew him through the plexiglass canopy, fracturing both of his arms and a collarbone. When Wetzel regained consciousness, Zaun was helping him remove the parachute harness. The silhouette of the H-3 hangars loomed nearby. Within an hour, the Iraqis captured both men.

The other pair of jets streaked across the target moments after Dash 2 went down. Dash 3 managed to drop its payload and escape, but Dash 4 took a missile that blew shrapnel through the fuselage and damaged an engine. The pilot jettisoned his bombs and nursed the plane back to Saudi Arabia for an emergency landing at Al Jouf.

At 2 A.M. Mike Menth hooked the arresting cable on *Saratoga's* deck, climbed from the cockpit, and walked stiffly to the debriefing room, where other pilots were planning the next night's attack. "We just can't do this anymore at low level. It's going to eat us alive," he reported. "The gun barrels did not melt."

*

*Sara*'s misadventures triggered recriminations and soul-searching. Within hours of Menth's return, an angry Riley Mixson sent a warning to his carrier admirals in the Red Sea:

> Low-altitude delivery is not the way to go in heavy AAA and associated SAM-6 environment that we have been experiencing in Iraq. Absolute minimum pullout altitude of 3000 feet is the norm. No lower than 3000 feet. 5000 feet or higher is better. Request you ensure your CAGs comply. No, repeat, no low-altitude deliveries without my express permission.

In the Persian Gulf aboard U.S.S. *Blue Ridge,* Vice Admiral Stan Arthur, the senior naval commander in Desert Storm, drafted a message of his own. Many in the Navy had grown accustomed to thinking of warfare in terms of quick raids and punitive jabs, like those launched against Syrian gunners in Lebanon in 1983 or against Libya three years later. Now, Arthur believed, it was necessary to think of war as a protracted campaign. To his admirals he cabled:

> Gentlemen — Far be it from me to dictate specific combat tactics, but I must inject my early observations relative to the age-old argument of low-altitude delivery versus high . . . We learned a hard lesson in Vietnam relative to AAA and later many told us we learned it wrong. I think not. There is a place and time for low-altitude delivery and it usually involves surprise. We can no longer count on surprise. That went away shortly after [3 A.M.] on 17 January 1991.

On *Saratoga,* planners immediately revamped their tactics for the next night's mission by converting a low-level attack against the Haditah television station west of Baghdad into a high-altitude strike. At Mixson's request, Menth and CAG Hendrickson flew by helicopter to *Kennedy* for further discussion. The admiral's most pressing concern was a suspicion — soon determined to be ill-founded — that Wetzel and Zaun had been shot down accidentally by a Navy F-14. But Mixson soon came around to the altitude issue. "I think we learned our lesson, didn't we?" he asked. "I still can't believe you guys went in low."

Hendrickson took full responsibility while pointing out that the plan had been sent to *Kennedy* for vetting. "I think it was ill-conceived," Mixson replied, "but you're right. It came through me. Let's go back to the basics we learned a long time ago and let's not deviate from them."

Never again would Hendrickson permit his pilots to contemplate low-altitude missions, not even when *Sara* began hunting mobile Scud launchers in earnest (an enterprise that accounted for a quarter of the carrier's attack sorties in the following six weeks). Henceforth, Navy bombs from all six carriers would be dropped mostly from sixteen to eighteen thousand feet. Other aviators tried to profit from *Saratoga*'s misfortune. Any pilot flying lower than the prescribed altitudes, Chuck Horner warned, "is going to be watching the war from the ready room."

Such edicts showed prudence and unquestionably saved lives. But the issue was hardly cut-and-dried. Had Iraq's fighter fleet displayed more pluck in challenging the allies, the higher altitudes would have offered dubious sanctuary. Both aircraft struck at H-3 were hit with missiles, not ground fire; three days later, another *Saratoga* plane would be lost, this time to a SAM at thirty thousand feet.

Furthermore, some munitions demanded low-level delivery. Four hours after the H-3 attack, an A-6 flying from U.S.S. *Ranger* in the Persian Gulf was shot down by ground fire while sowing mines around the Iraqi naval base at Umm Qasr. Both the pilot and bombardier-navigator, flying at five hundred feet, were killed.

As always in warfare, the calculus of benefit and risk, success and failure, life and death was profoundly intricate. Cheney's desire to eradicate the Scud launches against Israel had to be weighed against the military conviction that pilots seeking better visibility and accuracy placed themselves in greater jeopardy the lower they flew. As the Americans quickly discovered, trying to spot and then accurately bomb a mobile Scud truck in the dark from three miles up was extraordinarily difficult. Israeli pilots would soon claim, with irritating zeal, that they could more effectively root out the Scuds — not because they were better airmen, but because with Israel's fate at stake they were willing to fly as low and aggressively as necessary to find the missile launchers.

What was at stake for the Americans and the British and the French? Duty, courage, aviator élan — all played a part in pushing the pilots north. But the central question was more ambiguous and, in this war, could never be answered with certainty: at what point was a man's life a legitimate fee for the liberation of Kuwait?

Another mission that turned sour on the night of January 17 brought the point home. Fourteen Air Force F-15Es took off from Al Kharj within minutes of Mike Menth's launch from *Saratoga.* Veering east of the Iraqi air base at Jalibah, the bombers split up to hit three separate targets around Basrah. Four headed for a bridge; four veered toward a power plant. The remaining six streaked in single file toward an oil refinery southwest of the city. To avoid the many SAM batteries manned by Republican Guard units below Basrah, the six-ship formation flew at five hundred feet over the dark swampland of the Hawr al Hammar. As the attackers neared the refinery, already ablaze from an earlier Navy attack, ten Iraqi gun pits opened up from the south.

The inferno of the refinery clearly silhouetted the dark planes against the pall of smoke and low clouds.

One after another, the aircraft flicked their bombs and turned west to rendezvous for the flight home. The last plane, Thunderbird 6, failed to appear. The wreckage was never found by the Americans. After the war, the Iraqis returned the bodies of the two crewmen, Lieutenant Colonel Donnie R. Holland and Major Thomas F. Koritz.

Back at Al Kharj half a dozen survivors from the mission gathered in the tent of the two dead men, sharing a forbidden bottle of tequila and ignoring the siren warning of a Scud attack. Koritz had been a flight surgeon, Holland a tall, genial career airman with a passion for golf. On the table next to their cots, framed photographs of their wives and children stared at the pilots nursing their tequila. With brutal efficiency, an Air Force team would arrive in the morning to pack up the pictures and the flight suits and the shaving kits. Al Kharj, the air campaign, the war itself would go on as though the men had never existed.

The survivors drank until dawn, mostly in silence, then shambled off to a fitful sleep. *There's not a damned thing worth dying for in Iraq,* Buster Glosson had told them. And there was nothing more to say.

4

# The Left Hook

## Riyadh

On January 15, less than forty-eight hours before the war began, Norman Schwarzkopf had accepted an invitation from Buster Glosson to visit his strike planners in the basement of the Royal Saudi Air Force headquarters. Glosson wanted the CINC to pump a few hands, slap a few backs, and buck up morale before the long ordeal that would begin when the first bombs fell on Baghdad.

Schwarzkopf complied admirably. The air thickened with camaraderie. Toward the end of his tour, the CINC stood before a map with Glosson and Chuck Horner as Dave Deptula explained the sequence of attacks scheduled in the first hours of the war. Some twenty other officers lingered in the background, watching their commander-in-chief.

"Why," Schwarzkopf suddenly interjected, "are we not bombing the Republican Guard with B-52s?" Deptula, clad in his olive drab flight suit, hesitated before pointing out that the Tawalkana Division and other Guard units would indeed be bombed by B-52s on the night of January 17–18, about eighteen hours into the campaign. Subsequent raids had been scheduled against the Hammurabi and Madinah divisions.

Glosson jumped in. "Sir, because of the threat from surface-to-air missiles we never planned to bomb them until we've had a chance during the first day to work over all those SA-6s and SA-2s and SA-3s with our F-16s and F/A-18s." Glosson was aware of the CINC's fondness for B-52s; like many senior Army officers, he had seen the devastating consequences of Arc Light strikes in Vietnam. Glosson also knew of Schwarzkopf's preoccupation with the Republican Guard. Since late summer Schwarzkopf had directed that the Guard be attacked early and often during the air war. The CINC had often stressed that no allied ground offensive would be launched until coalition bombers had reduced Iraqi forces by at least 50 percent.

Now he flushed as he loomed over Glosson. The bonhomie vanished. "You've lied to me!" he barked.

"Sir," Glosson snapped back, "I've never lied to you. Period."

As other officers in the room exchanged glances, the CINC's voice rose. "I said I wanted them bombed from hour one and that's what I want! You people have been misleading me. You're not following orders. You're not doing what I told you to do."

Horner stepped in. "We don't have enough sorties to do it all. I —"

Schwarzkopf cut him off. "Goddam it! If you can't follow my orders, I'll find somebody who can!"

Glosson heatedly demanded, "You tell me how many B-52 air crew members you're willing to lose, and I'll tell you how many we can put in there. You tell me how many you're willing to have die."

"Goddam it, Buster! I've never been willing to have someone die unnecessarily."

"That's the issue," Glosson insisted.

"I don't want to discuss it anymore," Schwarzkopf said, turning away.

"Well —" Glosson began.

"I don't want to discuss it anymore!"

For a moment the four men stood as if in tableau. Horner assumed that Glosson would be ordered from the theater for impertinence. Deptula, astonished at the CINC's tirade, began calculating the chaos that would result from an abrupt reconfiguration of the first night's attack.

Schwarzkopf stalked from the room and up to Horner's office on the third floor, trailed by the two Air Force generals. When the door was closed, he jabbed his finger at them. "Don't either one of you ever again confront me like that in front of people. Goddam it, the guys we're going to have to fight are the Republican Guard. We've got the B-52s and we want to start pounding them right away." The color drained from his face and his voice softened. "I'm under a lot of pressure. You don't understand how much pressure I'm under."

Horner eyed the CINC curiously. Give me a break, he said to himself. Glosson began to reply, then thought better of it. We put the entire air campaign together by ourselves, he thought, with no help from you. Those are our pilots who are going out to die, and you have the gall to talk to us about pressure?

Norman Schwarzkopf was a terrestrial creature. For nearly forty years, since entering West Point as a seventeen-year-old plebe and on through company, battalion, brigade, division, and corps command, his professional life had revolved around the clash of ground armies. His heroes — Grant, Sherman, Creighton Abrams — were all, as he fondly called them, "muddy-boot generals."

He had never believed strategic air power likely to roust Saddam Hussein's army from Kuwait without the knockout blow of a ground attack. Although he accepted the centers of gravity laid out by John Warden and subsequently incorporated into the air campaign, his own focus was on Iraq's army, particularly the Republican Guard's three heavily armored divisions, which he sometimes referred to as "*the* center of gravity." In this he echoed Colin Powell, who had first drawn attention to the Guard in his meeting with Warden on August 11.

The Central Command that Schwarzkopf inherited in 1989 had focused on a Soviet thrust south into Iran, a hypothesis obviated by the collapse of central authority in Moscow. Almost immediately after assuming command, he began looking at other threats; in September 1989, with encouragement from Under Secretary Paul Wolfowitz, the CINC asked the Defense Intelligence Agency to scrutinize Iraq as a potential foe. By July 1990, he had assembled portions of a secret battle plan, Operations Plan 1002–90, *Defense of the Arabian Peninsula,* which was designed to parry an Iraqi invasion of Saudi Arabia through Kuwait.

Schwarzkopf's initial mission in Riyadh, on his arrival in August, had been to defend the Saudis from further Iraqi aggression. The forces at his disposal, under long-standing Pentagon war plans, included Horner's Ninth Air Force, the 1st Marine Expeditionary Force, and the U.S. Third Army. Using the basic blueprint of 1002–90, he intended to trade space for time in the event of an invasion of Saudi Arabia; he would defend enclaves around Jubail and Dhahran, grind down the Iraqis if they drove south, and counterattack when allied forces grew sufficiently strong. The war game in the summer of 1990, code-named Internal Look, showed that the strategy would work, albeit at a cost of nearly half the Americans' fighting strength. The plan as modified by Schwarzkopf was smart, necessary, and bitterly opposed by the Saudis, who favored a line of defense at the Kuwaiti border and resisted surrendering any territory, even temporarily, to the enemy.

As Desert Shield matured, Schwarzkopf began searching for an attack plan that would drive the Iraqi occupiers from Kuwait. OPPLAN 1002–90 contained no detailed counteroffensive scheme, so in late August he drafted the outlines of a four-phase offense: a three-part air campaign, built on Warden's Instant Thunder, followed by a ground attack in the fourth phase. Schwarzkopf told Bush he would need eight to ten months to assemble the necessary forces for a counteroffensive; with

Powell's help, the size of the force eventually doubled and the deployment time was halved.

Throughout the fall, Schwarzkopf stressed the importance of demolishing the Republican Guard. He personally dictated the mission of his main ground attack force: "attack deep to destroy Republican Guard armored-mechanized forces." But precisely where, how, and when the ground offensive should be launched remained a puzzle. A CENTCOM planning team, code-named Eager Anvil, initially concluded that conquering Kuwait with the force on hand "can't be done." Schwarzkopf knew as well as Powell or Cheney that a liberation of Kuwait at the cost of thousands of American lives would be a Pyrrhic victory, with dreadful political repercussions. Yet the history of armored attacks in the twentieth century was not reassuring; even the Nazi invasion of Poland in 1939 — a blitzkrieg against a surprised and technologically inferior foe — had resulted in 45,000 German casualties.

In mid-September, four Army officers arrived in Riyadh to help Schwarzkopf draft his plan. Led by a former helicopter pilot turned armored officer, Lieutenant Colonel Joseph H. Purvis, the quartet came from the School of Advanced Military Studies (SAMS) at Fort Leavenworth, Kansas. SAMS graduates were known informally — and, at first, derisively — as Jedi Knights, after heroic characters from the movie *Star Wars.* Steeped in military history, doctrine, and strategy, the SAMS Jedis served as a kind of brain trust within the Army. They embraced the same unofficial motto as the German general staff: "Be more than you appear to be."

On September 18 at 6 P.M., Schwarzkopf summoned the four men to his office, where he propped a small map of Kuwait against the couch. "Assume a ground attack will follow an air campaign," he said. "Tell me the best way to drive Iraq out of Kuwait given the forces we have available."

Under the guise of a Leavenworth team studying desert warfare, they moved into a small vault in the Ministry of Defense and Aviation (MODA) basement that had been occupied by Eager Anvil, now disbanded. Security teams frequently swept the chamber for listening devices; all working papers were locked in a vault at night, with a guard posted. Schwarzkopf initially limited to ten the number of officers "read into" the top secret operation. Even the Saudis remained ignorant of the planning operation.

For two weeks, Joe Purvis and his team pondered the daunting task of ousting an Iraqi army of occupation — which would grow, U.S. intelligence reported, to nearly half a million — with an allied force half its size. In southern Kuwait, the Iraqis had emplaced hundreds of artillery tubes, hundreds of thousands of mines, and a warren of trenches and barbed wire worthy of Verdun, including fire ditches filled with inflammable oil. Yet Iraqi generals appeared determined to defend vast tracts of barren desert, a strategy as daft as "a sailor fighting for a wave or an airman for a cloud," in the words of Major General Rupert Smith, a British division commander.

The SAMS group considered sweeping west of the fortifications, in a flanking maneuver similar to Guderian's Panzer attack through the Ardennes toward the English Channel in May 1940 or Rommel's run around the British at El Gazala two years later. U.S. Army doctrine called for attacking an enemy both deep behind his lines and at the point of greatest vulnerability. If the Iraqis expected an assault through southern Kuwait, doctrine and common sense suggested the attackers veer around the teeth of the defenses.

Yet in Purvis's view several factors militated against attacking west of the Wadi al Batin, the dry riverbed that angled from southwest to northeast to mark the western border between Kuwait and Iraq. Terrain in that sector of Iraq was *terra incognita.* Was the desert "trafficable," capable of supporting the thousands of tanks, trucks, and other vehicles needed in allied attack? Satellite photos, Bedouin reports, even historical records from British expeditions in World War I were inconclusive. A single armored division in heavy combat could burn half a million gallons of fuel a day and shoot five thousand tons of ammunition; the inability to drive fuel or ammo trucks into Iraq would be disastrous.

Equally troubling, a flank attack would require virtually all of the available allied ground forces, leaving no reserve. In the war with Iran, Iraq had displayed the ability to counterattack quickly, moving armored divisions a hundred kilometers in a day. If the American force was cut off and reduced to a fighting withdrawal, the military and political consequences would be horrific. Because neither American national sovereignty nor survival was at stake — a condition referred to in Riyadh as "the Israeli notion" — Purvis concluded that gambling on a flank attack was too risky without more combat and logistics power.

On October 6, Purvis presented Schwarzkopf with several options. The best course, he suggested, was to attack at night through western Kuwait about forty miles east of the Wadi al Batin. The border from the Persian Gulf to the wadi stretched for 130 miles; Saddam's forces could not be equally strong everywhere, and this sector appeared less fortified than the Kuwaiti bootheel. The allied spearhead would drive toward the high ground above Mutlaa Pass, west of Kuwait Bay, then cut the four-lane highway leading from the Kuwaiti capital to Basrah.

With luck, the Republican Guard divisions laagered north of the Kuwaiti border would roll south to blunt the attack, exposing themselves to allied air power. If those three elite divisions were demolished, Purvis believed, the rest of Iraq's army would be likely to capitulate. If the battle went badly, the Americans could dig in and fight, or retreat south without risk of being trapped and decimated. An amphibious landing by the Marines, he also concluded, would cost too many lives and would be more effective as a feint to tie down Iraqi troops on the coast.

Schwarzkopf stood at the map as Purvis spoke, his nose just inches from the thick black arrows. He agreed with Purvis that a flanking attack to the west was imprudent, given the forces at hand. Like many officers of his generation, the CINC was wary of underestimating the enemy. In Korea, Vietnam, even in Grenada, where Schwarzkopf had been the senior ground commander, American planners had underrated the foe, with results ranging from unfortunate to calamitous. "Do you think this will work?" he asked Purvis.

"It's very high risk," Purvis replied. "It may work."

"What would it take to guarantee success?"

"Another corps," Purvis answered promptly. Two or more divisions made up a corps. The Army's XVIII Airborne Corps currently in Saudi Arabia comprised four divisions, but only two were armed with heavy tanks.

Schwarzkopf nodded. "I agree."

Three days later Purvis flew to Washington with Glosson, Major Rick Francona, an intelligence officer, and Bob Johnston, the CENTCOM chief of staff. Bush and his advisers wanted to review Schwarzkopf's plans for both the air and ground attacks. The abrupt request to show a ground offensive plan irritated the CINC, who felt whipsawed between Colin Powell's desire to hold down the number of troops deployed in Desert Shield and the need to draft an adequate offense. "Goddam it," Schwarzkopf told Powell over the phone, "I told you over and over again we can't get there from here." But Powell insisted. The White House "is on my back," he explained. "They want to see what we can do."

Before Purvis and his team left Riyadh, the CINC gave them several last-minute instructions. Show them the proposal, Schwarzkopf said, but make the point that for us to feel confident of victory we need an additional corps. "We don't bullshit the president," he added. Any officer offering a personal opinion in Washington would be relieved and sent home in disgrace.

On Wednesday, October 10, the briefing was delivered first to the

Joint Chiefs, Cheney, and Paul Wolfowitz in the Tank, the chiefs' secure conference room in the Pentagon. Schwarzkopf's team repeated the performance on Thursday for Bush, Vice President Dan Quayle, Robert Gates, and Scowcroft in the White House Situation Room.

On both occasions, Glosson's review of the air campaign drew few critical comments. But Purvis's presentation of the ground attack — dubbed the One Corps Concept — made military officers and civilians alike uncomfortable. Scowcroft was particularly pointed in his questioning. "Why straight up the middle?" he asked. "Why don't you go around?" "Logistics," Powell replied. "We don't have enough force to go around." Johnston flashed two final viewgraphs labeled "CINC's Assessment"; one noted, "Planning still in conceptual state." To "go around," Schwarzkopf needed an additional corps.

The four officers left their only copy of the briefing documents with Powell and flew back to Riyadh, believing that they had adequately conveyed Schwarzkopf's message. In truth they had not. Schwarzkopf's desire for more forces was vaguely understood, but not the purpose those additional troops would serve. To Cheney, the idea was simply "a bad plan." Wolfowitz came away convinced that another corps would be used to reinforce failure in a frontal assault. Scowcroft was particularly appalled at a scheme he considered unimaginative, even foolhardy; he hectored Cheney for more creative alternatives.

Although Schwarzkopf's scheme was soon caricatured in Washington as moronic, it represented a reasonable attempt to make do with the forces available should Bush order an offensive in the next two months. Schwarzkopf and Purvis had carefully considered a one-corps flanking attack — both recognized the appeal of a grand sweep to the west — but rejected it for sound military reasons. If the CINC and his men had failed to convey those reasons, they nevertheless displayed courage in resisting the appeal of a more flamboyant plan that ran counter to their judgment.

Yet in the White House and some corners of the Pentagon the suspicion took root that the CINC's plea for more troops was simply the delaying tactic of a general hesitant to fight. Schwarzkopf's effort was dismissed by Scowcroft, Richard Haass, and others in the National Security Council with a curt "Thank you, General McClellan," a snide allusion to Lincoln's reluctant commander of the Army of the Potomac. And the One Corps Concept came to be known by the cruelly clever parody of a nursery rhyme: Hey diddle diddle, straight up the middle.

It took but two days for CENTCOM to realize how badly the presentations in Washington had been received. "Well," Powell told Schwarzkopf on the phone, "nobody's very happy with your ground campaign plan."

"It's not *my* ground campaign plan," the CINC replied heatedly. "I told you that. This is not what I'm recommending."

When Powell mentioned the McClellan jibe, Schwarzkopf grew even angrier. "Tell me who said that," he demanded. "I'll call the son of a bitch on the phone right now and explain the difference between me and McClellan if they're so stupid. If these guys are advising the president of the United States, they ought to know better than to make flip statements like that." Powell prudently kept the source to himself.

Cheney, hoping to galvanize Riyadh into more innovative thinking, began pushing his scheme to attack H-2 and H-3. The golden word *Inchon* — MacArthur's famed amphibious attack behind enemy lines in Korea — cropped up with frequency as the Pentagon groped for a more audacious plan. Schwarzkopf railed at Washington's discontent. "There is no damned Inchon," he thundered, slamming his fist on the table, "and somebody ought to come out here so I can show them there is no damned Inchon!"

Somebody did. Colin Powell flew to Riyadh on October 21. For two days he peppered Schwarzkopf and Purvis with questions. Most of the discussion centered on the One Corps Concept, but Purvis also showed the chairman his group's recent work on a two-corps attack.

Rather than a strike through Kuwait, the two-corps plan called for an attack into Iraq just west of the Wadi al Batin, angling toward — but not through — the Republican Guard divisions south of the Euphrates Valley. Because the SAMS team could not agree on a definition of "destroy," the initial purpose of the attack was to "defeat" the Guard. (A defeated enemy was defined as "no longer capable of putting larger than a brigade-sized force against us coherently.") The day before Powell's arrival, however, Schwarzkopf amended the plan. "I want you to draw an arrow through the Republican Guard," he told Purvis. "If we get two corps up there and we're still organized as a coherent force, we'll take that mass and destroy them." The Guard, the CINC subsequently decreed, was to be obliterated "as a military organization" in the sort of titanic battle Clausewitz had called *die Schlacht,* the slaughter.

But Powell considered this rendition of the two-corps scheme still too conventional: although intended to outflank the enemy, the attack lay too close to the wadi where Iraq would expect the allied thrust. The plan failed to exploit what Powell called the American "mobility differential," the agility purchased — at staggering expense — with M1A1 tanks, Bradley Fighting Vehicles, and a huge fleet of helicopters.

(Several proposals drafted by the Joint Chiefs' staff went so far as to suggest a Marine Corps attack from Jordan or a strike from Syria and southern Turkey.) The chairman agreed, however, with the necessity of a second corps. "Tell me what you need," he instructed Schwarzkopf. "If we go to war, we will not do this halfway."

On returning to Washington, Powell advocated almost a doubling of the American deployment. For a truly intrepid flanking attack from the west, the chairman argued, Schwarzkopf needed many more troops. Cheney readily agreed, having come to the same conclusion. On October 31, Bush concurred, although the decision was kept secret until November 8, after the national elections that week. Schwarzkopf would receive even more forces than he had requested: three additional aircraft carriers, a second Marine division, and, to bolster XVIII Airborne Corps, the heavily armored VII Corps from Germany, where it was no longer needed to deter a Warsaw Pact invasion.

In the same way that Instant Thunder had served as a counterpoint to the slow acceleration of the Rolling Thunder air campaign in Vietnam, so too did this massive buildup of ground forces signal a rejection of gradualism, of limited force, of the perceived strategic shortcomings that led to the quagmire in Southeast Asia. Encouraged by Powell, Bush embraced — in Cheney's infelicitous phrase — "the don't-screw-around school of military strategy." A force so formidable as to be invincible would mass in the Saudi desert, a force so huge that inevitably it contributed to the momentum propelling the nation toward war.

For the next two months, planners in Riyadh argued vigorously and at times bitterly over how far west the two Army corps should attack. But the basic concept of a flanking attack — eventually dubbed the Left Hook because of the resemblance to a boxer's roundhouse punch — garnered broad approval. During a two-hour session in mid-November with his enthusiastic division and corps commanders, the CINC forecast that the ground attack would be launched in mid-February. The Army spearhead, CENTCOM planners estimated, could reach the Basrah Canal from the Iraqi border in 144 hours. "We're not going to get any more troops and we're not going to get any more time," the CINC added, "so let's get on with it."

A month later, Cheney, Powell, and Bush approved the Left Hook. Among ten "operational imperatives," Schwarzkopf was to "accept losses no greater than the equivalent of three companies per coalition brigade," a formula that roughly translated into a cap of ten thousand friendly casualties.

Countless issues wanted resolution. Among the most difficult and consequential was how best to use the Marines. Purvis unilaterally allocated to them the task of protecting the Army's logistics tail in the west along the Wadi al Batin, a critical but prosaic mission suited to a force that possessed few heavy tanks. Predictably, the assignment provoked the wrath of the Marine commander, Walt Boomer, who wanted his troops both in the fight and closer to the Persian Gulf for resupply and support. Schwarzkopf agreed with Boomer. "Put them back on the coast," he ordered.

This in turn bothered the British, at the time intertwined with the Marines. London also had agreed to augment its force from a single brigade to a full division, the 1st Armoured, with 28,000 troops and 221 Challenger tanks. Under Schwarzkopf's new plan, the Marines and British would attack into the teeth of Iraqi fortifications in southern Kuwait but pull back if the defenses proved formidable; this "supporting attack" would serve as a diversion, tying down more than a dozen Iraqi divisions while the Army swung wide to the west in the main attack.

The British high command looked askance at the prospect of squandering the 1st Armoured Division in a diversionary expedition. Having shipped much of the British Army's firepower to Saudi Arabia, London now preferred to use it to good effect, preferably in tandem with the U.S. Army's main attack in the west. General de la Billière also fretted over high casualties if his forces were required to remain with Boomer's Marines. The Marines struck him as "mustard keen" to press the attack regardless of how stout the Iraqi defenses. A computer war game in London had predicted infantry casualties of 10 percent for every twenty-four hours in combat. ("Marines are gung-ho," Boomer observed icily, "but they're not stupid.")

At the same time, U.S. Army planners, aware of the Challenger fleet's prodigious combat power, slyly encouraged de la Billière to cast his fortunes with VII Corps. The British commander needed little such encouragement: he campaigned relentlessly in Riyadh and urged London to apply political pressure. Schwarzkopf and Boomer resisted. The Marines, beyond developing deep fraternal ties to the British 7th Armoured Brigade, the Desert Rats, also coveted their tanks and engineers. After weeks of haggling, the CINC capitulated. "Walt, I'm tired of talking with the Brits about this," he told Boomer. "We're going to move them over with the main attack. I'll find you another brigade."

In late December, Schwarzkopf formally shifted the British division to VII Corps. As a consolation prize, the Marines received a smaller U.S. Army unit, the Tiger Brigade from Texas. In a sorrowful, teary luncheon, the Marines and British bade one another farewell, their bands joining to play "This Moment in Time." As a parting gift, the Marines offered their erstwhile comrades a plaque inscribed with lines from the Saint Crispin's Day speech in *Henry V:*

> *And gentlemen in England now a-bed*
> *Shall think themselves accurs'd they were not here.*

## Incirlik, Turkey

Even as the war began, Turkey's role in the allied coalition remained uncertain. Many Turks resisted picking sides in the conflict, recalling the Ottoman error in permitting Germany to tug them into World War I, only to end up losing the war and their empire. President Bush had carefully nurtured the good will of the Turkish president, Turgut Özal, and now the president's solicitude paid off. Özal, hoping to remove a neighborhood despot while reaping aid from the United States and Europe, concluded that Turkey's involvement was a worthy risk; on January 17, he persuaded his National Assembly to authorize American attacks from the NATO base at Incirlik, roughly three hundred miles northeast of Iraq.

Shortly after midnight on January 18, the allies opened a second front with a wave of bombers from Incirlik that struck four radar sites in northern Iraq. Soon after, the first Tomahawks fired from the Mediterranean would also cross Turkey toward Iraq — and in some cases surreptitiously cut across the northeast corner of Syria.

For allied commanders in Riyadh, the aircraft from Incirlik permitted an extra hundred or so sorties a day, typically with two flocks of F-16s during daylight and a flock of F-111s at night, plus assorted Weasels, Ravens, and Eagles. Though very much a sideshow compared with the huge effort in the south, the Incirlik planes — code-named Proven Force — wreaked havoc on northern Iraq. In large formations that reminded some pilots of World War II raids, flights of eight to eighteen F-111s, flying at 25,000 feet and dropping a dozen bombs each, pelted rail yards, airfields, petroleum refineries, and ammunition dumps.

Proven Force also broadened the attacks against the Iraqi power grid. Relentless and wholly successful, that campaign steadily reduced Iraq almost to a pre-electrical state. John Warden's Instant Thunder plan had anticipated knocking out 35 percent of the country's electricity; instead, by war's end more than 95 percent of the generated power was gone and with it most of Iraq's capacity to pump sewage and purify water. Auxiliary generators were fickle, inefficient, and required fuel, supplies of which also were being destroyed wholesale. The blackout reportedly reduced some Iraqi surgeons to operating by candlelight.

Limited efforts had been taken before the war to restrict this demolition. In Riyadh target planners intended to leave generator halls largely intact, theoretically permitting a postwar reconstruction of the power grid in three to six months. On January 12, Buster Glosson wrote a memo notifying his wing commanders that "electric targets will be targeted to minimize recuperation time . . . Boilers and generators will not be aim points." But those instructions often failed to reach the pilots, who, with zeal and alacrity, attacked generators, transformers, switching stations, and transmission lines. Because of uncertainty about the effectiveness of the secret carbon filaments, most targets draped with Kit 2s were also hit with conventional bombs. Besides, the imprecision of dumb bombs in many cases made pinpoint targeting impossible. Even four months after the war, Iraq's generating capacity would be comparable with the country's electrical output in 1920.

One mission flown from Incirlik on the night of January 18 typified the dangers and difficulty of pinpoint bombing. A four-ship flight of F-111s flew east for forty-five minutes to avoid Syrian airspace, then turned south toward the Al Abbas dam on the Tigris River twelve miles north of Mosul, Iraq's third largest city.

The F-111s flying from Turkey, unlike more sophisticated models based in Saudi Arabia, could drop only dumb bombs, not precision-guided munitions. Despite Incirlik's balmy Mediterranean climate, the air crews wore heavy woolen survival suits in case they were forced to eject in the mountains of northern Iraq. Manning an aircraft nicknamed *Sheeba* was a young pilot, Captain Greg Stevens, and Major Mike Sweeney, the weapons systems officer (WSO, or "whizo"), both of whom had arrived in Turkey from their base in England only two days before.

Flight planners had promised the crews "an easy milk run" for their baptism of fire. But after crossing the mountains at ten thousand feet, *Sheeba*'s electronic warning system detected Iraqi SAM radar emissions in the heavily defended Tigris Valley (soon to be dubbed, sardonically, Happy Valley). Stevens and Sweeney dropped to four hundred feet, then to three hundred, lower than either had ever flown at night. As *Sheeba* and another F-111 angled toward the dam from the west, the second pair swung around to attack from the east.

Sweeney had been instructed to avoid hitting the dam while aiming for the hydropower station on the west side of the spillway. As the aircraft neared the target, he recognized the rectilinear shape of the transformer yard in his ground radar scope. Tracers stitched the sky over the dam. Stevens jinked right to avoid a stream of antiaircraft fire from a three-man gun crew clearly visible below. A moment later he was startled when, on a hillside just off the left wing, someone opened the door of a parked pickup truck and the dome light eerily illuminated the cab's interior.

"Accelerating to 510 knots," Stevens announced. "I'm going to range bomb now." "We're a little farther north than I want to be," Sweeney replied. Stevens nosed the plane slightly to the right. "Ten thousand feet to release," he reported. "Five thousand feet. Coming on the pickle button — now!"

*Sheeba*'s computer — calculating altitude, wind velocity, ground speed, and the distance to the radar cursors Sweeney had placed on the target — paused two seconds before pitching a dozen 500-pound Mark 82s into the night. The plane shuddered violently and began to wobble. "Shit, we're hit!" Sweeney shouted. "No, no, we're okay," Stevens answered. "It's just the bombs coming off."

Peering from the cockpit, Sweeney saw the dark mass of the dam to his left. In the reflection of the spillway the bombs blossomed red and orange, carving a six-hundred-foot channel of fire across the power station. Stevens pulled out of the valley and turned north toward Turkey. Pilots flying over the dam the following day confirmed severe damage to the transformers. Mosul reportedly was without power. The Air Force awarded Sweeney the Distinguished Flying Cross for valorous bombing under fire.

The valor was beyond question. The cumulative impact of such missions on Iraqi society, however, was devastating. More than two hundred combat sorties, plus the Kit 2 Tomahawks, would be launched against electrical targets. Glosson and Deptula had sound military reasons for picking those targets: disrupting enemy radars, communications, and computers; crippling defense industries; stripping away the power needed to open the heavy doors on airplane shelters and to refrigerate suspected biological weapons. They also hoped to bring the sting of war home to the Iraqi populace, thus adding to the pressures from within on Saddam's regime.

Yet certainly they underestimated the efficacy of the attacks and the pain of that sting. Schwarzkopf would declare on January 30 that some electricity had been left flowing "because of our interest in making sure that civilians [do] not suffer unduly." But Iraq was an industrial state; three quarters of its nineteen million citizens lived in cities and towns. With the world's second highest birthrate, the country had an exceptionally large number of children: 20 percent of the population was under five years old. The widespread loss of power was ruinous.

After the war, critics would charge that although civilian deaths directly attributable to the bombing campaign numbered in the hundreds or low thousands, tens of thousands more died from disease, degraded medical care, and the deprivation of adequate food and clean water.

In this issue no moral certitude obtained. The attack on Iraq was a far remove from the pristine, surgical pricking described by George Bush and his surrogates, yet it was equidistant from the fire bombing of Dresden and Tokyo. In the end the war was of a piece with all wars: unpredictable, cruel, and violent, damning the innocent and guilty alike.

## Washington, D.C.

After two early morning false alarms, several more Scuds fell on Tel Aviv shortly after dawn on Saturday, January 19, the Jewish Sabbath. Like the initial salvo, the attack lightly wounded several dozen civilians, killed no one, and triggered a rash of phone calls between Israeli and American officials. The most alarming, on the Hammer Rick line linking the Pentagon and Keriya, came from David Ivri to Paul Wolfowitz around 2 A.M. Washington time.

"We intend to respond because there has been no stoppage of the Scuds," Ivri warned in his heavily accented English. "We want to pass some details to you of where we plan to go so that you can get your forces out of the way." The Israeli air force planned a massive counterstrike, with two large waves of warplanes followed by helicopter attacks and the insertion of infantry commandos into western Iraq. (Curiously, Ivri was executing orders from Moshe Arens that he personally opposed; the director general had argued against an Israeli counterstrike, which he believed would engulf the Middle East in general war.)

Wolfowitz stalled. In nearly twenty years of service to Republican and Democratic administrations, Wolfowitz had weathered innumerable crises. A genial and exceptionally intelligent man, he had studied mathematics and chemistry in college before becoming an expert in foreign policy and arms control. Israel he knew firsthand from having lived there as a teenager. When he spoke, his voice rarely rose above a murmur. Now he sensed from the urgency in Ivri's voice that the Israelis were at the end of their rope and even closer to a retaliatory strike than on the previous night. "You can give us any information you want," Wolfowitz replied. "But there's been no decision on our part that we're going to get out of your way. All I can do is take down what you tell me."

The rebuff silenced Ivri, who hung up without providing additional details. Wolfowitz reported the conversation to Cheney, who flew a few hours later with Colin Powell to Camp David for a strategy session with the president. Tense and frustrated, Bush ordered Wolfowitz and Lawrence Eagleburger back to Israel as his personal envoys, and authorized a separate diplomatic mission to Jordan. Sensing the president's pique, Cheney leaned on Powell. "How the hell is it," the secretary asked, "that these guys keep launching Scuds?"

Powell returned to the Pentagon, where he had spent the previous two nights napping on the leather couch in his office beneath a portrait of West Point's first black graduate leading a cavalry patrol across the Great Plains. Using the big white phone console behind his desk, he again urged Schwarzkopf to turn his full attention to the Iraqi missiles. The CINC seemed oddly tone deaf in his public remarks about the Scuds. After the first barrage hit Israel, he had decried the "absolutely insignificant results" while adding, "We were delighted to see that the Iraqis did exactly what we thought they'd do." Such remarks hardly endeared him to the Israelis, who viewed the attack as neither insignificant nor delightful, or to Pentagon civilians. "The guy supposedly has read Clausewitz and knows wars are political, right?" Wolfowitz asked caustically.

Powell understood Schwarzkopf's preoccupation with the air campaign and the ground attack plan. He hesitated to tell the CINC how to run his war. No field commander, Powell knew, ever believed he was supported adequately in Washington, and Schwarzkopf needed reassurance more than most. In the main, the chairman thought the CINC's performance admirable. But on the nettlesome issue of the Scuds — the one aspect of CENTCOM's war plan that was not going well — he seemed willfully stubborn.

When Powell raised the issue, as he did often in the first days of the war, Schwarzkopf grew annoyed, then angry. "You know, you guys have completely lost your perspective," Schwarzkopf told him. "I appreciate your concern about Israel, but what about concern for us in Riyadh and Dhahran? We're getting Scuds rained on us, too." The CINC was particularly incensed at a Joint Chiefs' proposal to have Israeli target planners sit in CENTCOM's Riyadh headquarters; he also had pointed out that the first bomb dropped on Baghdad Thursday morning by itself contained the explosive power of ten Scuds. But Powell held firm, with an insistence that made several eavesdropping officers think of a man smacking a bull between the eyes with a two-by-four.

*

In the army of a political democracy, the French author Alexis de Tocqueville once observed, the most peaceful men are the generals. Colin Luther Powell proved the point. By constitution he inclined to diplomacy and compromise rather than confrontation and bloodletting. He was a devoted admirer of Dwight Eisenhower, not for the generalissimo's battlefield exploits, but because Eisenhower was a soldier who distrusted military solutions and advocated containment. An inveterate collector of aphorisms, Powell kept an epigram on his desk from the Greek historian Thucydides: "Of all manifestations of power, restraint impresses men most." For Colin Powell, war was not an abstract means to achieve national objectives but rather a brutal enterprise that produced dead soldiers, shattered lives, and smoking wreckage.

Born in Harlem to Jamaican immigrants, he had been raised in the polymorphic world of the South Bronx. His mother was a seamstress, his father a shipping clerk in New York's garment district. For fifteen years the Powell family lived in a four-story walkup a few blocks from the police precinct station later known — after drugs and violence had destroyed the neighborhood — as Fort Apache.

Powell enrolled at the City College of New York with intentions of becoming an engineer, an ambition promptly abandoned, he later joked, when he found himself baffled by the task of visualizing a cone intersecting a plane in space. With a lackluster college record and a degree in geology, he was commissioned a second lieutenant in the Army in 1958.

For a young black man of modest means, Powell wrote years later, the military was "a route out, a route up." Integrated only a decade earlier by President Harry Truman, the U.S. Army was undergoing a transformation from one of the nation's most segregated institutions to one of the most progressive in race relations. Powell soon developed, and would never relinquish, an emotional kinship with blacks who had served in uniform before him, including the so-called buffalo soldiers, all-black regiments that fought in the Indian campaigns after the Civil War.

In 1962, Captain Powell arrived in Vietnam as part of the growing cadre of American military advisers. He served with a South Vietnamese infantry regiment and returned home a year later with a Purple Heart earned after he stepped on a sharpened punji stake. Following several stateside tours and a year at the Army's Command and General Staff College — where he ranked second among twelve hundred officers in his class — he returned to Vietnam in June 1968 at the peak of American involvement in the war. There he served in the Americal

Division as executive officer of the same brigade that three months earlier, as he would learn much later, had massacred scores of civilians at My Lai. Even as chairman, Powell still wore the Americal's combat patch on his right sleeve.

Of the twenty-two years since returning from his second tour in Vietnam, Powell had spent seventeen in Washington. A credible if unexceptional career began to blossom in 1972 with his selection as a White House Fellow. During his one-year apprenticeship at the Office of Management and Budget, Powell's intelligence and stamina earned the admiration of the OMB director, Caspar W. Weinberger, and his deputy, Frank C. Carlucci. Bolstered by two powerful patrons — Powell subsequently referred to Carlucci as his "godfather of godfathers" — the young officer found himself on a very fast track to the top.

In the 1980s, after distinguished tours as a battalion commander in Korea and a brigade commander with the 101st Airborne Division, Powell served for three years as military assistant to Weinberger, who had become Ronald Reagan's secretary of defense. Now a major general, Powell bypassed the traditional way station of division command in 1986 when he received his third star and command of V Corps in Germany. Only six months later he was back in Washington as deputy to Carlucci, who had become the national security adviser; when Weinberger resigned in 1987, Carlucci moved to the Pentagon as defense secretary and Powell served the last two years of Reagan's term as national security adviser. In the summer of 1989, Richard Cheney, with the consent of President Bush, chose Powell as the new chairman of the Joint Chiefs. At fifty-three, he was the youngest man ever to hold the position.

Wherever Powell went, he exuded self-confidence, as though he were gliding above the shoals that snagged lesser mortals. (Another maxim displayed on his desk: "Never let them see you sweat.") His face had a quicksilver quality, shifting instantly from concentration to amusement, and his glasses magnified his eyes slightly to lend him an added intensity. Powell once estimated that as chairman he lost his temper on average ten times a day; but he was equally quick to guffaw, loosing a sharp, distinctive bray that filled the room and infected those present with his mirth.

Like Cheney, he lacked pretension. As a hobbyist, Powell enjoyed tinkering with old Volvos in the garage behind his Fort Myer quarters; as an epicure, his tastes ran to peanut butter sandwiches and ground beef. He was sentimental, loyal, and profane, with an agile mind and an appetite for work. Generous and good-humored, he could also be trenchant and sarcastic, as in his derisive reference to hawks in the

State Department as "the warriors of C Street." Powell claimed tongue-in-cheek that it was not until he was a young officer at Fort Benning that "I ever saw what is referred to as a white Anglo-Saxon Protestant." Even after becoming the ultimate Washington insider, encircled by WASPs, he cultivated the image of the outsider. He had become one of the seven men — known to Washington wits, depending on their degree of admiration or contempt, as the Seven Wonders or Seven Dwarves — who made up George Bush's inner council. Powell got along well with everyone in the group, but before going to a meeting at the White House he would joke about donning "a string of garlic to ward off the werewolves."

Perhaps more than any of the president's other councilors, Powell had resisted war with Iraq. From the moment Saddam invaded Kuwait until Bush's decision in late October to double the force in Saudi Arabia, he subtly sought to steer the United States away from a military solution. He preferred "strangling" Saddam with a United Nations blockade and economic sanctions. To Cheney, Bush, Scowcroft, and others, he had laid out the case for "grinding" down Iraq through containment — without overtly declaring himself a proponent of such a strategy. "There is a case for the containment or strangulation policy," he had told the president during an Oval Office meeting in early October. "This is an option that has merit. It will work someday."

Bush had not directly solicited Powell's recommendation, and the chairman did not offer it. His hesitancy would later earn Powell the sobriquet of "reluctant warrior," a phrase that did not entirely displease him. Yet by failing to confront Bush directly with his doubts, the chairman exposed himself to sharp questions after the war about whether he had shirked his obligations as the president's principal military adviser.

By early November, when Bush made clear his rejection of the strangulation option, Powell flung himself body and soul into preparations for war. In 1984, Caspar Weinberger had publicly laid out several criteria that had to be met before American combat forces were committed. Troops were to be employed only "with the clear intention of winning," with "clearly defined political and military objectives," with "the support of the American people and their elected representatives in Congress," and only as "a last resort."

Powell regarded the so-called Weinberger Doctrine — a distillation of the lessons of Vietnam — as a set of useful guidelines. The concept of "surgical" strikes and the incremental application of force repulsed him. If war was inevitable, then his task as chairman was to ensure that American victory was also inevitable. If that took half a million troops, two thousand planes, and fifty warships, so be it. George Bush had vowed that a war with Iraq would not be another Vietnam. In this, Colin Powell wholeheartedly agreed.

Powell was the most politically deft chairman since Maxwell Taylor, and he kept close rein on the four service chiefs without seeming to constrict them. Solicitous and confiding, he sought their counsel without recourse to formal votes, often meeting informally across the round mahogany table in his office. Powell referred to the chiefs, chairman, and vice chairman as "the six brothers." By common consent Powell became their proxy and mouthpiece; the service leaders nearly vanished from public view. Emasculated by congressional reforms in the mid-1980s that enhanced the chairman's power at the expense of their own, the chiefs played virtually no role in the decision to go to war, and they would be no more than bit players in the decision to stop it. In truth, they had all been reluctant warriors, badly scarred from Vietnam and wary of a fickle public that could cheer the armed forces off to war but turn venomous if things went badly.

The sole exception was the newest chief, General Merrill A. McPeak, appointed to command the Air Force after the dismissal in September of Michael Dugan. A tall, bony fighter pilot who had once commanded the Thunderbirds demonstration team and later the Air Force in the Pacific, Tony McPeak considered himself no different from his four-star brethren: a bureaucrat in uniform who had risen through the ranks by being a company man rather than a risk-taking entrepreneur. Yet he was dismayed by the foot dragging of the chiefs, whom he likened to "a bunch of fucking municipal bond salesmen" ready to dissuade Bush from war should the president show the slightest buckling in the knees. Shortly before the war began, McPeak observed that Winston Churchill had once said of his military chiefs that the toughest soldier, the staunchest airman, and the most intrepid sailor would collectively turn to mush. That's what we are, McPeak added: mush.

McPeak suspected that Iraq was weak. He believed the doubling of American forces in the gulf to be a mistake, driven by misguided fear of Iraqi military capabilities and the Army's desire to play a larger role in the conflict. He halfheartedly sought to convince Powell that Saddam could be defeated quicker and cheaper. "We won't get any style points for this," he warned. "The real test is to make it look like a no-brainer. Your friends admire you for showing grace as well as power, but all we're showing is power." Powell, determined to guarantee absolute success if war erupted, firmly rejected the argument.

McPeak pressed the issue no further, even keeping his own counsel in a private luncheon with Bush at the White House on January 14. As the junior member of the firm, he recognized a certain juvenile pugnacity in his own itch for a fight. A sensible reluctance among those commanding the world's most powerful military, he reasoned, probably served the country well. Moreover, McPeak was not certain he was right: the more he watched Colin Powell, the more he admired the man's sagacity. Tony McPeak would defer to the chairman's judgment, and not for the last time.

## Al Qaim, Iraq

The order to suppress the Iraqi Scud launches rocketed down the chain of command from Bush to Cheney and Powell, and thence to Schwarzkopf, Horner, and Glosson. On the late afternoon of Saturday, January 19, it reached Al Kharj Air Base, where twenty-four F-15E Strike Eagles were preparing to attack an ammunition depot in the Euphrates Valley. At 2 P.M., only six hours before take-off, the mission suddenly changed. American intelligence had discovered a complex of missile storage bunkers and assembly buildings — breathlessly described as "the motherlode of all Scud targets" — in western Iraq. Analysts also concluded that another missile barrage was planned against Israel later that night.

Air Force planners in Riyadh redirected the Strike Eagles to hit the Scud sites near Al Qaim, later known to American pilots — in deference to the many surface-to-air missile batteries girdling the city — as Sam's Town. Twelve bombers from the 335th Tactical Fighter Squadron were to hit the targets around 10 P.M., followed half an hour later by another twelve from their sister squadron, the 336th.

At Al Kharj, the abrupt change of orders was greeted with consternation. Although the pilots understood the political importance of destroying the Scuds, they resented not having time for meticulous planning. This diversion was viewed as Buster Glosson's hasty effort to demonstrate that the United States Air Force needed no assistance from Israel. While bomb loaders stripped the planes of five-hundred-pound Mk 82s and replaced them with cluster bombs and Gator mines, strike planners badgered Riyadh for more information about the new targets. Not until the air crews were ready to walk to their planes did a faxed list of specific bombing aim points arrive at Al Kharj. The required TOT — time on target — changed once, then twice.

No one was angrier at the haphazard way the mission was thrown together than the pilot assigned to lead it, Lieutenant Colonel R. E.

(Scottie) Scott, who had flown two nights before in the attack on the Basrah oil refinery that cost the lives of Donnie Holland and Thomas Koritz. Shortly before climbing into his cockpit, Scott confronted the director of operations for the 4th Tactical Fighter Wing, Colonel David W. Eberly, a slender, soft-spoken native of Brazil, Indiana, who also planned to fly in the attack.

"This thing is a goat rope," Scott declared. "It's the kind of mission that gets people killed." Eberly managed to affect an air of calm assurance. "We're going to do this, Scottie," he said. "We can do it." Privately, he shared the mission commander's misgivings. As he walked from the ready room to the flight line, Eberly confided to the wing commander, "This one is worth twenty years of flight pay."

The mission continued to deteriorate after take-off. The bad weather that had bedeviled *Saratoga* and *Kennedy* bombers two nights earlier still lingered over northern Saudi Arabia. Bouncing in the turbulent winds, the twelve Strike Eagles struggled to refuel from tankers orbiting above Ar Ar. Then Scott learned that the two F-4 Wild Weasels he had requested from a base in Bahrain had not been notified of the final TOT change; without the Weasels and their radar-killing HARM missiles, the Strike Eagles had no way to deal with the Iraqi surface-to-air missiles. Scott initially decided to wait for thirty minutes, then changed his mind, concluding that the attack would have to proceed without them. Eberly, flying eight miles behind Scott, agreed with the final decision; Glosson, he knew, placed great importance on the mission. Even so, Eberly's foreboding deepened.

The bombers flew west, turned out their lights, and crossed the border into Iraq. When the strike force neared Sam's Town, misfortune struck again. At Scott's request, two EF-111 Ravens had begun orbiting southeast of Al Qaim. Each aircraft, laden with three tons of electronic jamming equipment, began pumping thousands of watts of power at Iraqi ground radars. Through a tangle of antennae, including a large pod known as the football perched atop the vertical stabilizer, the Raven detected Iraqi search radar signals. The electronic warfare officer (EWO), sitting next to the pilot, then broadcast contradictory signals through a shallow bulge in the plane's belly called the canoe. More art than science, "casting 'trons" was intended to disrupt the "electronic ecosystem" and confuse the Iraqis just long enough to let the attack bombers slip across their targets. If the Raven was close enough and the EWO sufficiently deft, the Iraqi radar screens turned to snow.

But as these Ravens began their second orbit in a counterclockwise turn toward the Syrian border, a MiG-25 suddenly darted toward them at high speed. The Iraqi fired one air-to-air missile at the lead Raven and two at his wingman. The missiles flew wide, but the Ravens dived to escape and then, uncertain where the MiG was lurking, turned back toward Saudi Arabia.

Unaware of this drama the Strike Eagles pushed toward the targets, now without Ravens or Weasels. Scott led six jets northeast toward the Al Qaim highway (later known as Triple A Alley). The other six veered southwest. Having witnessed the consequences of low-altitude bombing at Basrah, Scott had instructed his pilots to attack from twenty thousand feet. Thirty miles from the target, the antiaircraft fire began; at ten miles the first SA-2s and SA-3s streaked skyward amid a barrage of flares. Scott and his wingman — Corvette 1 and Corvette 2 — roared over the Scud bunkers at 580 knots, dropped their bombs, jinked wildly to evade several missiles, and raced for the open desert.

Now came Corvette 3 with David Eberly and, in the back seat, his weapons officer, Major Thomas E. Griffith, Jr. Eight miles from the target, Eberly spied the bright orange flame of a SAM coming up on the right. The tiny fireball looked stationary on the aircraft canopy, an indication that the missile had locked on and was tracking them. Eberly pulled sixty degrees to the right and the missile shot past the airplane. Steering back to the left, he lowered the nose into an attack dive.

He never saw the missile that hit them. With a brilliant white flash and a violence that reminded him of an automobile slamming into a tree, Corvette 3 lurched up. The men were too shocked to yell. Eberly tried to scan the instrument panel, tried to make his eyes focus. He had the sensation that the aircraft had stopped in midair. The cockpit lighting was strange, although he couldn't tell why. He felt his gaze drawn to the fire warning lights on the left.

Eberly had always assumed that if this time came he would stay with his crippled jet until the last moment, but now his hands dropped instinctively to the ejection levers on either side of his seat. As he grasped the handles and tugged them upward, he felt the friction of metal on metal. In the back, Griffith inched his right foot toward the microphone pedal to broadcast a mayday call.

An explosive charge tore the plexiglass canopy from the fuselage. Subzero air ripped through the cockpit at more than five hundred knots. Then two small rockets detonated beneath the seats, blasting first Griffith, then Eberly, upward with a force thirteen times that of gravity. In less time than it takes for a human heart to beat, the men catapulted back over the bomber's twin tails and into the inky night, tumbling toward the scarlet tracers that leaped to greet them from Sam's Town far below.

*

He was on his knees. That much he knew and no more. He stared at the horizon. An orange glare blazed silently, as from a distant bonfire. He felt neither pain nor cold, heard not the slightest sound.

Fragments swam into his consciousness. His name: Eberly. David Eberly. He was a pilot, a colonel. He had a wife, Barbara, and a son, Timm. Perhaps he was in Nevada, on a Red Flag exercise. Or Oman? No, the wing had left Oman for Al Kharj. Then the war began. He had been flying with Griffith, flying far to the north. Then what? Kneeling, as though at a communion rail, he stared at the liquid glow, mesmerized. Perhaps he was dreaming.

He moved his right knee, and it made a noise, a scratching on the shale. Abruptly it came to him: I am in Iraq.

Eberly strained to hear the growl of an airplane, to hear any noise at all. The sky had cleared and was swollen with stars. A sliver of moon hung in the south. They've gone, he thought. The survival raft, automatically inflated during his descent, lay near the heap of his parachute. He had no recollection of unhooking the chute from his harness, no memory of anything after the friction of the ejection handles.

Now the cold came, seeping through his flight suit. He stood up, stumbling, and gathered the parachute around him like a shawl. He realized he was going into shock. Sitting in the raft, facing the southern moon, he slid the water flask from the calf pocket of his G-suit and took several swigs. Fifty feet away he saw the silhouette of a power stanchion with high-tension lines running toward the distant glow he now knew was Al Qaim.

The sudden sound of a truck forced him upright. Headlights scythed the desert. Eberly scrambled to his feet and scuttled to the southeast corner of the concrete stanchion. The truck stopped a hundred yards away, its lights falling just short of the pillar where he crouched, listening for the snarl of dogs and the metallic shuffle of soldiers grabbing their rifles. Instead, one man climbed from the cab and lit a cigarette. He stood and smoked, staring south, his eyes reflecting the crimson ember each time he inhaled. At last he flicked the butt away, climbed back in, gunned the engine, and drove northward.

Eberly knew he had to get away from the crash site. He bolted southwest, abandoning the raft and the remainder of the survival kit. His mind was clearer now. He thought of Scottie and the rest of the flight, now racing back to the tanker and the safety of Saudi airspace. He thought of his warm cot in the tent at Al Kharj, and of Barbara and Timm at home in North Carolina, but he pushed the images from his mind. Yea, though I walk through the valley of the shadow of death, he recited silently, I will fear no evil.

He remembered a photograph he had once seen of a World War I pilot with a parachute draped around his neck; he doubted that he looked as jaunty. Off to the right, another truck rumbled past, then another. From his vest he pulled the small survival radio and tried, without hope of success, to raise the AWACs plane far to the south. "This is Corvette three, on guard. How do you read?" No answer.

Suddenly the dark shape of an ejection seat loomed before him. He crept foward, fearful of finding Griffith's body still strapped in. The seat was empty.

A sharp crackle from the radio startled him. "Corvette three alpha," Griffith's voice called softly, "this is three bravo."

Eberly keyed the microphone. "Grif, where are you?" he asked before realizing how foolish the question was. What could Griffith answer? In Iraq? In the desert?

"Do you see the big power line overhead?" Eberly asked.

"Yeah, I see it."

"Can you see the headlights of the truck on the road? I'll give you a hack when the truck's abeam my position along the power line."

Fifteen minutes later Eberly heard the crunch of footsteps, and Griffith emerged from the darkness. Their reunion was almost curt, the exhilaration of companionship silenced by the instinct to keep moving, to find a wadi or a thicket in which to hide before dawn caught them in the open.

For several hours they stumbled southwest, following the power lines. All conversation focused on the flat terrain and the odds of being spotted from the road. Eberly felt rivulets running down his back and chest. He assumed it was sweat; not until daylight would Griffith see that it was blood oozing from a gash on the back of Eberly's head and a deep scrape on the left side of his jaw, injuries apparently suffered during the ejection.

At length they came on a shallow ravine choked with low shrubs. Collapsing to the ground, they huddled for warmth and tucked Eberly's parachute about them like a quilt. "It doesn't matter what happened," Eberly said softly. "We're alive, and if we can stay alive we'll get home." He pulled out the water flask. Each took a sip. He capped the flask and slid it back into his pocket. They slept.

## Seymour Johnson Air Force Base, North Carolina

Even before her husband left for the Middle East, Barbara Eberly had had premonitions. It was nothing she could put her finger on. But she had been a pilot's wife long enough to know of many sudden widows and fatherless children, and could not dismiss the misgivings out of hand. When David left in August, she played the plucky spouse. "You've waited twenty years to do this," she told him. Later, like Penelope on the cliffs of Ithaca, she waved farewell from the flight line as his plane took off in a pounding rain. But as Christmas drew near, she wandered among the clothing displays at a department store until she found a simple navy dress that she knew would be appropriate for a funeral.

She was a handsome woman, blond and delicate-featured, articulate, a Hoosier like her husband. On this first Saturday of the war, she was supposed to drive up to Virginia to see a friend, but she canceled the trip under the weight of her forebodings. "The idea of hearing that David has been shot down and then having to drive four hours to get home is too much," she explained on the phone. She lunched at the Officers' Club, did some errands in the afternoon, picked at Chinese takeout for dinner. Shortly before 10 P.M., as she was about to watch a video cassette that David had sent, the bell chimed on the side door.

Colonel James Wray, the vice wing commander who had remained in North Carolina to run the base, stood on the stoop in civilian clothes. He held a walkie-talkie. Behind him — in a breach of etiquette that signaled this was not a social call — stood his wife. Barbara opened the door and peered down the driveway, searching for the ominous fleet of blue cars that the Air Force dispatched whenever a pilot went down. The street was clear.

"I guess it's all right to let you in," she joked nervously. The Wrays stepped into the kitchen, smiling, and she knew immediately the smiles were of sympathy. She began backing up, through the kitchen and into the dining room. Wray held out his arms. "Barbara, we've got to talk." "No! No!" she protested, and her cries filled the house. "No, no, no!"

Half an hour later, Timm returned from the movies. He was eighteen, bigger than his father, blond like his mother. He walked into a house now filled with blue uniforms. A dozen faces turned toward him. Before a word was said, he lowered his head and burst into tears.

He refused to believe that his father was dead; she refused to hope that he was alive, and she snapped at her son when he persisted in asserting that somehow David had survived. Shortly before midnight the chaplain arrived and at her request led them through the rainy night to the base chapel. After they had prayed for David's soul and turned to leave, Timm lingered for a moment near the altar. "My father's alive, chaplain," he confided. "I know. I've seen him walking in the desert."

## Tel Aviv

On Sunday morning, January 20, a U.S. Air Force transport jet banked into its final approach to Ben Gurion Airport. The plane rocked lower and lower over the cobalt Mediterranean before crossing the broad white shingle that stretched from the domes and minarets of ancient Jaffa in the south — said to be the world's oldest port, founded by Noah's son Japhet — to the high-rise hotels clustered along Hayarkon Street in the north. Tiny puffs of smoke spurted from the jet's wheels as they touched the airport tarmac. The pilot taxied past the date palms bordering the squat, tatty terminal and cut the engines. The most important diplomatic mission of the war had begun.

Lawrence Eagleburger and Paul Wolfowitz, representing the departments of State and Defense, stepped from the cabin into the midmorning glare. Their instructions were simple. "Keep the Israelis from responding," James Baker had directed. "Do whatever you have to do to assure them that we have the Scuds under control." On the airport perimeter, the newly arrived crew of a Patriot battery — one of six that would be deployed across western Israel — worked feverishly to position their missile canisters and radar screens. No Scuds had fallen since Saturday morning; in that regard, the costly Strike Eagle attack the previous night had been successful. Yet the prospect of more attacks seemed certain.

It took only an afternoon for the two Americans to see that in the week since their last visit, Israel had become a nation under siege — but a nation learning to live with it. A public opinion poll found that 80 percent of Israeli citizens favored a continued policy of restraint even after the first two Scud salvos. Puppets on the television program *Kippy of Rechov Sumsum* — the Israeli *Sesame Street* — wore little gas masks during a call-in show intended to reassure anxious children. A group of ten senior rabbis passionately debated whether to cast an ancient Hassidic curse on Saddam Hussein, who, if truly evil, would theoretically die within thirty days. Visitors to the Holocaust exhibit at Yad Vashem, walking past photographs of Nazi crematoria, carried cardboard boxes containing their masks and atropine injectors.

On a tour of the shattered apartments of a Tel Aviv neighborhood hit by an earlier Scud, Eagleburger, clad in his boots and trademark red sweater, and clutching a cigarette in one hand and an asthma inhaler in the other, shouted to the cheering crowd, "Good for you! The people of Israel live!" He later slipped on his own gas mask in a hotel room, peered into the mirror, and gasped, "Gad! It's a short fat man from Mars!" To Wolfowitz he complained, "I can't smoke in this thing. I'd rather die on the roof breathing fresh air with a cigarette in my hand."

For four days, the Americans shuttled between Tel Aviv and Jerusalem in an endless succession of contentious, emotional meetings with Israeli government officials. ("A great kabuki dance," Eagleburger called it privately.) When Moshe Arens and David Ivri again requested an open corridor across Jordan or Saudi Arabia so that Israeli warplanes could attack the Scud sites, Eagleburger rebuffed them. "Look, we're not going to coordinate with you. We're not going to give you anything that lets you operate independently with some sort of strike."

Instead, the Americans agreed to summon a team of photo specialists from Washington to help the Israelis interpret satellite images and recommend targets for allied pilots. If the Israeli military at times seemed heedless of the sacrifices made by aviators like David Eberly and Thomas Griffith, they at last understood that taking out the Scuds would require not a few surgical strikes but a protracted campaign. They also hoped to turn a quick profit. When the Israeli foreign minister presented the Americans with a bill of $13 billion to defray war costs and build new settlements for Soviet immigrants, Eagleburger peered owlishly through his glasses and replied, "This is all very interesting and I'll take it back to Washington. But it's kind of expensive, wouldn't you say?"

Thirty minutes before a session with Yitzhak Shamir, harsh new instructions arrived from the White House. Tell the prime minister, the secret cable ordered, that if he firmly agrees not to attack Iraq, the United States will provide him with additional Patriot batteries. Eagleburger and Wolfowitz were appalled at the heavy-handed ultimatum, which seemed distressingly close to blackmail. They had scrupulously affirmed Israel's right to counterattack — the very cornerstone of Israeli military policy for forty years — while urging Shamir not to translate that principle into action. As always, Shamir had been enigmatic and prickly. Yet thus far he had steadfastly resisted pressure from his more bellicose cabinet ministers. The prime minister recognized the danger of giving Saddam an excuse to withdraw his forces from Kuwait and launch them instead in an Arab *jihad,* a holy war, against Israel. He also saw the potential calamity of embroiling Jordan, which had long served as a buffer between Israel and her more hostile neighbors to the east.

When Eagleburger relayed his new instructions to Shamir, the prime minister responded with scorn. "So if we don't do what you want us to do, you'll let people here get killed?" he asked. "This is not the way to speak to a friend. It leads to questions about what kind of relationship we will have in the future." Eagleburger privately agreed, and after the meeting he sent a sharp message back to Washington, reporting Shamir's displeasure and his own dismay.

Though never certain that Israel would remain on the sidelines, Eagleburger surmised that Shamir was seeking excuses not to act. That, Eagleburger believed, should be the basis for American diplomacy however long the war lasted. Washington should be patient, sympathetic, and firm, he concluded, allowing the Israelis to posture in their cabinet meetings while blaming the Americans for keeping them out of the war.

One final episode showed that the Americans were willing to permit Israel some latitude, if only to keep Saddam off balance. As Eagleburger and his delegation prepared to return to the United States, David Ivri pulled Wolfowitz aside and told him that Israel intended to test-fire a Jericho missile into the Mediterranean. Ivri promised that before any launch he would notify the Pentagon with a code phrase: *apple pie.*

Wolfowitz had been appalled at the White House suggestion on January 17 that a retaliatory attack on Iraq with Jerichos might circumvent concerns about Israeli fighters flying through Saudi or Jordanian airspace. Even before Moshe Arens summarily rejected the proposal, Wolfowitz had considered it "a dumb idea," likely to inflame the Middle East while leaving the Scud problem unresolved.

But a test shot with a dummy warhead was a different matter. On his return to Washington, Wolfowitz told Cheney of the planned launch. Two days later Ivri called the under secretary and reported, "Apple pie will take place." Wolfowitz offered no protest. Saddam should be reminded, the Americans had agreed, that even if Israel stood temporarily muzzled, she still had fangs.

As Eagleburger and Wolfowitz performed their kabuki rituals in Israel, another envoy flew to Amman for a different kind of dance with Jordan's King Hussein. Richard L. Armitage, a former Defense Department assistant secretary now working for Baker, had spent three tours in Vietnam after graduating from the Naval Academy. Son of a Boston street cop, he possessed a barrel chest, a voice like a rock slide, and the reputation of an official enforcer periodically dispatched for what he called "the wet work" of American diplomacy — thrashing a recalcitrant adversary without leaving any permanent scars. At 11 A.M., on Monday, January 21, he arrived at the royal palace for lunch and a blunt conversation.

King Hussein was playing a dangerous game. Although the Jordanian monarch was said to be a direct descendant of the Prophet Muhammad, Hussein ruled less by divine right than by guile and an ability to play off one opponent against another. The subjects of his small kingdom overwhelmingly supported Iraq. Four hundred newborn males in Jordan had been named Saddam since the invasion of Kuwait, and others now bore the name Scud. Barbers in Palestinian refugee camps reported a great demand for Saddam-style mustaches, and hundreds of women had tramped through the Jordanian capital chanting, "Saddam, our beloved, hit Tel Aviv with chemicals."

While privately assuring the Americans and Israelis that he was merely appeasing "my people in the street" by rhetorically supporting Baghdad, the king was also suspected of turning a blind eye to rampant smuggling across the Iraqi border in defiance of the United Nations embargo. During Desert Shield, the White House had warned Hussein that he risked forfeiting Washington's economic assistance if he failed to join the chorus condemning Iraq.

Armitage was not unsympathetic to the king's predicament. By the standards of Middle East royalty, Hussein was something of a populist; his support of democratic ideals had earned him rancor and snubs from the reactionary monarchs of Saudi Arabia and Kuwait. Armitage suspected that Hussein had endorsed Saddam's efforts to frighten the "Gucci sheikhs" of Kuwait before he realized that Iraq would invade and pillage the emirate. Now, Armitage believed, the king was like an errant youth who had succumbed to the pressure to join in a robbery, only to find himself an accessory to murder after the victim was shot.

Armitage found Hussein wearing an open-neck shirt and a gaudy cowboy belt, smiling nervously and gripping the arms of his chair as though awaiting the first truncheon blow of the inevitable wet work. Instead, the American played good cop to his own bad cop image. It is very much in Jordan's interest to refrain from attacking Israeli jets if they overfly your country, Armitage said. "We're going to get past this. We've got some immediate business to deal with now," he added, "but the United States respects Jordan." Hussein sighed and settled back into his chair as the court chamberlain handed him a cigarette. The king, Armitage realized, was utterly petrified.

As usual, Hussein tried to steer a middle course. The Israelis know the characteristics of our Hawk surface-to-air missiles, he said. "If they fly within the envelope of our missiles, I will have to shoot." To Armitage, the message seemed evident: tell the Israelis to veer around the Hawk batteries if they have to fly across Jordan.

"Be very clear in telling Saddam Hussein," Armitage advised, "that the use of weapons of mass destruction, particularly chemicals, will be dealt with in the harshest fashion." The king nodded and replied, "I do not think they will use chemical weapons."

As lunch ended, the Jordanian prime minister joined the two men. He looked ashen and exhausted, having returned only moments before from a trip to Baghdad to remind the Iraqis that Jordan had no wish to be a battleground in the *jihad* against Israel. Armitage inquired about the ten-hour drive on the Amman-Baghdad highway, which sliced past H-2 and H-3. "What did you see?" the American asked.

The prime minister said nothing; instead, he swiveled around and glanced upward, first over one shoulder, then over the other, the practiced gesture of a man who had just spent ten hours watching with dread for the American bomb that would blow him into the next world.

## Al Qaim, Iraq

The rising sun woke David Eberly and Thomas Griffith on Sunday morning, January 20. Eberly sat up stiffly, pushing away the bulky parachute. It was very cold — each breath sent a milky plume into the air — and fog shrouded the desert floor. Except for the power lines and a low hillock rising four hundred yards from the wadi where they had slept, the landscape was as bleak and empty as the open sea. Conversation was an effort, and the men said little. Griffith mopped Eberly's wounds. Then they gathered up the parachute and plunged into the fog toward the nearby hill.

When they had climbed along a goat path to the crest, the men found an oval depression, fifteen feet long, that offered a sheltered view of the road curving below them to the west. They took inventory. Eberly had the chute and his survival vest; his worldly possessions consisted of a radio, flares, a pistol, the water bottle — now nearly empty — and blood chits, promissory notes printed in English and Arabic that offered a reward for anyone delivering the pilots into allied hands. Griffith had his pistol, radio, and a survival kit with compass, flares, a map of western Iraq, two packets of water, and a small solar blanket. On the map they located the power line, the road, and then their hill, designated only by its elevation — 1181 feet. Their erstwhile target at Al Qaim lay to the northeast, and another hill — 900 — stood approximately two miles to the southwest. Less than ten miles due west was the Syrian border.

Turning off one radio to conserve the batteries, they monitored the other and periodically broadcast calls for help. "We're too far north," Eberly said after several futile attempts, "and nobody's going to be out here in the daylight." They dozed, then awakened a short while later to the sound of voices. Eberly crawled to the rocky parapet and peered over. A derelict oil truck sat on the side of the road, with a car parked behind it. Two men conversed briefly, then climbed into the car and drove south.

Eberly wondered whether the men could be rescuers, dressed incognito. Or perhaps it was a trap to lure the hiding airmen into the open. In hushed tones, he and Griffith debated walking back to the ejection seats to recover the other parachute and the water in Eberly's abandoned survival kit. But what if the seats had been discovered and were under surveillance? Or were booby-trapped? For now, they agreed, it was better to stay put. The day drifted by, and as the sun slipped below the horizon, the two men scraped at the rocks in their oval nest and slept again.

They woke once more to the familiar sound of airplanes high overhead. Another strike package of F-15Es was returning to bomb targets near Al Qaim. "This is Corvette three," Eberly called on the radio. "How do you read? If you read, come up twenty-eight, twenty-eight." A voice answered on the emergency frequency: "Corvette three, I read." Eberly instantly recognized the speaker as Major Gary Cole, a backseater from the 336th. "What's the hold-up?" he asked impatiently. "They're searching," Cole replied. "Well," Eberly urged, "tell them to hurry." And then the planes were gone, swept from range at nearly six hundred knots. Without success the two airmen tried repeatedly to raise the flight before again lying down in cold, frustrated misery.

*They're searching.* Seven combat search-and-rescue (CSAR) bases had been established at the start of the war, five in Saudi Arabia and two in Turkey. Schwarzkopf assigned CSAR responsibility to his special operations commander, Army Colonel Jesse Johnson, a move that soon stirred resentment among Navy and Air Force aviators. Rather than risk losing helicopter pilots and special forces troops in a random hunt for downed pilots, Johnson decreed that "reasonable confirmation" of a survivor's location and predicament be established before he would authorize a rescue mission. Only for the initial strikes of the war had several CSAR teams loitered over Iraq; thereafter it was deemed safer and more effective to stage from their home bases.

To prevent the Iraqis from luring rescue crews into a trap, each airman was to "authenticate" himself with code words kept in his file, personal trivia like his mother's maiden name or the breed of the family dog. To Jesse Johnson the procedures reflected a prudent balance between inaction and bravado, particularly since the Iraqis soon proved adept at "DF-ing" downed pilots — using direction-finding antennae to home in on American SOS broadcasts.

For several thousand aviators, however, the reasoning did not wash. Many came to believe that CSAR held a low priority for Johnson and his special forces, who were occupied with numerous other missions, from plotting secret strikes behind enemy lines to instructing Arab troops on the mysteries of soldiering. A vivid folklore persisted from Vietnam, where rescuers like the Jolly Green Giant helicopter crews had plucked thousands of survivors from crashes at sea and in the jungle with legendary feats of derring-do. From 1964 to 1973, Air Force rescue operations alone had saved nearly four thousand Americans at a price of seventy-one CSAR personnel killed; Navy teams recovered hundreds more.

By contrast, Johnson's effort seemed puny. For the thirty-five allied aircraft downed during the Persian Gulf War, only seven rescue missions would be launched and three pilots saved. At Al Kharj, as the hours ticked by without any sign that Eberly and Griffith would be rescued, cynicism took root among their fellow aviators. If you get shot down, they told one another, be sure you've got your walking shoes.

Such criticism, a natural consequence of frustration and anxiety, was unfair. Johnson *did* launch a CSAR team from Turkey shortly after Corvette 3 went down; the helicopter lingered near Al Qaim for seventeen minutes without making radio contact, then flew back to Turkey. Buster Glosson, closely monitoring the rescue efforts from Riyadh, learned that the Jordanians intended to dispatch a search team; for reasons unclear, the effort came to naught. Contradictory intelligence began piling up; one sketchy report noted that the two Americans had been captured and were being held for ransom. Although Cole, the airman who heard Eberly's transmission on the night of January 20, recognized the colonel's voice, the contact was too fleeting for him to pinpoint the position or authenticate Eberly's condition. Glosson considered asking Johnson to dispatch another CSAR team, but such an effort, he concluded, posed an unreasonable risk.

By midafternoon of their second day in the desert, Eberly and Griffith had decided to save themselves. Again they had picked up radio chatter, this time from a CSAR team searching for the crew of a Navy F-14 shot down during an attack on Al Asad Air Base, eighty miles east of Al Qaim. Without success they tried to break into the transmissions. Eberly crouched in their redoubt, scanning the empty sky to the southeast as the last transmissions faded and died. The time had come, he knew, to make a move. Neither he nor Griffith had eaten in forty-eight hours. Their water was reduced to a few sips. Soon they would be too weak to walk. The two aviators had hatched and rejected half a dozen schemes. One called for hijacking a car or truck by having one man lie in the road and the other hide in the brush. But what if the civilian driver resisted and they had to kill him? Moreover, where would they go? Iraqi soldiers surely had roadblocks near the Syrian border, and the rugged desert precluded cross-country driving. Pulling out the map, they plotted their course by foot: west by northwest to Syria, then onto the road that led to the Syrian town of Abu Kamal on the Euphrates River, perhaps thirty miles away.

With newfound energy they set to work ripping up the orange-and-green parachute and stuffing the strips into their flight suits for warmth. They also fashioned silk burnooses so that from a distance they resembled Bedouin. After burying the harness buckles and cords, they waited until sunset, then scrambled down the hill and set off at a measured pace.

The crescent moon rose like a grin, lambent on the desert pan. Headlights occasionally swept the nearby road, sending the men into a headlong sprawl for cover. The miles fell behind them. For the first time since being shot down, they chatted about home, about family — Griffith had four children — about the pleasures of a hot shower and a soft cot. They imagined being delivered to the American embassy in Damascus and then — where? Home to the States for a few days? Or straight to Riyadh, as each preferred, before returning to Al Kharj and the war?

Far away, they heard dogs barking. The noise drew closer. Alarmed, Eberly wondered whether Iraqi soldiers were looking for them with hounds. His flight suit was stiff with dried blood. As they slowed their pace, two Bedouin tents loomed ahead and from the shadows darted a dozen snarling mongrels. The dogs ignored Griffith and circled Eberly with fangs bared in the moonlight. Gathering his robe about him, he strode past the tents, certain they were being watched. As the dogs grew bolder, darting at his legs, he pulled the .38 revolver from his holster and cocked the hammer. The animals finally pulled back, retreating into the night.

Two hills rose before them like a camel's humps. Skirting to the north, they stopped to rest, exhausted from fighting their way through the tangled brush that seemed to grow thicker with each passing mile. Again they saw headlights, this time on a road to the west, which, they concluded, lay roughly a mile beyond the Syrian border.

Abruptly the radio came to life. "Mobil four one to Corvette three, how do you read?"

"This is Corvette three," Griffith replied. "We read you."

"What is your condition?"

"Corvette three A and B are okay. We're near the border, ten miles southwest of the target."

"Roger. Stand by."

Then, nothing. They assumed the call had come from a CSAR helicopter. (The voice was actually that of an F-15C pilot flying north out of Tobuk in Saudi Arabia.) Vainly they called again and again. The wind picked up, icy and relentless; sweaty from their brisk tramp, the pair huddled under Griffith's tiny foil blanket and dozed off, only to wake shivering uncontrollably. Behind them the moon had set. Staring to the east, they fancied the faint slap of helicopter blades. One hallucination embroidered another, and now Eberly imagined the winking red glow of aircraft lights. Numb with cold and despair, he and Griffith shared the last trickle of water.

Eberly proposed pushing on toward the Syrian road. Griffith suggested waiting for the rescue team, which surely would descend on them any moment. "I think we should ride out rather than walk," he said. Both men could barely speak, so violently did their teeth chatter. A few hundred yards away, Eberly saw the dim silhouette of a small building. Lightless, square, and flat-roofed, it looked deserted. "Let's at least go over there and get out of the wind," he proposed. They struggled to their feet and trudged toward the structure, Griffith twenty feet in trail.

Eberly had nearly reached the building when it exploded with gunfire. Muzzle flashes peppered the darkness, and bullets shredded the air with ripping noises. Other rounds kicked up small geysers of dirt at their feet. Above the staccato roar of automatic weapons fire, voices screamed at them in Arabic. Eberly, astonished not to have been hit yet, sank to his knees and hoisted his hands high over his head. "English!" he yelled. "English!"

The shooting stopped and was supplanted with an odd chant: "Iraq, Syria! Iraq, Syria!" Men hurried from the building, rough hands shoved the pilots toward the open door. The men steered the two Americans into a side room. Maybe we're in Syria, Eberly thought. Perhaps these are Syrian border guards. They'll get us to Damascus.

He and Griffith were pushed onto a bunk against one wall. Half a dozen uniformed men loomed over them. A blanket lay folded on the bunk and Eberly pulled the thick folds around him. Heat poured from a potbellied stove, and he reached toward it with his frozen fingers. "Cold," he explained, looking up at the men. "Water? Food? Please."

Then his heart sank. Glaring down from the opposite wall was the large photograph of a familiar face: that thick lower lip; the rounded jaw and heavy jowls; the brushy mustache; those black, dense eyes. Saddam Hussein.

# 5

# Delta

## Washington, D.C.

On the morning of January 22, just a few hours after David Eberly and Tom Griffith fell into Iraqi hands, a trim, fair-haired Army major general strode through the Pentagon's corridors to room 2D874 in the National Military Command Center. Wayne A. Downing had come to see Lieutenant General Thomas W. Kelly, director of operations for the Joint Chiefs, but he had kept this visit secret even from his boss; when Downing had called Kelly to ask for an appointment, all he disclosed of his purpose was "I want to talk to you about Scuds."

Downing headed the Joint Special Operations Command, the military's shadowy counterterrorist unit at Fort Bragg, North Carolina. Behind the closed door of Kelly's office, he pulled out a map and quickly sketched his proposal. According to intelligence analysts, the Iraqis appeared to be shooting Scuds at Israel from three different areas, or boxes, in western Iraq. The southern box straddled the Amman-Baghdad highway. The other two lay along the Syrian border, one near the town of Shab al Hiri, the third farther northeast, near Al Qaim.

Downing proposed setting up a field headquarters at Ar Ar, thirty miles south of the Iraqi border. From there he would insert a succession of large patrols by helicopter to clean out the southern box and then the northern pair. The patrols would attack not just missile launchers, but also communications sites, logistics bases, and other installations that supported the Scud campaign.

No matter how aggressively Air Force and Navy pilots attacked the Iraqis, Downing believed, they could not see from fifteen thousand feet what a soldier saw on the ground. Iraqi missile crews also appeared able to pack up and move within six minutes after firing, rather than the thirty minutes predicted by U.S. intelligence. To man the patrols, Downing suggested using JSOC's Delta Force, the elite commando unit usually reserved for counterterror missions. Formed in 1977, Delta comprised three squadrons, each with about 130 superbly trained and conditioned soldiers, who often wore civilian clothes and long hair when traveling abroad to disguise their military purpose.

For months before the war began, Delta Force and the other units in JSOC — including Army Rangers and Navy SEAL teams — had been frantically busy. Downing had flown to Saudi Arabia five times since early August. In October and November, under orders from Bush transmitted by Colin Powell, he and Buster Glosson designed an elaborate raid to free U.S. diplomats trapped inside the American embassy compound in Kuwait City. The mission, code-named Pacific Wind, involved eighteen F-15s, four F-117s, four Navy F-14 Tomcats, and several Raven and Prowler jamming aircraft. In the dead of night, a pair of Stealth fighters would bomb two power plants in Kuwait City to knock out the lights; another pair would flatten a high-rise hotel near the embassy to kill Iraqi officers living there and prevent enemy gunners on the roof from shooting at Delta's helicopters. Strike Eagles would then attack gun pits along the beach and bomb three boulevards leading to the embassy compound as Delta soldiers swooped in with helicopters to rescue the trapped Americans as well as diplomats from the nearby British embassy. Glosson and Downing planned to oversee the mission from a C-130 circling off the coast.

A Delta squadron had secretly flown to Nevada to practice operating in the desert. Others rehearsed in the sandy terrain of Fort Bragg's Sicily drop zone; F-15E crews practiced at Hurlburt Field in Florida. Pacific Wind would be the largest raid ever attempted by Delta — a "damned high risk," Glosson had warned. Powell worried that casualties would be high, among both the raiders and the diplomats. The plan to isolate the embassy by destroying a sizable portion of the neighborhood bothered Schwarzkopf, who also feared that such an attack could trigger all-out war. "If I direct you to conduct this operation," the CINC told Downing and Glosson in late autumn, "then I may need to go ahead and execute the rest of the war plan."

But in December, Saddam permitted the peaceful evacuation of the embassy and the raid plan had been shelved without Delta's deploying to the Middle East (except for thirty men assigned to set up logistics and communications and to serve as Schwarzkopf's bodyguards). JSOC then formed three task forces to combat an expected wave of terrorism around the world. One focused on southwest Asia, another on Europe, and the third on the United States. Thus far, however, the terror wave had not materialized, and JSOC was looking for work.

Downing knew that his Scud campaign plan faced several obstacles, perhaps the least of which was the Iraqi army. One hurdle was the deep suspicion held by many military commanders toward elite forces. "Armies do not win wars by means of a few bodies of super-soldiers but by the average quality of their standard units," General William J. Slim, Britain's great World War II commander, once declared. "Anything, whatever shortcuts to victory it may promise, which thus weakens the army spirit, is dangerous." That egalitarian sentiment persisted in the American military. JSOC had been created after the calamitous 1980 Iranian hostage mission in part to weave special operations more tightly into the fabric of conventional forces; its performance during the invasion of Panama, where Delta stormed a prison to free a CIA agent, won many plaudits. Still, by its very nature, Delta was unconventional. "Don't be doctrinaire," Downing urged his soldiers. "Think like a bank robber."

More formidable yet was a clash of egos. Downing's JSOC fell under the domain of General Carl W. Stiner, commander-in-chief of the U.S. Special Operations Command. Stiner was a force of nature, a craggy, exuberant Tennessean who had commanded XVIII Airborne Corps during the invasion of Panama. In early January, he swooped into Riyadh full of saucy ideas about how Schwarzkopf could best use special forces in the coming war. Conventional wisdom held that Schwarzkopf was wary of unorthodox warfare; perhaps more to the point, he distrusted anything that could subvert the precise timetable of his four-phase attack — such as a few hundred heavily armed commandos crashing through Iraq. "How am I going to explain," he asked Stiner and Downing, "what Delta Force is doing three hundred miles deep in Iraq?"

Worse, Stiner hinted at moving his own headquarters from Florida to Saudi Arabia. Alarmed at the prospect of another four-star bulling into his theater, Schwarzkopf reacted with disdain. Downing sensed that Stiner had become persona non grata in Riyadh and that any scheme bearing Stiner's imprint was unwelcome. For that reason he had come to see Tom Kelly behind Carl Stiner's back.

As Downing presented his plan, he was aware that one additional obstacle stood between Delta and the Scud boxes: the British were already there. The 22nd Special Air Service Regiment — Britain's commando force — had been training in the United Arab Emirates before moving to western Saudi Arabia in mid-January. The British theater commander, de la Billière, had once commanded the SAS, and he convinced Schwarzkopf that the unit could play hob in western Iraq by cutting roads and staging diversionary raids. Schwarzkopf allowed SAS troops to infiltrate across the border two days before the air campaign started. When missiles began to hit Israel, the regiment instead began hunting mobile Scuds.

That the SAS was permitted to fight while Delta remained in North Carolina bitterly offended those in the secret world of American special operations who knew of the arrangement. Downing admired the SAS, which was formed in North Africa in 1941. The regiment's exploits included daring attacks on German supply lines before the battle of El Alamein. But the unit was small and lacked the special helicopters needed to strike deep into Iraq.

Tom Kelly listened to Downing's proposal with mixed feelings, his blue eyes blinking deliberately behind his horn-rimmed glasses. The plan was dramatic and risky. Delta was viewed as the Joint Chiefs' private army; to have commandos captured or killed would be a political embarrassment as well as a military humiliation.

On the other hand, the Scud campaign was not going well. Bad weather had scuttled hundreds of aircraft sorties, and Colin Powell's displeasure mounted by the hour. "Goddam it," Powell had snapped at Schwarzkopf during one of their phone calls, "I want some fucking airplanes out there." Sending Delta would be a signal to the CINC of the importance Washington placed on rooting out the Iraqi missiles. Perhaps it also would silence the Israeli military, whom Kelly described as "arrogant little bastards who wouldn't last ten minutes on a European battlefield."

Kelly hailed from a long line of Philadelphia Linotype operators, and he had a folksy, unpretentious manner that Powell recognized as invaluable for the role of the Pentagon's daily spokesman during the war. Dry-witted ("I'm an Irish Catholic, and we enjoy pain") and, like Powell, given to aphorisms ("The leader is never equal to the led"), Kelly was frustrated by his inability to influence CENTCOM or pry timely information from Riyadh. The Panama invasion in 1989 had been run largely from the Pentagon; Schwarzkopf was determined that this war would be controlled in the theater. Kelly grudgingly conceded Schwarzkopf's prerogative, but the CINC's tight grip galled him.

Although Kelly and Schwarzkopf had been war college classmates, no love was lost between them. About the time of Downing's visit, Powell asked Kelly to assemble details of a recent Scud-hunting mission so that the chairman could present the information during a meeting at the White House. Kelly called Burton Moore, his counterpart in Riyadh, and asked for the data in an hour. Schwarzkopf, who overheard the call in his war room, flew into a profane rage and grabbed the receiver. "What the fuck are you doing? Did you pass this goddam order on?" the CINC demanded. "We can't do this in an hour." Kelly flushed, struggling to keep his temper. "Hey, Norm, the chairman asked for this," he replied. "You got a problem, call the chairman."

When Downing finished his pitch, Kelly gathered up the charts and maps. Despite his ambivalence, he considered the plan worthy of Powell's perusal. Downing left the office and, as requested, returned later in the day. Kelly handed back the charts. "Chairman says, 'Interesting, but not yet,' " he told the JSOC commander. "Just keep working on it." Downing tossed a salute and left the Pentagon to return to Fort Bragg.

That evening, Iraq delivered its most dramatic counterpunch of the war. A Scud warhead, slipping through the Patriot batteries, detonated in the Tel Aviv suburb of Ramat Gan. The blast damaged more than seventeen hundred apartments, wounded ninety-six people, and sent three elderly Israelis to the grave, victims of cardiac arrest. Search teams dug through the night for buried survivors. In the rubble of one apartment, rescuers found an eighty-year-old woman sitting in a closet with a memorial candle flickering in her hand, calmly awaiting deliverance.

As Tom Kelly had directed, Downing continued to refine his plan. He also called Carl Stiner and, with some trepidation, confessed to his visit to the Pentagon. Although Stiner at first seemed wounded by the disclosure, he quickly embraced Downing's proposal and soon became an advocate for the deployment of Delta Force as a sensible antidote to the Iraqi missile attacks.

George Bush had promised Israel, privately and then publicly, to suppress the Scuds by launching "the darndest search-and-destroy effort that's ever been undertaken." Yet how he expected to fulfill that pledge remained a mystery.

"Just the area of western Iraq alone is 29,000 square miles," Schwarzkopf told a television interviewer on January 20. "That's the size of Massachusetts, Vermont, and New Hampshire all put together, and you know there's not much point putting people on the ground to try and find nine, maybe ten trucks." In Washington this was interpreted as a reminder of the CINC's aversion to unleashing American special forces; the statement may also have been disinformation intended to obscure the activities of the British SAS.

Schwarzkopf further declared that thirty fixed Scud sites had been "neutralized" and "we may have killed as many as sixteen" of twenty or so suspected mobile launchers. Permanent sites had indeed been bombed; Iraq never attempted to launch a single Scud from those locations. The mobile tally, however, was fantasy based on notoriously optimistic pilot reports. Other U.S. intelligence estimates at the time confirmed only two mobiles destroyed. The CIA would not confirm that even one mobile launcher was actually demolished. (The agency's skepticism persisted to the end of the war.)

In a war celebrated for sophisticated weaponry, the Scud remained an annoying anachronism. The missile was a close relative of Nazi *Vergeltungswaffen*, weapons of retaliation like the V-1 and V-2. Germany in World War II had launched 9700 buzz bombs across the English Channel from France and Holland; British fighters and antiaircraft guns shot down many, but those which slipped through killed and wounded nearly 35,000 people. The Nazis also fired another 9300 at Allied-occupied cities on the Continent, particularly Antwerp. Although terrifying and lethal, the *Vergeltungswaffen* had little military effect, hampered by inaccuracy and German failure to concentrate their salvos on strategic targets, such as ports.

In the Iran-Iraq war in the 1980s, the two sides lobbed more than two hundred long-range missiles at each other as part of a slugfest that grew especially churlish during the so-called War of the Cities in early 1988. American intelligence gleaned a great deal from those exchanges, and from several Iraqi test-firings before the gulf war began. The Defense Intelligence Agency (DIA), which created "a Scud cell" on August 2, knew that Iraq had extended the Scud's range by adding extra fuel tanks and reducing the warhead payload. (The missiles fired in the gulf war were actually Al Husayns and Al Hijarahs, modified variants of the Soviet-made Scud.) Scud crews could aim only by estimating the distance to the target and adjusting the direction and elevation of the launcher. Once the missile was fired, it had no guidance system. With a "circular error probable" of more than three thousand meters, only half of the missiles could be expected to land within a two-mile radius of their targets.

Yet large intelligence gaps remained, particularly regarding mobile launchers and the fine points of Iraqi command-and-control — who could order the missiles launched and under what circumstances. Before the war began, some U.S. analysts estimated that Iraq possessed eighteen mobile launchers; Air Force target planners calculated that the entire Scud set — including fixed and mobile launchers, storage bunkers, and command centers — contained fewer than a hundred bomb aim points. (An aim point was the specific portion of any given target on which an attacking pilot was supposed to drop his munitions.) But the estimated number of mobiles soon increased, climbing to thirty-six and beyond, as intelligence experts realized that Iraq could jury rig a launcher simply by bolting a steel rail to the bed of a truck. In addition to TELs (transporter-erector launchers) made in the Soviet Union, the analysts soon concluded, Iraq had its indigenous MELs (mobile erector launchers).

Even more frustrating was the task of locating the elusive mobiles. Nuclear targeteers for several years had been thwarted in their efforts to track Soviet mobile missiles, mounted either on railcars or truck beds. The DIA director, Harry E. Soyster, had served as an Army lieutenant thirty years before on a Corporal guided missile crew. Soyster remembered how easy it was in exercises to hide the primitive, liquid-fueled Corporal among the bell towers and alleys of Italian villages. Then and now, Soyster cautioned, anyone hunting mobile missiles needed great luck to find them.

Such warnings came as no surprise to Buster Glosson and Dave Deptula, who had wrestled throughout the fall and winter over how much air power to allocate to taking out the Scud sites. Of advice they found no shortage. Israel offered a long list of locations where Scuds could be hidden, almost all of which had already been identified by the Americans. The 82nd Airborne Division, at Schwarzkopf's request, drafted a plan to parachute two brigades and insert a third by helicopter around H-2 and H-3. Although enthusiastically suppported by the division commander ("If you want to send a message to the world that you're serious about Scuds, drop the goddam 82nd Airborne on them," Major General Jim Johnson urged), the plan was deemed too risky. The Air Force and Navy ran tests by hiding mock Scuds in the California desert; pilots dispatched to bomb them reported that the "launchers" showed up well on radar — if the pilots knew where to look.

Throughout the fall Israel had sought American assurances that the destruction of Iraqi missiles would have the highest priority in allied war plans. In October, Cheney asked Glosson, "Okay, Buster, can I tell Arens that he doesn't have to worry about those Scuds pointing at him out of H-2?" "Yes sir, you can," Glosson replied. "I can guarantee there will not be a Scud launched from a fixed launcher after twenty-four to thirty-six hours. Not one. But I'm not going to tell you how long it will take to get the mobiles. I don't know, and neither does anybody else."

Glosson had concluded that one squadron of F-15Es — twenty-four planes — would suffice to suppress the mobiles. The estimate presumed that American intelligence would detect where the Iraqis stockpiled, fueled, and serviced their missiles, thus enabling allied warplanes to strangle the Scud campaign. Glosson also wanted to target Iraqi "scrapes," sites in the western desert that enemy missile crews were known to have surveyed as possible mobile launch locations. Deptula talked him out of it. "It makes no sense to hit those randomly. You'll drive yourself batty trying to anticipate where the launchers are," Deptula argued. "Also, you'll pull away airplanes from the primary effort." Glosson was convinced, and instead assigned bombers to remain on "Scud alert" until they received intelligence indications of an imminent launch.

But when the war began, the one-squadron estimate proved inadequate. To Glosson's chagrin, U.S. intelligence could not pinpoint the Scud logistic centers. Instead of wrecking the missile network, pilots found themselves crisscrossing much of Iraq in a hunt for eighteen-wheel launcher trucks. Many Iraqi decoys were so cleverly constructed — as United Nations inspectors discovered after the war — that they could not be distinguished from real missile launchers unless an observer was within twenty-five yards. Although allied aircrews would spot Scuds streaking through the sky more than three dozen times, they never found a vehicle that could be indisputably confirmed as a launcher truck.

As the Scuds continued to fall on Israel and Saudi Arabia, Glosson tripled the number of aircraft on Scud duty, adding squadrons of F-15Es, F-16s, and A-10s. Those on alert were shifted from their home bases to establish a permanent CAP — combat air patrol — over western Iraq. Glosson at one point even ordered the Strike Eagles to drop a bomb every half hour on any suspected Scud site to remind Iraqi missile crews that American planes were overhead. Cheney demanded detailed reports on the number of anti-Scud aircraft flying, the result of their sorties, and other intelligence regarding Iraqi missile activities. At Powell's subsequent request, Glosson sent daily accounts of all Scud-hunting activities to the Pentagon, which relayed them to Israel.

Still the Scuds fell. At wit's end, Glosson and Horner presented to Schwarzkopf late in January a plan for making sure that no more Scuds were launched toward Israel. It called for diverting nearly all allied aircraft — roughly two thousand planes — in a three-day campaign to flatten much of western Iraq, particularly the areas around Al Qaim, Rutba, and several other cities. Police posts, service stations, warehouses, and any other facility deemed remotely supportive of the Scuds would be destroyed. Bombers would sow mines on all roads in western Iraq and blow up more than sixty underpasses — favored mobile hiding places — on the Amman-Baghdad highway. It isn't carpet bombing, Glosson warned, but it's close.

"Will this detract from our main effort?" Schwarzkopf asked, referring to strategic targets in Baghdad and eastern Iraq.

"Yes, sir, sure it will," Glosson said. "We'll take all of the assets we've got and use them out there."

"Will it shorten the war?"

"Probably not."

Schwarzkopf asked several more questions before rejecting the scheme. The cost was exorbitant, both in terms of aircraft diverted and the number of Iraqi civilian casualties. Glosson privately disagreed, but he understood the CINC's reasoning. The American people, Glosson observed, would not abide certain acts of war, however justifiable from a military standpoint. Killing every living creature west of Baghdad was probably one of them.

## U.S.S. *Blue Ridge,* Persian Gulf

On the night of January 22 a wave of A-6 bombers from U.S.S. *Midway* attacked an Iraqi Al Qaddisiyah–class oil tanker anchored in the northern Persian Gulf, as well as a Hovercraft and a Zhuk patrol boat bobbing near the ship. Lookouts on the tanker, U.S. naval intelligence had concluded, were routinely warning Iraqi gunners ashore whenever a Navy strike force flew overhead toward targets in Kuwait or southeastern Iraq. This night, the A-6s ripped the ship and smaller boats from bow to stern, sinking the Hovercraft with cluster bombs, killing three crewmen, and igniting the tanker in an inferno that burned for weeks. The attack reflected an aggressive Navy campaign to clear the northern gulf of enemy ships and coastal missile batteries that menaced the allied fleet and threatened to thwart an eventual amphibious landing by U.S. Marines.

But the attack also violated Norman Schwarzkopf's orders. During Desert Shield, Schwarzkopf had decreed with fist-thumping emphasis that in the event of hostilities the Navy should sink "every goddam Iraqi ship in sight"; hours before the war began, however, he revised the "rules of engagement" by placing oil tankers and commercial facilities off limits. This reflected written instructions from Washington to avoid polluting the gulf. For reasons never completely determined, the modified order failed to reach the senior Navy commander, Vice Admiral Stanley Arthur, aboard his flagship, *Blue Ridge.*

Shortly after the attack, Schwarzkopf called Buster Glosson. "Did you give the Navy permission to sink a tanker?" he demanded. "Shit, no," Glosson replied. Schwarzkopf slammed down the receiver. The next call went from Schwarzkopf's J-3, Burton Moore, to Arthur's operations officer, Captain Robert L. (Bunky) Johnson, Jr. "What the hell are you guys doing?" Moore demanded. "You're out of control."

Johnson had grown accustomed to strident calls from Riyadh. Once, during a heated discussion with Moore over CENTCOM's request to have the Navy escort a cargo convoy into the gulf, Johnson had been surprised to hear Schwarzkopf suddenly come on the line. "What do you think this is, some kind of fucking debating society?" the CINC had roared. "You get those ships into the gulf and I don't want to hear another fucking word about it!" Johnson offered a feeble *aye, aye,* later reflecting that he had finally encountered a soldier who could teach a sailor how to swear.

But Bunky Johnson quickly realized that this call from Moore was more than routine intimidation. "We're considering court-martialing you," the J-3 continued. "We told you not to hit any oil tankers, and you've disobeyed orders. You —"

"Wait a minute," Johnson interjected. "What do you mean, you 'told' us?"

As Johnson began tracking down the misplaced order, Riyadh fired more warning shots. This latest episode aggravated CENTCOM concerns that the Navy during the first few days of the war had been excessively zealous in destroying targets of dubious military value. "The CINC is upset that you've violated the rules of engagement," the CENTCOM chief of staff, Major General Robert Johnston, advised in a call to *Blue Ridge*. "You've got a real problem on your hands."

A few minutes later Schwarzkopf called Arthur. Portly, genteel, and measured, Stan Arthur was widely admired as one of the Navy's premier flag officers. In Vietnam he had flown more than five hundred combat missions, winning eleven Distinguished Flying Crosses. Eventually rising to command the U.S. Seventh Fleet in Japan, he had arrived on *Blue Ridge* in December to lead the great armada gathering for the gulf war. But now Schwarzkopf berated him as though he were an errant ensign.

"The goddam Navy wants to blow up everything in the water that doesn't have an American flag on it. Use some sense and think about what it is you're doing instead of just violating my plan," the CINC demanded.

"I was doing my job as I saw it," Arthur offered. "I'll go back and review the bidding and give you a full report on what happened. But I'm the guy who did it. I okayed the attack."

Unappeased, Schwarzkopf let loose with a tirade that contained more threats of legal action. "Put somebody in charge of this," he added, "or I'll find somebody else who can follow my orders."

Arthur hung up, badly shaken. Following the CINC's order, in preparation for an eventual court-martial he began reconstructing the sequence of orders that had led to the tanker attack.

There was no court-martial, but the well had been poisoned. In Riyadh, the event reinforced convictions that the Navy was a renegade service incapable of teamwork and determined to fight the war as it saw fit. On *Blue Ridge,* Arthur imposed such strict restrictions on his subordinates — "tying their hands very tightly," in his words — that some feared the Navy was becoming overly timid. Neither perception was accurate, but without question the Navy's relationship with CENTCOM — and with her sister services — became even more tangled, beset with mutual admiration, envy, and suspicion.

"I simply have not got enough Navy to go around," Franklin D. Roosevelt complained in a famous burst of pique during World War II. In the Persian Gulf War, no such shortcoming obtained. Two very large fleets jammed two very small waterways, the gulf and the Red Sea. One hundred U.S. Navy ships, bolstered by fifty allied vessels, patrolled the region. Schwarzkopf wryly observed that he had more ships than water on which to float them.

By enforcing the blockade of Iraq, the combined navies had intercepted seven thousand ships since August and had boarded nearly a thousand of them in search of contraband. Six aircraft carriers now launched sorties around the clock. The perilous pace nurtured a culture of lunatic charm as pilots suited up to the strains of the Rolling Stones' "Satisfaction" (U.S.S. *Kennedy*) or "The William Tell Overture" (U.S.S. *Ranger*), or watched "motivational tapes" of comely, bare-breasted blondes while recalling their shore-leave expulsion from an exclusive restaurant for devouring all the exotic fish in the eatery's aquarium (U.S.S. *Theodore Roosevelt*).

As each ship developed a unique personality, so each fleet was distinctive, reflecting both the character of its commander and its accustomed region of operation. The Red Sea armada, composed largely of Atlantic Fleet ships used to working closely with other services in the North Atlantic or Mediterranean, was considered — in Riyadh and on *Blue Ridge* — cooperative and accommodating. The gulf armada, drawn from the Pacific Fleet, accustomed to steaming independently in the vast reaches of the western ocean, was viewed as more willful and autonomous.

Over this floating empire presided Stanley Arthur, searching for a balance between CENTCOM's grand war plan and the Navy's own requirements. Although he personally admired Schwarzkopf, he found his frustrations mounting steadily. In mid-December he had asked permission to sink Iraqi boats suspected of laying mines in the international waters of the northern gulf, itself an act of war. If an Army commander opened his tent flap and found someone planting a mine, the admiral argued, he would be justified in shooting the intruder. Schwarzkopf denied the request, reluctant to trigger all-out war while Washington was still seeking a diplomatic solution. "I feel more comfortable following our timeline," he told Arthur, "and I don't want to force the issue."

The Navy also considered Riyadh unsympathetic to the fleet's concerns about Iran, a hostile power with a large air force and ship-killing Silkworm missiles. When Iraqi fighters began seeking sanctuary at Iranian airfields soon after the war began, the Navy's anxiety soared; a sneak attack launched through the coastal mountains of western Iran would leave the fleet only ninety seconds to react. Arthur asked Riyadh, to no avail, for better intelligence and more fuel to position aircraft patrols near the Iranian coast. Also, because of incompatible computer systems, there were delays in the transmission of information from Riyadh to the far-flung allied forces, and when the war was under way, the Navy often found itself a day behind in learning of changes in CENTCOM's priorities.

Yet the Navy's relations with CENTCOM seemed downright amiable when compared to its blood feud with the Air Force. Wrote one British observer of his American cousins shortly after World War II, "The violence of interservice rivalry has to be seen to be believed, and was an appreciable handicap to their war effort." For two hundred years the American armed services had jostled one another in a relentless sibling rivalry; with the Cold War ended and the U.S. military budget contracting, the fractiousness intensified. Each service, more fearful of budgetary evisceration than of the Iraqis, sought to prove its value as a defender of the republic. Although few professional officers consciously put loyalty to service above the common good, subtle psychological factors came into play. This was the first major war in a generation, perhaps the last for another generation. War was the enterprise in which men at arms proved themselves. Glory, honor, self-esteem, promotions — all were at stake.

The Navy and Air Force wrangled about matters both foolish and urgent. The Navy resented aircraft rules of engagement — instructions to pilots on when they could and could not shoot, written in this war by the Air Force — which discriminated against Navy planes because they lacked redundant electronic means of distinguishing friend from foe. (Glosson, determined to avoid fratricide in a sky crammed with airplanes, offered a succinct rejoinder: "Tough shit.") The Air Force in turn berated Navy pilots in the gulf for often failing to broadcast the electronic signal that indicated they were indeed friend and not foe.

The Navy, largely reliant on Air Force tankers for refueling, threatened to boycott certain missions, convinced that the Air Force was stingy in the allocation of gas for some of the fighters protecting Navy bombers. (Glosson again: "Doesn't make a shit to me. I'll give the job to the Air Force or the Marines. Or the Saudis. Or the Brits.") The Navy also accused the Air Force of withholding stocks of JP-5, a jet fuel that, because of its relatively high flash point, was preferred on aircraft carriers for safety reasons. Some Navy officers on *Blue Ridge* even suspected the Air Force of manipulating battle reports to credit stealth fighters for damage actually inflicted by Tomahawks. In Washington, senior Air Force officers hinted that Navy pilots were having trouble hitting bridges and other targets.

If the squabbles seemed inane, they also spoke to the stress imposed on men at war. Issues that were momentous in the heat of battle turned to trivia in the cooler light of peace. Unlike the British comment about American internecine feuding in World War II, the tribal jealousies imposed no "appreciable handicap" on this war effort other than to make men grind their teeth and mutter imprecations at their brethren. Rivalry in the U.S. military was bred in the bone, and it would ever be thus.

## Tobuk, Saudi Arabia

*I'll give the job to the Brits,* Glosson had said. And the British would have accepted. Relentlessly game, only twice during the war did they ask Glosson to reconsider missions he had assigned them. But Royal Air Force losses had become a worry — a military one in Saudi Arabia, a political one at home. Five RAF Tornado GR1s already had been lost over Iraq, with ten crewmen either dead or captured. Among those killed, during an attack on Shaibah Airfield, was Wing Commander Nigel Elsdon, the highest-ranking allied officer to die during the war. At least two other British planes had limped home from combat full of holes.

The losses did not exceed prewar calculations; one Ministry of Defence study estimated that each of the three Tornado squadrons would lose a plane on the first night alone. But Britain had only forty-five Tornados deployed in the gulf among a hundred aircraft of all types.

Proportionately the losses far exceeded those of other allied aircraft — more than treble those, for example, of the F-15E like the one piloted by David Eberly.

Not since Operation Corporate, the British war against Argentina in the Falkland Islands in 1982, had the United Kingdom mounted an armed expedition beyond home waters. This enterprise, code-named Operation Granby after an eighteenth-century marquis, dwarfed the Falklands War. The British had amassed 45,000 troops in Saudi Arabia, second only to the number of Americans deployed.

In mid-December, London's contribution had been briefly overshadowed by a potential calamity when an RAF wing commander, David Farquhar, left his sedan unattended while browsing in an automobile showroom in west London. A thief broke into his Vauxhall and snatched a briefcase and a laptop computer containing detailed war plans used to brief Prime Minister John Major.

Police found the briefcase a few hours later in a parking garage, but the computer did not show up until it was anonymously mailed to British authorities three weeks later. Envoys from London flew to Washington and Riyadh to explain the mishap to their American colleagues. "Major loss of confidence all round, and Brits look stupid," General de la Billière fumed in a letter home. A court-martial found Farquhar guilty of negligence and stripped him of his seniority, but planners in Riyadh concluded that the allied attack scheme had not been fatally compromised.

The RAF, embarrassed but unbowed, flew from three bases in the gulf: Muharraq in Oman; Dhahran in eastern Saudi Arabia; and Tobuk, near the Red Sea. British training stressed very low altitude flying — 150 to 200 feet — in part, the Americans believed, because the RAF lacked the sophisticated electronic measures to defeat SAMs at higher altitude. (During one exercise in the Red Sea, Tornado crews "attacked" *Saratoga* by flying below the level of the flight deck, a feat that astonished even unimpressionable Navy pilots.)

Under NATO war plans Tornados were to attack Warsaw Pact airfields every twelve to twenty-four hours with the JP-233, a runway-cratering weapon that required pilots to fly directly over the target at roughly two hundred feet; in the gulf war, they drew the same assignment. A dangerous mission under any circumstances, it was particularly difficult against Iraqi airfields, which had long runways and numerous taxi aprons that gave Iraqi pilots several take-off options even if part of the field was damaged. (Some bases even had concrete plants to expedite repairs.) Moreover, British intelligence had concluded, wrongly, that Iraqi air defense crews mimicked their Soviet mentors by firing many surface-to-air missiles and relatively few antiaircraft guns.

In the first four days of the war the Tornados dropped more than a hundred JP-233s. Twenty feet long, the weapon resembled a large coffin fastened beneath the airplane. During a seven-second "stick," this dispenser spewed out more than two hundred mines while thirty runway-cratering bomblets floated to the ground on little parachutes. One explosion in the cratering charge penetrated the runway and a second heaved the concrete upward; the mines prevented Iraqi workmen from patching the holes. Yet the JP-233s were not as effective as the British had hoped. Intelligence analysts found that sand beneath the concrete tended to absorb the blasts, often lessening the amount of fracturing. Furthermore, few British crews had ever used the weapon before because of limited range space in the United Kingdom. Pilots on their first combat missions were alarmed to find that the munitions detonated like flashbulbs, brightly illuminating their low-flying aircraft.

As Tornado losses mounted, RAF pilots modified their attacks with such tactics as sending all planes across an airfield simultaneously rather than sequentially. But by January 22, 1991, British officers in Riyadh and at High Wycombe — the British command headquarters in Buckinghamshire — were reassessing their mission. The Americans had moved to higher altitudes, and Iraqi pilots appeared ever more reluctant to fly. No one was certain why British losses were so high; curiously, only two of the five Tornados appeared to have been lost while flying at low altitude and only one of those was dropping a JP-233. Theories ranged from the light paint on British planes — supposedly making them more visible to Iraqi gunners — to excessive risk taking by RAF pilots. "We have suffered a high rate of attrition in comparison with the other air forces. There is no denying that," Air Vice Marshal Bill Wratten, the senior RAF officer in Riyadh, told the press. "We have also been extremely unlucky. And bad luck doesn't last forever."

Inevitably, the losses took a psychological toll on the British pilots. Contrary to the Battle of Britain stereotype of an aviator climbing from the cockpit with a nonchalant grin and a jaunty wave, the pilots sometimes wept openly after an especially terrifying mission or the loss of a comrade. At Tobuk, where nearly half the Tornados were based, one pilot suffered a breakdown after seeing television footage of two captured British crewmen in Baghdad; the Tobuk commander, Group Captain Bill Hedges, considered banning further television viewing because of the effect on his crews' morale. (The distraught pilot, sent home to

Britain for hospitalization, returned to fly ten missions in late February.)

By the sixth day of the war, the British high command had had enough. The JP-233, no longer needed since Iraq was launching so few aircraft, would be shelved. Wratten called his subordinate commanders and announced, "We are going to medium-altitude bombing." On January 23, Tornados began dropping unguided bombs. Accuracy at times was dismal — pilots often missed their targets by hundreds of feet — and authorities in London proposed reviving the JP-233 missions. RAF officers in Riyadh stoutly resisted; accuracy improved, particularly after the arrival of new equipment that allowed pilots to attack with laser-guided munitions. British losses abated. Only one more Tornado would be lost, in mid-February. As Wratten had said, bad luck doesn't last forever.

No one in Riyadh, London, or Washington believed that the Iraqi air force had recused itself from the war simply because some of its runways had been perforated. Yet Iraq's strategy remained a mystery. Saddam possessed more than eight hundred fighters and bombers, which he reputedly referred to as his "angels." The fleet included forty-one state-of-the-art MiG-29s and seventy-five Mirage F-1s. In December, Iraqi aircraft had averaged 235 sorties a day; on January 17 Iraqi pilots flew 116 times.

But by the third day of combat the number was down to sixty sorties, and it dwindled steadily. Fifteen planes had been shot down or crashed in combat. (Seventeen allied planes were lost in combat the first week, a fifth of 1 percent of all combat missions flown.) The allies had achieved air superiority — defined as seizing sufficient control of the skies to attack without serious opposition — and were approaching air supremacy, the ability to fly virtually unopposed.

Some American pilots considered the Iraqis too frightened or too overwhelmed to challenge the invaders. Others thought they might concentrate against Proven Force, the small, less intimidating American host in Turkey. Chuck Horner believed Saddam wanted to husband his planes, ignite a ground war, inflict enough casualties to force the Americans to seek peace, and thus emerge with his air force intact.

Although allied fighter pilots longed for the glory of a good dogfight and air-to-air kills, most air power theorists recognized that the efficient way to destroy enemy planes was on the ground, as the Germans had done in World War II when they destroyed four thousand Soviet aircraft in a week. But Saddam's grounded aircraft appeared very well protected. Iraq had taken extraordinary measures to harden itself against attack, with fortified bunkers, redundant communications, and dozens of subterranean redoubts; some intelligence analysts assumed Saddam was preparing for nuclear war with Israel later in the decade. Among the most impressive strongholds were 594 aircraft shelters, many with steel-reinforced concrete roofs several feet thick, forty-ton doors, and water traps inside to foil incendiary attacks.

A few aircraft shelters were bombed early in the war because of suspicions that they harbored Scud missiles with chemical warheads. But as the first week drew to a close, the majority remained intact. Glosson and Deptula intended to attack them at an unspecified date later in the war after tending to more important targets.

In the Pentagon basement where Checkmate watched the war from afar, John Warden found himself returning again and again to the conundrum of the Iraqi air force. Warden had expected Saddam to request a cease-fire rather than risk the complete destruction of his regime. "What are these guys up to? Why aren't they flying?" he asked his staff. "What have they got planned?"

On January 21, Warden thought he had the answer to his own questions. The Iraqi planes represented a "fleet in being," he concluded, one that could be flushed ten or twelve days into the war, after the allies dropped their guard. Checkmate estimated that Iraq could launch three hundred planes simultaneously, half attacking Tel Aviv and the other half Riyadh. Such a counterstrike, although suicidal and of dubious military value, might provide Saddam with a psychological victory comparable to the Viet Cong's Tet offensive in 1968, a tactical failure that became a strategic victory because it undermined morale on the American home front. Similarly, Warden reasoned, Egypt had won an important political triumph by attacking Israeli forces with unexpected ferocity across the Suez Canal in 1973.

At 7 P.M. Riyadh time on the 21st, Warden called Dave Deptula. "This is the most important message I could pass to you in the war so far," he said gravely. "We've come to the conclusion that the Iraqis could launch an attack against Tel Aviv or Riyadh. It could be like the Egyptians crossing the Suez in 'seventy-three. You need to switch priorities to make the Iraqi air force your number one concern."

Deptula listened carefully. Some of Checkmate's brainstorms seemed silly, even addled, but Warden's passionate argument made sense and echoed similar concerns raised even before the war began. Deptula had been drafting the master attack plan for January 23 — two days in advance, as always — but after talking to Warden he scuttled the plan and started over. For ninety minutes he matched allied bombers to Iraqi airfields, neatly jotting the attack instructions in pencil. Twenty F-111s would hit aircraft shelters at Al Asad; twenty more would hit the H-2 shelters; sixteen F-15Es would strike Talil. Among other attackers and shelters targeted for the 23rd: Saudi Tornados and Navy A-6s to Mudaysis, Navy A-7s to Al Taqaddum, and forty Air Force F-16s to Al Jarrah.

As Deptula finished revising the attack plan, Glosson returned from "evening prayers," Schwarzkopf's regular seven o'clock meeting in the MODA basement. Deptula recounted Warden's call and explained the changes. "The Iraqis still have the ability to use these airplanes somehow," he said. "We need to take that away from them. Saddam may have something up his sleeve." Glosson concurred, thumping the wall map with his index finger to pinpoint the airfields now scheduled for attack.

Deptula carried the completed plan down the hall where other officers worked out radio call signs, ordnance loads, tanker routes, and a dozen other details. The finished air-tasking order, several hundred pages thick, was then dispatched by computer or courier to every squadron in the theater.

At midnight, Warden called again. Deptula told him the campaign against the aircraft shelters would begin on the early morning of January 23. "That's great, Dave!" he exclaimed. "That's just great!"

The destruction of the Iraqi air force began by trial and error. Air Force scientists in New Mexico had conducted tests that suggested a bomb landing near a hardened shelter would likely detonate with sufficient concussive force to buckle the floor and damage aircraft landing struts. But whether the thick roof could be penetrated without repeated blows remained uncertain. F-111 pilots, attacking a shelter at Ali al Salem Airfield on the first night of the war, had aimed for the steel doors and found the trajectory to be a shot requiring extraordinary precision.

When the attacks began on the 23rd, pilots discovered that a single 2000-pound bomb with a delay fuse would cut through the concrete roof if the angle of the weapon was perfectly perpendicular to the surface; otherwise the bomb would glance off, leaving only a charred divot. Ultimately, Air Force crews concluded that against many shelters they needed to drop two bombs at one-second intervals, the first to crack the roof, the second to penetrate the fissure and destroy everything inside.

Navy pilots from the Red Sea fleet who dropped Walleye bombs, which lacked the penetrating cases and fuses of Air Force I-2000s, found that their munitions bounced off the shelters or detonated without breaking through. Riley Mixson, the fleet commander, soon asked that his aircraft be excluded from shelter-busting sorties. Navy planes instead struck Iraqi jets parked in the open or other airfield facilities.

Stealth fighters also had problems with their munitions during the initial attack on Balaad Southeast, above Baghdad. On instructions from intelligence targeteers, the F-117 pilots carried GBU-10s, which could be dropped from a higher altitude than GBU-27s but also lacked some of the penetrating punch. More than a dozen planes dropped two bombs each. When pilots later looked at their gun camera videotapes, they saw tremendous explosions but no "smoodge," pilot slang for the extrusion of smoke indicating that a target had been gutted from the inside.

Planners in the Black Hole were furious at their intelligence counterparts. Deptula subsequently began specifying the ordnance to be carried on each F-117 sortie. The next night another wave struck Al Taqqadum — dubbed the Temple of Doom — with GBU-27s that sliced neatly through the shelters, blowing the doors off from the inside. Several nights later, again armed with GBU-27s, the stealth pilots attacked Balaad Southeast. The Iraqis, apparently concluding that the Balaad shelters were impregnable, had wheeled even more airplanes inside. This time the videotape recorded a horrific spectacle of flame, shattered concrete, and thick, coiling clouds of smoodge. Eventually 375 of the shelters would be destroyed or badly damaged.

Denied sanctuary, the Iraqis fled. More than a dozen Iraqi commercial airliners had flown to Iran at the beginning of the war. Now flocks of fighters followed them, seeking asylum on the soil of their former enemy. Within two days of the onset of the allied campaign against the shelters, more than twenty jets dashed east across the border. American pilots joked that Iraqi planes now bore bumper stickers that warned, "If you can read this, you're on your way to Iran."

Tehran assured the United States that the planes would remain grounded for the duration. Although skeptical, American officials were heartened by intelligence reports that Iran was stripping Iraqi markings from some aircraft and painting the wings and fuselages with the insignia of the Iranian air force.

## Washington, D.C.

Despite the low allied casualties, after a week of war the nation's enthusiasm had yielded to apprehension. The change in mood was a consequence of several factors: a native impatience; unrealistic expectations of quick victory; the emotional struggle to understand that war, so long in coming, had finally arrived. Richard Cheney had suggested publicly on January 21 that the war could last for months, a prediction that deepened the public's gloom.

To be sure, support for the cause remained very strong. Polls showed that four of every five Americans approved of the decision to attack Iraq. George Bush's approval rating stood at 84 percent, comparable with Franklin D. Roosevelt's after Pearl Harbor. Military recruiting offices reported a flood of inquiries from would-be warriors eager to enlist. An Arkansas entrepreneur peddled Iraqi flags, suitable for burning. Scattered antiwar demonstrations around the country — including 25,000 protestors who had gathered in Lafayette Park over the weekend to chant "shame, shame" at the White House — were countered with large prowar rallies in New York, Chicago, and elsewhere.

But the patriotism and bellicosity were tempered by anxiety, as though the country collectively had taken counsel of its fears. Half of those polled still expected at least five thousand American soldiers to die in combat. Only 20 percent now believed the fighting would be over within a few weeks, compared with 40 percent the day after the war began. Combat was confined to the Middle East, but this was a global war. Television and instant telecommunications allowed noncombatants around the world to experience air strikes and Tomahawk launches almost as they happened. That sense of immediacy had transfixed the nation. In the preceding week, tens of millions of Americans had become compulsive television viewers.

The war dominated church sermons, classroom lectures, and coffee shop chatter. Teachers and parents found themselves struggling to reassure children frightened at the prospect of terrorist attacks or bombs that could somehow fall at home rather than in the Persian Gulf. Signs of tight security became ubiquitous: bomb-sniffing dogs at airports, sharpshooters on the roofs of public buildings, policemen in bulletproof vests everywhere.

Public unease was reflected in the media and compounded by the press corps's own disgruntlement at home and in the war zone. Sixteen hundred journalists had massed in Saudi Arabia, roughly four times the number in Vietnam during the late 1960s. Unlike Vietnam, however, where reporters could roam unescorted into the field and file uncensored dispatches, in the gulf they were subject to controls similar to those imposed during the Korean War and World War II.

Such controls represented a legitimate attempt to manage the media throngs in Riyadh and Dhahran. Yet they also reflected a desire by the Bush administration and the military to shape the war's image in order to avoid giving aid and comfort to the enemy, maintain public support, and keep the allied coalition intact.

Under rules issued by the Pentagon on January 15, twelve categories of information — such as troop movements, tactics, and plans — were subject to censorship. All reporters were to be accompanied by military escorts during interviews. Coverage was organized by pools, small groups of reporters and photographers whose stories and pictures were then shared with all news organizations.

Reporters resented the escorts, who often intimidated those being interviewed and occasionally proved overzealous in trying to shape the coverage. Overt censorship never became a significant issue; only five dispatches among thirteen hundred filed during the war would be referred to the Pentagon for further arbitration, and only one was ultimately suppressed. But access to key individuals or those involved in sensitive combat roles, like special forces units or B-52 crews, was highly restricted. Moreover, the pool system would collapse once the ground war began, largely because of the Army's inability to get photographs, videotape, and written copy back to the clearinghouse in Dhahran quickly enough. Nearly 80 percent of the pool reports filed during the ground offensive took more than twelve hours to reach Dhahran, by which time the news was often stale or obsolete. One in ten reports took more than three days, far longer than the time needed for dispatches to reach New York from the battle of Bull Run in 1861.

For now, as the air war unfolded, reporters were mostly reliant on daily briefings in Riyadh and Washington, much as their journalistic predecessors had been on briefings in London before the invasion of Normandy in 1944. Although two or three such sessions were held daily, the information tended to be superficial and numbingly statistical, with arid updates on the numbers of sorties flown or Scuds fired. Few data on bombing missions were released, particularly regarding economic targets such as factories and oil refineries. Even less information was forthcoming on damage to civilian facilities struck inadvertently.

Thousands of feet of gun camera videotape of bombs missing their targets remained classified; only flawless missions displaying dead-on accuracy were released — with the audio recordings of cursing, hyperventilating pilots primly excised. Of 167 laser-guided bombs dropped during the first five nights of combat by F-117s, considered the most accurate aircraft system in the allied arsenal, seventy-six missed their targets because of pilot error, mechanical or electronic malfunctions, or poor weather. None of those was acknowledged by Riyadh or Washington.

If the information system constructed by the Pentagon had serious shortcomings, so too did the reportorial corps. Two decades had passed since the end of military conscription, and few journalists had had any personal military experience. Some of those pressed into service as combat correspondents knew little about weapons, tactics, or military culture. Their simplistic questions and fatuous news stories reinforced the military's stereotype of journalists as dilettantes innately hostile to those in uniform.

Even for those reporters who knew what they were about, the first week was exasperating. Hundreds of journalists in Riyadh and Dhahran saw little of the war beyond what they gleaned from television sets in their hotel rooms. Frustrated and restive, they began to question why, if the war was being prosecuted with near perfection, Iraq showed no sign of capitulating. This disaffection magnified the anxiety of a nation with half a million sons and daughters in harm's way.

Colin Powell had a more sophisticated view of the media than many Vietnam veterans who believed the press had lost that conflict by turning American opinion against both the war and the warriors. He did not view journalists — whom he privately called "newsies" — as enemies but as potential allies critical to maintaining public support. "Once you've got all the forces moving and everything's being taken care of by the commanders, turn your attention to television," he had advised younger officers in a speech at the National Defense University a year earlier. "Because you can win the battle [but] lose the war if you don't handle the story right."

Worried that impatience — among the public and in the White House — could upset Schwarzkopf's campaign timetable, Powell and Cheney appeared in the Pentagon E-Ring press briefing room at 2 P.M. on January 23 to describe allied progress one week into the war. Powell was determined to "handle the story right."

After brief comments by Cheney, the chairman took over. Using charts and maps propped on an easel, he acknowledged that unexpectedly bad weather over Iraq had scuttled many strike missions and hampered efforts to assess the bombing damage with reconnaissance cameras. Scuds remained a "vexing problem." The effect of air attacks on the entrenched and dispersed Iraqi army was uncertain.

"The Iraqi strategy, it seems to me, is [one of] hunkering down," Powell ventured. "They probably are questioning whether we can keep this up for an extended period of time and whether or not the political will and public support will be there to keep this up."

He wondered the same thing, although he kept the thought to himself. A secret Defense Intelligence Agency analysis suggested the war could indeed be long and bloody. Only 2 to 3 percent of Republican Guard armor — about twenty-five of eight hundred tanks — had been destroyed. Resupply of the occupying army continued with little disruption. Only half of the strategic targets bombed at the beginning of the war had been sufficiently damaged; they would have to be retargeted. Mobile Scud launchers, it was now evident, would never be completely eradicated.

Yet with patience would come certain victory. The air campaign, Powell believed, would probably last about three weeks. After that, barring an Iraqi collapse, a ground offensive would be necessary to prevent Baghdad from recapturing the initiative.

"We're in no hurry," Powell said. "We are not looking to have large numbers of casualties." Pointing on a map to the areas of Kuwait and southeastern Iraq where the Iraqis were entrenched, the chairman used a sentence he had carefully rehearsed. "Our strategy in going after this army is very, very simple: first, we're going to cut it off, and then we're going to kill it."

Powell had concluded that the line, although crude and blunt, conveyed the proper tone of confident determination. The Iraqi army, he continued, "is for the most part sitting there, dug in, waiting to be attacked. And attacked it will be. There is no question that this large force will become weaker every day. That's absolutely mathematical."

Pausing for a moment, Powell glanced up at the packed room of reporters and television cameras. Another aphorism on his desk declared: "You never know what you can get away with unless you try."

"Trust me," he now urged with the slightest smile. "Trust me."

# PART II

# MIDDLE MONTH

6

# Mesopotamia

## Baghdad

David Eberly heard cats mewing. The sound pleased him; cats would control the rats. A train of ants scurried across the filthy cell floor, but he had already learned to ignore those. He rolled up his absurdly long sleeves, then cuffed the trouser legs. The uniform, cut from yellow duck, was not as warm as his flight suit, which had been taken from him. The brown leather tag stitched to his breast pocket bore two letters: PW. Prisoner of war.

When he spotted a nail protruding from the wall, he worked it free and etched a vertical calendar in the plaster. His first sortie was on the morning the war started, January 17, a Thursday; on the night of the 19th he had been shot down; in the early morning of the 22nd he had blundered into the Iraqi border post. That meant today was Thursday, January 24, his first day in a Baghdad prison. He marked the date and carefully hid the nail, wondering how many times he would need it to scratch away another day. Hundreds of hatchmarks covered the wall, the disquieting runes of an earlier occupant.

For a few minutes after he and Griffith were captured on the 22nd, Eberly had clung to the hope that the Iraqis would let them go. Their captors at the border post had provided tea and food. "Thank you, *danke, gracias,*" Eberly replied, groping for the Arabic word.

When the interrogation began, he explained that the airplane had suffered an electrical malfunction and that the two of them had walked in the desert for three days. To mislead his captors, he told them he and Griffith were English pilots. The Iraqis took their pistols, the flares, the map, the radios; Eberly tried unsuccessfully to palm his radio battery so that the Iraqis could not use it. As they studied Griffith's blood chit — the promissory note — Eberly suggested that repatriating two able-bodied aviators could be worth a half-million-dollar reward. (Unlike RAF pilots, the Americans did not sew gold ingots in the lining of their flight suits.) The soldiers instead relieved him of his watch and a ten-dollar bill, then herded both men — cuffing and spitting at them — into the bed of a pickup truck. They lay with their hands lashed behind them as the truck jounced across a rugged track for twenty miles to another guard post, where they were questioned again, blindfolded, and eventually hauled to a military camp, where they spent the night handcuffed to a cot.

At noon the next day, the soldiers pushed them into the rear seat of a Toyota station wagon and drove them through a town in western Iraq. The small procession, led by a truck with a loudspeaker announcing the capture of two allied pilots, crept through the bustling marketplace, and a crowd began to swarm around the white Toyota. The men's blindfolds had been removed, but both remained handcuffed. Eberly briefly entertained the delusion that the throng was cheering them. Perhaps, he thought, the Iraqis had turned against Saddam, against the war. Maybe they would hide him if he could break free.

Sitting next to the unlocked right rear door, with Griffith and a guard on his left, he contemplated bolting from the car and plunging into the crowd. But as he studied the faces pressing ever closer, he saw that they were scowling and jeering, and he leaned forward to lock the door. At that instant a rock the size of a grapefruit smashed through the window behind his head, spraying glass across the back of the station wagon. Even the soldiers appeared frightened. The driver pressed the accelerator and sped through the town as the mob hurled stones and epithets at the fleeing car.

Later that afternoon — after several stops and the now-familiar ordeal of being pushed, questioned, and occasionally beaten by guards with black rubber tubes — they had reached Baghdad. The darkened city seemed deserted but for a few crazed motorists who careered through the streets, flashing their headlights and honking madly. Eberly was surprised that the city looked intact; he had imagined rubble and flames, something akin to Hamburg or Schweinfurt in 1945. The car threaded its way through a silent district of low buildings and stopped. He and Griffith were hauled from the rear seat and locked in an outdoor kennel, where they spent the night without food or blanket, huddled on a cold slab.

In the morning they were again shoved into the car, blindfolded this time. Twenty minutes later, the car stopped and guards led them through a doorway and down a steep flight of steps to an underground bunker. The facility, Eberly thought with alarm, might be a classic

F-117 target; he hoped the Iraqis would be sensible enough to get them out by sundown.

He heard the whirring of a camera and was cheered by the sound. He wanted his family and the U.S. government to know he was alive; a video record would make the Iraqis accountable for him. Surreptitiously he tugged at the zipper of his flight suit to display the bloodstained jersey underneath. A guard removed the blindfold.

"Your name and rank, please?" the interrogator asked.

"Colonel David William Eberly."

"What is your nationality, David?"

"From the U.S.A."

"What is the kind of your plane and which squadron?"

"F-15, from the Fourth Tactical Fighter Wing."

"What are your targets in Iraq in this mission?"

He hesitated briefly. Should he have disclosed his unit and the aircraft model? That seemed harmless enough. But what about the target? How much did the enemy already know about the mission? What would Griffith tell them? In theory he was obliged to reveal only his name, rank, service number, and date of birth. But perhaps, he reasoned, the Iraqis' propaganda could be turned against them; if nothing else, the Israelis should be reassured by American efforts to destroy the Scuds.

"Scud missile and associated chemical facilities."

"And how were you shot down?"

"I don't know."

The interview ended abruptly. Eberly was blindfolded again and led into a narrow corridor, where he sat for several hours. Already he had learned to make himself as unnoticeable as possible, pulling his legs up tight and trying to fade into the wall to avoid the kicks of passing guards. He also had begun to feign a shuffle, hunching over and slowing every movement to avoid antagonizing his captors.

When he needed food and drink, he asked timidly, and the Iraqis brought water and a meager bowl of rice. A guard led him to the toilet and for a few minutes removed the cuffs and blindfold. For the first time Eberly could examine his face and jaw — still swollen from the ejection injury — in a mirror. His ankles, apparently injured during the parachute landing and the long hike across the desert, had turned greenish-black. Worse were the hives now peppering his hands and body like a suit woven of red welts; during one stop en route to Baghdad, an Iraqi physician had swabbed the gash on the back of his head with a yellow liquid, probably penicillin, to which he was allergic. He carefully tore a pocket from his prison uniform and used one part as a handkerchief, the other as a washcloth to bathe his inflamed skin.

Shortly before dark he and Griffith were led from the bunker and driven to the prison. Trying to keep his bearings as they drove through the city, Eberly surmised that they had entered an army compound south of the capital. The Iraqis shoved him down a hallway and into a vacant six-by-six-foot cell next to Griffith's. The door gave a metallic screech as it swung shut. From the barred window he saw a dilapidated frame building in the fading light, the wood graying handsomely, like that of an old barn.

The night passed. In the morning, again blindfolded, he was hauled to the underground bunker for another interrogation, this time without the comfort of a video camera. He half-expected to be manacled to a dungeon wall, his back lashed with a bullwhip, but the inquisition was more annoying than brutal: a few kicks, a few whacks with a clipboard. He strained to hear the tread of a guard sneaking up behind him, which he now knew foretold a double swat to the head in an apparent effort to break his eardrums. The questions were surprisingly crude, either attempts to elicit information on subjects he genuinely knew nothing about — what are the American plans for a ground attack? — or unsophisticated demands for technical data that he found he could satisfy with information available in any library. Yes, he agreed, the Strike Eagle had two engines. No, the aircraft required a two-man crew.

Now, having been returned once more to the prison cell, he carved his calendar, then let his imagination conjure up a variety of shapes in the peeling plaster on the cell walls: Barbara's face; the silhouetted head of his Airedale, Ted; a procession of circus animals. Hunger gnawed. The Iraqis seemed to have forgotten his dinner. Without his flight suit he felt defenseless, stripped of his armor. The baggy yellow prison uniform was a constant reminder that he had failed, that he had gone toe-to-toe with the enemy and lost.

He pushed the thought from his mind. Such self-recriminations were futile, enervating. If his fate was to spend the war in prison, then he must be stalwart enough to survive. As he stood by the door, another prisoner shuffled past toward the toilet, a guard in tow. Eberly leaned toward the door and whispered, "I'm Colonel Eberly from the Air Force." The man said nothing, but paused long enough on his way back to answer softly, "I'm Lieutenant Zaun, U.S. Navy." "Hang in there," Eberly urged. "We're going to make it."

He lay on the dirty concrete as darkness slowly shuttered the prison and closed his eyes to prevent his mind from constantly groping for light. The bulkhead between sanity and madness, he had begun to realize, was thin and fragile.

Muted voices drifted down the cellblock. Eberly had no way of gauging how many aviators were imprisoned here, but he was probably the ranking officer. This was his command. The men must not lose hope or do anything foolish out of desperation. He struggled to his feet and groped toward the door, where he lifted his voice through the bars. "If you can hear me," he called, "let me remind you what we were told back in Saudi before the war started: there's nothing up here worth dying for."

His duty done for now, he again lay down in a fetal curl. His back pressed against the flaking plaster, he listened to the bawling cats and waited patiently for the freedom of sleep.

## Suqrah Bay, Oman

At 4 A.M. on January 26, wave upon wave of amphibious vehicles, each jammed with two dozen Marines, slid from the yawning bays of their mother ships and veered north toward the barren shingle several thousand yards distant. All but submerged in the dark Arabian Sea, the vehicles wallowed shoreward at nine knots. Harrier jets from U.S.S. *Tarawa* and U.S.S. *Nassau* roared overhead toward the coastline. In less than fifteen minutes, the vanguard of four battalions had breached the heavy surf and rolled across the sand. Behind them came jet-powered LCACs — landing craft, air-cushioned — skimming inches above the waves, and heavier landing craft loaded with tanks and artillery and armored personnel carriers. At first light, helicopters ferried additional battalions to reinforce the flanks of the four-mile beachhead.

Not since Inchon had such an amphibious force been hurled ashore. The operation, code-named Sea Soldier IV, embraced seventeen thousand Marines backed by eleven thousand sailors on thirty-four ships. Intended as a dress rehearsal for a landing in Kuwait, the operation was the fifth and largest landing on the Arabian peninsula since September. Three previous Sea Soldiers, like this version, had been launched on Oman's half-moon Suqrah Bay, one of the most desolate coasts in the world. The other exercise, Imminent Thunder, had attacked a Saudi beach in November. CENTCOM publicized the exercises to remind Baghdad of America's amphibious prowess. The message got through: Iraq had shifted several divisions and hundreds of heavy guns to reinforce Kuwaiti beaches.

Marine commanders watching their men crash ashore in Sea Soldier IV found themselves torn by ambivalence as they imagined the casualties that would result from a real beach assault under fire. After bobbing about in the Arabian Sea and Persian Gulf for months, the Marines longed to be part of the war. Marines had been storming beaches since March 1776, when two hundred men overpowered a British garrison to capture Fort Montague in the Bahamas; no Marine was oblivious of the Corps's exploits on Iwo Jima and Okinawa and a dozen other hallowed strands.

Amphibious operations demanded intricate choreography, uncommon valor, and luck. At Gallipoli in 1915, a British army plagued with misfortune and incompetent commanders impaled itself on the Turkish shore. The disaster gave rise to a conventional belief that in landings under fire the advantage invariably lay with the defender. After World War I, the Marines — motivated in part by interservice rivalry and a fear that the Corps would be reduced to a naval police force — diligently studied Gallipoli to develop new techniques, strategies, and equipment.

But landings early in World War II were sobering. At Dieppe in August 1942, six thousand commandos, mostly Canadians, waded onto the French coast, only to withdraw twelve hours later with 60 percent of the force dead, wounded, or captured. At Tarawa in the Gilbert Islands in November 1943, poor intelligence about local tides and coral reefs forced Marines to wade several hundred yards to shore under heavy fire. A thousand Marines died and more than two thousand were wounded in the assault thereafter known as Terrible Tarawa.

Although the Corps's subsequent successes elevated amphibious warfare to what has been called "the greatest tactical innovation of World War II," nuclear weapons seemed to render beach landings obsolete. "I am wondering," Omar Bradley said in 1949, "if we shall ever have another large-scale amphibious operation. Frankly, the atomic bomb, properly delivered, about precludes such a possibility." Even the brilliance of Inchon on the west coast of Korea in 1950, although it proved Bradley wrong at a cost of twenty-one American dead, came in the face of MacArthur's grim estimate that the odds against success were "five thousand to one."

An assault on Kuwait could be every bit as bloody as Terrible Tarawa. "I will tell you that if we proceed with this attack," the commander of the Marine amphibious force, Major General Harry W. Jenkins, Jr., had recently warned his superiors, "I intend to destroy everything in front of me and on the flanks to try to keep our casualties down." Jenkins fumed at the publicity given the Marine exercises, noting that every public mention of the American threat from the sea seemed to heighten Iraqi reinforcement. "This doesn't make a helluva lot of sense to me," he complained. "I'm in favor of informing the public and letting the media know as much as we can, but we're going to cut off our nose to spite our face."

Precisely how — if at all — to use the powerful host of Marines afloat had bedeviled American planners for months. Had Saddam carried his invasion into Saudi Arabia, Schwarzkopf could have landed Marines behind the enemy to cut them off, as MacArthur had done in Korea; but the Iraqis stopped at Kuwait's southern border. "If the Iraqis had come south, we could have gotten behind them and raised all kinds of hell," Jenkins told his staff. "Now the whole damned coast is fortified."

A gregarious, wry, thirty-one-year veteran who had commanded a Marine company in Vietnam during the long siege of Khe Sanh, Jenkins in September drafted a plan with ten options ranging from modest raids to full-blown amphibious landings. One plan proposed slipping two battalions onto the Al Faw peninsula at the northern end of the gulf to threaten Basrah as the Army attacked from the west. But Schwarzkopf feared that such an attack could inadvertently spill into Iran. Moreover, the two Marine divisions ashore had modified their proposed attack axis. Rather than drive north up the coast, Walt Boomer now wanted to strike fifty miles inland. He believed that an amphibious landing south of Kuwait City was critical, because it could provide the Marines ashore with a supply base and thwart an Iraqi counterattack to the west.

In tactics and equipment, amphibious combat had evolved considerably since the Pacific island campaign and the Korean War. To avoid enemy missiles, landing fleets had to disperse and zigzag constantly in "sea echelon areas" far from shore. Helicopters provided mobility and speed. The new LCAC craft — of which Jenkins had thirty-two — harnessed four jet engines, two to lift the ten-ton craft above the water and two to power a pair of huge propellers in the stern for thrust; the crew sat in a cabin that resembled an aircraft cockpit. And yet amphibious warfare in other respects had changed little: thousands of frightened, seasick men still wedged themselves into fragile landing craft to weave through gunfire and deadly obstacles toward a contested shore.

Perhaps no form of combat more affirmed the Clausewitz dictum that "three fourths of those things upon which action in war must be calculated are hidden more or less in the clouds of great uncertainty." Amphibious planners fretted over such arcana as hydrography, the ability of a beach to bear traffic, spring tides, neap tides, and harmonic analysis. Marines preferred to attack a convex coastline, such as a peninsula or promontory; concave coasts — in a bay, for example — permitted defenders to mass their fires on three sides. To pierce "the clouds of great uncertainty," planners drafted elaborate debarkation schedules, sea echelon plans, assault diagrams, and approach schedules replete with wave numbers, landing craft azimuths, and control officers to act as traffic cops under fire.

On January 16 Navy SEALs began reconnoitering Kuwaiti beaches at night, closing to within two hundred yards in rubber raiding boats before swimming the rest of the way with scuba gear. Their reports, supplemented by accounts from the Kuwaiti resistance, were not encouraging. At many sites the Iraqis possessed sweeping fields of fire along the waterline. The defenders had bricked up beachfront houses and apartment buildings to create a nexus of stout pillboxes, and had planted underwater mines, barbed wire, and steel dragon's teeth to rip the bottoms from landing craft. Also, poor hydrographic conditions characterized much of the coast. In some spots the slope was 1-to-1000, meaning only a foot of vertical depth for every thousand feet from the beach — the Persian Gulf equivalent of the reefs at Tarawa.

Amphibious planners looking for a suitable beachhead had finally settled on Ash Shuaybah, about twenty miles south of the entrance to Kuwait Bay. Here the hydrography ranged between 1-to-50 and 1-to-100, deep enough to bring the mother ships closer to shore and permit the landing craft to beach quickly — if combat engineers could demolish the obstacles.

But Ash Shuaybah was hardly ideal. Many high-rise buildings along the beach had been fortified. Equally troubling was a liquid natural gas plant south of the landing zone, abutting the port the Marines planned to seize after they had secured the beach. The gas storage tanks had enough latent explosive power, according to one estimate, to fling a two-pound chunk of shrapnel ten kilometers. Although Kuwaiti resistance fighters offered to drain the tanks before an assault, Navy and Marine planners began referring to the plant as "the nuke on the beach."

The gas complex and apartment buildings would have to be razed by naval gunfire and air strikes before any amphibious landing commenced, the Marines and Navy insisted. "I want battleship gunfire grid square by grid square, starting at the beachline," Jenkins declared. "I want it so thick that whoever survives will be in no mood to fight when the Marines get there." But Schwarzkopf, reluctant to inflict more damage on Kuwait than necessary, had thus far refused to authorize demolition of the plant or the buildings.

As Sea Soldier IV continued — with Suqrah Bay standing in for Ash Shuaybah — debate intensified over the wisdom of an amphibious landing, now code-named Desert Saber. On *Blue Ridge,* Stan Arthur and his staff worried about steaming into mined waters within range of Silkworm missiles. In Riyadh, Cal Waller argued adamantly against Desert Saber. "I couldn't live with myself if we made a decision to do an amphibious landing because the Marines want to do one," he told Schwarzkopf at one point. "There's nothing to be gained except a lot of Marines being killed while trying to get over those beaches."

Boomer was still convinced that Desert Saber was tactically vital; he assured Schwarzkopf, "I will never come to you and recommend an amphibious landing for any other reason than to help us win this campaign. I will never recommend an amphibious operation just because I think it will help the Marine Corps." In Washington the Marine commandant, General Alfred M. Gray, was believed to be the most ardent advocate. ("Al Gray just wants to build another Iwo Jima Memorial to all the Marines he's going to get killed," one of his fellow chiefs remarked bitterly.) Yet Gray had doubts. He had attended too many funerals after the 1983 bombing of the Marine barracks in Beirut to disregard the tragedy of more losses. "Do any of you know what it's like to lose an entire unit?" Gray had asked his senior generals during a session in Jubail. "Because you ought to think about it."

Still, Navy officers thought the Marines too casual about the threat of sea mines; they chafed at Boomer's frequent tinkering with his ground attack plan, which had a ripple effect on amphibious operation. The Marines in turn thought the Navy unprepared for the mine threat and ignorant of the value of amphibious operations. Both services regarded CENTCOM as unenlightened about the potency of their combined firepower in the gulf. Marine and sailor alike waited impatiently for Schwarzkopf to reveal his thoughts on the matter. For now, the CINC kept his counsel.

## Washington, D.C.

Iraqi Scud launches had neither stopped nor diminished; they had intensified. On January 25 ten missiles fell, most on Israel, where rescue dogs again sniffed through the rubble for survivors. Six more fell on the 26th. Schwarzkopf, demonstrating the nonchalance that infuriated both Israelis and Pentagon civilians, told an interviewer, "Saying that Scuds are a danger to a nation is like saying that lightning is a danger to a nation. I frankly would be more afraid of standing out in a lightning storm in southern Georgia than I would standing out in the streets of Riyadh when the Scuds are coming down."

On January 28, at the request of Moshe Arens — who caustically began referring to the Scud attacks as "more lightning storms" — Richard Cheney received three somber Israelis in his Pentagon office: David Ivri, the defense director general; General Ehud Barak, the deputy chief of staff; and Rear Admiral Avraham Ben-Shoshan, former head of the Israeli navy and now defense attaché in Washington. For the first time the Israelis revealed to Cheney, Powell, and Wolfowitz details of their plan for attacking the Iraqi launchers with a combination of air strikes and ground forces. Western Iraq was too vast for pilots alone to ferret out the missiles, Barak insisted; only soldiers could master the wadis and desert roads and subtle folds of the earth. Ben-Shoshan called it "reading Braille," a tactile sensing of the land and its many hiding places.

The Americans listened intently before Powell led Barak off for a more detailed discussion. Cheney repeated the familiar arguments against Israeli involvement, but he was intrigued by the idea of combining special forces. Perhaps, he suggested after the Israelis had gone, the Americans could develop a similar plan.

As Cheney soon learned, his commanders had been doing precisely that. At 1 P.M. on January 30, Wayne Downing, the JSOC commander, appeared in room 2C865, the Special Technical Operations Center. Here, the Pentagon's most secret "black" programs, revealed only to those with the highest security clearances, were run from an elaborate computer and communications complex known informally as Starship Enterprise. Downing quickly briefed Powell on the Delta Force proposal that he had sketched for Tom Kelly eight days earlier, including his plan to bivouac at Ar Ar and insert heavily armed patrols into the three Scud boxes. After thirty minutes, Powell said, "I want the secdef to hear this." Cheney joined them, and Downing repeated his pitch. "Just attacking the Scuds is like looking for a needle in a haystack," he said. "We need to attack their command and control and logistics, as well as the launch sites."

Powell and Cheney concurred. Although both men had agreed that no units would be assigned to Schwarzkopf against his will, the time had come for Washington to assert itself. "Our biggest concern is that the goddam Israelis are going to come into the war and the coalition is going to fall apart," the chairman told Downing. "We need to get over there and do something."

Downing offered three "force package" options, ranging in size from less than a squadron to two squadrons plus a Ranger battalion. As he expected, the secretary and chairman picked the middle course — Delta's 1st Squadron and pilots from Task Force 160, an Army helicopter unit, giving a combined force of four hundred men.

"How soon can you deploy?" Cheney asked.

"We're ready to go now," Downing replied.

"Then why," the secretary asked, "are you still sitting here?"

Downing excused himself and called Fort Bragg to alert his command before flying back to North Carolina. At 11 P.M., four C-141 and two C-5 transport planes lifted into the night and headed east.

Notwithstanding perceptions in Washington and Tel Aviv that Schwarzkopf was slow in suppressing the Scuds, the CINC had not been sitting on his hands. Allied aircraft now averaged more than a hundred counter-Scud sorties a day. B-52s flattened suspected sites at H-2 and H-3. U-2 and TR-1 reconnaissance planes crisscrossed western Iraq. Pioneer drones cut lazy circles above the eastern desert, scanning the roads and wadis with television cameras. Ten A-10 Warthogs flew to a makeshift base at Al Jouf and began patrolling in the west, their pilots peering through binoculars for any unusual movement. Buster Glosson assigned F-15Es from the 335th Tactical Fighter Squadron — with whom Eberly and Griffith had been flying — to permanent "scud-busting" duty at night using their radar and infrared pods.

Iraq, in contrast to its feeble efforts in other aspects of the war, proved tenacious in launching Scuds. Missile crews displayed ingenuity in the use of decoys, placing mock missiles among barrels of diesel fuel to simulate secondary explosions when hit, aluminum reflectors to emit confusing radar signatures, and heat generators to baffle infrared detectors. Occasionally the allied pilots duped themselves. One night in late January, an F-15E crew spotted a "hot" square in the infrared scope which looked like a camouflage net draped over a missile launcher. With flawless precision they dropped a 2000-pound bomb, only to discover that the suspected Scud was a flock of sheep.

Hundreds of allied intelligence analysts studied the missile firings for patterns. Saudi officials revealed that in the late 1980s they had financed a cartography project for the Iraqis. Unfortunately, despite a frantic search throughout the kingdom, the Saudis could find only four map sheets depicting just a small slice of eastern Iraq; the rest of the set had either been lost or given to Baghdad. Still, the map fragments were of help. Tiny triangular symbols, American analysts soon realized, represented precise survey points that Scud crews appeared to be using for some missile shots against Dhahran and other targets along the gulf coast.

Other analysts studied Iraqi radar in hopes of correlating electronic emissions with Scud firings. Missile crews needed weather data to adjust their trajectories, as well as assurances that no allied warplanes lurked in the launch area. Three types of radar — known to allied intelligence as End Tray, Ball Point, and the Chinese-made Fan Song — provided such information. Although U.S. intelligence concluded that Fan Song emissions in particular seemed to foretell a Scud shot, converting that deduction into a successful counterattack proved difficult. The radar usually emitted for only a few minutes, allowing the allies little time to pinpoint the location. Moreover, radar and mobile launcher might be miles apart. Efforts to intercept Iraqi radio broadcasts — with eavesdropping aircraft, satellites, or the listening posts strung from Tobuk to the gulf — were frustrated by Iraq's vast network of coaxial and fiber-optic land lines. Hard to cut and harder to tap, the lines kept radio traffic to a minimum.

Perhaps the most intriguing Scud SIGINT — signals intelligence — was deciphered by a tall, skeletal warrant officer from the XVIII Airborne Corps named James E. Roberts, Jr. Roberts was classified in the Army as a 98 Charlie, a signals intelligence analyst. He had won a Bronze Star in Panama for parachuting with a Ranger battalion onto Rio Hato with a satellite radio lashed to his back. Three days after the first Scud fell in the gulf war, Roberts believed he had cracked an Iraqi code that would enable him to predict the location of most launches. By manipulating certain intercepted radio messages and applying an arcane formula to a jumbled sequence of letters and numbers, he put together what appeared to be an eight-digit grid coordinate.

Most other intelligence analysts thought Roberts was daft: his formula seemed incomprehensible, more guesswork than ingenuity. Officials at the National Security Agency, the nation's mammoth eavesdropping bureaucracy, derided his "crystal ball." Even so, enough senior officers — including Schwarzkopf's intelligence chief — considered the predictions plausible, and they positioned aircraft near the anticipated launch sites. Sometimes the Iraqis fired; sometimes they did not. Only thrice could Roberts divine a precise time for the launches — a shortcoming that hamstrung efforts to ambush the Scud crews — and his prognostications never proved accurate nor timely enough to allow the allies to cripple the shooters.

But he did have a triumph on January 26, when he predicted that a missile would be launched toward Riyadh at 10:50 P.M. Sitting in the

Army's headquarters building outside the Saudi capital, Roberts anxiously watched the screen of a computer that was tied into the missile early-warning network. At 10:48 P.M. a message flashed on the screen: "Scud launch." Roberts leaped to his feet. "I told you!" he crowed. "I fuckin' told you!" and he danced across the room in a joyful splay of arms and legs, for a moment at least the happiest man in Saudi Arabia. After the war, the Army pinned another Bronze Star on Roberts for "meritorious achievement."

On the morning of February 1, Wayne Downing and Delta Force arrived at King Fahd International Airport. While his troops flew west to their new base at Ar Ar, Downing headed for Riyadh to see Schwarzkopf, whom he had known for many years. The men had served together at Fort Lewis, Washington, where Downing commanded a Ranger battalion and Schwarzkopf a brigade in the 9th Infantry Division. During a war game in Alaska in the mid-1970s, Rangers had attacked Colonel Schwarzkopf's command post one night when the colonel was sitting in the latrine. The CINC had forgiven that episode, but he was hardly overjoyed to see Downing now. "The Scud is a pissant weapon that isn't doing a goddam thing. It's insignificant," he grumbled. "I also want you to be sure you know who you're working for, and it isn't General Stiner. You work for me, you son of a bitch." Schwarzkopf was still wary of Carl Stiner's efforts to infiltrate the theater.

"Another thing," Schwarzkopf said. "If you personally go into Iraq, I'm going to relieve you."

"You don't have to tell me that," Downing protested.

"I know you," the CINC said, one eye pinching into a squint. "I don't want you going across that border and getting yourself captured or killed. One, because of the embarrassment, and two, because you know too much."

Before leaving Riyadh, Downing also visited the British special forces commander. Colonel Andy Massey immediately flung open the curtain shrouding his map of western Iraq and for an hour regaled his Yankee comrade with the travails of the 22nd SAS Regiment. The British unit, Massey explained, had been operating from Al Jouf for about two weeks. Lacking long-range helicopters with sophisticated night navigation equipment, the 250 commandos concentrated on inserting their patrols into the southern Scud box along the Amman-Baghdad highway. Some teams traveled through Iraq on foot; others used Land-Rovers mounted with Milan antiarmor missiles; others used motorcycles.

The Bedouin were a ubiquitous nuisance, Massey warned. If a Bedouin spotted a commando, the odds were one in two that he would alert the Iraqi army. Wretched weather in the high desert also was dangerous. Two men had frozen to death, one after swimming an icy river, the other after hiding from an enemy patrol in a watery ditch. Unprepared for the frigid nights, SAS teams had even been forced to kindle small fires beneath their Land-Rovers to thaw frozen fuel lines. (By war's end, the SAS would suffer four dead and five captured.) Eight men in a patrol currently in Iraq had been scattered; Massey was not certain of their whereabouts. Despite formidable Iraqi resistance, the British believed they had begun to push the Scud crews into the two northern boxes. There, Massey suggested, was where Delta could accomplish the most. The SAS, he added, was now calling that region Mesopotamia, in ancient times the name for the land between the Tigris and Euphrates.

Downing liked the name and he took it with him to Ar Ar, along with Massey's advice. Soon Delta and Task Force 160 — known as the Joint Special Operations Task Force — began infiltrating deep into Mesopotamia, focusing on several hundred square miles around H-2 and H-3 code-named Area of Operation Eagle. A typical patrol consisted of twenty to forty commandos sent out for ten to fifteen days at a time. Working closely with Buster Glosson, Downing synchronized the missions so that infiltrations and exfiltrations occurred at night while bombing raids distracted the Iraqis. Then a flock of Pave Lows, Blackhawks, and vehicle-carrying Chinooks flew deep behind the lines from Ar Ar, dropped off the patrols, and returned before dawn.

Although Blackhawk gunships occasionally struck enemy targets, the commandos' principal mission was reconnaissance. Hiding on ridgelines and in wadis, moving only at night, they watched for military traffic and, with their radios, summoned air strikes. As Downing had once encouraged his men to think like bank robbers, now he urged them to think like Scud crews. "If you were driving a truck from Al Qaim to Shab al Hiri," he would ask, maps spread at his feet, "what route would you most likely take?" But surveillance required a patience and restraint that taxed even the most dedicated soldier. "Sir, I'm not going back out there unless you tell me I can kill the next Iraqi I see," one sergeant complained to Downing. "I can't stand to see them and not do anything about it. I just want to shoot these people up." After a few days' rest, the soldier sheepishly asked to join the next patrol, promising to keep his pugnacity in check.

As the British had warned, Bedouin roamed the western desert. The Americans soon learned to gauge whether the sparse vegetation in a particular area would support grazing, a certain symptom that herdsmen and their flocks were nearby. The closer the commandos edged to Al Qaim, which lay on the Euphrates, the lusher the grazing, the more common the Bedouin, the stouter the defenses, and the richer the enemy targets.

In contrast to the passivity of their besieged comrades in the east, Iraqi soldiers in Mesopotamia remained bellicose and tenacious. Several times they pursued the Delta patrols for miles across the desert. Once, after a discovered commando was badly wounded in the neck and leg, an armored column chased the patrol for fifty miles before the Americans succeeded in covering their trail, doubling back, and losing the Iraqis in the dark. In early February, nine Iraqi armored vehicles drew so close to a fleeing patrol that an F-15E overhead could not distinguish friend from foe. The pilot courageously turned on his lights and swooped in low to scatter the pursuers while helicopters circled north to rescue four Delta troopers stranded on foot near Al Qaim. Two nights later, five Iraqi helicopters chased another compromised team, which fled, unwittingly, toward two hundred enemy soldiers waiting in ambush. Again an F-15E dropped to two thousand feet, destroyed one helicopter with a laser-guided bomb, and caused enough confusion for the patrol to escape.

Schwarzkopf kept close rein on the operation, personally approving every mission sent across the border. After Downing's repeated pleas, the CINC permitted another Delta squadron and a Ranger company at Ar Ar, increasing the force to eight hundred. (At peak, two hundred American commandos prowled through Mesopotamia.) Although combat casualties were very light, seven men died in one of the worst accidents of the war when a TF-160 rescue helicopter, returning with a soldier who had injured his back in a fall from a cliff, crashed in a snowstorm less than two miles from Ar Ar.

Few direct Scud kills could be confirmed — and CIA analysts still refused to count any mobile launchers as destroyed. But the harassment campaign clearly confounded the missile crews. Again with Glosson's cooperation, Delta and F-15Es sowed hundreds of Gator mines on roads, overpasses, and suspected hide sites to channel the Iraqis into areas more easily watched and saturated with bombs. Scuds continued to fall, but less frequently. The daily average of five missiles during the initial ten days of the war dwindled to one a day for the balance of the conflict. By late February, U.S. intelligence would conclude that the mobile launchers had been pushed into a narrow corridor only ten miles in diameter on the outskirts of Al Qaim. Downing and Andy Massey, conferring by phone several times a day, congratulated each other for "establishing Anglo-American dominion over western Mesopotamia."

*

Although higher authority had foisted on him Delta's private war in the west, Schwarzkopf remained undisputed master of the other special operations units in Saudi Arabia, and he rejected nearly all requests to let them steal across the border and join the fray. Nine thousand Navy SEALs, Army Special Forces soldiers, Air Force gunship crews, and other unconventional troops took part in the war. They provided services ranging from search-and-rescue missions to marksmanship training for Kuwaiti refugees. Scores of Green Beret advisory teams, in an undertaking reminiscent of their celebrated role in Vietnam, joined Syrian, Egyptian, Saudi, and other allied forces to provide instruction and to present Schwarzkopf with an appraisal of how the Arabs were likely to fight.

But the CINC and his special operations commander, Colonel Jesse Johnson, stoutly resisted most "direct action" suggestions to insert men behind the lines. Of more than sixty direct action proposals advanced by the Army's 5th Special Forces Group, for example, all but a handful were turned down. To sever a fiber-optic line connecting Baghdad with Basrah, for example, Special Forces commanders suggested slipping through one of the manhole covers that provided access to the cable. The proposal was rejected, as was a plan to blow up a microwave tower southwest of Baghdad. When U.S. intelligence concluded that Iraq was possibly shipping Scud missiles by rail from a military plant at Taji to H-2 and H-3, a Special Forces team — with the CINC's blessing — practiced ambushing trains and tearing up track in Saudi Arabia. But when further intelligence showed that the purported Scud train kept an erratic schedule and occasionally carried civilians, the plan was scuttled.

Though his caution was immensely frustrating to those chafing for action, Schwarzkopf had sound reasons for resisting most of the special operations proposals. He was reluctant to upset the timing of his ground campaign or suggest to the Iraqis that an allied attack would come from any direction other than through Kuwait; he wanted forces available for deep reconnaissance and other unconventional missions as the ground war drew closer; and he recognized that precision-guided bombs and allied air supremacy obviated the need for many traditional commando operations, such as blowing up bridges.

Several approved schemes failed. In early February, three Special Forces A-Teams unsuccessfully tried to find mobile Scud launchers near the Iraqi town of As Salman. Two of the teams were compromised and extracted almost immediately; the third pulled out after thirty futile hours behind the lines.

In another mission, intended to smuggle secure radios to the Kuwaiti resistance, SEALs provided ten Kuwaiti commandos with wet suits, ferried them in rubber boats along the coast south of Kuwait City, and managed to get them to the water's edge undetected. But on spotting Iraqi guards on the beach, the Kuwaitis balked and the mission collapsed.

On yet another direct action operation a week into the war, three American Green Berets — a medic, an engineer, and a radio expert — joined a commando team from the British Special Boat Service (SBS) and flew by Chinook helicopter to a road along the upper shore of Bahr al Milh, the large, shallow lake forty miles southwest of Baghdad. There they dug a six-foot trench in search of a fiber-optic cable that reportedly connected Baghdad and Karbala.

Finding no cable, they dug another trench perpendicular to the road and uncovered five lines — none of them fiber optic. The raiders nevertheless chopped a section from each cable, crammed several hundred pounds of explosives into the trench with a delay fuse, and flew back to Saudi Arabia, where they presented Schwarzkopf with an Iraqi road marker as a souvenir. If nothing else, the mission reinforced American convictions that the British were permitted a great deal more latitude in their enterprises — and were having a great deal more fun.

## Andover, Massachusetts

The hunt for Scud launchers unfolded as the United States and Israel faced another crisis, which triggered something close to panic: the stockpiles of Patriot missiles were dwindling. Although Tom Kelly had assured the Pentagon press corps on January 22 that "we don't have any great concerns" about Patriot stocks, a hard mathematical reality had since eroded that confidence. At the time of Kelly's statement, Patriot crews possessed 499 PAC-2s, the variant designed to shoot down ballistic missiles. Raytheon Company and its subcontractors, having surged to three shifts a day, could build about four new PAC-2s every twenty-four hours. But the twenty-seven Patriot batteries in Saudi Arabia and Israel were shooting up to ten missiles at each incoming Scud. At that rate, the U.S. Army calculated, the supply could be exhausted around the third week of February, just as the ground war was set to begin.

Raytheon engineers and Army missile experts now believed they understood the curious behavior that had first been observed in the earliest Scud attacks, including the tendency of incoming missiles to multiply in flight. To extend the range of their Soviet-made Scuds, Iraqi technicians had lightened the payload and lengthened the motor section, apparently by welding additional fuel tanks to the missile body. This had been accomplished with consummate ineptitude. The aerodynamic stress of a fall through the atmosphere at four thousand miles an hour caused the welds to break, shattering each Scud into several pieces. By dint of shoddy workmanship, the Iraqis had created a warhead with a low radar signature and an accompanying flock of decoys — not unlike the sophisticated re-entry vehicles on Soviet and American intercontinental ballistic missiles.

A Patriot crew was supposed to fire two missiles at each Scud, a redundancy intended to improve the probability of destruction. But the Patriot radars, seeing the swarm of incoming objects at Mach six, automatically launched two Patriots at each chunk of warhead, fuel tank, and motor. American missilemen had only the murkiest idea what happened after that, since the Patriots lacked on-board data recorders to permit a detailed reconstruction of each engagement. On the basis of flimsy evidence, the Army would claim a kill if a Patriot merely happened to detonate in the vicinity of a Scud. Some missile experts — particularly in Israel — had begun to doubt the Patriot's efficacy. No one, however, was prepared to challenge the public perception that the weapon was nearly flawless in parrying the Scuds.

Further complicating matters was the unnerving propensity of the Patriot to fire at false electronic targets. Patriots had been designed for deployment in a relatively isolated location, not — as was the case particularly in Israel — in the middle of a city or an airfield surrounded by electronic distractions. Pilots testing their electronic countermeasures, for example, unwittingly bombarded the missile battery with stray electrons, apparently prompting false launches. On January 25, when Iraq flung seven missiles at Israel, the combination of disintegrating Scuds and electronic clutter initiated the launching of thirty-one Patriots, some of which reportedly bored into the ground.

Searching for solutions, the Army vice chief of staff, General Gordon R. Sullivan, flew from Washington to Andover, Massachusetts, on January 26 to meet with Raytheon executives. Sullivan — who would become the Army chief after the war — was easy to underestimate. Bald, laconic, unprepossessing, he rarely raised his voice. Yet he could be tough and demanding; subordinates never left the vice chief's presence wondering who was in charge. "Hope," Sullivan often warned, "is not a method." Told of the alarming ratio of Patriots to Scuds, Sullivan responded with the terse command "Make the rate go down."

Several possible solutions were examined during the meeting at Raytheon and subsequent brainstorming sessions. New software could be developed to help the Patriot ignore the electronic clutter and distinguish warhead from debris. Because the disintegrating motor and fuel tanks tended to trail behind the more streamlined warhead, missile crews also could manually restrain the Patriot from shooting at those fragments. But prospects were not bright for accelerating PAC-2 production; even if more missile bodies could be built, manufacture of the intricate fuses could not be speeded up. The Joint Chiefs, who personally decided whether new PAC-2s went to Israel or Saudi Arabia, were to expect no surge in supply.

Touring the Patriot plant, seeing rows of women in smocks diligently threading wires through electronic breadboards, Sullivan was reminded of photographs depicting World War II defense plants, where Rosie the Riveter and millions like her toiled for the common good. "You can probably understand me because I come from Quincy," Sullivan said, exaggerating his New England accent in an impromptu speech to a thousand workers. "In the Civil War, William Tecumseh Sherman wrote a letter to Ulysses Grant. 'I knew wherever I was that you thought of me,' Sherman told him, 'and if I got in a tight place you would come.' That's the nature of the United States. That's what you're all about. I thank you for what you're doing for our country."

The workers cheered wildly before returning to their breadboards, eager to help a friend in a tight place.

On the return flight to Washington the vice chief stopped in Delaware to inspect the U.S. military mortuary at Dover Air Force Base. Construction crews swarmed over two new hangars being erected to shelter the legions of dead expected from the gulf once the ground war began. Here the fallen would return to American soil, like so many before them. More than twenty thousand corpses from Vietnam had passed through this charnel house, as had the 241 Marines murdered in their Beirut barracks in 1983 and the 248 paratroopers killed in a plane crash in Gander, Newfoundland, in 1985. Cold and drear, Dover was thick with ghosts, with grief, with the anguish of countless final homecomings.

Sullivan listened as an officer explained in detail where the bodies would be hosed clean, where their teeth would be X-rayed for identification, where the gunmetal-gray caskets would be stored to await shipment to the soldiers' hometowns. All would be done with assembly-line efficiency. Like the making of PAC-2s, body bag production had been accelerated, and there was a stockpile of more than sixteen thousand "human remains pouches" — as the Pentagon insisted on calling them. Each sack was seven feet ten inches long and thirty-eight inches wide, with six stout handles and a heavy zipper running from head to toe.

Dover's goal was to identify every corpse, to have no unknown soldiers from the Persian Gulf War. The mortuary staff had quadrupled in size, and the building expansion was nearly complete. But with so many bodies anticipated, the mortuary doubted it could provide a proper honor guard to receive each fallen soldier with a dignified ceremony. Thus, the Pentagon on January 21 had declared Dover off limits to the media, which would inevitably want to shoot demoralizing television footage of America's heroes being unloaded with a forklift.

After two hours Sullivan boarded his plane and flew back to Washington, somber yet satisfied that this tragic and necessary task was properly organized. Only once had he balked, when the briefing officer asked whether he wanted to view the body of a naval aviator who had been killed in an accident. The vice chief shook his head. There was no reason to intrude. He had seen enough.

## Al Ahmadi, Kuwait

U.S. intelligence had no doubt that Iraq possessed both means and will to impose a devastating scorched earth policy on Kuwait. The emirate's oil industry was particularly vulnerable. Shortly after the August invasion, an estimated three dozen Iraqi engineers and a thousand troops began packing most of Kuwait's 1080 oil wellheads with C-4 plastic explosives. The soldiers also dug extensive trenches to shield the web of detonation wires from allied bombs. Sabotage had been carried out on oil fields in other wars, such as the fires set in the Baku wells during the first Russian revolution in 1905. But Kuwait's oil patch offered an unprecedented target. The Burqan field alone — second largest in the world — contained fifty-five billion barrels, twice the total of all U.S. reserves compressed into an area smaller than Rhode Island.

Six wells reportedly blew up in December near Al Ahmadi, south of Kuwait City, apparently as part of Iraqi experiments on how best to configure and trigger the charges. More oil facilities blew up after the war began, with smoky fires reported in the Al Wafrah field on January 21. Blazing storage tanks at Ash Shuaybah and Mina Abdullah were detected a day later.

Yet those incidents paled when compared with the sabotage first detected by reconnaissance cameras in the early hours of January 25. Several oil tankers berthed at Al Ahmadi since October — holding a combined cargo of more than a million barrels — now rode high in the water. U.S. intelligence concluded that much of their crude had been pumped into the Persian Gulf. Worse, Iraqi engineers had opened the pipeline leading from the Ahmadi Crude Oil Storage Terminal to the Sea Island Terminal, a loading buoy eight miles offshore used to fill deep-draft supertankers. A river of oil swept through the underwater pipeline to form an immense, oblong slick in the northern gulf.

American analysts never determined with certainty what the Iraqi saboteurs had in mind. Baghdad's presumed intent was to disrupt any amphibious operations and perhaps clog the Saudi desalination plants that provided much of the kingdom's fresh water. George Bush angrily denounced the environmental terrorism as "kind of sick" and symptomatic of Saddam's desperation. Televised images of oil-fouled cormorants provided Washington with a public relations boon and yet another opportunity to condemn Baghdad's moral bankruptcy.

The Pentagon spokesman Pete Williams predicted that the oil slick was "likely to be more than a dozen times bigger" than the 262,000 barrels dumped into Alaska's Prince William Sound by a reckless Exxon tanker in March 1989. Even that estimate proved conservative. The spill ultimately was calculated at double or even triple the size of the record 3.3 million barrels dumped into the Gulf of Mexico during the Ixtoc I disaster of 1979. (At least some of the eleven million barrels that ultimately spewed into the gulf came from other sources, including the tanker attacked by Navy jets, leakage from Iraq's offshore terminal at Mina al Bakr, and run-off from wells subsequently sabotaged ashore.)

For nearly twenty-four hours, military planners in Riyadh and Washington debated how best to stanch the flow from Sea Island. Intelligence analysts queried petroleum experts in the United States, as well as exiled Kuwaitis in Taif. About ten hours after the spill was detected, an engineer from Kuwait's oil industry arrived in a white Mercedes at the MODA building in Riyadh, blueprints of the Ahmadi complex in hand. By Friday evening, January 25, a two-pronged attack had been planned. The first step involved incinerating some of the pollutants by igniting the fountain of crude bubbling up at the Sea Island Terminal. This was accomplished that night, inadvertently, during a Navy gunfight with an Iraqi patrol boat suspected of laying mines.

The second and more difficult attack would target two inland manifolds, shedlike structures housing the valves that controlled pipeline pressure. Schwarzkopf, ever mindful of the danger of appearing wantonly destructive, paraphrased an infamous aphorism from Vietnam in reminding his lieutenants that "we are not in the business of destroying Kuwait while we are liberating Kuwait."

As CENTCOM contemplated the proper response, Buster Glosson was ready to act. He had called John Warden in Checkmate and the Tactical Air Command at Langley, Virginia. "What would be the impact of exploding a bomb on those manifolds?" Glosson asked. Within hours, Checkmate and TAC both assured him that the attack would pinch off the leak. Glosson then ordered five F-111s carrying precision-guided GBU-15 bombs to prepare for a pinpoint strike that was to avoid adjacent storage tanks and the nearby town of Al Ahmadi.

But as the bombers taxied onto the runway at Taif Friday night, Glosson took a call from Colonel B. B. Bell, Schwarzkopf's aide. The Kuwaiti government-in-exile had complained to Washington that the attack could demolish a good portion of the oil field. "The CINC is concerned that bombing those things will make it worse," Bell said.

"I've already checked that out," Glosson replied.

"People are telling the CINC —"

Glosson cut Bell off. "They're telling the CINC bullshit. I've already checked it out."

Bell persisted. "The CINC said don't hit the target until he personally approves it."

Glosson shrugged and hung up, then ordered the F-111s to stand down pending further orders. "We're letting oil flow into the Persian Gulf for no reason," he complained to Horner.

A thick overcast rolled over the Kuwaiti coast, causing additional delays, because GBU-15s required clear skies to work effectively. For nearly twenty-four hours the pilots waited, refining their attack plan as additional advice poured in from Riyadh and oil poured out of Al Ahmadi. "Schwarzkopf himself called," the wing commander, Colonel Tom Lennon, told the crews. "He said, 'Don't fuck this one up.' " At last the weather cleared, the Kuwaitis were sufficiently reassured, and the attack was rescheduled for 10:30 P.M. Saturday. The planes took off at dusk and headed east. One aircraft turned back after developing mechanical trouble. Shortly after 10 P.M., the remaining quartet crossed the Saudi coastline and veered north over the gulf.

The two manifolds lay five miles inland and three miles apart. Because of dense Iraqi air defenses along the coast, the plan of attack was unusually complex. Aircraft number three and number four would drop the bombs, which would then be guided by aircraft one and two. Each GBU-15 carried a small television camera in the nose, permitting a weapons officer to steer the falling bomb with radio commands relayed from a joystick in the cockpit.

As planned, number three darted toward the coast at fifteen thousand feet, released its weapon eight miles from the target, then turned south to evade the antiaircraft fire. The weapons officer in number one, flying parallel to the beach sixty-five miles away, began steering the bomb. But only a few seconds into the weapon's two-minute flight, the electronic data link to the GBU-15 abruptly broke. The officer glumly radioed the code word that indicated he no longer had control of the plummeting bomb: "Goalie, goalie, goalie."

"I've got it! I've got it!" replied Captain Bradley A. Seipel, the weapons officer in aircraft number two. Seipel and his pilot, Captain Mike Russell, were also flying parallel to the coast, several miles behind the lead plane. "I think I've got the pump house," Seipel called as he stared at the blurry image on the cockpit television screen. The picture began to drift, an indication that Seipel's data link also was beginning to fray. "Come on, you piece of shit!" he urged.

"Sixty seconds to go," Russell called.

"Come on! Come on!" said Seipel. The drifting stopped momentarily, began again, and then the picture held steady. "It's intermittent, but I've got it."

"Only thirty seconds to go."

"Okay, I've got the target. A little to the right. Come on!"

The manifold appeared to rush forward on Seipel's screen before vanishing in a spray of white as the bomb detonated. Russell banked the plane sharply for another pass parallel to the coastline.

Two minutes later number four made its run to the coast and released the second GBU amid a barrage of 100mm antiaircraft fire. Seipel took control of this bomb, too. "I've got the general target area. I've got the tank farm."

"One minute, Brad," said Russell.

"Is that the building? Yep, I've got the target. Good picture. Good data link control. I've got it."

"I love it. Forty seconds to go."

"I'm going terminal."

"Twenty seconds."

"Coming on in."

"Five seconds to impact."

"Looking good. Looking sweet. Come on, baby!" Seipel let out a jubilant yell. Again the screen went white as the second manifold blew up.

The effect of this attack would remain in dispute long after the war. Oil continued to seep from the thirteen-mile length of pipe that connected the manifolds and Sea Island Terminal, adding to a slick already thirty-five miles long and ten miles wide. Some analysts believed that the American bombs were less important in stanching the flow than were several Kuwait Oil Company workers, who reportedly took advantage of the confusion after the air attack to slip into the tank farm and close the valves manually.

Nevertheless, on Sunday night, twenty-two hours after the raid, Schwarzkopf held a press conference to announce that the F-111 attacks had reduced the flow "to a trickle." If it was intended to thwart an amphibious landing, the CINC declared, the Iraqi tactic would fail. Further enemy sabotage, he acknowledged, was "entirely conceivable . . . [But] I certainly don't accept the fact that we should not fight a despot [because] we might for some reason pollute a shore. And I don't take that lightly, believe me. I'm a lover of the outdoors. I'm a lover of the environment. I'm a conservationist.

"Unfortunately," the CINC observed, "war is not a clean business."

7

# Khafji

## Washington, D.C.

At 9 P.M. on January 29 an expectant buzz filled the House of Representatives, where Washington's highest and mightiest had gathered for one of the republic's oldest rituals: the State of the Union address. Senators and representatives, cabinet secretaries and military chiefs mingled on the crimson and cobalt blue carpet of the House well, swapping gossip and banter beneath the packed gallery.

Not for a generation, not since Richard M. Nixon in the sad, waning months of Vietnam, had a president delivered the State of the Union when his nation was at war. Not for two generations, not since Franklin D. Roosevelt, had a commander-in-chief faced a Congress and a country more buoyed by good tidings from the front or more unified by a common martial purpose. For the moment at least, *pro patria* held sway, a thousand intractable ills receded, and the glory of the American century revived in a starburst of optimism, pride, and can-do confidence — the very signatures of the national character. However fleeting, the hour was golden, and the golden hour belonged to George Herbert Walker Bush.

"Mr. Speaker!" the doorkeeper's voice rang above the din. "The president of the United States!"

He fairly bounced into the chamber, too tall and angular for elegance but not without a statesman's carriage: the head cocked back in a confident tilt, the patrician nose, the thin lips drawn into the familiar, lopsided grin. Down the center aisle he moved, propelled by applause, murmuring greetings right and left, pumping hands, pointing at those whom he could not touch.

None suspected — and he was too circumspect to reveal himself on this wide stage — that the president was thoroughly irritated. Moments before leaving the White House for the short ride down Pennsylvania

Avenue to the Capitol, Bush had learned of a diplomatic gaffe at the State Department that threatened to steal his thunder. James Baker and the new Soviet foreign minister, Alexander Bessmertnykh, had concluded three days of talks by drafting a joint communiqué on the gulf war intended to quell speculation that Moscow was straying from the coalition. In private, Bessmertnykh had expressed dismay at the growing Iraqi casualties and urged the Americans to end the war quickly. But Baker considered the public statement drafted by the two men to be so innocuous — a rehash of a pronouncement issued in Helsinki the previous September — that he failed to clear it with the White House; the secretary instead repaired to Blair House for a sandwich before Bush's speech. Bessmertnykh, however, eager to demonstrate that Moscow was still an influential player in the Middle East, emerged from the State Department's C Street entrance and immediately read the communiqué — first in Russian, then in English — to a battery of waiting television cameras.

The gulf war could end, the statement decreed, "if Iraq would make an unequivocal commitment" to vacate Kuwait and take "immediate, concrete steps" to comply with various United Nations resolutions. The two-page document called for redoubled efforts to solve broader Middle East problems and bring "a real reconciliation for Israel, Arab states, and Palestinians." Notwithstanding Baker's conviction that the announcement contained little that was newsworthy, reporters detected a shift in tone from Bush's adamant declaration, as recently as January 23, that there could be "no pause now that Saddam has forced the world into war." Moreover, after insisting for months that broader Middle East issues would not be linked to the gulf war, the superpowers addressed both in the same document. Bessmertnykh finished reading, shoved the papers into his suit pocket, and rode away in an embassy car.

At the White House, Brent Scowcroft was just concluding a background briefing for journalists on the imminent State of the Union speech when a reporter asked about the communiqué. "Does this mean that you don't insist anymore that Iraq withdraw completely from Kuwait, that they only have to pledge that they'll do that?" Scowcroft, veiling his surprise, hesitated before replying with a terse "No. They have to leave Kuwait."

Hurrying back to his office, the irate national security adviser called Baker at Blair House to voice his displeasure. He then set out — a copy of the document now in hand — to warn Bush. Scowcroft found the president emerging from the White House make-up room, face thickly powdered for the television cameras.

"This is not good," Scowcroft said. "This creates the impression that we've blinked. It implies a linkage between the war and other problems in the Middle East." Scowcroft climbed into the limousine after the president, and the two men huddled over the statement as they sped toward the Capitol. At first, Bush observed mildly that "it sounds like Jim got a little carried away over there." But Scowcroft's irritation was infectious: by the time the motorcade pulled up beneath the massive white dome, Bush too was upset. The statement would anger the Israelis and puzzle the allies. And the timing was atrocious, upstaging the president before one of the most important speeches he would ever deliver. Both men agreed that little could be done other than to issue a White House statement declaring no change in U.S. policy.

Now, standing at the House podium with the speaker and vice president on the dais behind him, Bush concentrated on the task at hand. Surveying his audience, he spoke firmly, confidently. "Halfway around the world, we are engaged in a great struggle in the skies and on the seas and sands," the president began. "We know why we're there. We are Americans — part of something larger than ourselves. Together, we have resisted the trap of appeasement, cynicism, and isolation that gives temptation to tyrants."

After quickly recounting the end of the Cold War and skimming over his domestic legislation proposals, Bush paused to gather himself before returning to the war against Iraq. "Almost fifty years ago, we began a long struggle against aggressive totalitarianism. Now we face another defining hour for America and for the world. There is no one more devoted, more committed to the hard work of freedom, than every soldier and sailor, every Marine and Coast Guardsman — every man and woman now serving in the Persian Gulf."

The president was barely halfway through his speech, but the chamber erupted in a standing ovation. For fully a minute, Republican and Democrat, congressman and spectator, applauded in a rousing, emotional tribute to the sons and daughters they had sent to war.

If his generals had been molded by Ia Drang, Khe Sanh, and the pathetic images of Saigon abandoned in a pell-mell rout, George Bush had been shaped by Pearl Harbor, Normandy, and the viceroy MacArthur aboard the *Missouri* in Tokyo Bay. The Second World War had been the defining experience of his youth. As a young Navy pilot, shot down and rescued in the South Pacific, he had absorbed the prevailing belief in just wars that pitted the good and the selfless against the evil and the rapacious.

Bush's world was largely white and black; shades of gray baffled and annoyed him. The war against Saddam provided him with a clear-cut moral cause, animating his presidency with a sense of purpose that had been absent during his first two years in the Oval Office. The Persian Gulf crisis forced Bush for the first time — and, as it happened, the last — to rise above the limitations of his character, vision, and political philosophy to become, briefly, an extraordinary man.

Few would have expected it. He was often derided as a "wimp" or a "lap dog" or someone "who reminded every woman of her first husband." When unsure of his footing, Bush appeared frenzied, screechy, and inarticulate. He habitually dropped personal pronouns and wandered into syntactical thickets that implied intellectual befuddlement. He was ridiculed for his shallow enthusiasms and an unseemly willingness to shift positions — on economic policy, abortion, civil rights — for political expediency.

He was also thoughtful, gracious, and remarkably well versed on virtually every public issue of the day, a man of self-effacing wit even political opponents found difficult to dislike. Notwithstanding his upper-class background — Phillips Academy in Andover, Massachusetts, Yale University, a $300,000 grubstake from family and friends to enter the oil business in Texas — Bush had a genuine if narrow plebeian streak. He owned a bowling ball, liked horseshoes and country music, and watched too much television. He enjoyed walking on the South Lawn for impromptu chats with tourists through the iron fence. He was an inveterate practical joker, whether teeing up exploding golf balls made of chalk, wearing a George Bush rubber mask, or fooling waiters at the Chinese embassy by tugging a twenty-dollar bill across the floor with a string.

Politically, he was cautious and reactive. Distrustful of government as an agent of change, he embraced a deeply conservative view of the presidency as a caretaker's appointment on behalf of the status quo. "We don't need to remake society," Bush had declared in 1988; "we just need to remember who we are." His highest ambition — "the vision thing" — was "to see that government doesn't get in the way." He viewed White House inaction as a political virtue, although his metabolism kept him in perpetual motion. As the nation's forty-first president he sought to distance himself from Ronald Reagan, whom he had loyally served as vice president for eight years, largely by rhetorical artifice, notably his promotion of philanthropy, voluntarism, and "a kinder and gentler" America. The "country's fundamental goodness and decency," Bush believed, would assert themselves without meddling from Washington.

Bereft of innovative ideas and strong convictions, but possessing a complacent satisfaction with the world as he found it, he limited his domestic agenda to tinkering "at the margins of practicable change," in the words of his budget director, Richard Darman. Like Reagan, Bush advocated tax cuts for the rich. Otherwise, he ceded legislative initiative to the Democrats, who controlled both House and Senate. In keeping with his minimalist approach to government, his chief contribution to lawmaking was a lavish use of the presidential veto, which he had successfully exercised twenty-one times in the preceding two years. If he had trouble articulating what he was for, George Bush had no difficulty demonstrating what he was against.

In foreign affairs, clearly his first love, he was more active. Preserving the status quo abroad required eternal vigilance and a willingness to practice big stick diplomacy. Bush was vigilant and he was willing. "Domestic policy can get us thrown out of office," he once said, quoting John Kennedy, "but foreign policy can get us killed." Irrepressibly gregarious, he had spent more than two decades — as ambassador to the United Nations and China, as CIA director, as vice president, and now as president — cultivating personal ties with world leaders. After the invasion of Kuwait, those ties served him very well. With a sense of mission and with leadership skills he rarely displayed on domestic issues, Bush had rallied the world to his side.

He also rallied America, although not without a struggle. "In the life of a nation," Bush had said in announcing the first deployment of U.S. troops in August, "we are called upon to define who we are and what we believe." During the next five months, Bush sought to define what the country should believe in and to explain why half a million Americans should be placed in harm's way to restore the throne of a sybaritic emir in a tiny country most Americans could not find on a map. The muscular assertion of vested national interests — such as denying an avaricious despot control of 40 percent of the world's petroleum — would seem crass and ignoble if espoused by the Oval Office. Searching for a loftier objective, he had hopped from rationale to rationale. Redeeming Kuwait's right to self-determination, thwarting Saddam's quest for nuclear weapons, preserving jobs in Western democracies — Bush had tried all of these lines without convincing his nation that collectively they represented a legitimate *casus belli*.

He spoke — in his State of the Union speech, in fact — of "a new world order where diverse nations are drawn together in common cause to achieve the universal aspirations of mankind." However elevated, the aspiration rang hollow, like the abstract musings of a blind man struggling to describe sight. For no rhetorical flourish could obscure the reality that this was a war on behalf of the old world order, a war waged for the status quo of cheap oil and benign monarchies. It was only when he framed the conflict as a moral crusade that George Bush found his voice. The British would take credit for stiffening his spine, claiming that Margaret Thatcher, before stepping down as prime minister, had "performed a successful backbone transplant." Yet in truth his resolve was bred in the bone, nurtured half a century before in the war against fascism. "How can you say it is not moral to stop a man who is having children shot on the streets in front of their parents?" Bush had asked in extemporaneous comments to the Republican National Committee on January 25. "Was it moral for us in 1939 to not stop Hitler from going into Poland? Perhaps if we had, hundreds of Polish patriots would have lived, perhaps millions of Jews would have survived."

Three days later, he held, in a speech to the convention of National Religious Broadcasters, that although the gulf conflict was not a religious war, "it has everything to do with what religion embodies — good versus evil, right versus wrong, human dignity and freedom versus tyranny and oppression . . . It is a just war. And it is a war in which good will prevail." His anger was genuine: in the privacy of the White House, he referred to Saddam not only as another Hitler but also as "that lying son of a bitch."

Bush would be accused in postwar jeremiads of waging a private war, of cynically manipulating the nation to further his political ambitions and distract the electorate from his dismal domestic record. Such interpretations caricatured the inner man. Certainly he was aware that a successful war could enhance his political standing, just as another Vietnam would destroy him. But his motivation — and his success in rallying the country behind him — came from a belief in an American archetype: the great national myth of a peace-loving people who find no glory in war yet periodically rouse themselves to lead a military expedition against infernal forces.

In casting the war as crusade, he had fallen back on a conviction that lay close to the heart. This was, perhaps, as near to vision as George Bush could come. He proselytized a nation eager to believe again in its own goodness, a nation reluctant to accept that it could or should trade blood for oil, a nation much like its leader in preferring a world painted in blacks and whites. After twenty years of wary introspection following Vietnam, the country was again ready to see itself as an armory of virtue.

Two hazards awaited the man who would mount this crusade. The first was the risk of failure. The second, less obvious, was the risk of success. By relentlessly demonizing the enemy and defining the struggle in moral terms, Bush had dug himself a trap. He could win the war — the splendid little war in which victory was defined as defeating the Iraqi army and liberating Kuwait — and still lose the crusade. For crusading fervor could never be wholly sated by such limited though sensible objectives any more than Lieutenant George Bush would have been satisfied with Hitler in power, the Japanese warlords afoot, and Axis fascism extant. He had drawn the terrible swift sword without knowing when or how to sheathe it again.

The president had nearly finished his speech. He spoke deliberately, with the assurance of a man who knew his listeners were hanging on every word. In the gallery overhead and on the House floor below, the audience sat rapt, all eyes fixed on their commander-in-chief.

"The world has to wonder what the dictator of Iraq is thinking," Bush said. "If he thinks that by targeting innocent civilians in Israel and Saudi Arabia he will gain advantage — he is dead wrong. If he thinks that he will advance his cause through tragic and despicable environmental terrorism — he is dead wrong. And if he thinks that by abusing the coalition POWs, he will benefit — he is dead wrong."

Bush paused as the final refrain rang through the chamber: dead wrong.

"Among the nations of the world, only the United States of America has had both the moral standing and the means to back it up," he continued. "This is the burden of leadership and strength that has made America the beacon of freedom in a searching world. This nation has never found glory in war. Our people have never wanted to abandon the blessings of home and work for distant lands and deadly conflict."

With evangelical fervor, Bush brought the speech full circle. "Our cause is just, our cause is moral, our cause is right. Let future generations understand the burden and blessings of freedom. Let them say, we stood where duty required us to stand. Let them know that together we affirmed America, and the world, as a community of conscience."

He finished with a broad grin. The room thundered with long, loud, self-righteous applause.

No one clapped louder than the exotic figure in white robes rising from his chair behind the Joint Chiefs and Supreme Court justices. Prince Bandar bin Sultan, the Saudi ambassador, thought the president's speech simply first-rate. Bush, the ambassador would later gush, had revealed himself as a historical figure on a par with Churchill and Roosevelt. *We stood where duty required us to stand.* The words made the hair stand up on the back of Bandar's neck beneath his burnoose.

Propriety demanded that Prince Bandar sit with the other foreign emissaries and distinguished guests. Had protocol reflected political reality, however, Bandar would have been sitting up front with the cabinet, of which he was virtually a member ex officio. As the link between the American government and the Saudi royal family, Bandar had perhaps the pre-eminent role in helping two alien cultures bridge their differences and join in common cause against Saddam Hussein. Beginning with King Fahd's agreement in August to admit coalition troops into Saudi Arabia, Bandar had been instrumental in convincing the Saudis that the Americans would make good on their promise to defend the kingdom, expel Iraq from Kuwait, and then leave when the job was done.

In a capital full of intriguers, fixers, and grand characters, none eclipsed Bandar for savoir-faire. He was a grandson of his country's first king — a less than unique distinction, since Abdul Aziz ibn Saud had sired forty-three males. His uncle was the current king, his father the defense minister. For seventeen years Bandar had been a fighter pilot, ultimately flying F-15s from Dhahran Air Base after taking flight training in Texas and a master's degree in international relations from Johns Hopkins University.

In 1983 Bandar became Riyadh's ambassador to Washington, immediately cutting a wide swath. He favored beautifully tailored blue suits — except when the occasion demanded traditional Saudi garb — and immense Cuban cigars, which he discreetly hid from view whenever a photographer appeared. In a Virginia mansion above the Potomac River, he lived with his wife, six children, and Welsh bodyguards; his elegant, wainscotted embassy office overlooked the Watergate Hotel in Foggy Bottom. Ubiquitous, melodramatic, charming to a fault, the prince described life in the American capital with fighter-pilot slang: "the closest to pulling Gs I ever came outside a cockpit." The prince, *Newsweek* magazine once observed, "understands the game of not exactly lying, but not telling the whole truth either."

To illustrate Saudi Arabia's position in the world, Bandar liked to recount an anecdote about Queen Victoria and Benjamin Disraeli. Does Britain have no permanent friends? the queen had asked her prime minister. "Britain has no permanent friends or permanent enemies," Disraeli replied. "It just has permanent interests." Under Bandar's prodding, Riyadh's interests more often than not coincided with Washington's. During Iraq's war with Iran, when the United States saw Baghdad as a bulwark against Tehran's militant Islamic fundamentalism, Bandar had served as a secret middleman between Saddam and CIA Director William J. Casey. He helped provide Iraq with top secret U.S. satellite information about Iranian troop movements; he had personally signed a contract for sophisticated communications equipment made by American companies and secretly shipped to Baghdad.

Bandar's interests also coincided with those of George Bush. He had carefully cultivated Bush for years, even joining the then vice president on a fishing expedition and hosting a lavish party for him in 1985. A frequent visitor to the White House, Bandar, acting as the president's emissary, made a clandestine visit to Saddam in April 1990. In October, two months after the invasion of Kuwait, he had assured Bush that Iraq was a paper tiger that could be defeated in two weeks of war.

Several hours before the State of the Union speech, Bandar had visited CIA headquarters in Langley for a secret intelligence briefing. Though he admired U.S. technical competence, the prince thought his American friends often naïve and misinformed about what was really happening in the Middle East. The exodus of Iraqi aircraft to Iran was a case in point. The CIA now counted more than ninety planes on Iranian soil, two thirds of them high-performance fighters. Some analysts feared a double-cross, with Tehran permitting those planes to launch a sneak attack. To Bandar that was nonsense. Iran, he believed, was the world's most cynical nation. Worried that Iraq would sue for peace and emerge from the war with its air force intact, Tehran had cunningly encouraged Baghdad to keep fighting by offering sanctuary to aircraft it had no intention of returning.

In Bandar's view, Washington's anxiety about the fragility of the allied coalition also was misplaced, a consequence of know-nothing American Arabists who persisted in viewing the Arab world either patronizingly — as a collection of illiterate tribes — or romantically — as warriors in white robes on rearing chargers. Syria, he was certain, had no intention of helping archrival Iraq even if Israel entered the war. The Syrian strongman Hafez Assad, Bandar had observed, "would sell his mother to get Saddam." Saddam's threat to "make the fire eat up half of Israel" now looked like bluster, much to the delight of Assad, who could tell other Arab leaders that Iraq had simply provided Tel Aviv with world sympathy and American Patriot missiles. Bandar also envisioned a new world order, though one somewhat different from Bush's. The postwar Middle East, he believed, would be dominated by a Saudi-Egyptian-Syrian axis, perhaps with the Turks eventually joining the coalition.

Bandar had no doubt that Iraq would soon fold. Iraqi defectors — immediately spirited away by Saudi border troops, much to the Americans' irritation — drew a portrait of abysmal morale and beastly conditions in Saddam's battered ranks. Tens of thousands awaited the chance to defect. Informal communication back and forth across the border also suggested that some of the Iraqi generals, perhaps even a corps commander, might be willing to turn against Saddam in a coup d'état. (The Saudi army, Bandar had learned, also planned to permit the capture of fourteen Saudi soldiers carrying phony maps and documents to reinforce Iraqi suspicions that the main allied attack would come through the Kuwaiti bootheel in the eastern half of the country.) Bandar had slightly revised his earlier assertion that the conflict would last only two weeks. Now — as he told King Fahd, Colin Powell, and anyone else within earshot — he predicted a ground phase of "not more than a week to ten days." The war, he believed, would end by late February.

Bandar watched as the president, his speech finished, stepped from the podium and plunged into the crowd of well-wishers on the House floor. The prince had very much enjoyed this spectacle. Everyone — the administration, the Congress, the media — was trying to do the right thing, fearful of poisoning the country as the country had been poisoned in Vietnam. It was, Bandar had told a friend earlier in the day, "a beautiful thing to watch."

Gathering his robes, he shuffled from the chamber with the rest of the crowd. Bandar was eager to hear further reports from home. Shortly before leaving for the Capitol, he had spoken by secure satellite phone to General Khalid's staff in Riyadh. Reports were sketchy, he was told, but the Iraqis apparently had launched a modest offensive. Shooting had been reported along the western Kuwaiti border as well as in the east, in a scruffy Saudi border town known as Ra's al Khafji.

## Observation Post 4, Saudi Arabia

Since early January the twelve hundred Marines of Task Force Shepherd had formed a thin screen running northwest to southeast in Saudi territory just west of the Kuwaiti bootheel. Lightly armed, the task force — actually two mingled battalions, now commanded by Lieutenant Colonel Cliff Myers — served as a trip wire to protect the 1st Marine Division laagered around Kibrit, more than thirty miles to the south. Myers's mission was to watch for enemy movement. His men also reinforced three small reconnaissance teams manning the Saudi police posts that straddled narrow cuts in a fifteen-foot sand berm erected years before to deter smugglers. The most important post, astride the shortest route from Kuwait to Kibrit, lay in the extreme corner of the bootheel. Known as Observation Post 4, the site was dominated by a seedy, fly-infested building that resembled a Beau Geste fort because of its rooftop stone battlements. About thirty Americans and Saudis occupied OP-4, directing air strikes and spying on the Iraqis with sophisticated optical and listening equipment.

Except for sporadic rocket fire, the Iraqis had mostly ignored the Marines during the first two weeks of war. Four Army psychological operations teams attached to the task force routinely broadcast surrender appeals through huge loudspeakers, which also were used to annoy the enemy with rap music played at decibel levels approaching the inhumane. (The Iraqis often replied with long bursts of gunfire.) In a rain and hail storm on the night of January 25, the Marines slipped two artillery batteries close to the berm at OP-6 — near the "elbow," where the border again turned to an east-west axis — and fired several dozen rounds at targets near Al Jaber Air Base. While pulling back from the raid, a sleepy driver plowed into the rear of the armored vehicle in front of him, killing three Marines — the first blooding of Task Force Shepherd.

The earliest symptom of trouble on January 29 came at 6:30 P.M., when Iraqi troops began electronically jamming Marine radios. They achieved modest success, disrupting UHF, VHF, and HF communications, but failed to jam the Marines' SINCGARS system, designed automatically to shift frequencies several times a second. Myers, who had set up his command post near OP-4, put his four companies on full alert and then tried futilely to reach division headquarters. A few minutes later Charlie Company, arrayed twenty-five miles north near OP-6, reported enemy armor moving south on the Kuwaiti side of the berm. "Flash, flash, flash!" Myers radioed. "Tanks, tanks, tanks."

Delta Company, commanded by Captain Roger Pollard, waited five kilometers northwest of the reconnaissance team at OP-4. Delta's firepower included seven LAV-ATs (light armored vehicles, antitank), with TOW missiles and thermal sights for night vision, as well as thirteen LAV-25s, each armed with a 25mm cannon but no thermals. Because the desert sloped from west to east, the Marines could see into Kuwait across the berm, which ran north to south. The night was clear and cool, with a full moon so brilliant that the glare ricocheting off the desert floor actually hampered visibility. Around 8 P.M., Delta gunners spotted the fleet of Iraqi tanks rolling south. Pollard radioed Myers that at least fifty armored vehicles appeared to be closing in on OP-4.

The reconnaissance team, dug in near the fort and now thoroughly alarmed, was armed only with machine guns, rifles, and a few light antitank weapons. A seven-man squad darted out in front of the fort to set up firing positions among several mounds of sand. When the Iraqi column was barely three hundred yards away, the Marines opened up, hitting one T-55 tank and temporarily stalling the attack. A star cluster flare, the signal requesting immediate help, now burst above the observation post in a vivid shower of sparks.

Pollard spotted the flare and ordered his company forward toward the berm, with first platoon on the left, second platoon on the right. Lacking thermal sights and TOWs, the LAV-25s could neither see nor fight the Iraqi tanks effectively at this range, so the seven LAV-ATs took the lead. Through their thermals, which registered heat emanations, the AT gunners counted more than sixty crimson dots massing around OP-4 but still in Kuwaiti territory. At Pollard's command four of the ATs — moving abreast roughly twelve hundred meters apart — rolled forward on the right at ten miles per hour. They closed to within three kilometers of the berm.

In the vehicle designated Green 1, which was positioned farthest left of the four ATs, Sergeant Michael G. Wissman, Jr., called the AT commander, Sergeant Nicholas V. Vitale, to report that his gunner had spotted what appeared to be a tank at "three thousand meters, right front." Vitale was skeptical; that would mean the Iraqi was on the west side of the berm. The enemy appeared not to have spotted D Company, and the Marines had yet to take any Iraqi fire.

After asking Wissman to reconfirm the target, Vitale talked directly by radio to the gunner. (Only the gunner in an AT could see through the thermals.) Yes, the gunner insisted, he had a tank in his sights at 2500 to three thousand meters. Now two other ATs reported targets within range. Vitale called his company commander, Pollard, and notified him that they were about to fire.

Almost simultaneously, orange flame licked from the TOW launchers on the three ATs. Two TOWs, controlled by their gunners with thin wires that unspooled behind the streaking missiles, raced due east only to embed themselves in the berm short of the Iraqi tanks. The third TOW, fired by Green 1, darted south by southeast. The missile faded first to the right, then to the left as the gunner made several quick corrections to keep his cross hairs centered on the small crimson box in his sights.

Yet the gunner was confused, tragically misoriented. He had targeted not an Iraqi T-55 but Green 2, a sister LAV-AT to his right with four Marines inside: Lance Corporal Daniel B. Walker, Lance Corporal David T. Snyder, Private First Class Scott A. Schroeder, and the vehicle commander, Corporal Ismael Cotto. Only four seconds after Green 1 fired, the TOW warhead sliced through the rear troop hatch of Green 2, detonating fourteen TOWs stacked in a storage rack and seventy-one gallons of fuel. Green 2 exploded in a monstrous fireball, flinging the troop hatch a hundred yards away and killing Cotto, Walker, Snyder, and Schroeder.

Pollard, spotting several muzzle flashes from Iraqi tanks, assumed that the AT had been destroyed by enemy fire. Others knew the truth — "I think you shot Cotto," Vitale radioed Green 1 immediately — but kept silent in shock, shame, or uncertainty. Not for many weeks would the truth out. The men at OP-4, seeing that the explosion had momentarily diverted the Iraqis, dashed for the four Humvees and a five-ton truck they had hidden behind the berm. West across the desert they sped, past the pyre of Green 2 burning beneath the harsh moon. As Delta Company braced for an attack, the Iraqi column rumbled forward, sweeping around the Beau Geste fort and through the gap in the berm.

## R'as al Khafji

Perched on the thin littoral between the gulf and the endless desert, Khafji was an unpretty border town with a small port, an oil refinery, and the misfortune of lying within range of Iraqi field guns in southeastern Kuwait. An artillery barrage on the first day of war had forced the town's fifteen thousand souls to flee — breakfast rolls and half-empty pots of tea still littered the white tablecloths at the Khafji Beach Hotel — and the abandoned streets now belonged to stray dogs and grazing camels.

As in the west, allied surveillance teams manned several border posts just above Khafji, which lay seven miles south of Kuwait. The easternmost post, OP-8, sat on the beach a few hundred meters from the border; two kilometers farther south, a twenty-man group of U.S. Marines, Navy SEALs, and Army intelligence soldiers occupied the Khafji desalination plant. But the main force of Marines and Arab troops remained thirty miles or more to the south, reflecting Schwarzkopf's strategy of trading space for time should Iraq attack.

Life on the border had settled into an uneasy routine, with opponents swatting at one another across a no man's land of barbed wire and minefields. Iraqi rocket crews occasionally flipped on their truck headlights long enough to infiltrate south through their own minefields, where they blindly fired a few Frog or Astro rockets. Marine ANGLICOs — spotters trained to coordinate air and naval gunfire — retaliated with bomb or napalm strikes. Invariably the Iraqis got the worst of such exchanges. On January 26, for example, Marine Harrier jets dropped cluster bombs on a makeshift barracks north of Khafji; ninety minutes later, as an Iraqi burial detail pulled bodies from the rubble, the Harriers attacked again. Watching through a television camera mounted on an airborne drone, the Marines counted more than a hundred enemy casualties.

Despite such attacks — or perhaps because of them — the Iraqis grew bolder. During the day on January 29, U.S. surveillance teams had detected increased enemy movement across the border, including unusual daylight repositioning of tanks and armored personnel carriers. Harriers and Air Force A-10s responded by hammering targets in the so-called Wafrah Forest, a patchwork of Kuwaiti orchards and cultivated fields twenty-five miles west of the gulf. Between attacks, the Iraqis labored to clear wreckage from the road leading out of the forest to the Kuwaiti coast. Although wary, the Americans anticipated no immediate attack. Three enemy soldiers — later judged to be part of an enemy ruse — defected to OP-8, where they assured the Marines that their officers had all fled north to escape the bombing.

In truth the Iraqis had assembled several battalions in the Wafrah Forest. At 8:30 P.M. Lieutenant Andrew Hewitt, watching the forest through thermal sights from OP-1, spotted the lead Iraqi platoons emerging from the trees. Within fifteen minutes, Hewitt counted nearly a hundred vehicles in column on the road paralleling the border. The lieutenant tried frantically to summon an air strike, but all available pilots had been diverted farther west to help the besieged Marines at OP-4. By 9:15 the enemy had moved south to within two thousand yards of OP-1. Frustrated and frightened, Hewitt and the other ANGLICOs piled into their vehicles and fled from the border, five minutes before Iraqi T-55s rolled over OP-1. Ten miles east, the Marines at OP-7 held their ground long enough to direct several strikes by Cobra helicopter gunships; then they too headed south.

On the beach at OP-8 around 9 P.M., ANGLICOs heard the distinctive creak of armor tracks echoing among the houses on the Kuwaiti side of the border. Crouched in a sandbagged bunker carved from a dune, the Marines radioed a warning to the men in the desalination plant. When the sounds drew closer, an ANGLICO poked his head from the bunker to observe the coastal road. An Iraqi soldier atop a personnel carrier barely fifty meters away opened fire with his machine gun — and missed. The Marines fired back with a grenade launcher, and also missed. The ANGLICOs and a four-man SEAL team crawled across the sand to their Humvees and raced south through the dunes, small arms and tank rounds whizzing overhead.

At the desalination plant, Iraqi illumination shells began bursting above, followed by the distinctive *clumpf* of mortar rounds and then a barrage of red, yellow, and green star cluster flares. The Marine officer in charge, Lieutenant Colonel Rick Barry, ordered his men to fall back into Khafji. Saudi border troops and national guardsmen also fled south, leaving their helmets but taking their flak jackets, which they used as prayer rugs. (Some Marines also abandoned their gear, including classified maps and cryptographic equipment; either ignored or overlooked by the Iraqis, most of the kit was recovered several days later.)

With the border abandoned, Barry regrouped in a Saudi "safe house" next to the water tower that loomed over southern Khafji. "We're not retreating one more fucking step," he told Captain Jon Fleming, who had helped coordinate the border ANGLICOs from the desalination plant. Hoping to find a better vantage point from which to watch the four-lane road leading into Khafji from Kuwait, Fleming prowled around the base of the water tower in search of a stairway. He discovered instead several Saudi soldiers in a basement, barefoot and drinking tea with dazzling insouciance.

"Don't you know the war's started?" Fleming asked, gesturing toward the distant sound of gunfire.

One of the Saudis shrugged. "The Marines," he answered, "are here to protect us."

After discovering a metal door leading into the water tower, Fleming and another Marine officer bounded up eight flights of stairs. Hyperventilating with excitement and exertion, they peered north through a narrow window. A dozen round-turreted T-55s and at least as many armored personnel carriers had crossed the causeway leading into northern Khafji. The Iraqi gunners sprayed machine gun fire at rooftops and upper windows, apparently to suppress any snipers; soon tracer rounds streaked toward the water tower.

"Hey, Marines!" a sergeant's voice called from the ground below. "C'mon!"

Bullets pinged off the tower as Fleming and his companion galloped back down the stairs and leaped into a Humvee. Chasing after Barry's receding taillights, they raced south toward Al Mishab through the ceremonial archway welcoming visitors to Khafji.

The Iraqi army had conquered itself a Saudi town.

Unbeknown to the Iraqis, however, more than dogs and camels remained in Khafji. A pair of six-man Marine reconnaissance teams had been trapped by the sudden enemy thrust. One team, inserted on foot the night of January 28, took refuge on the rooftop of a four-story building along the town's main street. Spotting the convoy of American vehicles racing south from the desalination plant shortly after 9 P.M., the team leader had radioed Task Force Taro, the main Marine force in Al Mishab. "Is there something going on up north that we don't know about?" he asked.

The second team, commanded by Corporal Lawrence Lentz, had been in Khafji nearly a week, watching for infiltrators and providing early warning of Iraqi Scud and rocket attacks. Lentz, a three-year veteran from Concord, North Carolina, had moved his men into a two-story house still under construction in northeastern Khafji, about two kilometers from the other team's eventual location. From a second-floor window he too watched as the friendly forces fled south: first the Saudi guardsmen in their Toyota trucks, then the ANGLICOs and SEALs in Humvees. Behind them Lentz saw tank and machine gun fire as the Iraqis fired on the telephone company building north of Khafji. Soon the Iraqis pressed south and entered the town in a fusillade of wild firing.

Lentz checked his defenses. An eight-foot concrete wall around the half-acre compound offered some protection. The team's two Humvees, one of which carried a fifty-caliber machine gun, were hidden against the house. Lentz set up an M-60 machine gun near the front door to cover the compound gate; he emplaced another light machine gun upstairs and claymore mines outside near the Humvees. He checked the most potent weapon of all — the radio — and found he had good communications with an artillery unit, the other team, and Task Force Taro.

About 10:30 P.M. the Iraqi shooting seemed to ebb. Lentz called his captain in Al Mishab. "I'm leaving it up to you," the captain said. "If you think you should get out, we'll try to help you."

Lentz paused before answering. If the Iraqis discovered his hiding place, the compound walls and machine guns would be no match for tanks and enemy mortars. On the other hand, the two Marine teams were ideally placed to direct air and artillery strikes in a counterattack. The men, he knew, were very frightened; Lentz was terrified himself. "No," he told the captain, "we'll stay."

## OP-4, Saudi Arabia

The size, speed, and tactical sophistication of the enemy attack caught the American high command flat-footed. Several brigades from the Iraqi 5th Mechanized Infantry Division, perhaps supported by other units, had combined to launch three distinct spearheads — at OP-4, OP-7, and Khafji — with two more to follow on January 30. The enemy objective was — and would remain — something of a mystery. Allied intelligence later concluded that the attack had been nearly two weeks in the making, undoubtedly with Saddam's authorization. Baghdad's presumed intent was to parlay a modest military achievement into a major political victory, perhaps by inflicting enough damage on Saudi and American forces to dishearten the home front. The incursions may also have been an attempt to draw the coalition into a ground war prematurely, and to capture more American prisoners.

Without air power to cover their troops in the open desert, the Iraqis' attack was doomed to fail — an inevitability that exculpated the surprised U.S. ground commanders. Until allied air superiority could assert itself, however, CENTCOM and the Marines faced several tense hours trying to fathom Iraqi intentions while blunting the offensive. Fearful that the enemy would drive toward the port and logistics compound at Al Mishab, engineers planted a thousand pounds of plastic explosives beneath the highway five miles south of Khafji, where encroaching *sabkahs* — nearly impenetrable desert badlands — formed a natural chokepoint.

A more immediate anxiety was the lightly defended forward logistics base at Kibrit, inland and thirty-five miles south of the Wafrah Forest. "My God," exclaimed the 1st Marine Division operations officer, Lieutenant Colonel Jerry Humble, late Tuesday night. "They're after Kibrit!" (The anxiety was heightened by several inaccurate reports during the night that placed the Iraqi tanks that had come through OP-4 much farther south than they actually were.) Knowing that a successful attack on the huge ammunition dump at Kibrit would devastate Schwarzkopf's plan for a ground offensive, the Marine logistician, Brigadier General Charles C. Krulak, marshaled clerks, typists, and other troops into a hasty perimeter defense north of the dump, and called for reinforcements from the Army's Tiger Brigade.

Walt Boomer, the senior Marine commander, also worried about Kibrit; he had gambled by putting the logistics base so far forward in anticipation of an eventual allied ground attack. Of equal concern to Boomer, however, was the recapture of Khafji. Though the town might be tactically worthless, politically it was invaluable. Khafji could not remain in Iraqi hands.

But diplomatic niceties intruded. Khafji lay in the sector controlled by three brigades of Saudis and their Arab allies. Many Americans suspected the Saudis incapable of serious fighting, much less ousting the Iraqi army from an occupied town; Saudi soldiers were viewed, like those lounging by the Khafji water tower, as indolent, barefoot tea drinkers relying on the Marines for protection. Now, Boomer realized, that stereotype would be put to the test.

At OP-4 the Iraqi juggernaut had stalled, thanks to the timely arrival of Marine and Air Force warplanes and the gritty reluctance by Captain Pollard's Delta Company to give ground. After the destruction of Green 2, Pollard reorganized his force into a firing line three thousand yards southwest of the overrun reconnaissance post. With battlefield ingenuity, the Marines used the thermal sights on the LAV-ATs to spot enemy armor for the LAV-25s, which then fired at the Iraqis. Although the cannon rounds would not penetrate the tanks, the tracers and flashing ricochets provided beacons for the pilots overhead. Two tanks thus damaged by Air Force A-10s barreled into the berm and were abandoned by their crews.

Fearful of being flanked by Iraqi armor moving to his right, Pollard pulled the company back a thousand yards, unleashed a volley of TOWs, then moved back another thousand yards and waited for an air strike by a second pair of A-10s that had come on station. Twice the Marine gunners marked a T-55 with cannon tracers; twice the A-10s reported difficulty in finding the Iraqi tank.

For the two pilots peering down at the desert more than a mile below, the battlefield was a welter of vehicles and intertwining gunfire. Flying at night was unusual for A-10 pilots, practiced infrequently in peacetime. This unit, the 355th Tactical Fighter Squadron, had been converted to night flying during Desert Shield in anticipation of precisely this kind of predicament — nocturnal, close air support of troops locked in battle. No pilots in the Air Force were more eager to help their brethren on the ground than those flying the A-10 Warthogs.

Unlike the F-15E and the F-111, however, the A-10 had no sophisticated infrared equipment; instead, the pilots navigated and picked out targets through the relatively crude infrared lens in the nose of a Maverick missile carried under the wing. "Hot" objects — like a tank or an LAV — showed up as tiny white dots on a six-inch cockpit screen. Although the 355th also had some practice using flares for illumination and as reference points on the ground — a common tactic in Vietnam — a supply shortage had limited their training.

To give Delta Company's air controller just such a reference point, one of the A-10 pilots popped a magnesium flare. It floated to earth like a second moon, landing directly behind the second-to-last vehicle on Pollard's left flank. Concerned that the bright glow silhouetted the

LAV for Iraqi gunners, a Marine leaped from the back of the vehicle and tried to bury the burning flare with a shovel.

The second A-10, guided by the Delta air controller, finally picked out the Iraqi T-55 and rolled in for the kill. Slewing the cross hairs of his missile onto the target, displayed as a white rectangle on the cockpit screen, the pilot pressed the trigger, the "pickle button." With a whoosh and a rattle the Maverick leaped from its rail, correctly yanking free the data cable and blanking out the screen.

What happened next, however, was not correct. Instead of locking automatically on the target selected by the pilot, the missile apparently "went stupid" — malfunctioned — and dived directly toward the earth. It hit not the Iraqi tank but an LAV-25. The warhead easily punched through the thin armor roof, spraying molten metal into the troop bay; the consequent explosion flipped the LAV turret twenty meters across the desert and knocked the driver clear out of his vehicle. He was discovered five hours later, cut and burned, but miraculously alive. The blast killed seven other Marines.

Pandemonium swept Delta Company, the radios frantic with a gabble of voices. For the first time since the fight began, Pollard was frightened. The Iraqis, he believed, had swept behind him and were picking off his troops with tank fire. But a quick search of the desert showed no one on the company's flank or in the rear. He calmed the men and soon understood what had happened. "I just lost an LAV-25," he radioed the battalion commander, Myers. "I think it was a friendly missile." The company pulled back another thousand yards.

It was nearly midnight. In four hours of combat Delta had suffered not a scratch from Iraqi fire. But eleven Marines were dead.

The fight in the west persisted another nine hours. At 1:30 A.M. Iraqi infantry, covered by several dozen armored vehicles moving behind a thick screen of smoke, occupied the vacant OP-6. Incessant air attacks and TOW missiles fired by Task Force Shepherd's Charlie Company drove them back into the Kuwaiti desert by dawn. At OP-4, where Myers relieved Delta with Alpha Company, fifteen Iraqi tanks massing near the fort were attacked by A-10s and helicopter gunships. A flight of Marine Cobras destroyed three tanks with TOW missiles, missed two others, then hit a sixth from the rear with a TOW that caused the round turret to flip in the air and land upside down in the turret ring. By 9 A.M. Wednesday, the Iraqis were in retreat. Myers moved his companies onto the berm, and for two hours they called in air and artillery strikes on the fleeing enemy.

Someone raised an American flag over the shattered hulk of the LAV-25, where a doctor and a graves registration detail began the grim task of collecting arms, legs, and torsos. Two kilometers away, not enough of Green 2 was left even to post the colors. Myers was struck by a certain hardening that had overtaken his weary Marines, an abrupt aging in their eyes. Many had the "thousand-yard stare" common to combat veterans. No callow men remained in Task Force Shepherd: innocence, too, had been charred beyond salvage. Around the neck of a dead Iraqi, slumped grotesquely in his shattered tank, a truculent sign was hung: "Don't fuck with the United States Marine Corps."

## Khafji

Confusion, war's constant companion, held sway in the east. An estimated Iraqi battalion, perhaps six hundred men and several dozen armored vehicles, occupied Khafji. Beyond that, little was certain. Radio Baghdad, making the most of its prize, crowed, "O Iraqis! O Arabs! O Muslims who believe in justice! Your faithful and courageous ground forces have moved to teach the aggressors the lessons they deserve!" A Qatari tank battalion, maneuvering west of town, took a barrage of artillery fire that seemed to come from American guns. Two U.S. Army trucks inexplicably wandered into the captured town; one escaped under fire, but Saudi national guardsmen found the other crashed into a cinder-block wall with the engine still running and two soldiers, including an enlisted woman, missing, apparently captured. Farther west, near OP-1, a Saudi national guard battalion watched as an Iraqi battalion crossed the border (with turrets traversed, the Saudis later asserted, in a gesture of surrender). Following two hours of negotiation a firefight erupted and the Iraqis pulled back into Kuwait.

After five months of misery in the Saudi desert, many Marines were eager to vent their frustrations with a quick, violent counterattack to recapture Khafji. But there remained the issue of Arab control over this sector of the war zone. At least one senior Saudi commander sensibly proposed encircling the town and waiting for the besieged Iraqis to surrender. The dozen trapped Marines, however, complicated matters. The team on the rooftop in the center of town had nearly been discovered when several Iraqi soldiers entered the building lobby. Peering down the stairwell, the Marines saw the enemy helmets bobbing below, then burned their secret radio codes and other sensitive documents in anticipation of being overrun. A quick artillery strike distracted the

Iraqis, and the Marines remained undetected. Their security, however, was tenuous.

On the late afternoon of January 30, the commander of Marine Task Force Taro, Colonel John Admire, drove north from Al Mishab to confer with Colonel Turki al Firm, brigade commander of the 2nd SANG (Saudi Arabian National Guard). "We have two reconnaissance teams in the city," Admire said. "I think they can continue their mission for thirty-six hours or so, calling in artillery and air strikes for us. After that, their position will probably be compromised." Turki took the hint. After a moment's pause, he replied simply, "We attack."

At midnight several armored companies of Saudis and Qataris, accompanied by a few Marines, pressed into Khafji from the south. Their mission was to probe the Iraqi defenses, determine enemy strength and disposition, find the besieged Marines if possible, then press forward to liberate the entire city. The probe, however, had all the finesse of a cavalry charge. The Arab troops careered through the streets of southern Khafji for several hours, shooting at enemies, real and imagined, as well as at one another. The Iraqis fired back with equal indiscrimination, and for a few hours Khafji resembled Beirut. So many rounds missed high that the pilots overhead wondered whether they were taking antiaircraft fire. Just after 4 A.M., the probe withdrew.

The two reconnaissance teams, seeing the streets painted with tracers from all points of the compass, prudently remained hidden. Lawrence Lentz and his squad, concealed in their unfinished house, had spent the day calling in artillery strikes. One salvo destroyed a mobile rocket launcher near a brick tower north of town. Another ripped through a ten-man foot patrol marching on the causeway, flinging bodies into the air.

When the Arab probe began Wednesday night and a TOW streaked past the compound wall, Lentz moved his men to the northeast corner of the house to get as far as possible from the firefight. "Do you know what would happen if one of those things hit this house?" a Marine whispered. "Please," Lentz replied, "don't tell me."

After the shooting ebbed, a new wave of Iraqi tanks and personnel carriers rumbled down the causeway and a parallel beach road. Both reconnaissance teams called for additional artillery. Nearly fifty rounds whistled in on Khafji, the shells cracking open above the earth to spray hundreds of shrapnel bomblets on the battered city. The barrage lightly wounded one Marine hiding on the rooftop and shredded three tires on Lentz's Humvees.

As in the west, the greatest killing was done from the air. Other Iraqi units had emerged from their revetments to marshal for a follow-on attack into Saudi Arabia; on the night of January 30 allied pilots pounced on the enemy columns, hammering the Iraqis with thousands of cluster bombs, mines, rockets, Mavericks, TOWs, laser-guided munitions, and old-fashioned dumb bombs. "It's almost like you flipped on the light in the kitchen late at night and the cockroaches start scurrying, and we're killing them," observed a Marine pilot involved in the attacks. "They're moving in columns, they're moving in small groups and convoys. It's exactly what we've been looking for." The skies over southern Kuwait were so thick with aircraft — from B-52s to helicopters — that pilots worried most about midair collisions.

Not all the killing was intentional. Four Saudis were killed in an errant attack by a mixed flight of Air Force F-16s and Qatari F-1s. (The Americans subsequently claimed the Qataris had actually dropped the fatal bombs; the Saudis, typically reticent, kept quiet about the episode.) A similar fate nearly befell the 2nd Marine Division when eight cluster bombs narrowly missed a battalion command post. The blasts sprayed shrapnel as close as fifteen yards from the Marines.

Nor was all the killing done by the allies. Among the planes flying north of Khafji that night were three AC-130 Spectre gunships, slow-moving special operations aircraft first unveiled in Vietnam in 1967. Armed with machine guns, cannons, and a 105mm howitzer, the AC-130 needed the cloak of darkness to compensate for its lumbering pace. At 6:19 on January 31, as dawn spread over the Persian Gulf, an AWACs plane ordered the third and final gunship still on station — tail number 69–6567 — to return to base. Instead, the crew of five officers and nine enlisted men loitered above OP-8 in search of a Frog missile battery spotted earlier by the Marines. At 6:23 the gunship again was ordered to break off the attack. "Roger, roger," the co-pilot replied casually.

A few seconds later the AWACs picked up a weak, strangled cry: "Mayday, mayday." An Iraqi SAM had struck between the fuselage and inboard left engine, shearing the wing off. Spectre 69–6567 heeled over in a fatal helix. Spiraling to the left, nose down at a seventy degree angle, the aircraft smashed into the gulf a mile from the desalination plant in water barely ten feet deep. Fourteen men died.

On the early morning of January 31 Colonel Turki's 2nd SANG again plunged into Khafji, this time intent on recapturing the town. Two Qatari tank companies moved north to block Iraqi reinforcements with help from Marine artillery and air units. Again the attack resembled a Wild West shoot-out, although with automatic weapons fire and tracers rather than six-shooters. The Iraqis stubbornly held their ground, and Turki summoned reinforcements. Despite the artillery rounds falling nearby, the advancing troops paused to pray. Walt Boomer, listening to reports of the battle from his field headquarters in the south, vented his exasperation. "I don't think the Saudis have even approached Khafji yet," he declared in anger. "*This* is the outfit that's going to conduct the breach attack on our flank?"

Brave but impetuous, with little thought to clearing the city block by block, the Saudis darted haphazardly through the streets, firing over one another with heavy machine guns. The town filled with the sounds of shattering glass, whining bullets, explosions, ricochets, and tank rounds bursting against the battered water tower — a bedlam punctuated by Saudi officers screaming surrender demands through loudspeakers. The attackers had pushed about a third of the way into the city, past the used car lots and groceries in southern Khafji, when an antitank missile fired from a three-story building destroyed a Saudi personnel carrier, killing six and wounding three. The attack faltered and then resumed as the officers urged their men forward.

By late morning, Lawrence Lentz learned that the other Marine reconnaissance team had managed to sprint back through the lines to safety. "There are still isolated pockets of resistance," an officer advised Lentz by radio, "but if you're going to get out, this is the time to do it." As the squad prepared to flee the compound, an Iraqi sniper opened up from a building a hundred yards to the west. The Marines stitched both sides of the sniper's window with M-16s and fired two rocket-propelled grenades; the shooting stopped. At 1 P.M. Lentz and his men drove through the compound's front gate, flat tires flapping on the pavement. Picking their way cautiously through the side streets of eastern Khafji, the team soon reached the checkpoint on the road to Al Mishab, where their fellow Marines provided a jubilant welcome.

By late afternoon on the 31st most of Khafji was liberated. Scores of surrendering Iraqi soldiers emerged from their hiding places, waving and grinning as Saudi guards marched them south to a prison compound. Bodies littered the battered streets. The charred corpse of a Saudi soldier sat seared to the driver's seat in a smoldering personnel carrier. A few yards away, a dead Iraqi soldier lay wrapped in a blue, blood-soaked blanket, arm draped over a face frozen in a final grimace. Near another fallen Iraqi stood six undamaged vehicles stuffed with food, medicine, and ammunition. Throughout the town the detritus of war greeted the liberators: dead camels, blackened tanks, walls and streets pocked with artillery and small arms fire.

By Saudi count, thirty Iraqis had been killed and 466 captured, thirty-seven wounded among them. Nineteen Saudis and Qataris had died, and thirty-six were wounded; an uncertain number of these had fallen to friendly fire. American losses, including those in the fighting out west, were twenty-five dead, nearly half from fratricide.

Schwarzkopf, who had monitored events from Riyadh, harrumphed that Khafji was "about as significant as a mosquito on an elephant." (Among the CINC's contributions was dissuading the Saudis from flattening the town with air strikes rather than suffer the disgrace of Iraqi occupation.) If not the desert Stalingrad that some participants claimed — and Saudi bards immediately set to work composing paeans to their famous victory — the two-day scrap held portents of the coming allied ground attack.

To the immense relief of the Americans, Saudi guardsmen had demonstrated that they could fight with zeal and courage — if not with tactical prowess. Braced by his success against the vaunted Iraqi legions, Khalid became more insistent on a larger role for Arab troops in the ground campaign. Instead of following in trace as the Marines pushed north through eastern and central Kuwait, the Arab forces were determined to breach the minefields and man their own attack sector along the coast. Schwarzkopf and Boomer assented. The configuration would free the Marines to focus farther west on more significant tactical objectives, such as Kuwait International Airport and the high ground overlooking Kuwait Bay.

More important, the battle cut the Iraqis down to size. A credible battle plan had been executed badly by enemy troops who lacked air power, the ability to adjust artillery fire, even the wherewithal to avoid their own obstacle belts when forced to retreat. Allied warplanes riddled a number of tanks and trucks trapped between two minefields north of Khafji. Schwarzkopf later estimated that 80 percent of the Iraqi 5th Mechanized Division had been obliterated. The enemy also lacked the fire in the belly required, in military vernacular, to close with and destroy their foe.

Although the boldness of the attack impressed some American officers — "This is an enemy who is not going to go down easily," warned Colonel Ron Richard, operations officer of the 2nd Marine Division — a new confidence stole into the allied war councils. "The Iraqis will quit and they'll quit early if you hit them hard," predicted John Admire, the Task Force Taro commander. In Riyadh the CINC's intelligence chief, Brigadier General John A. Leide, told Schwarzkopf on the 31st: "Sir, I hate to say this but these guys aren't worth a shit. They can't put it together in a cohesive way and they can't operate coherently above a brigade level."

The battle also illuminated several allied shortcomings. Failure to clear sea mines and destroy Iraq's ship-killing missiles had kept the battleships *Missouri* and *Wisconsin* beyond range in a fight ideally suited to their sixteen-inch guns. (*Wisconsin*'s skipper, Captain David Bill, was nearly apoplectic with frustration. "Why in the hell aren't we there?" he asked in a stinging message to the *Blue Ridge.* "The world wants to know.") And no one could be reconciled to the friendly fire deaths. The power and range of modern munitions, launched in the dark by confused, weary, and frightened men, at times overwhelmed the inadequate measures intended to control them.

If air power had been the decisive arm in the two-day fight, it also engendered a new wariness. With sardonic humor, some Marines painted American flags on the top of their helmets. Syrian troops, learning of the Saudi deaths by allied fire, dropped their stubborn refusal to mark their vehicles with recognition symbols. In the 1st Marine Division the grisly dismemberment of eleven men resulted in a new order: henceforth all Marines would wear one dog tag around the neck and another tucked into the laces of the left boot.

Despite Schwarzkopf's dismissal of Khafji's strategic importance, he remained clear-eyed about war's horror. In a television interview as the battle ended, the CINC declared that the combat deaths were "sobering to the American people, and I don't think that's unhealthy."

## Dover, Delaware

Thanks to the grim efficiency that General Gordon Sullivan had witnessed at Dover a week earlier, the military mortuary succeeded in promptly identifying the dead Marines so that the bodies could be released for burial. None would be cursed with that epitaph of anonymity so common in earlier wars: KNOWN BUT TO GOD. Within a week or so, this first clutch of native sons to fall in ground combat would be home again.

Corporal Stephen E. Bentzlin, commander of the LAV-25 destroyed by the errant Maverick missile, left three young boys behind. "He was fighting for peace," his mother said, "and he found it." Before his funeral in Minnesota, his widow, Carol, asked that the casket be opened for her to have a final moment alone with her husband. He lay wrapped in a green blanket, his uniform draped over him. She kissed him goodbye. In her last letter before his death, Carol Bentzlin had written: "Good night, Steve. I miss you. I love you. You're a hero."

Lance Corporal Dion J. Stephenson was sitting next to Bentzlin when they died. When his body arrived in Salt Lake City from Dover on a commercial airliner, baggage handlers and refueling crews stood at attention in the cold, foggy night as the plane taxied to Gate 5. Six Marines in dress uniforms hoisted the casket from the conveyor belt into a hearse; on the street in Bountiful, Utah, where his parents lived, the neighbors kept vigil along the curb, singing "America the Beautiful" as the cortège arrived. After a funeral Mass at the Cathedral of the Madeline, he was laid to rest at Bountiful City Cemetery. Standing among the mourners near Stephenson's father — who had served with the Corps in Vietnam — was General Al Gray, the Marine commandant. Four Apache helicopters flew in tribute overhead, one veering away in the "missing man" formation as a bagpiper played "The Marine Hymn."

Lance Corporal Daniel B. Walker came home to Whitehouse in the piney woods of east Texas, where the flags in front of city hall flapped at half mast and black bunting trimmed the yellow ribbons. Dan Walker had been riding in Green 2. He was twenty when he died. "It paralleled his life," his mother, Robin, said. "When he was a kid, whenever he did something wrong, he always got caught the first time. That's how he was. You know how some kids can do things forever without getting caught? He always got caught." He had been a high school dropout, busing tables and tossing pizzas and listening to heavy metal music at such volume the speakers once caught fire. Then the Marines took him and sent him to war, perhaps not yet a man but certainly no longer a boy. A week before he died he had called his mother from Saudi Arabia, promising to take her out for dinner and a beer in his dress blues when he got home. Instead, three doleful strangers in dress blues appeared at Robin Walker's door at 3:30 A.M. on January 31.

Nearly half of Whitehouse attended the memorial service. They packed the high school gymnasium, young and old, black and white, friends and strangers, craggy veterans in their blue American Legion caps and pink-cheeked boys in their wash-and-wear shirts. "One of our own has fallen," a Navy chaplain told the mourners. "There is something within our biological structure that screams out and says it is morally wrong for the old to outlive the young. This is one of the times when God doesn't seem to make sense. This is the worst that life gets."

Sitting in the front row, Robin Walker sobbed and clutched her son's dog tags. His sister held a white rose. A family friend read a eulogy written by Dan's father, Bruce, who was too broken-hearted to read it himself: "Daniel went about his life with purpose, resolve, and an impeccable heart. The glory of his spirit shines like an oh-so-bright star in the darkness of my despair . . . Go proudly into your next life, son, and know we loved you very much."

A three-mile procession rolled down Farm Road 346 to the country cemetery where generations of Walkers slept. Marine pallbearers carried the flag-shrouded casket to an open grave beneath the boughs of an ancient oak. "Fire three volleys," a lieutenant commanded. The shots cracked and faded, leaving only the smell of cordite and the wail of a startled baby. The Walkers held hands as a bugler played Taps. The lieutenant presented the flag, folded into a trim triangle, to Robin Walker. She clutched it and wept. *This is the worst that life gets.* Her son's casket, crowned with a single white rose, vanished into the waiting earth. Dan Walker was home, and he had brought the war with him.

8

# The Riyadh War

## Riyadh

With Khafji recaptured and the Iraqis repulsed, the American military services could return to battling their more implacable foes: each other. The blunting of Baghdad's feeble attack had again displayed allied air hegemony. But as the allies' own ground offensive drew nearer, a struggle over who would control the huge air armada intensified.

In basic terms, the Army and Marines wanted the air attack focused on the enemy forces they would soon have to fight in the Kuwaiti Theater of Operations. They wanted aircraft trolling the northern horizon in a relentless campaign against Iraqi armor, troops, and particularly artillery in Kuwait and southeastern Iraq. The Air Force preferred to prosecute the strategic campaign against the twelve target sets laid out before the war, one of which happened to be the Republican Guard. These divergent points of view soon hardened into acrimony, and a classic groundpounder versus flyboy conflict erupted. No one was happy.

Since man first bolted bomb to wing, such battles had been waged. While many ground commanders during and after World War I envisioned airplanes as a useful weapon for attacking enemy troops on the battlefield, an air power champion like Billy Mitchell insisted that "the hostile main army in the field is a false objective and the real objectives are the vital centers" far from the front lines. In April 1943, George Patton complained bitterly to his superiors that insufficient air support had slowed the advance of his II Corps in Tunisia; when one air officer suggested that inexperienced troops were the real reason for II Corps's problems, an enraged Patton demanded a public apology. Later in World War II, during planning for the assault on Europe, air power enthusiasts argued that Allied bombers should continue hammering strategic targets in Germany. They were overruled, and the bombs instead fell on

France in support of the successful Normandy invasion — but the cost was a three-month delay in attacks on the petroleum industry that fueled the Third Reich. As Checkmate's John Warden wrote in *The Air Campaign*, "Powerful forces are pulling the ground commander one way and the air commander another."

Schwarzkopf's ultimate ambition in the gulf war, expressed to his division commanders as early as November 1990, had been succinct and sanguinary: "I want every Iraqi soldier bleeding from every orifice." In prosecuting the air attack, however, he had given the Air Force what amounted to carte blanche. "There's only going to be one guy in charge of the air: Horner," the CINC had told his subordinates in the fall. "If you want to fight your interservice battles, do it after the war." Thus empowered as commander of all allied air forces, Chuck Horner concentrated his planes where he thought they best supported the CINC's overall war objectives.

At the heart of the matter lay a subtle shift in the balance of power between sister services. Air power in previous wars had usually served a supporting function, subordinate to the ground "scheme of maneuver." But in this war, fought in a theater ideally suited for aircraft now technologically capable of precision attacks, the roles were reversed. To Horner, Buster Glosson, and Dave Deptula, the air campaign was going very well despite the diversion of three squadrons to Scud hunting and weather so atrocious that four of every ten sorties through late January had been canceled. (On the basis of historical meteorological records, planners had anticipated overcast blanketing Iraq 13 percent of the time; in fact, clouds obscured the targets three times more frequently.)

By attacking Iraq's centers of gravity — leadership targets, communications, road and rail networks, bridges, and the like — the allies had weakened Saddam's isolated army, as Glosson had promised Bush when outlining the proposed air campaign at the White House in October. "I can guarantee you that the Iraqis won't be able to feed, resupply, or move their army because I'll have all the bridges down and I'll take their resupply away from them," Glosson boldly assured the president. "Over a period of time they'll shrivel like a grape when the vine's been cut." Colin Powell, listening carefully to Glosson's presentation, had stepped in. "You cannot guarantee the president that the Iraqi army will pull out of Kuwait." "That's right," said Glosson. "But I can guarantee him that if he'll wait long enough, I'll destroy it in place."

Already the Iraqis were restricted to moving mostly at night, and even then at risk of attack. A commander who could reposition or resupply only during darkness, a German Panzer officer once observed, was like a chess player allowed but one move for every three made by his opponent. Allied intelligence claimed that by January 30 Iraqi military supplies into Kuwait had shrunk from twenty thousand tons per day to two thousand tons.

Nor was the Iraqi army ignored: pilots routinely bombed both Republican Guard and front-line units. Several hundred sorties a day were devoted to tactical targets in the Kuwaiti theater, complementing the hundreds flown north of the Euphrates against strategic targets. Horner and Glosson also carved Kuwait and southeastern Iraq into grid squares called "kill boxes." Each box measured thirty by thirty miles, and was further subdivided into four quadrants. Pilots unable to hit a target farther north might, for example, be diverted to box AH-4 in the Kuwaiti bootheel to drop their bombs on any tanks or artillery spotted there. Airborne "killer scouts" patrolled the boxes from above to pinpoint targets for their heavily armed partners, known as "killer bees."

But to limit the strategic campaign now by further concentrating the air fleet against the Iraqi army was folly, Glosson and Deptula believed, a strategy as ill conceived as an Army directive in the early days of manned flight that limited an aircraft's flying radius to the distance covered by ground troops in a day's march. The Iraqi army would shrivel and then be destroyed in place — but in due time.

The Army had a different view. Corps and division commanders, while admiring the efficacy of the strategic campaign, wanted a voice in directing which targets would be hit south of the Euphrates, where enemy troops were concentrated. In European war games, ground commanders always weighed in with recommendations of what should be bombed fifty or even a hundred miles in front of their lines; yet in the gulf war Horner and Glosson — with Schwarzkopf's blessing — controlled all targeting beyond the Saudi border. The Army could nominate targets by asking, for instance, that aircraft attack an artillery battery five miles outside As Salman in southcentral Iraq.

But the Air Force believed that Army targeting data were often obsolete: pilots complained that in many cases Iraqi forces had repositioned or the targets had been struck already. Nominated target locations were supposed to be pinpointed on the map within a hundred meters and the position revalidated by intelligence just four hours before an air strike — requirements the Army often found impossible to meet. Moreover, the diversion of fighters to Scud hunting meant fewer aircraft available for operations in the Kuwaiti theater.

Consequently, of an average 110 Army nominations on any given day, only a couple of dozen might appear on Glosson's air-tasking order (ATO), used to orchestrate daily attacks. Those which *were* attacked often had been low on the Army's priority list. Other targets might be hit as part of the kill box strikes, but the Army frequently had no way of knowing what had been destroyed in those sorties. The Army also disliked the kill boxes because the initiative for target selection rested with pilots rather than with ground commanders. Frederick M. Franks, Jr., the VII Corps commander who would be responsible for the Left Hook flanking attack into the heart of the Iraqi army, grew agitated in his calls to Calvin Waller, the deputy CINC. "Cal, I'm not getting my share," Franks complained. "I need your help."

Walt Boomer and his Marines shared the Army's dismay, although the personal admiration Boomer and Horner held for each other ensured that the dispute would remain civil — at least on the three-star level. Unlike the Army, the Marines also possessed their own air force in the theater, the 3rd Marine Air Wing (MAW). Some Marine planes flew under Glosson's control as part of the strategic campaign. But the Marines reserved others for attacks against Iraqi forces in southern Kuwait, and eventually kept all of their F/A-18s under Marine Corps control. This in turn angered Glosson, who complained of Marine recalcitrance to Schwarzkopf.

"Buster, do you really need them? Does it really make a difference?" the CINC asked. "You can send the whole Marine air wing home as far as I'm concerned," Glosson snapped. "Okay," Schwarzkopf said, "if it doesn't make a difference, then let's just leave it that way."

For ground commanders preparing to send their troops into the teeth of Iraqi defenses, the issue was not academic. Schwarzkopf had decreed that Iraqi armor and artillery should be reduced by half before the ground offensive began. That would give the allies the overwhelming strength — force ratio, in military parlance — preferred by an attacker against an entrenched enemy. Army and Marine generals wanted to "shape the battlefield" not only by directly causing attrition of Iraqi forces but also by battering the enemy's will to resist and his ability to maneuver and resupply. Should the offensive begin before the Iraqi strength was reduced, higher American casualties could be expected.

Glosson, whose swagger alternately bemused and infuriated his military brethren, became a lightning rod for Army discontent. Among other prewar boasts he had asserted that if all allied combat aircraft were simultaneously flung against the Iraq army of occupation, the enemy force could be halved in ten to fourteen days. In late January, when it became apparent that the Iraqi army remained largely intact, the XVIII Airborne Corps commander, Lieutenant General Gary Luck, began asking his staff, "Where's Buster?" The quip became a taunt: "Where's Buster?"

In early February the Army took its grievances directly to Schwarzkopf, along with a scheme designed to peel some authority away from Horner and Glosson. Because the senior Army commander, John Yeosock, instinctively avoided confrontations with the CINC, his operations officer, Brigadier General Steven L. Arnold, was designated as point man.

As he rode to the MODA building from the Army's headquarters south of Riyadh, Arnold felt unvarnished dread. No officer in Saudi Arabia had taken more abuse from Schwarzkopf than Steve Arnold. Small-framed, genial, and self-effacing, winner of two Silver Stars and two Purple Hearts in Vietnam, Arnold had been given responsibility for translating the broad concept of the allied ground offensive — the Left Hook — into a detailed battle plan. Consequently, for three months he had been a prime target of the CINC's wrath.

"This is dumb! This is stupid!" Schwarzkopf would roar. "Yes, sir," Arnold would dutifully agree. "I didn't brief that very well. I didn't make the point properly. Let me try again." Arnold tried to roll with the blows; if the CINC needed to vent his frustrations by berating his subordinates, then one of Steve Arnold's contributions to the war would be to serve as a punching bag. But the battering had worn him down. He had yet to emerge from a meeting with Schwarzkopf feeling good about his efforts or himself.

To ensure the destruction of half the Iraqi armor and artillery, the Army plan called for an immediate increase in sorties flown against such targets — and greater Army influence over target selection. To guarantee that those targets would indeed be struck, Arnold proposed an arbiter who could properly apportion the aircraft needed to do the job: Cal Waller. Arnold knew that the deputy CINC sympathized with his Army colleagues. Waller agreed that reducing the Iraqi army from the air would take time and had to begin in earnest. "It's like trying to stuff spaghetti up a wildcat's ass: you don't achieve much and you get your hand covered with scratches," he had observed. Waller also shared the Army's wariness of Glosson. For weeks he had watched Glosson's adroit handling of Schwarzkopf — "playing the CINC like a Stradivarius," as he later put it. Listening to Glosson "hum and woof and throw out all this pilot talk," Waller on more than one occasion had buried his head in his hands and muttered, "Holy Christ!"

Arnold spent thirty minutes making his case in the small amphitheater down the corridor from the MODA war room. "As we get closer to G-Day it's very important that we have more leeway about where we put the air," he urged.

Waller, sitting at Schwarzkopf's elbow, endorsed himself as a new air battle manager. "You, as the land component commander, should be directing Horner and telling him what to emphasize," he advised the CINC. "They're just pounding the living daylights out of the strategic targets, but we ought to be devoting as much effort to targets in front of the corps to shape the battlefield. Since you don't have the time to do this, you ought to put out a message authorizing me to do it."

Schwarzkopf listened with uncharacteristic reserve, his face drawn with exhaustion. The CINC had been reluctant to tinker with the air campaign; both Waller and Arnold believed that he hoped air power alone could defeat Saddam, obviating any need for a potentially costly ground attack. But to Arnold's surprise, no temper tantrum materialized. Instead, the session ended with Schwarzkopf concurring with the Army proposal. The Iraqi army, he agreed, should be attacked with great vigor. And Cal Waller could arbitrate the Army nominations and make certain that the airplanes flew where they were supposed to fly.

"If any flights are not attacking the Iraqi land army," Schwarzkopf subsequently told Glosson, "I want to know why." The strategic campaign would continue, but the CINC made it clear that his focus was shifting in anticipation of the ground war. "The corps commanders say we're not meeting their priorities," the CINC also told Glosson. "I've got to get them thinking that we're listening to them and supporting whatever they need." Waller would "adjudicate" target nominations from the ground commanders "so they can't blame you for it," Schwarzkopf added.

Buster Glosson was too loyal — and too prudent — to ignore his commander-in-chief. But in his journal he scribbled, "This is a sad day . . . because we've shifted our focus prematurely from what we'd been asked to accomplish to preparations for a land campaign."

After the initial surge of twelve hundred daily sorties flown against strategic targets in the first few days of the war, Glosson had planned to keep at least five to seven hundred sorties a day flying strategic missions for three weeks. Already pilots flew three hundred daily sorties in close air support missions, another three hundred against the Republican Guard, and hundreds more against other targets in the Kuwaiti theater. (On January 30, for example, twenty-eight B-52 strikes alone dropped 470 tons of explosives on the Guard.)

Glosson privately railed against what he described, with some hyperbole, as "this foolish and lightweight" abandonment of the strategic campaign. He had long suspected that the cabal of "green suiters" running the war — Army officers like Powell, Schwarzkopf, and

Waller — had at least tacitly inflated the Iraqi threat to justify the presence of two enormous Army corps in Saudi Arabia. All of the services, he knew, were trying to prove their worth in anticipation of savage budget cutting now that the Warsaw Pact had collapsed. "You can take their wives," Horner once joked, "but don't take their budgets."

Sorties against strategic targets plummeted to about 250 a day, fewer than half the number Glosson wanted. Many of them, moreover, were flown by electronic warfare planes and fighter escorts rather than bombers. By the first week of February, only the F-117 stealths and one package of twelve to fourteen F-111s from Saudi airfields consistently hit targets north of the Euphrates.

"Your objectives in the strategic air campaign have not been met," Glosson warned Schwarzkopf. Bad weather, inaccurate bombing, and the Scud diversion meant that some command centers, military factories, and nuclear, biological, and chemical sites remained intact. But the CINC held firm. Instinctively, his attention was drawn to the coming ground war; most strategic targets had been damaged to his satisfaction. "We are progressing enough," he told Glosson. "We're having an impact."

Rather than fight the green suiters head-on, Glosson resisted obliquely. Schwarzkopf's targeting directives, handed down in his nightly seven o'clock meeting, might require "interpreting" — the results often at odds with Army expectations. Or airplanes would suddenly be diverted to strategic targets because of last-minute intelligence information. Or Glosson might convince Schwarzkopf of the need to batter a particular target above the Euphrates one more time. "Okay, do it," the CINC would say, and another attack package slated for strikes in the Kuwaiti theater would instead head farther north.

Army commanders became more upset. Fearful of being outfoxed, they believed — wrongly — that little had changed in the targeting priorities. But control over the targeting remained largely in Air Force hands. Finally, Waller had had enough. The DCINC confronted Glosson one night. "Henceforth, now and forever," Waller warned, "if anybody diverts aircraft without my knowledge, I'm going to choke your tongue out." Glosson protested his innocence and promised to conform to the DCINC's wishes. Ultimately, of 3067 targets nominated by the Army for the air-tasking order during the war, slightly more than a third would be flown.

Differences both in philosophy and culture lay beneath this squabble. Glosson and his Air Force colleagues sought to defeat the nation of Iraq with attacks on the heart of the enemy's political and military infrastructure. The Army and Marines were more concerned with destroying the Iraqi army of occupation. Army commanders concluded that the air planners were waging a separate war out of a conviction that victory could be achieved through air power alone, contrary to the Army belief that it would take the "synergy" of combined air and ground power to eject Saddam's forces from Kuwait.

Many Army and Marine strategists believed the Air Force was trying to prove its worth at a time when one of the service's primary missions — maintaining the U.S. fleet of intercontinental bombers and land-based nuclear missiles — appeared increasingly obsolete. If air power defeated Saddam, the Air Force could still claim a position as first among equals in the post–Cold War restructuring of America's military.

Also at play, of course, was decades-long sibling rivalry. Each service nurtured officers loyal to their uniform, proud of service tradition, and steeped in schools of strategic thought — variously stressing land, sea, or air power — that were not always reconcilable. "The military services," a RAND Corporation study observed in 1987, "have acquired personalities of their own that are shaped by their experiences and which, in turn, shape their behavior." (Interservice bickering after World War II drove Eisenhower to a four-pack-a-day cigarette habit.) Cooperation and camaraderie certainly outweighed sniping and dissent during the gulf war, but mutual suspicion cast its long shadow.

One other episode in early February contributed to Glosson's conviction that the strategic campaign was unraveling. Alerted by a spy in the Iraqi capital, American intelligence detected three radio-controlled MiG-21 drones at Al Rashid Air Base in southern Baghdad. Each carried a metal tank of the type used to drop chemical weapons on enemy troops during the Iran-Iraq war. Concerned about the risk of drones with chemical payloads getting airborne, Glosson directed that the aircraft be attacked. F-117s destroyed one MiG in its hangar, but bad weather prevented prompt attacks on the other two.

After talking to Schwarzkopf, Glosson ordered Tomahawk missiles launched against the Al Rashid hangars. The attack called for six TLAMs at noon on February 1. Navy planners on *Blue Ridge,* working under a tight deadline, initially scheduled three missiles from the Mediterranean and three from the Red Sea. On discovering that the former would have to cross Syrian airspace in daylight, however, the planners shifted all six shots to U.S.S. *Normandy* in the Red Sea.

As planned, the missiles streaked toward Baghdad in a neat line from west to east at midday. But instead of attacking almost simultaneously, they crossed the capital at roughly sixty-second intervals in full view of Western television cameras. The Navy later concluded that five missiles struck the hangars, with a sixth possibly shot down; the Iraqis claimed that at least two crashed or were destroyed in flight. For whatever reason, the attack appeared to cause civilian casualties. Reporters saw substantial damage in Baghdad's Karada neighborhood, where eighteen people — including several children — were reportedly killed or wounded. "It was so powerful that my entire house is gone!" cried an Iraqi merchant, standing before the charred wreckage of his home.

Colin Powell had shed much of his antipathy toward the Tomahawk, but he now considered the weapon to have outlived its usefulness in this war. The Navy had fired 288 missiles, about 50 percent of its theater stockpile. Slightly more than half had struck their intended targets; hit or miss, each cost roughly $2 million. The chairman also had begun to suspect that attacks on the Iraqi capital were reaching a point of diminishing returns. The televised images of marauding missiles juxtaposed with maimed civilians were unsettling. Like Schwarzkopf, Powell found his gaze drawn to the enemy army south of the Euphrates. "Jesus Christ," he told the CINC, "every time you pull the trigger it's another two million dollars." Schwarzkopf in turn talked to Glosson. "I don't want to see any more damned TLAMs flying around Baghdad in the daytime," he said.

"Well, you either have TLAMs flying in the daytime or you don't have anything," Glosson replied. "Because I am not sending airplanes in there during the day to get shot down."

"The chairman's upset," Schwarzkopf pointed out, "so we'll have to watch that." A day later the CINC imposed an even more rigid restriction: "Don't launch any more TLAMs unless I tell you to."

Glosson and Horner considered the Tomahawks an important complement to the nightly air raids on the Iraqi capital. Together they kept both enemy leadership and populace in a state of unease. The concept was similar to a strategy presented to Winston Churchill during the Allied conference at Casablanca in January 1943: "By bombing the devils around the clock, we can prevent the German defenses from getting any rest." A computer printout tacked to a wall in the Black Hole expressed the current Air Force view: THE WAY HOME IS THROUGH BAGHDAD.

But the Tomahawk war was over. The Al Rashid shots would be the last of the war. Several times Glosson appealed for additional TLAM missions, but Schwarzkopf held firm, convinced that any target slated for destruction could be bombed by manned aircraft. For the last month of the war Baghdad was granted a daytime respite from allied attack. That reprieve, Glosson believed, was a mistake.

The Joint Chiefs' operations officer, Tom Kelly, always on his toes when facing a pack of journalists, offered an ingenious reply when asked about the TLAM damage in Baghdad. "We have been doing everything we can to avoid collateral damage and we've worked very hard on that. If you've been watching television reports coming from Baghdad, you've seen a lot of neighborhoods that certainly were not struck."

But Western audiences failed to see many neighborhoods that *were* struck. Contrary to the Pentagon's antiseptic portrait of a war fought with nearly flawless precision, bombs went awry, targets were misidentified, civilians died. By the end of the war air strikes killed nearly 2300 and injured six thousand Iraqi noncombatants, according to figures provided by Baghdad to the United Nations and believed to be reasonably accurate. The toll was remarkably close to that predicted by General Tony McPeak, the Air Force chief, who had warned Bush in December, "You're going to kill two thousand people you're not mad at."

Without question the Americans and their allies took great pains to minimize "collateral damage," for reasons both humane and political. Most senior officials recognized that civilian carnage could inflame the Arab world and undermine the coalition's moral standing. American planners believed, with some justification, that they worried more about Iraqi civilian casualties than Saddam did. Pilots designed bombing runs to avoid unintended targets, even to the point of putting themselves in greater jeopardy. A ten-page "no fire" list proscribed mosques, hospitals, schools, archaeology digs, and dozens of other sites. No consideration was given to inflicting the sort of destruction that had been visited on Dresden or Tokyo in World War II.

In the main those efforts succeeded. During the Normandy invasion in 1944, one civilian died for approximately every four tons of bombs dropped; in Vietnam in 1972, one died for every fifteen tons. In the gulf war, if the Iraqi civilian tally is accurate, there was one noncombatant death for roughly every thirty-eight tons. And some of the damage may have been caused by descending surface-to-air missiles and antiaircraft fragments.

Yet in an attempt to deprive Baghdad of propaganda fodder, American civilian and military leaders resorted to absurd overstatement. Bush pronounced the bombing "fantastically accurate." A few days later he added, "This war is being fought with high technology." Tony McPeak's attempts to release a single snippet of gun camera video showing a laser-guided bomb hitting the wrong target in Baghdad were blocked in the Pentagon. In early February the CENTCOM chief of staff, Major General Robert Johnston, declared, "I quite truthfully cannot tell you of any reports that I know of that would show inaccurate bombing . . . I cannot tell you of any that I know that have grossly missed the target."

Johnston was an honorable officer, but his claim was ridiculous. Of 227,000 allied air munitions dropped during the war, 93 percent were dumb bombs, usually dropped from high altitude and in high winds that buffeted the attacking planes; imprecision was inevitable. Delivery systems had come a long way from the old "TLAR" method — "that looks about right" — yet they were far from flawless.

If darkness, confusion, or enemy fire caused an attacking F-16 pilot to fly twenty knots too slow, his first bomb would hit sixty feet short of the target; a dive angle that was five degrees too shallow would leave the bomb 130 feet short. An F-111 crew out of Turkey, dropping 500-pounders from 23,000 feet, typically released fourteen bombs that detonated on a path eighteen hundred feet long; an error of only three hundred feet in calculating the exact altitude above the target would alter the bomb trajectory enough to cost two hundred feet in accuracy. And B-52s, hardly instruments of surgical precision, would drop 72,000 bombs during the six-week war. As a survivor of a B-52 raid in Vietnam wrote, "One lost control of bodily functions as the mind screamed incomprehensible orders to get out."

Dumb bombs, according to postwar estimates, hit their targets only 25 percent of the time. (Even that was a significant improvement from the early days of World War II, when only one British bomber in five laid its payload within five miles of the target. "Precision bombing," the historian and critic Paul Fussell wrote, became a "comical oxymoron relished by bomber crews with a black sense of humor.") Bad weather, smoke, and haze forced Navy A-6 pilots in the gulf to forsake laser-guided targeting equipment for far less accurate radar bombing on a third of their missions.

Even the F-117s with their precision-guided munitions were bedeviled enough by clouds, enemy gunfire, and pilot error to miss their targets with at least one bomb out of four — and more than that on some missions. The three waves of stealth fighters flown on the night of January 30 were not atypical. Wave one — dispatched against bridges, communications facilities, a telephone exchange, and Ali al Salem Airfield — reported nine hits and five misses. Wave two struck more bridges, three airfields, and communications targets at Basrah and Umm Qasr in southeastern Iraq, with sixteen hits and twelve misses recorded. The final wave involved seven planes — three others aborted because of technical problems — that hit ammunition dumps and suspected chemical and biological facilities at Salman Pak and Abu Ghurayb; these tallied eleven hits, one miss, and two "no-drops" because of foul weather.

The consequences of inaccuracy could be horrific. In Najaf, south of Baghdad, allied bombs damaged fifty houses on January 20; in one flattened building examined by reporters, thirteen of fourteen family members reportedly died. In Al Dour, northwest of the capital, twenty-three houses were demolished; the correspondent Peter Arnett interviewed one weeping woman sitting amid the wreckage of her home where her three brothers, their wives, and eight children died. In Diwaniyah, the bombers that destroyed the telephone exchange on January 17 also shattered an adjacent hotel and apartment building, killing fifteen. An errant attack on a bridge in Nasiriyah on February 4 killed fifty. Bombers inadvertently struck a hospital near Kuwait City, killing two Indian medical workers, the wife of an Egyptian doctor, and the two-year-old son of a Filipina nurse.

Basrah was especially hard hit. According to evidence compiled by William M. Arkin of Greenpeace, who conducted the most thorough on-site survey after the war, the Ma'qil neighborhood was struck three times at a cost of 125 dead. The Athiriya neighborhood lost thirty-five dead on January 29. Another thirty-five were killed in the Hakimiya neighborhood on February 4. In some cases targets hit purposely may have been misidentified. Bombers demolished warehouses containing food and consumer goods in Kut, Samawah, Basrah, and other cities. A four-acre plant in Baghdad's western outskirts was destroyed in raids on January 20 and 21. Allied intelligence believed the facility was used for biological weapons research; the Iraqis claimed that the plant manufactured infant milk formula.

"Its mystery," the British air power analyst J. M. Spaight wrote of the bomber in 1930, "is half its power." The other half was terror, havoc, and death. By exercising reasonable care, the allied leadership avoided wanton killing, but thousands died nonetheless, directly or indirectly — perhaps tens of thousands. The sanitary conflict depicted by Bush and his commanders, though of a piece with similar exaggerations in previous wars, was a lie. It further dehumanized the suffering of innocents and planted in the American psyche the unfortunate notion that war could be waged without blood, gore, screaming children, and sobbing mothers.

Two weeks into the war, Schwarzkopf appeared before the Riyadh press corps to offer his appraisal of the conflict. He delivered a bravura performance. Wielding a pointer like a rapier, the CINC nimbly summarized the allied strategy and tallied the targets attacked — airfields, bridges, headquarters, convoys. With Glosson at his side, he also narrated several gun camera tapes of successful air strikes. One snippet showed an Iraqi truck driver crossing the Mufwultadam bridge moments before it was demolished with a 2000-pound bomb — the closest the public would come to seeing a human being in the cross hairs of an American weapon. Schwarzkopf pronounced the driver "the luckiest man in Iraq."

The CINC closed his briefing with a brief homily on Baghdad's mendacity. "With regard to Saddam Hussein saying that he has met the best that the coalition has to offer," Schwarzkopf added, "I would only say: the best is yet to come." And so, too, the worst.

## Al Kharj, Saudi Arabia

The exodus of Iraqi aircraft to Iran came as an unpleasant surprise to Buster Glosson, who before the war had postulated that any fleeing enemy pilots would head west toward Jordan. In anticipation, Glosson adopted a code phrase to use with his fighters: Horner's buster. ("Buster," by an appropriate coincidence, was pilot slang for full throttle.) "If I give you a Horner's buster call, you will shoot down that fleeing airplane even if you have to run out of gas to do it," Glosson ordered. "I'll pick you up and get you another airplane, but you *will* shoot it down."

Instead, the Iraqis fled east by the score. To thwart the migration, Horner and Glosson in late January inaugurated a tactic known as a "bar CAP" (barrier combat air patrol). Three squadrons of F-15C Eagles began flying round-the-clock surveillance over routes most commonly used by the escaping Iraqi pilots. The various CAP sectors were designated by women's names: Carol, Charlotte, Elaine, Emily, Cindy.

Although flying bar CAP was tedious — and some pilots found themselves disconcerted by the prospect of shooting a fleeing enemy in the back — most welcomed the chance to down a MiG or a Mirage under any conditions. Iraq's refusal to challenge the allied fighters had frustrated the Americans and intensified the competition between sister units. (Glosson's staff in Riyadh proposed assigning the radio call sign "Whiner" to the 1st Tactical Fighter Wing, which seemed excessively eager to push aside rival squadrons in an effort to tally more MiG kills.)

The war had not produced a single ace, defined as a pilot destroying at least five enemy planes. Historically, fewer than 3 percent of all American fighter pilots were aces, but they accounted for 40 percent of all enemy planes shot down in air-to-air combat. In past wars air combat had featured glorious, swirling dogfights. Over North Korea in September 1952, for example, thirty-nine U.S. F-86s fought seventy-three MiG-15s; four American planes and thirteen Chinese were lost. Nothing remotely similar unfolded over Iraq.

Of the bar CAP sectors, Cindy CAP offered a fair chance of snaring enemy planes because of its proximity to the large air bases around the Iraqi capital. Cindy lay in a narrow swath of airspace between the Iranian border and Baghdad's eastern surface-to-air belt. Typically, four Eagles at a time from the 53rd Tactical Fighter Squadron patrolled the sector. Two fighters would orbit counterclockwise in a thirty-mile oval while the other pair shuttled south to refuel. Each mission lasted six to seven hours — Cindy CAP lay seven hundred miles north of the squadron's base at Al Kharj — as the pilots threaded the narrow channel between Iraqi SAMs and Iran, which repeatedly threatened to shoot down any American straying into Iranian airspace.

At 2 P.M. on January 30, Cindy CAP was manned by four aircraft led by the squadron commander, Lieutenant Colonel Randy Bigum. It was Bigum who had caused such perturbations in his squadron a month earlier by dividing the forty pilots evenly into day and night fighters. Two weeks of war had widened the schism. The night shift, including the redoubtable Killer Miller, had more grievances than ever: difficulty sleeping during the day; more intense (or at least more visible) Iraqi SAM and antiaircraft fire; greater worries about midair collisions; and, most aggravating of all, no opportunities to shoot down a MiG because the enemy refused to fly at night. For its part, the day shift had become vexed — and perhaps a bit guilty — at hearing their nocturnal comrades complain.

Bigum was aware of the night shift's unhappiness, and he also continued to fret over drug use in the squadron. By sharply restricting the distribution of go and no-go pills, he believed he had resolved most of the problems. But some men seemed at least psychologically addicted to the no-go tablets — unable to sleep without the sedatives — and there were rumors of bootleg pills at Al Kharj. Bigum had grounded two pilots for medical reasons, one indefinitely and the other for three days. The squadron commander himself felt frayed from the endless combat missions and the effort to hold his unit together. "Sir," one of his captains had recently confided, "I think you're overstressed." The suggestion so alarmed Bigum that he visited the flight surgeon. "You're the guy in charge of me," he told the doctor. "If you tell me that I'm slipping, I'll take corrective action — maybe even step aside and go home. I'll relieve myself of command if I have to."

In what had been a routine afternoon sortie, Bigum and his wingman, Captain Lynn Broome, had refueled down south and were returning to Cindy CAP to spell the other two Eagles when AWACs radioed a warning of enemy aircraft: "Xerex three-one. Snap, two seven zero. Bandits, three zero zero. Ninety miles." Bigum rolled west ninety degrees, searching toward Baghdad with his radar. AWACs called again: "Skip it. Skip it. Bogus targets." Bigum and Broome steered back north, irritated at the false alarm.

Eighty miles south of the CAP, AWACs called a third time: "Bandits west, seventy miles. High. Fast." This time it was real. A pair of MiG-25 Foxbats, flying at 42,000 feet and an astonishing one thousand knots — faster than an F-15's top speed — streaked from the Iraqi capital toward Cindy CAP. The two Eagle pilots on CAP, flying under call signs Vegas and Giggles, turned to face the enemy fighters. Giggles, slightly in front of his wingman, fired two Sparrow air-to-air missiles at the lead Foxbat, which in turn fired at Vegas. The Foxbat banked north in a sweeping turn at twice the speed of sound, outrunning both Sparrows.

Vegas peeled south to avoid the enemy missile. He then re-entered the fight and fired three Sparrow missiles at the second Foxbat, but for reasons never determined, none of them left the Eagle's wings. Vegas, alarmed, broke south. Giggles fired a final, futile missile at the fleeing MiGs and turned to protect his wingman.

Randy Bigum watched this drama unfold on his radar scope. The Iraqis, he realized, had tried to ambush the planes patrolling Cindy CAP; they were not simply fleeing to Iran. Having failed, both Foxbats now curled back west with their afterburners lit, evidently heading toward Al Taqaddum Air Base on the far side of Baghdad. Bigum turned to give chase. If he and Broome angled south of the capital, Bigum calculated, they might cut off the Iraqis.

The race began. Bigum kept his eyes on the radar scope; Broome was trailing by thirty miles over his left wing. In their war with Iran the Iraqis occasionally tricked enemy pilots into giving chase, only to destroy them with a sudden attack from below by Mirage F-1s. Bigum was so intent on avoiding such a trap and watching the Foxbats that he failed to note a 140-knot southwesterly wind pushing him far to the north. Only when he glanced out the cockpit hoping to spot the Iraqi contrails did he see his mistake. There below lay the presidential palace, the sun-spanked Tigris, and the office buildings of downtown

Baghdad. At the same time the Eagle's electronic warning gear detected emanations from SA-2 and SA-3 tracking radars. "Oh, my God," Bigum muttered. From AWACs came a gratuitous radio call: "Heads up for SAMs."

But the SAM batteries failed to launch, probably afraid of hitting the Foxbats. Bigum again concentrated on the enemy fighters, now twenty miles away. Each had performed a split S — an acrobatic half loop — and dropped almost to the ground. Broome fired two Sparrows at the trail Foxbat; neither hit. Bigum angled down to twenty thousand feet and glanced up long enough to see the twin runways of Al Taqaddum ten miles dead ahead. The lead Iraqi had slowed from a thousand knots to under three hundred, drifting into his final approach to land from the northwest.

Now Bigum fired. The Sparrow darted from under his plane and climbed sharply before knifing back toward the ground, a sign that the missile had locked onto its target. Bigum watched as the first Iraqi landed and rolled down the runway. "Come on, bitch!" he urged the missile. "Come on, bitch!" But the Sparrow never made it. The Foxbat had slowed to a forty-knot taxi, and the radar-guided missile could no longer distinguish between aircraft and ground clutter.

Then the trail Foxbat floated into view a mile from the western end of the runway, landing gear down. Bigum squeezed off another missile. Again the Sparrow climbed and dived. By this time Bigum had descended to eight thousand feet, directly over the airfield. Only concern at hitting the MiG, he guessed, had kept the Iraqi gunners from firing at such an easy target. As he banked left to escape, the second Foxbat touched down. Bigum saw the curve of the pilot's helmet and puffs of smoke spurt from the tires. Ten feet from the Foxbat's left wingtip, the Sparrow plunged into the runway and exploded. The Iraqi taxied unscathed toward the flight line.

The Eagle pilots had fired ten missiles to no effect. A week later Vegas and Giggles would destroy four Iraqi fighters fleeing toward Iran. But for Bigum, the chance had come and gone, never to return. If fortune had robbed him of two kills, it also had permitted him to fly without penalty across downtown Baghdad and Taqaddum in midday. The lesson was not lost. In the squadron ready room Bigum tacked up a sign: "Don't let your eagerness to get a MiG cause you to be our first casualty."

The narrow escape also proved cautionary for the Iraqis. In the final week of January, enemy fighters had averaged more than twenty sorties a day, including escapes to Iran. Now the flights stopped completely. In the first few days of February, not a single Iraqi aircraft left the ground. Saddam still owned Kuwait, but Glosson and his pilots owned the sky.

## Riyadh

"A general," Napoleon once observed, "never knows anything with certainty, never sees his enemy clearly, never knows positively where he is." The maxim was as true for Norman Schwarzkopf in the Persian Gulf — despite the proliferation of spy satellites and reconnaissance aircraft — as it had been for a field marshal squinting through the smoke at Austerlitz with a field glass. Many questions troubled the CINC during the war, yet none more than the difficulty of seeing the enemy clearly and determining where he had been weakened and where he remained strong. In modern warfare the art of estimating the harm inflicted on the enemy is known as battle-damage assessment — BDA — and from the moment the first bomb detonated in the Persian Gulf War it was a source of controversy, squabbling, and vexation.

One episode involved Glosson. When he appeared with Schwarzkopf in the CINC's televised press conference on January 30, he brought a gun camera tape taken by an F-15E during a night attack on a convoy of Iraqi vehicles. Glosson had asked several experienced photo intelligence analysts in Riyadh to review the footage, and they had assured him — one jokingly said he'd bet $1000 — that the destroyed trucks were mobile Scud launchers and support vehicles. Eager for evidence to confirm success in the campaign against Iraqi missiles, Schwarzkopf and Glosson proudly rolled the tape for Riyadh reporters and a worldwide television audience. "We knocked out as many as seven mobile erector launchers in just that one strike," the CINC declared.

Among those watching were analysts at the CIA's headquarters in Langley, Virginia. "My God," one exclaimed as cluster bombs ripped through the convoy, "those are oil tankers!" Some analysts at the Defense Intelligence Agency thought the vehicles might even have been milk trucks. Word of those opinions reached Riyadh within minutes. Schwarzkopf normally flushed when he was upset; this time he whitened with anger. Glosson, after an outburst of disbelief — "You've got to be shitting me!" — summoned the photo analysts to his office. "How could this happen?" he demanded. "How could we mislead the American people like this?"

The incident soon blew over — the "misled" American people remained ignorant of the contretemps — but other BDA disputes continued. Gauging how badly Iraq had been hurt was not an end in itself but a means of determining how much fight the enemy had left. Schwarzkopf's most important decisions, such as when to launch a ground attack, required a detailed assessment of Iraqi strength.

To help the CINC, the United States had cobbled together an immense intelligence operation. Twenty-three different types of aircraft eventually flew over Iraq and Kuwait gathering information; among them were the high-flying U-2s, which shot more than a million feet of film during the war. A half-dozen satellites wheeled overhead, each taking hundreds of images with telescopic, infrared, or radar sensors. The so-called Keyhole satellites could discern objects only six inches in diameter; the Lacrosse, designed to track Warsaw Pact armor, could see through clouds with sophisticated radar. Two other satellites, the Magnum and Vortex, intercepted Iraqi communications. Electronic spy planes, with code names like Rivet Joint and Senior Span, also eavesdropped on the enemy.

Hundreds of analysts scrutinized the images on light tables at the National Photographic Interpretation Center in the Washington Navy Yard. Hundreds of others worked in the Pentagon's Joint Intelligence Center (JIC) or the makeshift research center set up at Bolling Air Force Base outside Washington. Five hundred more labored in CENTCOM's own JIC, built in the MODA basement, or the JIPC, the Joint Imagery Production Complex, at Riyadh Air Base. By war's end, there were two hundred tons of intelligence "product" — including countless sheaves of satellite and U-2 photos — in Saudi Arabia. The Americans commandeered Saudi bread trucks to haul the stuff around.

The sheer volume of this prodigious effort created problems. Analysts could not process satellite and reconnaissance photos fast enough to keep pace with an air war featuring two to three thousand sorties a day. On the other hand, unusually heavy cloud cover sometimes thwarted satellites and reconnaissance planes. Overcast prevented the allies from reconnoitering most strategic targets in Baghdad and elsewhere until five days after the bombing began, a delay that put analysts behind from the outset. The Iraqi SAM threat also kept spy planes from flying over portions of the battlefield in the first weeks of war; even when the planes could fly, BDA missions often lagged several days behind the attacks, giving the Iraqis time to reposition equipment or otherwise confuse the American analysts.

Unlike the saturation bombing of World War II, when destruction of a ball-bearing plant or aircraft factory could be gauged by the depth of the rubble or square footage of roof demolished, the damage wrought by precision-guided munitions was often hard to assess. A 2000-pound laser-guided bomb might punch a hole in the roof and vaporize the contents of a building, but BDA analysts, limited to photographs of the penetration hole, would report, "Possible damage to roof." The system was so cumbersome and ineffectual that Glosson and Dave Deptula jury rigged their own BDA operation in the Black Hole; they scanned gun camera footage and interviewed pilots by phone to determine which targets required additional strikes.

Schwarzkopf's intelligence chief was the earnest, intense one-star Army general John Leide. Leide, who held a law degree from Syracuse University, had commanded rifle companies in the Dominican Republic and Vietnam, as well as a Special Forces battalion at Fort Bragg. In 1989, he was serving as the U.S. military attaché in Beijing when the Chinese government brutally crushed democratic dissidents in Tiananmen Square. None of those assignments, Jack Leide soon concluded, had been as stressful as trying to make sense of the intelligence picture in Riyadh.

Exhausted by a long succession of twenty-hour workdays and repeatedly savaged by Schwarzkopf's temper, Leide was so close to collapse that Waller feared he would suffer a nervous breakdown. (In one typical exchange, Schwarzkopf, after being told that a particular Iraqi unit was believed to be "sixty-four percent combat effective," snapped, "Not sixty-five percent? Not sixty-three? Goddam it, you don't know what the hell you're talking about, do you?") "Jack, don't promise the CINC the moon," Waller advised. "There will be times when you'll have to tell him, 'Sir, I just don't know.' "

Under Schwarzkopf's instructions to "use sound military judgment," Leide and his staff tried several innovations to get through the BDA morass. Rather than rely wholly on empirical evidence of damage done to the twelve target sets, they began offering subjective estimates based partly on intuition of how effectively the Iraqis could function. Each target set was illustrated with a chart that displayed levels of damage ranging from slight to moderate to severe. Objective damage was marked with an O, subjective estimates with an X.

In February, for example, damage to Iraqi airfields was rated objectively at moderate; subjectively, since the Iraqis had stopped flying except for the occasional dash to Iran, the damage was rated closer to severe. In like fashion, objective damage to Scud targets was rated between slight and moderate. But because the missile launches had dwindled steadily, the subjective X was placed at moderate. Schwarzkopf, always quick to appreciate common sense, liked the system.

Tallying the destruction of Iraqi armor and artillery was more complicated. Schwarzkopf assigned responsibility for estimating attrition in the Kuwaiti theater to the U.S. Army. At first the Army intelligence chief, Brigadier General John F. Stewart, Jr., decided to use only SIGINT — signals intelligence from intercepted radio communications — and photographic evidence.

The system quickly proved inadequate. For one thing, Iraqi troops rarely talked on the radio. Furthermore, the satellite cameras on which the Americans initially relied could take wide-angle shots that were too blurry for accurate BDA, or high-resolution photos of individual targets — a tank here, an artillery tube there — that precluded a comprehensive understanding of how the enemy tank or artillery battalion had fared as a whole. Stewart lamented that the task was like trying to make sense of a televised football game in which the cameras focused only from afar on the entire stadium and surrounding city, or up close on only one linebacker.

In late January, Stewart began adding pilot reports to the mix, but he counted no more than half the reported kills — on the assumption that pilots overestimated their prowess — and only those from A-10s. Warthog pilots, Stewart reasoned, flew slower and lower to the ground, and they had wingmen who could help confirm the destruction.

In early February the formula was adjusted yet again to count 100 percent of U-2 photographs — the spy planes had begun flying directly over Iraqi positions with high-resolution H-cameras — as well as 50 percent of the gun camera videos from F-111s. A-10 pilot reports were devalued to count only a third of their estimated kills. If the system seemed haphazard, at least it established a baseline for estimating enemy losses. On February 1, for example, the Army estimated that 476 Iraqi tanks had been destroyed; by February 6, the number was 728.

A rift soon developed between the CIA and Riyadh. To avoid the "light at the end of the tunnel" optimism that had characterized reports from the field in Vietnam, the CIA charter called for autonomous analysis that gave the president an impartial and independent view of the war. But in providing timely BDA, the agency was limited for the most part to satellite images. (Pictures from U-2 and RF-4 spy planes, for example, often arrived in Washington four or five days late.) Cautious and inherently conservative, CIA analysts used elaborate "keys" to translate visible destruction into damage estimates. Often their estimates ran counter to those in Riyadh, where the assessors had benefit of pilot reports — known as "ego BDA" — defectors' accounts, gun camera footage, and a dozen other sources of information that helped provide them with a feel for the battlefield.

The schism was deepest in regard to the question of damage to Iraq's tanks, personnel carriers, and artillery tubes. As the number of sorties against such targets grew, the gap between CIA and CENTCOM damage estimates became so wide that Riyadh asserted, for example, that four times more tanks had been destroyed than the CIA could confirm. In the CIA view, CENTCOM was dangerously optimistic. In Riyadh, conversely, CIA reliance on satellites was likened to the physical limitations of the "one-armed paper hanger." Among other problems, satellites could cover less than 20 percent of the tactical targets struck on any given day; furthermore, unless a tank's turret or tracks had been blown off, imagery analysts had difficulty determining whether the target was destroyed.

Underlying the dispute was a struggle for control between two bureaucratic organizations, a turf battle exacerbated by the inevitable friction between a field headquarters and Washington. Also at play was the preoccupation with what Cal Waller called "the dead-things count." The effort to quantify Iraqi losses — part of the CINC's desire to destroy half of the enemy's equipment before launching the ground war — assumed a life of its own. Schwarzkopf sensibly took pains to avoid the obsessive body-count mentality so prevalent in Vietnam. But in the gulf war, the tank count became its equivalent.

Using John Stewart's estimates and other intelligence, Jack Leide came up with one additional innovation. He installed a map of the Kuwaiti theater in a corner of the MODA war room. Each Iraqi division in Kuwait and southeastern Iraq was represented with a small paper sticker. A green sticker marked a division judged to be 75 to 100 percent combat effective, with most of its equipment unscathed and the unit's fighting capacity largely intact. A yellow sticker represented a division 50 to 75 percent combat effective; a red sticker meant the division was less than 50 percent intact and no longer considered a serious threat.

Leide, who knew that Schwarzkopf was happiest when intelligence was presented in simple packages, referred to the map as the Cartoon. The CINC stared at it with ferocious intensity and soon memorized the status of all forty-two divisions. With a thick forefinger he would jab at a particular sticker, directing Horner and Glosson to sharpen their air attacks on the enemy's 14th Infantry or 12th Armored or Tawalkana Republican Guard division. Slowly but inexorably the Cartoon became a vivid mosaic, as green began to give way to yellow, and yellow to a bright and bloody red.

## U.S.S. *Blue Ridge*, Persian Gulf

After more than a week of gentle prodding from Vice Admiral Stan Arthur, Schwarzkopf agreed that the time had come for a decision on Desert Saber, the proposed amphibious landing at Ash Shuaybah. Shortly after 11 A.M. on February 2, Arthur's fleet helicopter, *Blackbeard Zero One*, touched down on the *Blue Ridge* landing pad. As the engine died and the twirling rotors spun to a stop, the CINC emerged from the passenger bay into the briny glare on the ship's fantail. Arthur and Walt Boomer, the Marine commander, welcomed him aboard. The admiral led the way forward to the ship's conference room on the second deck.

The Navy had a problem, and Arthur wasted no time making certain Schwarzkopf understood the obstacles facing Desert Saber. Before the Marines could be put ashore, minesweepers would have to clear a channel from the middle of the Persian Gulf to the Kuwaiti coast. But to protect the sweeper force of boats and helicopters, Iraqi missile batteries and shore guns had to be obliterated first. Arthur estimated the latter task would take a week. The minesweeping would take another eighteen days. Finally, three to five additional days of naval gunfire would be necessary to suppress Iraqi gunners who could pick off Marines landing along the Ash Shuaybah coast.

All told, the timetable called for at least twenty-eight days of preparation before the first Marine set foot on a Kuwaiti beach. That would push an amphibious landing, timed to coincide with the ground offensive by Army and Marine divisions, into early March — *if* the minesweeping went smoothly.

After the war, Schwarzkopf would criticize the Navy minesweeping force as "old, slow, ineffective, and incapable of doing the job." The CINC had a point. Sweeping carried none of the glamour of nuclear submarines, naval aviation, or Tomahawk missiles; so for decades it had received short shrift. Two ships, U.S.S. *Pirate* and U.S.S. *Pledge*, had been sunk by mines in the Korean War. Crude but lethal mines planted by the Vietcong in the Mekong Delta waterways again revealed shortcomings in American minesweeping capabilities. Naval battle plans for World War III had called for the United States to amass a fleet of large capital ships like carriers, cruisers, and submarines against the Soviet Union, while other NATO navies supplied the minesweepers. Scant attention, however, had been paid to the mine threat posed by potential Third World adversaries, and more than one allied naval officer wondered whether the Americans knew what they were doing in the gulf war. The British naval commander in the gulf, Commodore Christopher Craig, was so unnerved by an early version of the mine-clearing plan — under which Royal Navy mine hunters were to sail within a few miles of Iraqi shore batteries — that he considered the scheme "totally unworkable" and "tantamount to suicide."

U.S. intelligence estimated that Iraq possessed one to two thousand mines. Most were old-fashioned contact mines of World War I vintage, dozens of which had been spotted adrift on the gulf's currents. But the inventory also included sophisticated "influence mines," triggered by a ship's magnetic field or the acoustic throb of propellers passing overhead. Where those mines had been laid was unknown, in part because Schwarzkopf — wary of provoking the Iraqis — had restricted surveillance by Navy ships and planes before the war to an area at least fifty miles south of the Kuwaiti border. The American minesweeping force in Desert Storm consisted of two new ships (*Avenger* and *Guardian*, designed primarily for deep-water missions), three antiquated boats, and six helicopters, none of which could effectively detect mines in less than thirty feet of water. In addition to five British mine hunters, there were several sweepers provided by the Saudis, a trio of Belgian ships working primarily in the Gulf of Oman, and several others.

Mine-infested waters typically required three separate sweeps. First, helicopters dragged a long metal blade through the sea to slice the tethers holding moored mines, which were destroyed with gunfire or explosives after they bobbed to the surface. Then helicopters pulled a hydrofoil sled that simulated the acoustic and magnetic signature of a passing ship. Finally, minesweeping ships hunted any remaining mines with sonar.

Sweeping was tedious and uncertain. Turbidity, changes in water temperature, and the ingenuity of mine designers — some influence mines had sophisticated counters that prevented them from detonating until many ships had passed by — complicated the operation. The Navy calculated that two full days of sweeping a single five-square-mile swatch of gulf yielded only a 60 percent certainty that all the mines had been detected.

As he sat in the *Blue Ridge* conference room, Schwarzkopf began to get the picture. Boomer, listening for the first time to the Navy's proposed timetable, realized with dismay that Desert Saber would likely delay the allied ground attack for weeks. Nor was minesweeping the only snag. The Navy also worried about Iraq's antiship Silkworm missiles dotting the coastline, as well as potential suicide attacks by enemy aircraft bearing Exocet missiles. During the U.S. campaign against the Japanese on Okinawa, 250 of nineteen hundred kamikazes had penetrated American defenses and sunk twenty-five ships; the memory lingered nearly half a century later. As Arthur sat slumped in a chair, arms folded across his chest, one of his staff officers, Captain Gordon Holder, unveiled a collage of charts and reconnaissance photographs depicting Iraq's defenses along the Ash Shuaybah shingle. The pictures clearly showed dozens of gun pits and pillboxes now bristling from coastal apartment buildings, as well as the natural gas plant — the "nuke on the beach" — farther south.

"Every high-rise between the beach and the four-lane highway to the west is going to have to come down," Arthur warned, "or the Marines making the landing will be too exposed. I'll have to level the buildings with naval gunfire and air strikes."

Schwarzkopf looked nonplused. For months he had stressed the importance of minimizing damage to nonmilitary structures, a point that Colin Powell also had raised repeatedly. "I can't destroy Kuwait in the process of saving Kuwait," the CINC said.

"Well," Arthur said, "we also can't accept the casualties we'd take if we do a landing there without knocking down those fortifications."

Holder pointed out the natural gas plant. That too would have to be destroyed, or the Marines would be at risk from an inferno.

"You can't destroy the industrial infrastructure of this country," the CINC repeated. He turned to Boomer. "Walt, can you conduct your attack without the amphibious assault?"

Boomer paused a full thirty seconds before replying. Tall and whippet-lean, with prominent ears and a pleasant drawl, the Marine commander had long assumed that his two divisions ashore would be bolstered by their seventeen thousand comrades at sea. Fear of an amphibious landing kept at least three Iraqi divisions riveted to the coast. Without Ash Shuaybah's port, moreover, logisticians would have to push even greater stockpiles of ammunition and fuel overland to support the Marine attack through the Kuwaiti bootheel. But delaying the allied offensive in order to launch a potentially bloody amphibious assault no longer seemed justified. "Yes, sir," Boomer replied at last, "I can do it. But we'll have to continue the deception of a full-blown landing. That has to be a high priority. We've got to keep those three Iraqi divisions tied up on the coast."

Schwarzkopf had heard enough. Dead Marines awash in the Kuwaiti surf was a hard image to shake. The CINC had slept badly in recent nights, tormented by the dread of making a bad decision that would kill his men. Dark, violet pouches had pooled beneath his eyes.

He turned to Arthur. "Okay. I want you to hold the pressure on them and make the Iraqis believe you're going to go with the amphibious landing — right up to the last minute. But you'll go only if Boomer gets in trouble." Otherwise, the landing would be a feint. "Without destroying everything," Schwarzkopf directed, "I want you to continue with your battlefield preparation, using the battleships to shell targets. Make it look as if you're coming with the invasion as scheduled. But don't tear down the countryside unless we absolutely know it's something we've got to do."

Planning for Ash Shuaybah would continue, the CINC added. But the new focus of amphibious operations would be a more modest attack against the 2500-man Iraqi brigade entrenched on Faylaka Island, east of Kuwait Bay.

Arthur agreed. His staff, listening from the back of the conference room, gave a nearly audible sigh of relief.

At 2:15 P.M., Schwarzkopf climbed back into *Blackbeard Zero One* for the short flight to the mainland. If at times the measure of a commander was his willingness to accept destruction and death in pursuit of a larger objective, at other moments he could be gauged by his intuitive sense of when to draw back from the abyss. An amphibious invasion would have wiped out a significant stretch of the Kuwaiti coast; at least some Marines probably would have died, needlessly. Given the Navy's subsequent troubles with Iraqi mines, it is not inconceivable that one or more ships would have been lost.

Schwarzkopf's intuition and military judgment aboard *Blue Ridge* served him well. He had arrived at precisely the right decision, one worthy of a commander-in-chief. Desert Saber would be one less worry to roil his rest.

9

# The Desert Sea

## Baghdad

The sirens began yet again with a low growl that rose in pitch until the entire city seemed wracked with a single howl. David Eberly stirred from the floor of his cell. He sat up, bracing his back against the cold wall and tugging the green wool blanket over his head as a shield against falling debris. In the distance a bomb exploded, powerful enough to send tremors through the prison. Eberly felt the vibration from the floor carry up his spine. Although air raids had become a nightly ritual, the attacks had never come close enough to shake the cellblock. For the first time he seriously reckoned with the possibility of being killed by his own countrymen. He drew no comfort from the irony.

A narrow window, eight inches wide by five feet long, was cut into the outer wall. Partly shuttered with steel louvers and girdled in chicken wire, it afforded him a pinched and fractured glimpse of the sky over Baghdad. Directly outside, an antiaircraft gun now opened fire with a tremendous din. He saw the bright flick of tracers, crimson against the velvet night. Staring up beneath the cloak of his blanket, he thought abruptly of Francis Key watching the British bombardment of Fort McHenry. The lyrics came to him first, then the melody, threading lightly through his head despite the screaming sirens and booming syncopation of the Iraqi guns. From this he did draw comfort, amid the rocket's red glare and the bombs, the bombs bursting in air.

On the night of January 31, nine days after his capture, Dave Eberly had been rousted from his cell in the camp south of Baghdad. He was blindfolded, handcuffed, and chained. He assumed the Iraqis were staging another interrogation, but instead they shoved him onto a small bus. He was frightened and cold, disoriented at being manhandled in the middle of the night. After a drive through the city his captors led him into a building and up an elevator to the second floor. As he shuffled down a corridor, straining for a familiar sound or recognizable smell, he guessed that he was in a hospital or other medical facility. He wondered whether the Iraqis conducted experiments on their prisoners, subjecting them to cold or carrying out other research, as the Nazis had done. A heavy door swung open. A rough hand tugged off the blindfold and shoved him inside a cell. The door slammed shut.

At dawn, when light seeped through the narrow window, he saw that the building was no hospital. His cell measured nine by six feet, with red brick walls and a massive steel door. A broken toilet squatted in one corner. Plastic sheeting covered a broken water faucet. Compared with his previous cell, this one looked clean. Heavy traffic noises drifted through the window during the day; at night, the keening sirens. Occasionally a generator chugged outside. Eberly would later learn that the compound housed the Iraqi Intelligence Service regional headquarters, located east of the Tigris in central Baghdad; for now, allied prisoners called it the Biltmore.

Survival in the Biltmore, he soon discovered, boiled down to a few irreducible necessities: try to keep warm; eat anything remotely edible; use the toilet down the corridor at every opportunity. For the first two days he received no water and only a single ladle of broth served in a plastic bowl. Always slender, he was fast becoming gaunt. The outbreak of hives from his violent reaction to the penicillin tormented him. His skin cracked and festered. His fingers swelled like sausages. Eberly learned to sit with his hands in plain view but slightly cupped to protect them from the periodic slashes of his captors. He also learned to distinguish the tread of different guards — that of the sadists from that of the benignly indifferent. Sometimes he sat with his ear pressed to the steel door, straining for a few syllables of English amid the barking of the guards and the radio blare of Arab music, and, bizarrely, the occasional lilt of the theme song from the television show *Bonanza*.

Small triumphs cheered him. On the floor he discovered a few dried flakes of soap, which he scraped into a teaspoon-sized ball and hid in hopes of someday being permitted to wash himself. Working a screw loose from the drain plate, he used it as an awl to scratch another calendar on the wall; he stopped at February 28 rather than permit himself to contemplate another month's imprisonment. For exercise and warmth, he paced the cell: three steps lengthwise, two across, three steps back, two across, again and again. With the blanket wrapped around his head and shoulders to trap his body heat, he imagined that he resembled E.T.

Dehydrated and weak from hunger, he fell prey to delusions. During the night of February 2, the sound of a cleaning crew clattering down the corridor with mops and buckets awakened him. He listened as steel doors opened and shut. Finally, the door to his cell — number twenty-six — swung wide. A smiling Arab in a red and white kaffiyeh held out two pieces of warm bread and a small bottle of water. Eberly stared in disbelief. His mind raced. The Saudis have found us, he thought. They're masquerading as a cleaning crew. Maybe they'll come every night with food. Maybe they'll help us escape. The man gave him the bread and water before shutting the door, and Eberly fell back asleep, warmed by the happy belief that deliverance was at hand.

It was not. Soon his spirits flagged. He had succeeded in keeping thoughts of home from weighing too heavily, but his mind conjured up the scene — in remarkably accurate detail — of the wing vice commander knocking at the door in North Carolina with news that his plane had gone down. He saw Barbara, wide-eyed with alarm. He saw her backing away, turning, terrified. He felt utterly helpless. The brick cell pressed in like the walls of a tomb. He paced — three steps, two steps, three steps, two steps — in a gyre of despair.

On the morning of February 7 the cell door opened and a guard beckoned. "Come." He made a shaving motion. "Clean." Eberly shook his head and gathered the blanket around him. "No. Warm." Another guard, one of the bullies, pulled him into the corridor, where he was handed a dull razor, a shard of mirror, and a marble bowl of filthy water. Wincing at every stroke, nicking himself a dozen times, he scraped at the stubble. The left side of his jaw was still swollen and scabbed from the ejection injury. When he finished shaving, the guard plastered his bleeding face with strips of dirty newspaper and shoved him back into the cell.

An hour later he was blindfolded and led out to the elevator. When the blindfold was removed, Eberly found himself in front of a television camera inside a small auditorium. He shivered from the cold. A middle-aged Iraqi was perched in an adjacent chair, nattily dressed in a gray suit and red necktie. Several soldiers sat in the amphitheater seats. As ordered, Eberly stripped off the yellow duck prison garb and slipped into one of the American flight suits lying in a pile on the floor. He surreptitiously pulled the bandage from his head and wound it around his neck so that it formed a bloody ascot. The Iraqis stared disdainfully at his battered face.

The camera whirred. "Do you have any advice for your fellow pilots?" asked the man in the suit. Eberly shook his head. "I have nothing to say." The Iraqi tried again. "Is there something you want to tell your family?" Again he paused, trying to look pathetically confused in the hope that his captors would consider it useless to interrogate him further. Once more he shook his head and replied, "No."

The interview ended. He changed from the flight suit into the prison uniform, and a guard led him back to the cell. The door slammed, its metal clang echoing against the bricks. For once, he welcomed his solitude and the chance to pace in peace.

## Riyadh

As the first week of February slipped by, the MODA war room concentrated almost solely on the shape and timing of the ground offensive. The broad concept had been fixed since November: a sweeping attack by two Army corps — the so-called Left Hook — which would swing west of the Wadi al Batin to strike the Iraqi flank while the Marines and Arab forces plunged north into Kuwait.

But critical details — subtle points of position, logistics, and synchronization — remained issues of contention. Every proposal seemed to beget two counterproposals; the bold strokes of arrows on a map engendered a thousand putative Napoleons, each with a field marshal's certainty on how best to steer the grand march of armies. Schwarzkopf fretted that he would be pressed by his superiors in Washington to launch the attack prematurely, before his forces had gathered and his plan was complete. "I need you to help me on this," he told Carl Vuono, the Army chief of staff, early in February. "We're on track. I need to keep to my schedule."

In the east, the scuttling of the amphibious landing meant the two divisions of Marines ashore could expect no direct bracing from their seventeen thousand comrades at sea. Schwarzkopf and Colin Powell repeatedly stressed that the Marines' thrust into Kuwait was a "supporting attack," intended not to overrun the numerically superior Iraqis but to keep the enemy fixed in place while the Army struck from the west. U.S. intelligence estimated that 170,000 Iraqis occupied the Kuwaiti bootheel; Walt Boomer commanded about seventy thousand troops, including the Army's Tiger Brigade, which had joined the Marines in January after the British moved over to the Army's VII Corps.

Khafji had provided the first clear evidence of Iraqi weakness. Subsequent intelligence — for instance, a number of intercepted radio communications from the Iraqi 3rd Corps headquarters in Kuwait — reinforced the picture of an enemy in disarray. Boomer wanted to strike where the enemy seemed weakest, and thus he amended his attack plan five times before the ground offensive began. Although some of

Schwarzkopf's lieutenants thought the permutations bespoke indecision, the CINC left the Marine commander to his own devices and avoided the interference that had so aggrieved field commanders in Vietnam. For Schwarzkopf's patience, Boomer would be forever grateful, although he deplored the CINC's volcanic outbursts on other issues.

The conundrum facing Boomer in early February was how best to align his two divisions. The Marines at one point had anticipated attacking twenty miles inland from the Persian Gulf, then plowing north on a route parallel to the coastal highway. The point of attack subsequently migrated west along the bootheel; the coastal corridor was assigned to a force led by Saudis and Omanis, while the Egyptians, Syrians, and others took a sector in western Kuwait.

The 1st Marine Division had urged Boomer to fling his attack through the elbow, where the Saudi-Kuwaiti border above OP-6 returned to an east-west axis. From a purely tactical standpoint, the idea had merit, since the Al Jahra heights commanding Kuwait City lay only twenty-five miles from the elbow. Yet logistical complications and thickening enemy defenses argued against the proposal, and by mid-January it was dead. On January 22 Boomer approved the Southwest Option, which called for both Marine divisions to bull through a single breach in the Iraqi defenses east of OP-4, right at the Kuwaiti heel.

But the single breach troubled Boomer. The scheme required the 1st Division to cut a gap through two obstacle belts of Iraqi mines and barbed wire, then stand aside while the 2nd Marine Division pushed through to attack north. At best, shoving five thousand vehicles of one division through five thousand vehicles of another promised a thirty-mile traffic jam; at worst, if the Iraqis used chemical weapons, as the allies expected, the breach could become a slaughterhouse while the two divisions untangled themselves.

In early February the 2nd Division commander, Major General William M. Keys, offered Boomer another choice. A thickset Naval Academy graduate with bushy brows and a pugilist's mashed nose, Keys had won the Navy Cross and a Silver Star in Vietnam; from his staff he earned the sobriquet Mumbles because of his occasionally incomprehensible diction. But Boomer's trust in Keys — with whom he had served as a *co-van,* an adviser to Vietnamese marines, more than twenty years earlier — was absolute.

In a meeting at the 2nd Division command post thirty miles south of the Kuwaiti border, Keys came right to the point: "I have to tell you I think we're doing a dumb thing with one breach." Enough additional mine-clearing equipment — some of it provided by Israel — had recently arrived in Saudi Arabia to permit a second penetration of the Iraqi lines. Keys proposed looping his division some eighty miles northwest to attack near OP-5 through the Umm Gudair Oil Field. The Iraqis would not expect an assault into the foul-smelling wasteland of pipes, vats, and power lines.

"Christ, are you sure?" Boomer asked.

"I guarantee you," said Keys, "that I can do it with this division."

Boomer took a day to ponder the proposal. Many 1st Division staff officers, though troubled by the single breach, preferred that option because it would keep Marine combat power massed. If the double breach was adopted, the division believed the assaults should be side by side to reinforce each other, and staggered so that the 1st Division's attack began several hours before the 2nd's.

But Keys carried the day. Boomer decided to split his force. The two divisions would attack almost simultaneously: the 1st through the southern border near the bootheel; the 2nd, fifteen miles away through the western border. He would accept Keys's guarantee, trusting the judgment of his fellow *co-van.* The decision was a reminder that in the United States Marine Corps, *semper fidelis* was an assurance never far from the heart.

In the west, Schwarzkopf took a direct hand in deciding how to align the two oversized Army corps — a total of 255,000 troops — that would carry his main attack. One proposal in November had the XVIII Airborne Corps breaching the Iraqi defenses just above the border, then spinning to the northwest as VII Corps followed and veered northeast. But, as with the Marines, the prospect of mingling thousands of tanks, fuel trucks, and sundry other vehicles soon killed the idea.

Instead, Schwarzkopf segregated the two corps. In the west, XVIII Corps would be responsible for severing the Euphrates Valley to cut off reinforcements from Baghdad and isolate the Kuwaiti theater; VII Corps, with its vast tank fleet, was instructed, in the CINC's secret mission statement, to "attack deep to destroy Republican Guard armored/mechanized forces. Be prepared to defend [in the] vicinity of the northern Kuwait border [to] prevent Iraqi counterattacks from reseizing Kuwait."

Yet the issue of how far west to push the attack sparked debate well after the basic plan had been devised. Steve Arnold, the Army's operations officer responsible for fleshing out the offensive plan, originally aimed the XVIII Corps at Jalibah, the large Iraqi air base south of the Euphrates Valley and seventy miles west of Basrah. After Jalibah was captured and the adjacent Highway 8 severed, Arnold believed, the war would in effect be won. Iraqi forces in the Kuwaiti theater would be trapped, unable to escape through the marshes and lakes north of the highway or the sand dunes to the south. The enemy could stand and fight, surrender, or retreat into Basrah.

Schwarzkopf, however, kept nudging the attack axis farther west until eventually XVIII Corps commanders found themselves studying street maps of As Samawah, a Euphrates River city nearly a hundred miles west of Jalibah. The 101st Airborne Division drafted a plan for its 2nd Brigade to seize bridges and roads in three locations on the outskirts of As Samawah, and the division's 3rd Brigade was to attack near Nasiriyah, another river town, forty miles west of Jalibah.

Schwarzkopf kept his cards close to the vest, leaving Arnold and other planners in Riyadh uncertain of his rationale for pushing the attack out so far. They suspected he felt pressure from Cheney and Powell to develop an innovative flanking attack, one far removed from the discredited hey-diddle-diddle-straight-up-the-middle plan proffered in October. The allies also hoped that the credible threat of an attack on Baghdad would frighten Saddam politically and keep him from sending his troops to reinforce the south.

But the farther west the attack migrated, the more difficult it became to sustain logistically. CENTCOM planners anticipated a demand for fifteen million gallons of fuel a day once the ground war began, all of which would have to be trucked to the fighters. Four hundred miles separated the port of Al Jubail, on the Saudi coast, from XVIII Corps headquarters far to the west in Rafha; As Samawah lay another 160 miles north. Despite the frenetic construction of two huge logistics bases along the road to Rafha, the soldiers in XVIII Corps were near the end of their tether. Butch Neal, Schwarzkopf's deputy operations officer, spoke for many when he said, "Jesus, I'd rather have them in closer where they can influence the action. We've got them eighty million miles out there in the west where it's a logistics nightmare."

Colin Powell, studying his maps in the Pentagon, also worried about the political consequences of American soldiers embroiled in house-to-house fighting in As Samawah and Nasiriyah, and the civilian casualties that would surely result. Other officers alluded in hushed tones to "a bridge too far," the disastrous overextension, in 1944, of Allied forces at Arnhem bridge in Holland that caused nearly twelve thousand American casualties.

Schwarzkopf eventually reeled the Army back in — although not without delivering another tirade. Arnold was giving a briefing in the MODA headquarters when the CINC suddenly interrupted: "Why the hell are you going so far west? What are you doing?"

Arnold, too stunned to point out that the attack axis reflected the CINC's own orders, said nothing. Schwarzkopf soon worked himself into a booming monologue. Glaring at John Yeosock, he accused the Army commander of failing to think through the attack. The Republican Guard was supposed to be the focus of the attack, the center of gravity, the CINC reminded him; sending thousands of combat troops toward As Samawah and Nasiriyah risked involving both American soldiers and Iraqi civilians in bloody urban warfare. Moreover, if Iraqi armor counterattacked XVIII Corps, VII Corps would be too far away to provide reinforcements.

Gesturing toward the map, Schwarzkopf pointed at the planned attack by Arab forces across Iraqi defenses in western Kuwait. The Egyptians, whom the CINC considered an "indispensable" spearhead for other Arab troops, had persistently requested attack helicopters and mine-clearing equipment from the Americans to support their offensive. Cal Waller had joked that "what the Egyptians are facing, two sick prostitutes could handle," but Schwarzkopf could not afford to be so cavalier. "How do I tell the Egyptians that half the U.S. Army's combat power is going way to the west, halfway to Baghdad?" Schwarzkopf demanded. "How am I going to explain to them that they'll be going into the meat grinder when our Army is way out here? I am not at all comfortable with this. I want you to go back and look at the plan again."

Arnold and Yeosock dutifully complied, instructing XVIII Corps to avoid As Samawah and Nasiriyah. Instead, the 101st would sever Highway 8 with a helicopter assault into the Euphrates Valley midway between the two towns. The French 6th Division and U.S. 82nd Airborne would screen the corps from any counterattack out of the west by pushing up to the Iraqi logistics base at As Salman. The 24th Mechanized Infantry Division, the corps's heavy armor unit, would attack farther east to seize Jalibah Air Base before threatening Basrah.

The focus of VII Corps remained the Republican Guard, still dug in above the northern Kuwaiti border. Four heavy divisions and an armored cavalry regiment would plunge into Iraq on roughly a fifty-mile front between the Wadi al Batin and the eastern edge of XVIII Corps, angling north and then east to strike the enemy in the flank. Reviewing the battle plan yet again, the CINC seemed satisfied at last. The blitzkrieg was nearly ready.

Behind the dance of arrows on Schwarzkopf's map lay three distinct phenomena that would converge in a battlefield trinity and complete the rout of Iraq: desert warfare, the tank, and the Army's search for a new combat doctrine. Since the age of Napoleon, commanders had gradually freed themselves of the traditional obligation to arrange their forces in a continuous line to avoid being outflanked. But in none of the nine major wars previously fought by the United States had the combination of topography, weaponry, and fighting methodology offered such an opportunity to avoid the shackles of linear warfare. Schwarzkopf saw the chance and seized it.

If the horrors of modern combat permitted any lingering romance about the art of war, it could be found in the clash of armies across the open desert. Battle captains like T. E. Lawrence, Erwin Rommel, Field Marshal Bernard Montgomery, all tugged at the imagination of the commanders now massing their forces in the Saudi sand. Schwarzkopf once described donning Arab robes and posing before a mirror "just like . . . Lawrence of Arabia." Scores of officers carried copies of *The Rommel Papers* in their kit bags, and Correlli Barnett's lyrical history, *The Desert Generals,* was recommended reading in VII Corps.

Most American colonels and generals had spent at least a year or two as young officers hunting an elusive enemy in the primeval confusion of Vietnam's triple-canopy jungle. By contrast, the desert seemed to offer visibility, clarity, and endless fields of fire. "The desert suits the British, and so does fighting in it. You can see your man," wrote the World War II field marshal William Slim in *Defeat Into Victory.* The quotation was cited widely by the Americans, who, no less than the British, liked being able to see their man.

Desert warfare inspired metaphor. The North African campaign in the early 1940s "was fought like a polo game on an empty arena," Correlli Barnett wrote. He also described the desert "with its agoraphobic vastness and emptiness and sameness . . . as naked and overwhelming as a bare stage to a green actor." But the prevailing simile, repeated fervently by U.S. commanders, likened the desert to the sea. In the same way that oceans are uniform, desert terrain offered few constraints. Forces moved like fleets steaming in formation, with visibility uninterrupted to all points of the compass. Advantage was gained through maneuver rather than by seizing this hill or that road junction. Although a desert commander could hardly ignore *sabkahs* and wadis and other topographical idiosyncracies any more than a sea captain could ignore shoals and islands, freedom of movement was the rule, not — as in Europe and Southeast Asia — the exception.

Like a fleet, Lawrence once wrote, desert armies were singular "in their mobility, ubiquity, their independence of bases and communications, in their ignoring of ground features, of strategic areas, of fixed directions, of fixed points." In VII Corps, where the nautical analogy held most firmly, the armored divisions preparing to set out toward the Republican Guard were likened to aircraft carrier battle groups. To make the point during a visit by the corps commander, Fred Franks, officers in the 2nd Armored Cavalry Regiment gently rocked the military van in which Franks sat while serenading him with "Anchors Aweigh."

If the lack of landmarks and natural barriers afforded opportunity, it also presented hazards. The wise commander worried incessantly about his exposed flanks. Unlike Europe, where a thousand church steeples and hilltops provided aiming stakes for artillerymen, the desert challenged the most savvy gunner to distinguish north from south. (The Defense Mapping Agency accordingly printed thirteen million maps for the gulf war.) Desert warfare also placed a premium on competent logisticians. An army may move on its stomach, but it does not move far without immense stocks of fuel, ammunition, spare parts, and water. Horatio Herbert Kitchener, the British hero of Khartoum, was a middling tactician who parlayed his expertise in logistics and transportation into victory. In World War II, the battle for North Africa became a struggle between supply officers. The Germans and Italians found themselves perpetually short of fuel, thanks to attacks by Allied aircraft and submarines based on Malta; Rommel, his supply lines stretched a thousand miles from Tripoli to El Alamein, had to divert fuel from the Luftwaffe to keep his tanks moving, and thereby diminished his air cover. The British smartly kept their men and machines stoked by building a railroad and water pipeline behind the army as it moved. (The Italians, catering to a different need with *la dolce vita* flair, included a motorized brothel in their supply trains.) In the gulf war, no significant movement could or would be made without consideration of the logistical consequences.

One other characteristic distinguished desert warfare historically: battles tended to be decisive. Battered armies, like stricken fleets, could not rely on topography to shield a retreat. Rearguard action in the open was suicidal. Without forests, mountains, or rivers to slow a pursuing opponent, withdrawal often turned into rout.

Bones littered the world's deserts to prove the point. The vain and ambitious Roman proconsul Marcus Licinius Crassus led nearly forty thousand soldiers into the Parthian wasteland at Carrhae in 54 B.C.; only five thousand escaped. Two thousand years later, during Operation Compass in December 1940, the British destroyed ten Italian divisions in North Africa and captured more than 130,000 prisoners at a cost of five hundred British dead. At El Alamein, Rommel lost 55,000 dead, wounded, or captured, as well as 450 tanks, in a fight that marked the beginning of the end for the Third Reich.

"Just as the desert is incapable of compromise, battles fought therein result in total victory or total defeat," wrote the historian Bryan Perrett. Desert combat possessed a grim purity, the unencumbered clash of force on force in what Admiral Horatio Nelson described before Trafalgar as "pell mell battle." Clausewitz referred to the direct swap of fire between two opposing armies as a "cash transaction"; in the desert, such commerce often left but small change.

In the gulf war, as in North Africa, the armored tank served as the sharp point on the spear. Developed by the British during World War I as a "landship" — again the nautical parallel — capable of breaking the stalemate of trench warfare, the tank was so named because early versions of the secret weapon left the factory beneath tarpaulins deceptively labeled "water tank." (The first combat model, however, was dubbed "Mother.") The chief of staff of the Royal Tank Corps during the Great War, Colonel J.F.C. Fuller, imagined armored contraptions of great speed and range capable of penetrating the enemy front and overrunning his command centers. It took only a quarter-century for this vision to be realized: in World War II the German Panzer commander Heinz Guderian effectively massed his tank fleets, punched through the enemy, and then swept behind the lines in a devastating encirclement. A generation later, Israel used Guderian's blitzkrieg tactic to near perfection during the Six Day War, in 1967.

Now another generation had passed and the American Army boasted a battle tank unlike anything employed by Fuller or Guderian. The M1A1 Abrams, first developed in the 1970s, was a sixty-seven-ton behemoth with thermal sites that permitted its four-man crew to spot targets through smoke and haze at ranges of two miles or more by detecting heat emanations. Capable of thirty miles per hour cross-country and nearly twice that on hard-surface roads, the Abrams combined speed, power, and lethality. In World War II, a stationary American tank had to fire an average of seventeen rounds to kill another tank at a range of seven hundred meters; the Abrams, which could fire on the move, improved that efficiency to one round at two thousand meters. Abrams gunners fired either HEAT rounds, which injected a 3000-degree jet of burning gas into the target, or sabot rounds, which struck the target with a three-foot-long dart of depleted uranium comparable, an Army document proclaimed, to "the force of a race car striking a brick wall at two hundred miles per hour, but with all of its energy compressed into an area smaller than a golf ball."

More than two thousand M1A1s in Saudi Arabia awaited the signal to attack. (The two sides, combined, had approximately ten thousand armored vehicles, compared with two thousand at El Alamein and the eight thousand assembled by the Soviets and Germans at Kursk, where the greatest tank battle in history was fought, in July 1943). A third of the M1A1s had been fitted with additional armor plates, which made them nearly invulnerable. American tank commanders believed that no Iraqi tank round — not even one fired from the Soviet-made T-72 — could penetrate the frontal sixty degrees of the Abrams, where the armor was two feet thick. The Abrams outweighed the T-72 by roughly twenty tons, and nearly all of the extra weight was in the form of protection for the crew. If the M1A1 had an Achilles heel, it was its fuel gluttony. Burning roughly six gallons for each mile traveled, the Abrams's 500-gallon tanks required refilling every eight hours of operation, which again made logistics paramount. Nevertheless, the M1A1 was a machine of exceptional killing capacity. Like the shark, it had a dreadful, evolutionary beauty.

Also evolutionary was the Army's idea of how best to wage war. Even before the fall of Saigon, in April 1975, the service had entered an era of creative introspection. Focus shifted from the lost cause in Southeast Asia to a prospective war against the Warsaw Pact in Europe. The Army's best thinkers, led by an irascible general named William DePuy, began contemplating a new doctrine — the blueprint for forces in combat — that could exploit the firepower of modern weapons and counter the numerical superiority of the Soviet military. DePuy's effort bore fruit in a revised version of Field Manual 100–5, unveiled in 1976. The doctrine stressed an "active defense," in which a division commander quickly shifted six or eight battalions to thwart an attack by twenty or more Soviet battalions. Geared almost entirely to Armageddon in Central Europe, FM 100–5 envisioned hammering the enemy with repeated blows to slow his advance. Strategists spoke of "battle calculus," which identified enemy targets to be "serviced" at a quantifiable "kill rate."

Active defense proved ephemeral, a necessary way station on the road toward a more comprehensive and creative doctrine. Critics, pointing to the Army's traditional "offensive spirit," chafed at the stress on defense and a lack of subtlety in "meeting the strength of the Soviet attack head-on." They decried a mechanistic concept of war that ignored the confusion of battle, devalued the human variables of courage, leadership, and endurance, and concentrated on Europe to the exclusion of probable flash points elsewhere. The doctrinal stress on a paramount first battle seemed to ignore the idea of an extended campaign, in which thrust was met by counterthrust.

In 1982, after six years of perfervid debate, the Army published a new doctrine in yet another edition of FM 100–5. It urged battle commanders to "look deep," a hundred miles or more behind the front lines, in order to disrupt enemy echelons with strikes by air, artillery, and special forces. The fight would be joined simultaneously across the width and depth of the battlefield, from platoons exchanging direct fire to fighter pilots attacking enemy supply lines and reserve units far to the rear. The doctrine celebrated deception and maneuver in the spirit of Robert E. Lee and Stonewall Jackson. It also cited, among many historical battles, those of Vicksburg in 1863 and Tannenberg in 1914 to illustrate the efficacy of speed, surprise, and mobility, which "protects the force and keeps the enemy off balance."

At the same time, the Army overhauled training methods and raised personnel standards to attract smarter soldiers, who were easier to instruct and less inclined to buck the rules. In keeping with its stress on the link between ground and air forces, the new doctrine was called AirLand Battle. Unlike active defense, AirLand was greeted with enthusiasm throughout the Army. Initiative, agility, synchronization, and depth — the basic tenets of AirLand — soon became part of every officer's vernacular. With minor modifications, the doctrine remained intact as the Army's catechism for Desert Storm. The Marines, admiring the revised version of FM 100–5, incorporated much of it into their 1989 field manual.

War would always be fought, in Matthew Arnold's haunting image, "on a darkling plain swept with confused alarms of struggle and flight, where ignorant armies clash by night." George C. Marshall, whose 1941 war manual served as a model for the 1982 document, had warned that mobile combat was "a cloud of uncertainties, haste, rapid movements, congestion on the roads, strange terrain, lack of ammunition and supplies at the right places at the right moment, failures of communication, terrific tests of endurance, and misunderstandings in direct proportion to the inexperience of the officers."

Yet, by serendipity, AirLand was perhaps best suited to armored warfare in the open desert. No battlefield on earth allowed a commander to look deeper, move quicker, or seize the initiative faster. Here, unlike Vietnam, the enemy had few hiding places. The terrain magnified both the effects of air supremacy and the Americans' technological advantages. In Iraq, the Army had found the perfect killing field.

## Al Qaysumah, Saudi Arabia

Like so many features of the modern military, the army corps was a legacy of Napoleon Bonaparte. By placing two to four French divisions — each with about ten thousand soldiers — beneath a corps commander, the emperor increased his ability to maneuver massive troop formations first to confuse and then to crush his opponents. In World War II, the British and particularly the Germans recognized corps operations as a means to concentrate combat power in North Africa and prevent the desert's vastness from scattering the divisions. "One point was very firmly fixed in my mind," Montgomery later wrote. "Desert warfare was not suited to remote control."

Under Schwarzkopf's war plan, VII Corps was the linchpin of the allied ground attack. "The mission of VII Corps," the CINC had decreed in November, "is to destroy the Republican Guard." First formed in France in 1918 and permanently based in Germany for most of the Cold War, the unit was now the largest corps ever mustered by any army: 146,000 soldiers and more than fourteen hundred tanks.

At first blush, the man in charge of this juggernaut seemed an unlikely candidate for the task. Short and slender, with a trim white mustache and blue-gray eyes behind gold-rimmed spectacles, Lieutenant General Frederick M. Franks, Jr., looked less like George Patton or Erwin Rommel than a grandfather — which he was — or a college professor — which he had been. At West Point, class of 1959, Franks had captained the baseball team; as a graduate student in English at Columbia University, he concentrated on John Milton and Cromwell's influences on seventeenth-century literature before returning to teach at the academy. He was meticulous, quiet, and even-tempered. At times, his quiet manner and pensive pauses made him appear hesitant or indecisive.

But Fred Franks possessed formidable inner grit. In May 1970, as a thirty-three-year-old cavalry officer during the American invasion of Cambodia, he was interrogating captured prisoners when a North Vietnamese soldier heaved a grenade from a bunker. The explosion ripped off much of Franks's left leg. After six months in a Valley Forge hospital, where recurrent infections melted fifty pounds from a frame that was already spare, he agreed to amputation. For twenty years since, including command tours with a cavalry regiment and the 1st Armored Division, he had walked through life with an artificial limb strapped on at the knee.

In many ways, Franks personified the American Army. Maimed in Southeast Asia, made whole again through force of will during a painful recuperation, he also had endured the dark days after Vietnam, when racial strife, drugs, and indiscipline nearly destroyed the Army. As deputy commandant at Fort Leavenworth, Franks became a staunch advocate of AirLand doctrine. His affection for ordinary soldiers bordered on reverence. He spoke with emotion of the "bright blue flame" that still burned from the war, of young men on the amputee ward at Valley Forge so traumatized by public disdain toward the Army that they pretended to have lost their limbs in car wrecks or factory accidents. My God, he had told himself, we can't let this happen again.

Franks also offered a study in contradiction. At heart he remained a young cavalryman, admiring the spontaneity, agility, and brio that had characterized the cavalry since the days of Jeb Stuart. As a division and corps commander, however, he could be cautious to the point of inaction, a man — in the words of one subordinate — "who couldn't make a decision to pee if his pants were on fire." Combining a low tolerance for uncertainty with an insatiable appetite for more information, Franks often waited until the last instant before issuing orders — to the dismay of his staff and division commanders.

In part this reflected a faith in intuition, a belief that it was a commander's "feel" for the battlefield that gave him the sense of when to act. As a corps commander in Germany he had been a notorious procrastinator, but in Saudi Arabia those closest to him — like Colonel Stan Cherrie, the corps operations officer and a fellow amputee who had first met Franks on the ward at Valley Forge — thought his tendency to delay had become less pronounced. Franks's orders in the gulf war, Cherrie noted with surprise, seemed unusually crisp, clear, and timely. Others, however, were not so complimentary. "If you're expecting VII Corps to be aggressive," Cal Waller warned Schwarzkopf, "it ain't gonna happen. It just ain't gonna happen."

Through December and January, Franks's most pressing concern was getting his forces from Germany to Saudi Arabia in fighting trim. According to CENTCOM's timetable, the corps was supposed to arrive by January 15. Instead, as the war began, thousands of tons of equipment, vehicles, and ammunition remained scattered from Bremerhaven to Dammam. Even in early February, ships bearing much of the combat punch of the 3rd Armored Division had yet to make port.

Unlike XVIII Airborne Corps, VII Corps had no contingency plans for deploying more than a battalion to another theater. Not since 1958, when two divisions left Germany for Lebanon, had a sizable American force moved out of Europe for combat duty elsewhere. Starting from scratch in November, Franks and his planners frantically searched for ways to move eighty-three separate units, 37,000 pieces of equipment, and tens of thousands of soldiers from one continent to another in two months. Planners at corps headquarters in Stuttgart even studied the possibility of shipping some equipment by train across the Soviet

Union and into Turkey before they remembered that Soviet railroads have a different track gauge from most of western Europe. By mid-February, the corps would employ 465 trains, 312 barges, 578 airplanes, and 140 ships to haul itself to the war.

A thousand nagging difficulties slowed the migration. Transporters needed specially designed "low-boy" rail cars to carry tanks and other heavy equipment to the ports. But only three hundred low-boys could be found in all of Europe, and many had been reserved to carry ammunition to Turkey for Proven Force. German rail workers, accustomed to peacetime working hours, had to be persuaded to labor past 4 P.M. and on weekends; American officers carried baskets of wine and bread to station masters, pleading with them to keep the rail yards open a few more hours. The *Bundesbahn* — the German rail system — possessed intricately detailed manuals on how to load and secure heavy equipment, which the Germans, predictably, followed to the letter. Thus, every tank gun barrel was lashed down with a two-inch cotton rope before a train would budge — even though the Abrams carried a heavy iron bar designed to do the same thing.

Three thousand shipping containers overwhelmed the German container-handling capacity. In an effort to keep the huge metal crates moving, the Americans decreed that every container be loaded and shipped within seventy-two hours of arrival at Bremerhaven, Rotterdam, or Antwerp. Some, in consequence, were sealed up with nothing inside but what was referred to as "sailboat fuel" — air. Despite attempts to ship each brigade's equipment in orderly "unit sets," most commanders counted themselves lucky if just a battalion's kit arrived coherently in Saudi Arabia. The equipment of one unlucky battalion ended up on twenty different ships.

Trains arrived at the wrong European port; foul weather upset shipping schedules; ships broke down on the high seas with alarming frequency. The star-crossed *Jolly Sreraldo*, for example, flying a South American flag while hauling a battalion's worth of tanks and half a brigade of helicopters, blew an engine twelve hours out of Rotterdam, then blew a main bearing off the coast of France, then stopped in Malta to treat a crewman's appendicitis, then stalled in Piraeus when the crew balked at entering the war zone after the air campaign began. When it finally arrived, more than two weeks late, the ship suffered one last indignity: the unloading ramp jammed at the dock.

Confusion and congestion in Saudi ports multiplied the snafus. In addition to the units coming from Germany, thousands of British troops and the 1st Infantry Division from Kansas poured into the country to join VII Corps. Because stevedores stacked shipping containers willy-nilly, their labels could not be read to reveal what was inside. When the Scuds began to fall, four ship captains — three in Dammam and one in Jubail — pulled out to sea before being wooed back to complete their unloading. Thousands of tanks, armored personnel carriers, and other vehicles required repainting to cover the forest green camouflage scheme of Central Europe with desert tan.

An early plan to build a rail line to the west, as the British had in North Africa, never materialized, so huge truck convoys ferried the equipment on the two-lane Tapline Road, soon dubbed the Trail of Tears. (Two hundred Third World truck drivers bolted when the war began; like the ship captains, most eventually were enticed back to work.) To save wear and tear on their armored vehicles, commanders wanted to use sturdy flatbed trucks, known as heavy-equipment transporters, or HETs. But VII Corps had fewer than two hundred HETs. The Americans eventually acquired more than thirteen hundred, many of those leased or borrowed from the Saudis, Egyptians, Germans, and Czechs. A few HETs, designed for lighter Soviet-made tanks, blew their tires when loaded with a hefty Abrams.

Yet for all its shortcomings the deployment was a masterpiece, a tribute to perseverance, ingenuity, and what came to be called "brute force logistics." In less than three months, a force larger than the combined Union and Confederate armies at Antietam arrived in Southwest Asia, bringing with it nearly as many tanks as had fought at El Alamein. Slowly but inexorably, VII Corps massed in the desert.

Fred Franks had a habit of sitting by himself for hours in front of three maps of southeastern Iraq, each set to a different scale. Unconsciously he moved his hands with small, graceful gestures, like a conductor leading a symphony. His staff, watching surreptitiously from a corner of the tent, realized their commander was fighting the battle against the Republican Guard in his imagination.

A corps commander sought clairvoyance. During combat, he disengaged himself from the immediate mayhem to look ahead twenty-four to seventy-two hours: his job was not to kill the enemy, but to provide killing opportunities for his tactical subordinates by putting the enemy in harm's way. Now, as his logisticians and transporters muscled the corps into the theater, Franks tried to anticipate Iraqi intentions and position his own forces accordingly.

American intelligence initially assumed the enemy would extend the so-called Saddam Line — the fortified barrier of minefields, trenches, and barbed wire that began along the gulf coast — a hundred miles or more west of the Wadi al Batin to a point in the Iraqi desert where broken terrain formed a natural tank trap. Franks therefore planned to have the 1st Infantry Division breach the barrier west of the wadi while the rest of the corps plunged north through the gap before swinging northeast toward the Republican Guard. But as Iraqi engineering efforts petered out, Franks changed his scheme of attack so often that some wags began referring to "the Dance of the Fairy Divisions."

His ultimate plan, however, was a good one. The 1st Cavalry Division, assigned the role of theater reserve by Schwarzkopf, would feint into the wadi to reinforce Iraqi suspicions that the attack was coming up that easy avenue into Kuwait. Farther west, the 1st Infantry Division would breach the Saddam Line, followed by the British 1st Armoured Division, which would then turn east toward the Iraqi armor units entrenched north of the border.

Still farther west, beyond the Iraqi barriers, the 2nd Armored Cavalry would lead two armored divisions across more than a hundred miles of open desert in an enveloping attack toward the Republican Guard above Kuwait. Those two heavy divisions — the 1st and 3rd Armored — would plunge abreast into Iraq on a front only twenty miles wide, a scheme Franks approved once the division commanders convinced him their units could squeeze themselves into such narrow confines.

Franks expected three fights: the first against Iraqi infantry in the front-line trenches, the second against enemy armor reinforcements moving out of Kuwait, the third and most important against the Republican Guard. The key to crushing the enemy and minimizing American casualties, Franks believed, was keeping his divisions together rather than letting them bolt across the desert in a gaggle. "I'm not going to attack with five fingers. I'm going to attack with a fist," he told his commanders. "We will not hit the enemy piecemeal. We'll hit him with mass."

Several VII Corps planners, including operations officer Stan Cherrie, believed that before hitting the Republican Guard the corps should pause for a day or so at Objective Collins, an area ninety miles north of the Saudi border. A pause would allow the logistics trains time to catch up with the armored spearhead and give the combat troops a chance to rearm, refuel, and rest. But Franks refused to hear of it. An Army corps, he believed, was like an ocean liner — or an aircraft carrier: hard to stop and harder to get moving again. He also was influenced by a recent series of war games at Fort Leavenworth, which suggested that the corps, once halted, might take up to twenty-four hours to recapture its momentum.

In the war plan sent to his division commanders, Franks wrote, simply, "No pauses." So adamant was he that staff officers in the corps headquarters hesitated to raise the issue again within his earshot. They referred instead to "the P-word." Not until the ground attack began would the P-word be uttered openly — when Fred Franks himself decided to bring his corps to a halt.

## U.S.S. *Wisconsin,* Persian Gulf

If desert warfare seized the imagination of American Army commanders, the battleship held similar sway over many a sea dog. Her strategic value had long been eclipsed by nuclear submarines and aircraft carriers. As a tactical weapon she appeared doomed to follow the crossbow and blunderbuss. Yet, for traditionalists, no maritime silhouette better symbolized the American thalassocracy: the pugnacious, jutting bow; the looming superstructure; the trio of triple-barreled turrets, each heavy as a frigate. When employed as a gun platform, she remained nonpareil, capable of tossing a shell with the heft of an automobile more than twenty miles. In the gulf war, her hour had come round at last.

For more than a hundred years, naval theorists had periodically consigned the battleship to obsolescence, ostensibly doomed first by mines and torpedos, next by air attacks, finally by guided and cruise missiles. The golden age of dreadnoughts seemed to end at Jutland, then again at Pearl Harbor. Yet new wars erupted, and the battle wagon inevitably took her place on the firing line. In Korea, American battleships sent more than twenty thousand sixteen-inch shells ashore; in Vietnam, another five thousand. In the Persian Gulf, *Wisconsin* and *Missouri* — two among a battleship quartet pulled from mothballs in the 1980s — had hoped to sweep Iraqi targets from the Kuwaiti coastline. American uncertainty over mines frustrated that ambition, and a grand opportunity to use the big guns in defense of Khafji had gone begging.

But on February 3, finally assured of clear steaming by the minesweepers, *Missouri* moved twenty miles off the northern Saudi coast and hammered enemy bunkers in extreme southeastern Kuwait. On February 6 *Wisconsin* relieved her, throwing eleven shells at an artillery battery in the ship's first combat shoot since 1952. A day later, at dusk on February 7, *Wisconsin* unlimbered for a more extensive mission against Iraqi forces reported by naval intelligence to be sheltering at Khawr al Mufattah, a small marina twenty miles north of Khafji.

"Everything in the naval world is directed to the manifestation at a particular place . . . of a shattering, blasting, overbearing force," Winston Churchill, then First Lord of the Admiralty, declared in 1912. With respect to naval gunfire, the dictum was still valid. High above *Wisconsin*'s teak deck, in the cramped crow's nest known as Spot 2, stood the man who knew more about sixteen-inch gunnery than anyone in the U.S. Navy. Master Chief Steve Skelley had converted a boyhood love of battleships into a lifelong passion for shooting. Short and voluble, with wiry hair and a staccato laugh, Skelley could talk for endless hours about great shoots of yore: of *West Virginia*'s brilliant gunnery in the 1930s, of the 973 rounds dumped by the Japanese on Henderson Field at Guadalcanal, of *Massachusetts* decimating a Vichy French destroyer squadron with eight hundred rounds at Casablanca.

Skelley had stood in Spot 1 aboard U.S.S. *Iowa* north of Puerto Rico on April 19, 1989, when turret number 2 exploded, killing forty-seven sailors. That accident, still not fully explained, cast a pall over the Navy's battleship fleet. Many felt the gulf war would be the dreadnoughts' last hurrah — again. *Iowa* and *New Jersey* had been decommissioned already, victims of a shrinking military budget, and *Wisconsin* was expected to be next. Like his eighteen hundred shipmates, Skelley wanted the end to come in a blaze of glory.

At 7:31 P.M. a warning klaxon hooted through the ship, and with a single round from the left gun of turret number 3 aft, the marina shoot began. An orange fireball boiled from the muzzle, followed by a concussive shock that hammered the throat and rattled in the chest. Sailors on the bridge had lowered the windows to prevent them from shattering; the blast even jolted men in the main conning station, encased in seventeen inches of armor plating. Traveling at more than two thousand feet per second, the shell soared into the night with a sound like ripping silk. Fifty seconds later it struck Kuwait.

In Spot 2, Skelley read the instruments that precisely measured muzzle velocity — critical to accurate shooting — and watched the television monitor that carried an infrared image of the target taken from a Pioneer drone orbiting overhead. The marina basin, barely a hundred yards across, contained four piers. More than a dozen boats, including several Kuwaiti yachts supposedly commandeered by Iraqi special forces, bobbed in their berths. Estimating that the first shell had landed short of the marina by a thousand yards, Skelley phoned adjustments to the gunners below. Two more rounds burst from turret 3. One fell short, the other carried long. The best shooters in World War II spoke of "finding the target in eight or ten rounds," an imprecision that Skelley described as "the price of admission."

Sixteen-inch gunnery melded equal parts science, art, and alchemy.

Many of the twelve hundred shells in *Wisconsin*'s magazines bore manufacturing dates from the late 1930s; much of the powder was World War II vintage, requiring careful blending with a more recent concoction to restore its propellant punch. A dozen other factors played into good shooting, including barrel wear, air temperature, sea state, and ship speed. The Pioneer drones, launched from the fantail with a small rocket and controlled with radio signals, augmented the ancient practices of eyeing the "fall of shot" from the ship or using human spotters close to the target. But gunnery also required an intuitive sense of ballistics gained only by experience. In the eight years since battleships had emerged from mothballs, Skelley had fired more than three thousand rounds, hoping, as he put it, to "resurrect disciplines fallen into disuse."

Shortly after 8 P.M., following eleven individual shots used to "walk in the round," *Wisconsin* fired her first three-gun salvo. A one-second interval separated each of the three blasts to allow the turret to recover from the torque. Seventy-five sailors manned each turret, which rested on bearings the size of basketballs seven decks below. After firing, each barrel dipped toward the main deck, raising the breech for reloading. A hoist lifted fresh bags of propellant from the powder flats below; another elevator raised a new shell from the projectile deck. The gun captain and rammer shoved in powder and shell. Another sailor attached the primer — a small cartridge the size of a .22-caliber bullet — and the breech snapped shut. In the gun plot room, a gunner pressed the twin brass triggers. Three sharp beeps echoed through the ship, followed immediately by the blast. The recoil kicked each barrel back four feet. Compressed air automatically blew residual gases from the barrels, which then dipped again for reloading. A nimble crew could fire and reload in forty seconds.

The rounds began to find their mark. In the claustrophobic confines of Spot 2, Skelley watched on the video monitor as shells burst across the marina, nineteen miles northwest of the ship. The Pioneer's camera zoomed in and out, clearly showing the hail of shrapnel sweeping the basin. Each 1900-pound shell burst into two or three thousand steel shards that weighed from less than an ounce to fifty pounds. Against "soft" targets, the killing radius extended about sixty yards, with an effect equivalent to the scything of a football field with several thousand machine gun rounds. No Iraqi soldiers could be seen in the picture, but several boats began to list and fires raged. Skelley phoned more adjustments below. *Wisconsin* loosed a nine-gun broadside, then another. One shell neatly sliced a pier in half. More boats burst into flinders or sank in their slips.

After seventy-five minutes and fifty rounds the salvos ceased, leaving the ship smothered in the acrid smell of gunpowder. The bombardment had destroyed or badly damaged sixteen boats and had so lacerated the basin that Iraq soon abandoned Khawr al Mufattah. Always excited when the big guns boomed, Skelley now felt ecstatic, virtually dancing beneath the plexiglass dome of Spot 2. In gunnery vernacular, the marina had been a "beautiful shoot."

Stan Arthur, the Navy commander on *Blue Ridge,* had spoken boldly before the war of firing the sixteen-inch guns "until their barrels melt." Lack of opportunity and a belief that the shells would be needed to support an amphibious landing led to an initial conservation of ordnance. But Schwarzkopf's cancelation of Desert Saber allowed the two battleships to unlimber with greater vengeance. So much steel raked the southern Kuwaiti coast that the CINC finally told Cal Waller, "Get hold of Stan Arthur and tell him to quit wasting ammunition hitting the same area over and over again."

Waller called *Blue Ridge.* "Cal, I understand your concern," Arthur replied, "but what the hell do you want me to do? The battleships are about to be decommissioned. I can take all these sixteen-inch shells and shoot them at the enemy, or I can take them home and put them in a museum." Waller laughed, as did Schwarzkopf, who issued a counterorder: "Have at it." Until war's end the firing would continue, with more than a thousand shells eventually falling on the Kuwaiti littoral. Soon the dreadnoughts would vanish, but Chief Skelley and his fellow crewmen had resurrected their abandoned disciplines just long enough to leave a scar.

## Taif, Saudi Arabia

Buster Glosson's pique at having to shift the air campaign from strategic to tactical targets was nearly matched by his frustration over the inability of American pilots to kill more tanks. Despite a steady rise in CENTCOM's "dead-things count" — an increase disputed by the CIA — Glosson believed the tally of Iraqi armor was not commensurate with the number of sorties now devoted to the task.

An allied ground offensive, to his chagrin, appeared inevitable. Every enemy tank left intact posed a threat to the soldiers and Marines who would surge across the border sometime in the next few weeks. Yet Glosson saw few good options for accelerating the destruction. Iraqi commanders often camouflaged their armor beneath sandbags, rendering them nearly invisible from two miles up. Even when A-10 or F-16 pilots spotted the tanks or personnel carriers, thick earthen berms shielded them from all but direct hits. Glosson could permit his pilots to drop down to two thousand feet or lower, where they could more easily find their targets and bomb with better accuracy, but aircraft losses would soar proportionately. Or he could find a new technique.

On February 3, he called Colonel Thomas J. Lennon, commander of the sixty-six F-111 Aardvarks in the 48th Tactical Fighter Wing at Taif. Lennon, whom Glosson had known since Vietnam, was autocratic and controversial. Many pilots disliked him, privately calling him Slif — Short Little Ignorant Fucker. Despite four hundred combat missions in Southeast Asia and decades of flying experience, Lennon had flown only two hundred hours in the F-111 — fewer than most of his young captains and majors. He upset his pilots and weapons system officers (WSOs) by frequently shuffling crews when their performance displeased him. He encouraged rivalry — goading the competitive instincts characteristic of most airmen — by erecting a large scoreboard in the wing headquarters, with mission results updated daily.

In Glosson's view, however, Lennon was the best wing commander in Saudi Arabia: a man of iron will and solid accomplishment, with a shrewd sense of tactics and a brazen bent for confronting higher authority on behalf of his men. Contrary to prewar computer simulations that predicted the loss of one in every five F-111s, not a single Aardvark had been — or would be — shot down.

"We've got a problem," Glosson told him. "The Iraqi tanks are buried and they've got sandbags on top of them. I want to know if you can see them at night and if you can take them out." The two men briefly debated the appropriate munitions. Glosson initially suggested CBU-87 cluster bombs, which Lennon denounced as "not worth a shit because we can't hit anything with them." Ultimately they agreed to try the GBU-12, a 500-pound laser-guided bomb.

The Aardvark was a Vietnam-vintage airplane designed as a long-range bomber against bridges, airfields, command bunkers, and other strategic targets — not as a tank killer. It carried an infrared targeting system, called Pave Tack, that sensed the faint radiation emitted by warm objects and then converted the infrared signal into a visible image on the WSO's cockpit scope. Before the war, during an exercise code-named Night Camel, Lennon's crews tried with little success to find American tanks hidden in the Saudi desert.

But those tanks had been widely dispersed and were cold, their engines shut down. As Lennon would discover, the Iraqis sometimes kept their tanks running at night for warmth and power. Also, even beneath sandbags, metal absorbed heat from the sun during the day — a phenomenon called solar loading — and released it at night; the cooling-rate differential between metal and the surrounding desert produced a heat pulse that was detectable by Pave Tack. Also, the Iraqi practice of digging protective revetments disturbed the cooler sand beneath the surface in ways that left a distinctive infrared signature.

On February 5 Lennon and his wingman — respectively Charger Zero Seven and Charger Zero Eight — took off from Taif on the first tank-hunting mission. At 9 P.M., flying at eighteen thousand feet, they found the Madinah Republican Guard Division entrenched just above the northern Kuwaiti border. Lennon's WSO, Steve Williams, slipped the rubber hood of the video scope over his head. The tanks showed up as tiny white boxes in the Pave Tack screen.

"Oh yeah! Oh yeah! I can really see them," Williams reported. "Come on, turn! Turn! Hurry up!" Lennon, eyeing the desultory antiaircraft fire now climbing toward them, leveled the plane as Williams marked a white rectangle with the laser beam used to guide the GBU-12. Thirty seconds after the first bomb dropped from the plane, the rectangle vanished in a fiery bloom.

Within minutes Charger Zero Seven and Zero Eight had emptied their bomb racks and turned back toward Taif. Lennon immediately studied the gun camera videos. Of eight GBU-12s launched, it looked as though all but one had destroyed either a tank or a personnel carrier. He called Glosson in Riyadh. "Seven for eight," Lennon told him. Glosson paused, then asked dryly, "Why didn't you get eight for eight?"

In truth, Buster Glosson was elated. Now he was certain the Iraqi army would literally die in its tracks. On February 7, he launched forty-four F-111s on the first mass tank-killing raid.

Glosson kept the new technique secret lest the Iraqis devise a countermeasure to what soon became a wholesale slaughter of armored vehicles. The Aardvark crews became adept in interpreting chiaroscuro, finding carefully camouflaged enemy forces by reading the subtle shades of whites, grays, and blacks in their Pave Tack scopes. At Taif, the thirty-by-sixty-mile attack sectors came to be known as "tank boxes." From dusk until dawn, aircraft saturated the principal boxes in northern Kuwait and adjacent Iraq. When Schwarzkopf objected to the pilots' cold-blooded slang for the attacks — "tank plinking" — Horner relayed the CINC's qualms to his air crews, thus guaranteeing that the term would forever be fixed in their lexicon.

Within a week the Aardvarks — occasionally joined by F-15Es equipped with similar infrared systems — were claiming a hundred or more armored targets destroyed every night. Horner established nightly quotas, "like an insurance salesman." Lennon added tank plinking to his headquarters' scoreboard. By the end of the war the top crew would be credited with thirty-one kills. A few airmen even complained of boredom, of missions that too much resembled the proverbial shooting of fish in a barrel.

But with its salutary consequences for the allied cause, tank plinking had one unfortunate effect: it widened the gap between CENTCOM and the CIA over battle-damage assessment. Without ready access to F-111 and F-15E gun camera footage, the CIA had to rely on satellites. The agency also suspected pilots of inadvertently inflating the tally by rekilling tanks and other armored vehicles destroyed in previous attacks. The changes in BDA methods used by the Army's John Stewart and adopted by CENTCOM's Jack Leide were seen as arbitrary, illustrative of the confusion in Riyadh. Occasional miscalculations by Schwarzkopf's intelligence officers further inflamed the rivalry. At one point, for example, the Tawalkana Republican Guard Division was considered more than 50 percent destroyed and thus combat ineffective; when analysts in Saudi Arabia realized they had counted armored vehicles from other divisions among the Tawalkana tally, they resurrected the Tawalkana at three-quarters its strength.

Eventually the disparity between Washington and Riyadh would reach stunning proportions. When CENTCOM estimated that fourteen hundred Iraqi tanks had been destroyed of the 4280 believed deployed in the Kuwaiti theater, the CIA confirmed only 358 of those. In an effort to "proof" Riyadh's figures, John Stewart selected three bellwether Iraqi units: the 12th Armored Division, the 10th Armored Division, and the Tawalkana. Using their complex formula of pilot reports, gun camera videotape, and U-2 footage, Stewart's analysts estimated that armor and artillery in those units had been reduced, respectively, to 60 percent, 55 percent, and 75 percent. During the first week of February, U.S. reconnaissance planes again flew over the units, snapping hundreds of frames with high-resolution cameras. The new estimates were nearly identical: 61 percent for 12th Armored, 55 percent for 10th Armored, and 70 percent for the Tawalkana.

Yet soon after tank plinking began, the CIA made note of the BDA schism between Langley and Riyadh in the President's Daily Brief (PDB), a secret intelligence summary distributed to senior government offices, including the White House. Reflecting the agency's mission to provide the chief executive with independent intelligence, the PDB memorandum disavowed CENTCOM's estimates and alerted the president to the agency's inability to confirm more than a fraction of the reported damage.

In Riyadh, the PDB sent Schwarzkopf into a steaming rage. Soon he was to make the most difficult decision of his military career: when to launch the ground attack. Soldiers would live or die on the basis of his judgment. The CIA was seen as cynically hedging its bets, providing an alibi in case the Iraqis managed to inflict heavy casualties on the allies. "Here I am trying to make a decision," Schwarzkopf fumed, "and they're doing this to me." In the CINC's view, when he most needed support from the home front, the Central Intelligence Agency had walked away.

## Riyadh

On February 8, Richard Cheney, Colin Powell, and Paul Wolfowitz flew to Saudi Arabia for a review of Schwarzkopf's ground offensive scheme. None of the three was alarmed by the BDA brouhaha; each had committed himself to support the CINC unless his judgments seemed patently addled. But Cheney especially insisted that the plan pass what he called "the sanity test." Unless Schwarzkopf and his subordinates could explain the attack in terms a civilian like Cheney could understand, the secretary would not feel comfortable advising Bush to authorize it.

Powell took advantage of the long flight to Riyadh to offer Cheney another tutorial in the mysteries of war. "Iraqi artillery will be useless against our attack," the chairman predicted. "One, it's mostly towed and can't be moved quickly. Two, because it's in fixed positions, it won't be able to adjust to our movement. Three, by now Iraqi artillerymen are nowhere near the guns at night if they can help it for the very simple reason that they're not stupid. They're getting bombed at night, and they haven't been training or registering their guns."

Iraqi gunners would be unable to spot their targets effectively, Powell added. "We can blind them. Artillery is useless if you can't see the target electronically or visually. With our total dominance in the air, and with our ability to see them, they'll shoot one round, and all hell will descend on them."

On Saturday, February 9, Powell and the two civilians settled into the big leather armchairs in the MODA war room for a full day of briefings. The secretary, wearing an open-necked shirt and blue blazer, sat with Schwarzkopf on his left and Powell on his right. The CINC had instructed his staff to give the visitors "absolutely every piece of information we can" on subjects ranging from logistics and intelligence to terrain and weather.

One after another, Schwarzkopf's senior generals offered a status report on his forces and described their role in a ground attack. Chuck Horner promised more than a hundred sorties over the battlefield every hour, ready to respond to calls for air support once the attack began. "The ground war is going to move so fast the men on the ground aren't going to know what they need a day or two ahead of time," Horner predicted. "They probably won't know what they need even several hours ahead of time." Stan Arthur quickly sketched his amphibious options, and spoke of the Navy's readiness to resurrect the landing at Ash Shuaybah if necessary. Powell peppered the admiral with questions about potential casualties and damage to the Kuwaiti coastline. Like Schwarzkopf, the chairman considered Desert Saber so risky that it should be resorted to only in an emergency.

After listening to Walt Boomer explain how the 1st and 2nd Marine Divisions would attack into the Kuwaiti bootheel, Powell again urged caution. "Your mission is secondary. Don't get decisively engaged," he warned. "If you start to take a lot of artillery fire, just stay out of range. If you force the Iraqis to shoot at you, then they're not shooting at Freddy Franks and VII Corps, and you've accomplished your mission. And if they don't shoot at you, then you can move and go for broke. Just remember: you are the supporting attack."

John Stewart, the Army intelligence chief, predicted that the Iraqis would most likely remain entrenched in a "positional defense," although a few units might gather themselves for a mobile counterattack. Powell and Cheney asked Stewart more than a dozen questions about his BDA estimates. How did you arrive at your calculations? What kind of imagery are you using? Why are you using pilot reports? How accurate are they? Are the Iraqis using tank mock-ups and other deceptive measures? Could the enemy be fooling us?

Stewart fielded the queries for nearly forty-five minutes and pointed to a graph that plotted air sorties and estimated Iraqi attrition over the next two weeks. If allied aircraft could double their sortie rate against Iraqi armor and artillery, Stewart predicted, then the "crossover date" — the point at which half of the enemy's strength was destroyed — should arrive on February 21. Schwarzkopf turned to Horner. "Can you do that?" "I'm sure we can do it," Horner replied. "We could even lift it higher than that."

Schwarzkopf devoted the afternoon session to reports from three Army field commanders. Fred Franks led off with a detailed review of the eleven Iraqi divisions confronting VII Corps's four divisions. The corps, Franks explained, planned to bypass five enemy infantry divisions in the front lines, roll over the Iraqi reserves, then strike the Republican Guard in the western flank. Anxious to avoid a premature attack, Franks wanted his units to have at least three weeks of desert and gunnery training once all troops and equipment had arrived in the country.

Franks was scheduled to speak for thirty minutes, but he droned on for an hour while Schwarzkopf fidgeted in his chair and glared at Cal Waller, who had been charged with keeping the briefings on schedule. In November, Schwarzkopf had been delighted at George Bush's announcement that VII Corps was being dispatched to Saudi Arabia, but he had been skeptical of Franks since the 1980s, when both served in the Pentagon. Now he was even more certain that the corps commander was a pedant — "not four-star material," in the CINC's cutting assessment of Franks made to Carl Vuono. To Yeosock, Schwarzkopf had derided VII Corps's "slow, ponderous, pachyderm mentality," which he contrasted to the "audacity, shock action, and surprise" needed in the Army's attack. (Later, outside the war room, Waller confronted Franks. "Freddy, goddam it, why did you spend twice your alloted time up there?" "I didn't realize I went that long," Franks replied. "I just wanted the secretary and chairman to understand." Waller shook his head. "Oh, Freddy, come on.")

Major General Ronald Griffith, commander of the 1st Armored Division, followed Franks and quickly flipped through thirteen viewgraphs, all prominently stamped "Secret." One chart, labeled "Scheme of Maneuver," showed the division's planned attack axis as the far left wing of VII Corps. Powell pointed to the Iraqi crossroads town of Al Busayyah, where 1st Armored would pivot east toward the Republican Guard. "Ron, what if you get into a fight down south and can't get your division up to Al Busayyah?" the chairman asked.

"Sir, I don't think that's going to happen," Griffith answered. "But if we can't get there, I've already talked to the 3rd Armored Cavalry [from XVIII Airborne Corps]. They'll come over and take Al Busayyah if I can't." Another of Griffith's slides listed four potential supply problems: ammunition, spare parts, fuel, and chemical warfare protective suits.

The last man on the agenda revived a room of listeners grown somnolent. Of the Army's four hundred generals, Barry R. McCaffrey was perhaps the most flamboyant and controversial. Winner of two Distinguished Service Crosses and two Silver Stars in Vietnam, McCaffrey had the chiseled good looks of a recruiting poster warrior: hooded eyes; dark, dense brows; a clean, strong jawline; hair thick and gunmetal gray.

In 1969, as a company commander on his second tour in Vietnam, he was leading his men into an enemy bunker complex when a machine gunner opened fire from only fifteen feet away, killing three and wounding sixteen. One round nearly blew off McCaffrey's left arm; two bones protruded from the skin and a fountain of blood arced three feet through the air. As he skidded on his nose in the dirt, clawing his way from the line of fire, other rounds shot the pistol from his hand and the canteen off his hip. A medic dashed into the maelstrom, grabbed him by the web gear, and yanked him to safety. McCaffrey successfully pleaded with the surgeons not to amputate, and for two years Army doctors worked to rebuild the arm with bones from his hip, skin grafts, and a metal rod used to connect wrist and elbow.

To his many admirers, McCaffrey represented the officer par excellence: articulate, fearless, charming, intelligent. To his detractors — some of whom certainly were stricken with envy — he was a self-aggrandizing showman. The Army's youngest division commander, he now led the 24th Mechanized Infantry Division — Schwarzkopf's old unit — and was a personal favorite of the CINC's. Since early fall McCaffrey had urged Army planners not to overestimate Iraqi military prowess. He had predicted, if Bush ordered an attack with just one American corps, the encirclement of Kuwait City in three days and the defeat of the Republican Guard in a week.

As the only heavy division in XVIII Airborne Corps, McCaffrey now explained, the 24th Division would dash north to sever the Euphrates Valley between Talil and Jalibah air bases. The 26,000 soldiers would then turn east toward Basrah as VII Corps was swinging east into the main body of the Republican Guard. In reaching Jalibah, he predicted, the division would shoot seventeen thousand tons of ammunition and burn 2.5 million gallons of fuel. The attack plan stressed speed and violence.

"I think the division will take five hundred to two thousand casualties," McCaffrey added. "I'm pretty sure it will be at the lower end."

"How do you come up with two thousand?" Cheney asked.

McCaffrey turned full face toward the secretary. "I've got ten battalions with four companies each," he replied. "I think each company will be in a couple of fights and have an average of twenty-five soldiers killed or wounded in each fight — if the Iraqis resist effectively. My estimates come from having been a company commander, and getting twenty-five soldiers hit in a fight is about the price of doing business."

After a few questions from Powell, Cheney said to McCaffrey, "I appreciate your briefing, General. But what I want to know is: what are you really worried about?"

"I'm a very cautious person," McCaffrey replied slowly. "I've been wounded in combat three times. My son is in the 82nd Airborne and his life is at stake." He paused, then leaned toward Cheney with dramatic effect. "But I'm not worried about a thing."

The men in the room stirred. "The force is fully modernized," McCaffrey continued. "The logistics are all in place. The planning has been done, thousands of hours of planning. The troops' spirit is tremendous. In my judgment, we're going to take the Iraqis apart in ten days to two weeks. I feel uneasy telling you this, but I personally don't believe the enemy is very good."

Cheney, clearly impressed, offered a lopsided grin. "Thank you very much, General."

The session ended. Cheney, Powell, Wolfowitz, and Schwarzkopf repaired to a small office across the hall. Cheney, having listened to the CINC's lieutenants for nine hours, believed that the plan passed the sanity test. He and Powell agreed that Schwarzkopf had come a long way since the first faltering proposal unveiled in Washington the previous October. After discussing the day's briefings for a few minutes, the men turned to the question of the ground offensive's launch date.

The president does not want to wait any longer than necessary, Cheney said, but you can take as much time as you need to prepare. Schwarzkopf assured the secretary that the force was ready to fight. February 21 appeared to be the optimal date for attack, but with "three or four days of latitude" to ensure clear weather for air support. "I think we should go with the ground attack now," Schwarzkopf recommended. "At the rate we're consuming munitions, I'm not sure how much longer we could keep up the air attack."

The CINC's aggressiveness impressed Wolfowitz, who recalled a hesitant Schwarzkopf in December half-jokingly refer to a "365-day bombing campaign." On more than one occasion the CINC had hinted that he would call Bush or even resign if pressed to strike before his ground forces were ready. But now he seemed eager to finish what he had begun.

"Would anything be gained by continuing the air campaign?" Wolfowitz asked, deliberately offering Schwarzkopf a chance to delay the offensive.

The CINC curtly dismissed the notion. "We're ready to go," he replied.

"Start your preparations," Cheney ordered. "I'll take those dates back to the president." When they joined the other commanders in the CENTCOM auditorium, Powell observed, "This is the way the world's only remaining superpower is supposed to behave."

The Pentagon trio left Riyadh on Sunday, February 10, stopping briefly in Khamis Mushait to visit the F-117 wing. Cheney and Powell autographed a laser-guided bomb with a black marking pen. The secretary scribbled: "To Saddam, with affection. Dick Cheney, Secretary of Defense." Then the chairman: "To Saddam — You didn't move it and now you'll lose it."

On the return flight to Washington, Wolfowitz raised several concerns. Had the air campaign really gone on long enough? Had the bombing accomplished all that it could? Both secretary and chairman strongly agreed that it had. What about the Marines, Wolfowitz wondered. Walt Boomer seemed to have reservations about attacking into Kuwait in less than two weeks. Powell pointed out that the Marines' mission was to occupy the enemy, not overrun Kuwait. If they get in trouble, the chairman said, we'll just pull them back. "That's easier said than done with Marines," Wolfowitz noted with a smile. "It's hard to give them a mission order that says, 'If this gets tough, turn around.' " Powell insisted he had the matter in hand.

On Monday, the secretary and chairman met privately with President Bush in the Yellow Oval Room upstairs in the White House residential quarters. Powell briefly described Schwarzkopf's plan and told of giving the CINC a "window" for the attack, beginning on the 21st. Bush asked few questions. Cheney thought the president seemed resolved, even eager to press ahead with the ground offensive. "Sounds great," Bush declared. "Let's do it."

# 10

# Al Firdos

## Riyadh

"Command is a true center of gravity," John Warden had written in *The Air Campaign,* "and worth attacking in any circumstance in which it can be reached . . . The utmost attention should be given to a concerted attack on the enemy's command system."

Heeding his own advice, Warden and the Checkmate staff had stressed the importance of smashing the Iraqi high command as well as the communications network used by Baghdad to control its military forces. In August, Warden had placed "Iraqi leadership" innermost among the five concentric circles on a diagram denoting his targeting priorities. In succeeding months, planners at CENTCOM and in the Black Hole had likewise devoted "the utmost attention" to pinpointing Iraqi leadership targets. Thirty-five bunkers, command posts, and headquarters made up the target set on January 17, a number that would increase to forty-six by war's end.

Killing an enemy's leaders, a strategy known in contemporary military parlance as decapitation, was an ambition as old as warfare itself. Rarely in modern combat, however, had commanders possessed either the intelligence or the means to find and destroy a foe's senior commanders. Among the few exceptions in the twentieth century was the code-breaking triumph that permitted U.S. pilots to ambush and shoot down Admiral Isoroku Yamamoto, architect of the Japanese attack on Pearl Harbor, over a Pacific island in April 1943.

In drafting the Persian Gulf campaign, American planners hoped that the combination of sophisticated intelligence and precision weapons would permit the decimation of Iraq's high command. Foremost among enemy leadership targets, of course, was Saddam Hussein. Since August 8, when Checkmate officers first scribbled "Saddam" at the top of their target list, the Iraqi commander-in-chief had figured prominently in the ambitions of the air campaign. Saddam was seen as the *casus belli;* ergo, his removal would restore the peace. General Michael J. Dugan, who was then the Air Force chief of staff and Warden's patron, had publicly disclosed during a trip to Saudi Arabia in September 1990 that Saddam was "the focus of our efforts." Dugan believed that Saddam's elimination would likely lead to a prompt Iraqi withdrawal from Kuwait, a sentiment widely shared in the U.S. government.

Richard Cheney fired Dugan for his indiscretion. Thereafter, American commanders maintained the public fiction that no Iraqi individual was targeted. This charade satisfied the letter if not the spirit of a U.S. law prohibiting assassination of foreign leaders. American officials could, and did, claim that their bombs were intended to destroy various command posts, not any particular person.

Of equal weight, the rhetorical artifice prevented Saddam's demise from becoming an unachievable war aim. During the invasion of Panama in 1989, a posse of 25,000 American troops combed the country for four days before finally snaring the deposed dictator, Manuel Noriega. That embarrassing memory lingered in Washington's corporate memory, among civilians and military officers alike. "Let's not get ourselves in another situation like Panama where the whole goddam operation depends on finding one guy in a bunker," argued the Army chief of staff, Carl Vuono, during a Desert Shield strategy session in the Pentagon. Vuono's plea drew vigorous nods from his fellow chiefs.

Although Schwarzkopf had rated the chances of killing Saddam as "high," no one in Washington or Riyadh underestimated Saddam's elusiveness. He reportedly had survived five assassination attempts in 1981 alone and countless plots since. (More oblique but equally unsuccessful was George Bush's secret order, in August 1990, authorizing the CIA to help Iraqi dissidents overthrow Saddam.) On the eve of the war Tony McPeak, Dugan's successor as Air Force chief, estimated the prospects of killing the Iraqi leader with air strikes at no better than three in ten.

Nevertheless, from the opening minutes of Desert Storm, the Americans had tried to beat the odds. Among more than four hundred sorties against Iraq's "national command authority," bombers attacked the presidential palaces in Baghdad, a command complex and bunker at Abu Ghurayb west of the capital, Saddam's purported retreat at Taji on the Euphrates River, his summer home in Tikrit, and other potential hiding places. Several times U.S. intelligence believed it knew where Saddam had spent the previous night.

But the allies never possessed sufficient "real time" intelligence to lay the kind of ambush that had killed Yamamoto. Eventually the

Defense Intelligence Agency concluded that Saddam slept only in private houses, rarely if ever spending two consecutive nights in the same bed. Other reports indicated he sometimes traveled in a caravan of American-made recreational vehicles; one air strike on just such a convoy reportedly killed a number of Saddam's bodyguards but left the Iraqi leader unscathed. After three weeks of war, with increasing numbers of allied aircraft being pulled from the strategic campaign to prepare for the eventual ground attack, the hope of ending the war by decapitation had been reduced to a waning faith in blind luck.

Since killing Saddam was at most an unstated war aim — more aspiration than formal objective — this failure was nettlesome for allied strategists but not particularly troubling. Even if the Iraqi leader remained alive, the continued bombing of his command network was seen as providing useful benefits for Schwarzkopf and others anticipating the ground campaign. Shattering the means with which Baghdad communicated orders would leave the Iraqi army isolated and adrift. American commanders presumed that in a rigidly authoritarian society, where decision making was highly centralized, attacking the enemy's command-and-control could cripple its forces in the field. Saddam might remain in power, but he had been reduced to running a modern war with a command system — in Warden's analogy — "not much more sophisticated than that used by Wellington and Blücher at Waterloo in 1815."

For Buster Glosson, Dave Deptula, and other officers in the Black Hole, the continuing attacks offered another and potentially greater benefit: the incessant hammering of Iraqi intelligence, secret police, military, and other agencies could loosen Saddam's grip on his country and perhaps pave the way for a coup or an insurrection. If the pillars that supported and protected Saddam were weakened, he could be brought down. "When taken in total," an Air Force operations order predicted, the result of strategic air strikes would be "the progressive and systematic collapse of Saddam Hussein's entire war machine and regime."

As Clausewitz had warned, with a sentiment loudly echoed in Warden's writings, only "by constantly seeking out [the enemy's] center of power, by daring to win all, will one really defeat the enemy." The path to victory, Glosson and his men believed, still wound through Baghdad rather than through Iraq's army of occupation in Kuwait. Even if allied bombs failed to kill Saddam outright, they could cause his downfall — but only if the allies dared to win all.

*

The commander-in-chief and his senior military advisers walking toward the Oval Office after a private strategy session in the White House living quarters. Left to right: Richard B. Cheney, secretary of defense; President George Bush; and General Colin L. Powell, chairman of the Joint Chiefs of Staff.

An aerial photo of central Baghdad used by the U.S. Air Force's Checkmate planning group for initial targeting.

Colonel John A. Warden, head of the Checkmate staff, at Andrews Air Force Base en route to Saudi Arabia, where he would present his strategic air campaign plan on August 20, 1990.

Lieutenant General Charles Horner, commander of allied air forces during the Persian Gulf War, had his own ideas of how the strategic campaign should unfold.

Brigadier General Buster Glosson (left) and Lieutenant Colonel David A. Deptula review Iraqi targets in the Black Hole, the top secret vault in the basement of the Royal Saudi Air Force headquarters in Riyadh.

A Tomahawk cruise missile bursts from its launcher box aboard U.S.S. *Wisconsin.* Few American military planners knew that missiles fired from the Persian Gulf crossed Iran en route to Baghdad.

The Baghdad international communications center, known to allied planners as the AT&T building, was bombed by the first wave of stealth fighters at 3 A.M. on January 17, 1991.

Colonel David W. Eberly in the cockpit of an F-15E Strike Eagle before his ill-fated mission in western Iraq.

General H. Norman Schwarzkopf in the Black Hole moments before erupting in anger at the Air Force for allegedly disobeying his orders to pummel the Iraqi Republican Guard with B-52 strikes. Left to right: Brigadier General Buster Glosson, Lieutenant General Charles Horner (behind Glosson), Schwarzkopf, and Lieutenant Colonel David A. Deptula.

General Colin Powell points out Iraqi air defense concentrations during a briefing for the Pentagon press corps on January 23, 1991.

President George Bush delivers his State of the Union address on January 29, 1991, as the battle of Khafji unfolds in northeast Saudi Arabia. Behind the president are Vice President Dan Quayle and Speaker of the House Tom Foley.

Lieutenant General Frederick M. Franks, commander of the U.S. Army's VII Corps, and his senior commanders. Seated, from left: Brigadier General Robert P. McFarlin, 2nd COSCOM; Major General John Tilelli, Jr., 1st Cavalry Division; Major General Ronald H. Griffith, 1st Armored Division; Franks; Major General Tom Rhame, 1st Infantry Division; Major General Paul E. Funk, 3rd Armored Division; Major General Rupert Smith, British 1st Armoured Division; and Colonel L. Don Holder, 2nd Armored Cavalry Regiment.

An M1-A1 Abrams tank with its turret partially traversed during the 24th Infantry Division's attack into the Euphrates Valley.

Lieutenant General Fred Franks (left), VII Corps commander, and Major General Tom Rhame, 1st Infantry Division commander, discuss the breach of Iraqi defenses on the first day of the ground attack, February 24, 1991.

Major General Barry McCaffrey, commander of the 24th Infantry Division and one of the Army's most flamboyant generals, addresses his troops beneath a camouflage net.

Lieutenant General John Yeosock, commander of U.S. Army forces, in his headquarters at Eskan Village shortly after learning that the ground attack would be accelerated by nearly a day. Behind Yeosock is his operations officer, Brigadier General Steve Arnold.

Schwarzkopf confers with Lieutenant General Gary Luck, commander of XVIII Airborne Corps, outside the corps headquarters at Rafha, Saudi Arabia.

Major General Wayne A. Downing, seen here after receiving his third star, commanded the Joint Special Operations Task Force, which included hundreds of Delta Force soldiers who roamed through western Iraq.

Lieutenant General Walt Boomer, shown here after his promotion to four-star rank, commanded the I Marine Expeditionary Force and was responsible for attacking into the heart of Iraqi defenses in eastern Kuwait.

Engineers in Bahrain inspect the huge hole blown through the hull of U.S.S. *Tripoli* by an Iraqi mine.

The wreckage of an Iraqi SU-25 Frogfoot bomber at Jalibah Air Base in the Euphrates Valley.

Iraqi forces dug tens of thousands of sand revetments in a largely unsuccessful attempt to shield tanks, trucks, and troops from the allied attack.

Gathered in the Central Command war room on February 9, 1991, final review of the allied ground war plans are, from left, Lieutenant General Calvin A. H. Waller, deputy CENTCOM commander; Powell; Cheney; Schwarzkopf; and Paul Wolfowitz, under secretary of defense.

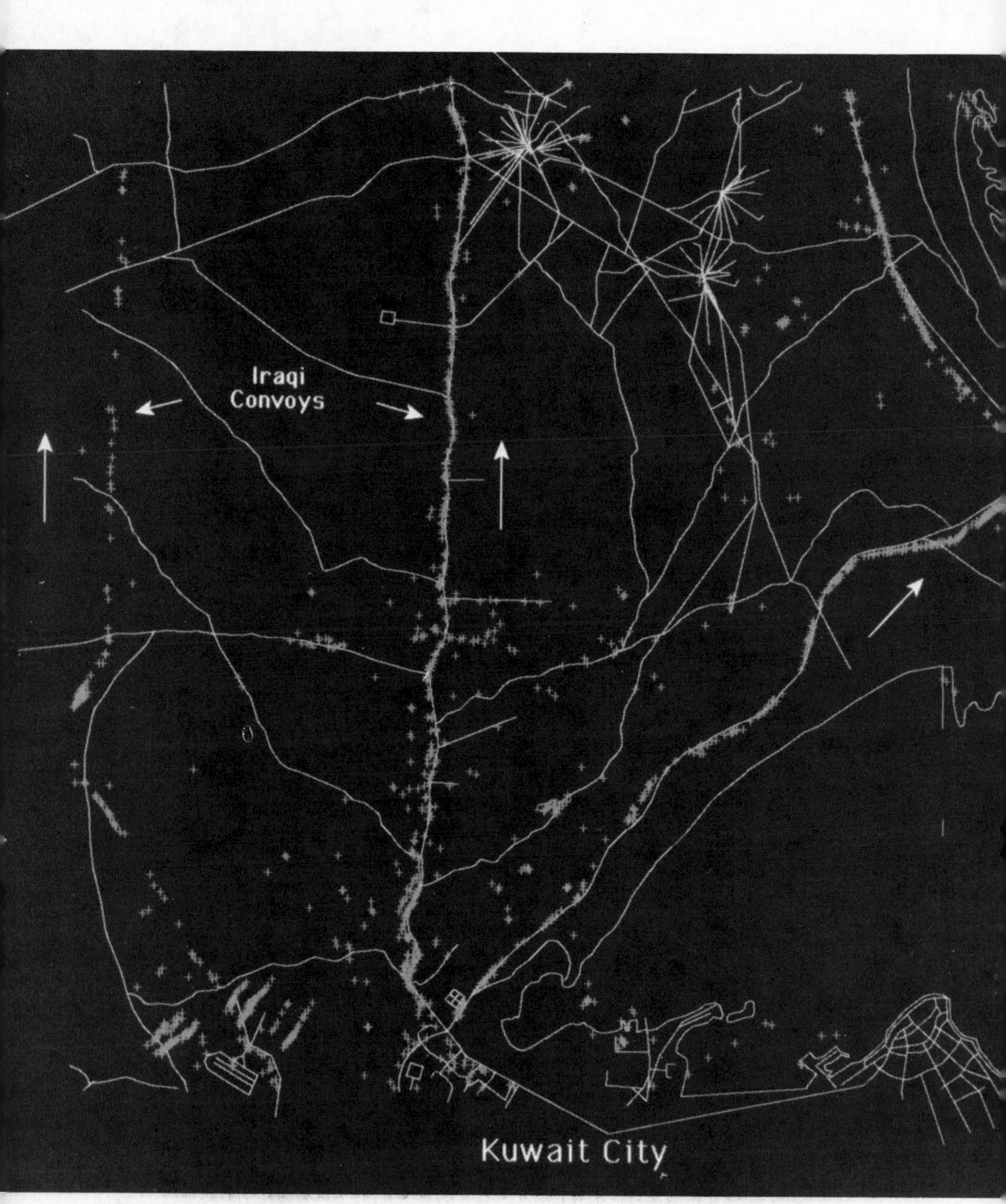


This image displayed on the Joint Surveillance Target Attack Radar System (JSTARS) radar scope shows thousands of Iraqi vehicles fleeing north from Kuwait City as the war reached its climax. Each "+" represents a vehicle or group of vehicles on the move.

Using plastic explosives, Iraqi engineers blew up hundreds of Kuwaiti oil wells as part of a scorched-earth policy in the occupied emirate.

Iraqi regional intelligence headquarters in Baghdad, also known as the Biltmore, where allied prisoners of war were inadvertently bombed on February 23, 1991.

Iraqi prisoners captured by the Marines' Task Force Ripper on the first day of the ground attack. By the end of the war, more than seventy thousand Iraqis would be in allied custody.

Schwarzkopf holds up a bottle of sand from the beach at Kuwait City after the emirate was liberated.

On the last full day of war, the wreckage of civilian and military vehicles litters the so-called Highway of Death from Kuwait City to Basrah after allied air attacks.

Schwarzkopf and Lieutenant General Khalid bin Sultan, commander of the Arab forces during the war, sit across the armistice table at Safwan from Iraqi military delegates, including Lieutenant General Sultan Hashim Ahmad, second from right. The Iraqi officer at far right is unidentified.

Colonel David Eberly is reunited with his son, Timm, and wife, Barbara, at Andrews Air Force Base after his release from an Iraqi prison.

Schwarzkopf and his senior staff officers lead the victory parade down Constitution Avenue in Washington, D.C., June 8, 1991.

One network of potential command posts that had heretofore escaped destruction began to attract renewed American interest during the first week of February. In the early 1980s, Finnish contractors built twenty-five large air raid shelters in greater Baghdad. In 1985, ten of those had been extensively hardened with reinforced concrete ceilings and protection against the electromagnetic pulse emitted by nuclear detonations. American intelligence possessed detailed drawings of the ten hardened shelters and presumed them to be incipient command bunkers. In November, Glosson had even interviewed European contractors who worked on the 1985 renovations.

During the late fall, CENTCOM planners had placed the ten bunkers on a preliminary Desert Storm target list, with plans to attack them by at least day three of the war. But in early January, Dave Deptula took a closer look. Satellite imagery indicated the Iraqis had surrounded the bunkers with fencing, but otherwise they appeared unoccupied. Air Force intelligence in Riyadh also detected no activity.

"From my perspective," Deptula told the intelligence officers, "I don't see any need to target them until we get some corroboration that they're being used as command and control facilities." Returning to the Black Hole, where all strategic targets had been listed in neat columns on a wall chart, Deptula drew a thick X through each of the ten bunkers with a felt marking pen.

By early February, three of the bunkers showed signs of life. One appeared to be used by the information ministry, another by administrators from the Iraqi defense ministry. The third seemed less innocuous. Located in the middle-class suburb of Amariyah, near Jordan Street in southwest Baghdad, the square facility occupied nearly an entire city block. Two large ventilation shafts protruded from the roof. Iron grates covered several arched doorways, one of them the entrance to a ramp that led forty feet under ground. Single-story houses dominated the neighborhood to the north and south; to the east, a school stood across the street and, beyond it, a mosque. A sign outside the building proclaimed in Arabic and English: "Department of Civilian Defense — Public Shelter Number 25." The Americans called it the Al Firdos bunker, after an adjacent section of Baghdad.

At the CIA, interest in Al Firdos had been piqued by January 17 satellite photos showing that fresh camouflage paint had been applied to the roof. Later photos showed increased activity around the building and military vehicles parked outside. Beginning on February 5, U.S. intelligence began collecting SIGINT cuts — signals intelligence from radio emanations in the vicinity of the shelter. The CIA, which believed the building had been used as a command bunker during the Iran-Iraq war, now concluded that it had been commandeered by the Mukhabarat, or General Intelligence Department, a successor to the Ba'ath Party's secret police. Mukhabarat headquarters near the Baghdad Horsemanship Club had been bombed early in the war; CIA analysts believed the organization had found a new home in Public Shelter Number 25.

On February 10, Charles Allen, one of the CIA's senior national intelligence officers, visited John Warden in Checkmate's basement warren in the Pentagon. Allen, who had become a close Warden confederate, laid out his evidence regarding the bunker and recommended that it be put back on the target list. The CIA, he added, saw no sign of the Iraqis using the facility as a civilian bomb shelter. On February 11, Warden sent Deptula a message suggesting that the building had become "an alternative command post because of damage to other facilities . . . Baghdad may believe that coalition forces will not attack the bunker because of potential [damage] to the nearby school and mosque."

Schwarzkopf's intelligence chief, Jack Leide, also had been examining the shelter. A fresh satellite photo sent by Rear Admiral Mike McConnell, intelligence director for the Joint Chiefs, again showed military vehicles parked in the compound. Leide sent a query to the Defense Intelligence Agency: Is it possible that this is being used as an air raid shelter? Told that no evidence had been found to that effect, Leide also recommended targeting the bunker.

Buster Glosson, however, considered the evidence purely circumstantial. The camouflaged roof, the fence, the satellite photos of military vehicles, the electronic intercepts — none of it seemed conclusive. In a police state as compulsive about security as Iraq, camouflage and fencing were common; even flour mills and water treatment plants were splotched with camouflage paint. Glosson considered Checkmate's analysis irrelevant. "That assessment," he later commented, "isn't worth a shit."

But in the second week of February each of the most senior officials in the U.S. government — fewer than a dozen — received a sealed envelope containing the latest report from one of the few effective spies operating in Iraq, a top official in Saddam's government who was secretly working for the Americans in Baghdad. The spy, whose information had previously proven accurate, warned that Iraqi intelligence had begun operating from Shelter Number 25. McConnell called Glosson to make certain he had received the letter. He had, and he believed it corroborated the circumstantial evidence regarding the bunker. The target, he concluded, was now worthy of attack.

On February 11, the bunker was added to the master attack plan. The raid was scheduled for the early morning of February 13 in the time block marked 0130 to 0155 Zulu — 4:30 to 4:55 A.M. in Baghdad. Six other targets, including the Iraqi Internal Security Directorate, would be hit before dawn by this third and final wave of stealth fighters to enter Iraq that night.

For Shelter Number 25, the attack plan listed precise grid coordinates as well as the aircraft and munitions to be employed — a pair of F-117s each dropping a 2000-pound laser-guided bomb. In the space reserved for "target description," the plan read: "Al Firdos bunker. Activated, recently camouflaged command-and-control bunker."

## Washington, D.C.

Nearly a month into the war, the so-called Israeli problem was considerably less vexing in Riyadh and Washington than it had been in January. Delta Force and the SAS combed western Iraq with a fury that chased Scud missile crews into an ever-shrinking patch of desert near Al Qaim. Seventy-five to 150 counter-Scud sorties flew overhead every day. The Iraqis rarely fired now except when bad weather enhanced the concealment. From an average of five daily Scud launches during the first week of war, the attacks in the subsequent two weeks tapered off to fewer than one a day. Patriot launches diminished proportionately, and anxiety over the stockpile of PAC-2 missiles abated. (The Pentagon sent 120 new PAC-2s to Israel before resuming shipments to Saudi Arabia, much to CENTCOM's irritation.)

Version 34, a new software package intended to correct Patriot's proclivity for firing at false targets, arrived in Israel and Saudi Arabia on February 4. Israel, which commanded the six Patriot batteries on her soil, four of them operated by U.S. Army crews, had ordered all launchers switched from automatic to manual firing modes after the January 25 barrage in an effort to target only incoming warheads instead of debris. American air defense experts considered the order ill advised: men could not react as swiftly as computerized machines. Even so, the dispute seemed inconsequential, since the average warning time of incoming missiles, flashed to Tel Aviv from the Pentagon over the Hammer Rick line, had doubled from two or three minutes to five or six. In the American view — at least as it was publicly expressed —

Patriot was flawless. "The Patriot's success, of course, is known to everyone," Schwarzkopf told the press. "It's one hundred percent so far. Of thirty-three [Scuds] engaged, there have been thirty-three destroyed."

In Tel Aviv and at the Israeli embassy in Washington, this gasconade was known as "the Patriot bullshit." The Israelis had a point. American claims were based largely on faith and wishful thinking. None of the twenty-one Patriot batteries in Saudi Arabia possessed "embedded data recorders," digital instruments capable of measuring precisely what occurred when a Patriot intercepted a Scud that had disintegrated at Mach six fifteen kilometers above the earth. That was because U.S. commanders feared the recorder equipment would interfere with missile operations. The Israelis, by contrast, had three recorders and a computerized system used to track all Scud fragments and analyze each engagement. Imperfect though they were, the Israeli data strongly suggested Patriot shortcomings.

This infirmity was hard for the Pentagon to admit. Having proclaimed Patriot a war hero, the Army was reluctant to pull its champion off the pedestal. (Not until more than a year after the war would the Army acknowledge that it had "high confidence" in only ten of the Scud kills proclaimed in Israel and Saudi Arabia; a subsequent study found that as few as 9 percent of engagements resulted in confirmed "warhead kills," although in other cases Patriots apparently knocked the Scuds off course or "dudded" the warheads.) Yet even the Israelis recognized the political and military utility in lauding the missile. When an Israeli officer suggested publicly disclosing qualms about Patriot, Avraham Ben-Shoshan, the military attaché in Washington, snapped, "You shut up. This is the best weapon we've got against the Scuds because it's the only weapon. Why tell Saddam Hussein that it's not working?"

In private conversations with the Americans, however, the Israelis showed no such reluctance. On February 4 at the defense ministry headquarters in Tel Aviv, Major General Avihu Ben-Nun, commander of the Israeli air force, confronted Colonel Lew Goldberg, head of the Pentagon's Patriot management cell. "The Patriot doesn't work," Ben-Nun declared bluntly. Of twelve Scud-Patriot engagements analyzed by the Israelis, at most only three warheads had been destroyed. Some Patriots had apparently caused damage by chasing Scud debris into the ground. "You ought to stop producing the Patriot until you fix it," the Israeli general added.

"Sir, would you rather have nothing until we can produce something that's perfect?" Goldberg replied. "Or would you rather have this, which is doing a large part of the job?" Ben-Nun responded with a scowl and a shrug. Two days later, back at the Pentagon, Goldberg presented the Army's top generals with a briefing entitled "Israeli-Patriot Issues" and classified "Secret–Close Hold." On his first page, Goldberg quoted Ben-Nun: "System doesn't work."

As if to prove the point, another Scud hit Tel Aviv on Saturday morning, February 9, damaging more than two hundred buildings and wounding twenty-seven Israelis. One flaming chunk of debris landed in the middle of a third-floor apartment; the tenant dialed the police and complained, "The missile's in my living room and it's burning my home down."

Defense Minister Moshe Arens was among those Israelis unhappiest with Patriot's performance. Notwithstanding the decrease in Scud launches, the Israeli military strongly believed that Saddam would authorize chemical warheads in desperation as the allied ground attack neared. Arens feared that the odds of successfully parrying such an attack were slim. Behind his grievances regarding Patriot was the defense minister's ulterior motive: he believed a "window of opportunity" was opening for Israel finally to strike back at Saddam. "We want to act," he told an Israeli colleague. "It's been four weeks, and we're tired of being sitting ducks."

Israel's forbearance, Arens believed, had dangerously eroded forty years of deterrence built on a policy of swift, sure retaliation. Israeli strike planning had continued apace, with several provisional dates set for an attack. (Some Israelis even contemplated direct action against Saddam himself; in 1980, during the American hostage crisis in Tehran, Israeli planners in a training exercise proposed snatching Ayatollah Khomeini from the Iranian city of Qum and using him to barter the Americans' release.)

By early February, Arens had concluded that an Israeli strike was unlikely to break up the allied coalition — Iraq's defeat seemed imminent. But the attack would slap Saddam's face, demonstrate Israel's ability to destroy mobile missiles, rebuild deterrence in the Middle East — and signal the Americans that Israel was not completely in Washington's pocket.

With Scuds continuing to fall, Arens believed he had almost convinced Yitzhak Shamir, his prime minister. Almost. As politically cunning as he was stubborn, Shamir balked at authorizing a specific date for the Israeli strike. Instead, he sent Arens to Washington for what was known in the Israeli embassy as "the cold shower."

At 11:30 A.M. on Monday, February 11, Arens, dressed as though for a funeral in dark suit and dark tie, walked into the Oval Office to face

Bush, Baker, Cheney, Quayle, Haass, and several other senior Americans. The defense minister quickly recounted the damage inflicted by the most recent Scud attacks. "We're very close to having to take action," he warned.

Unaware of either Arens's retaliation theory or details of the Patriot's purported flaws, Bush said, "There's been a dramatic decrease in the number of Scuds launched, so why are you so upset? Why are you so determined to do this?"

"Every time a Scud lands, it causes terror in Israel, Mr. President," Arens replied. "We see sights of destruction in Israel that have not been seen in Western countries since World War Two."

From the breast pocket of his suit jacket, Bush pulled an index card printed with results from a recent public opinion survey in Israel. The polls, the president said, show that Shamir's policy of restraint is very popular.

"That's a very brittle level of support," Arens retorted. "God forbid, if a chemical warhead lands or a conventional warhead does considerable harm, that support will be wiped out."

"What can you do that we can't do?" the president asked.

Arens, unaware that two Delta squadrons and a Ranger company had laagered at Ar Ar, alluded to the Israeli plan to use helicopters and ground troops in western Iraq. The minister then returned to the Patriot issue. "In my opinion, the Patriots have had only about a twenty percent rate of success," he said. "That's a very rough estimate."

Puzzlement etched the president's face. "What do you mean by that?" Bush asked.

"Out of ten Scuds that we try to intercept with Patriot," Arens said, "maybe two are successfully destroyed."

Cheney, though angry at the Israeli's insolence, suppressed the urge to accuse him of ingratitude. "There's a fundamental disagreement," Cheney interjected, "about how effective the Patriot is."

After thirty minutes the session broke up with Arens and the Americans thoroughly irritated with one another. The president and his men had anticipated a warm, cordial chat between strategic allies; instead, the confrontation revived the hostility that had long characterized relations between Bush's administration and Shamir's Likud government. Arens, Richard Haass observed sourly, "crapped all over us."

The defense minister drove to the Pentagon for a private meeting with Cheney and Wolfowitz. Apparently recognizing that his brusque manner had offended his hosts, Arens turned conciliatory. He thanked Cheney for both the Patriots and Hammer Rick. He also asked to be alerted before the allied ground attack began.

Again raising the issue of Israeli retaliation, Arens asked whether the Americans would stand aside in western Iraq during a strike by the Israeli Defense Forces. Cheney shook his head. Will you open a corridor through Saudi Arabia so that our aircraft do not have to fly over Jordan? Arens persisted. Cheney again refused. Arens paused for a moment, then asked, "Have you thought about putting in special forces?" Cheney assumed his best poker face and shrugged. "That's something," he said, "we can't talk about. I'm sure you understand."

From the Pentagon, Arens drove to the State Department. In Baker's suite, he asked for $1 billion to defray the cost of keeping Israeli warplanes on alert and other expenses. As Baker and Lawrence Eagleburger were pressing for an itemization of those expenses, Arens was summoned to the telephone in the office outside the secretary's sitting room. Another Scud had struck Tel Aviv, detonating just five hundred feet from Arens's house near Ben Gurion Airport. The minister called home to determine that his wife was unhurt, then coolly picked up the thread of his conversation with Baker.

Within hours, the account of Arens's aplomb had circulated through Washington, stirring sympathy and admiration for the Iceman's poise. If not forgotten, the morning's prickly confrontation in the Oval Office was at least forgiven. Once again the Iraqi genius for poor timing had served to solder cracks in the allied alliance.

## Baghdad

At the exact moment that Arens's mission to Washington was ending, another diplomatic venture began, more than six thousand miles away. Shortly before midnight in Baghdad on February 11, a convoy bearing the Soviet envoy, Yevgeni H. Primakov, wheeled into the driveway of the Al Rashid Hotel.

Primakov's journey had been singularly unpleasant. To forestall the indignity of being shot down by American fighters, Primakov flew from Moscow to Tehran, then transferred to an Iranian plane for a flight to Bakhtaran in western Iran. At the Iraqi border he joined a motorcade whose cars had been smeared with mud for camouflage. Casting nervous glances at the night sky, the drivers raced to Baghdad with their headlights off.

Primakov found his room at the Al Rashid lit with a kerosene lantern and stacked with jerry cans of water. Even so, his accommodations surpassed those he would have found at the Soviet embassy, where the ambassador and a dozen diplomats were bivouacked in large steel tubes embedded in the garden as makeshift bomb shelters. Once Primakov settled in for the night, the usual nocturnal explosions rocked greater Baghdad. Unfazed by the envoy's well-publicized presence, the Americans launched attacks in the early morning of February 12 against Saddam International Airport, military intelligence headquarters in Baghdad, the information ministry, and other targets.

Thickset and triple-chinned, with a gift for political survival and self-promotion, Primakov had first met Saddam in 1969 while working as *Pravda*'s correspondent in the Middle East. The U.S. government considered him a meddlesome intruder who represented the old-school Arabists in the Soviet bureaucracy loath to abandon Moscow's patron-client relationship with Iraq. In October, he had urged that Kuwait present Saddam with "face-saving" incentives to withdraw by ceding Iraq two disputed Persian Gulf islands and Kuwait's portion of the Rumaylah oil field. Nevertheless, Primakov's visit appeared to have the blessing of Mikhail Gorbachev, whose fundamental support for the American-led crusade against Iraq was tempered with a reluctance to forsake Soviet influence in the Middle East.

Primakov's mission began Tuesday morning with a tour of bomb damage in Baghdad. Four weeks of war, he discovered, had reduced the Iraqi capital to nineteenth-century privation. The city lacked running water, telephone service, garbage collection, and electricity. Every night hundreds of motorists desperate for gasoline queued up at service stations for a seven-gallon ration intended to last two weeks. Under a water-rationing program adopted in late January, tanker trucks circulated through neighborhoods every three days, mobbed by residents toting tubs, bottles, and pitchers. Without power, Baghdad's two sewage-treatment plants no longer functioned, and millions of gallons of raw waste poured into the Tigris. Despite the filth and a shortage of fuel for boiling tainted water, the river served as both communal well and public bath.

Food was adequate, although prices were sometimes extortionate: a kilo of rice sold for as much as fifteen dollars, compared to sixty cents before the war. Urbane Iraqis found themselves slaughtering and plucking chickens as their rural grandparents once had. Hospitals reported shortages of X-ray film, blood-typing kits, drugs, and many other accoutrements of modern medicine. Some physicians boiled dressings from one patient to use on another; when generators failed, surgeons operated by lantern, candle, and flashlight. Cases of sepsis, dysentery, hepatitis, and other diseases soared. When morgues overflowed, health workers reportedly buried the dead in hospital gardens.

The impact on Iraqi morale was difficult to judge. Frightened children leaped into their parents' laps when the nightly sirens howled. Many residents noted the difference between the relentless, nightly raids by the Americans and the sporadic, desultory attacks by Iran during that long war. (The allies flew more air combat sorties in a single day than Iran had flown in eight years.) If most neighborhoods escaped bombardment, the steady destruction of the capital's most prominent landmarks was evident to everyone: the graceful July 14 suspension bridge, now half-submerged beneath the Tigris; the Palace of Congresses convention center, shattered by two laser-guided bombs; the telecommunications centers, where customer accounts swirled in the wind like confetti. Every morning the city skyline was altered, reduced. Yet equally evident was the American focus on military targets rather than indiscriminate bombing of the population. Although certainly not normal, daily life assumed a wartime equilibrium.

Baghdad had announced in early February that seventeen-year-old students, previously exempt from conscription, would now be drafted. Protest against the measure, like disenchantment with the war itself, was limited to grumbling and to graffiti scribbled furtively at night on walls in the downtown souks. Many Iraqis huddled by their shortwave sets at night, listening to news broadcasts from the BBC or Voice of America, although rumors periodically swept the city like epidemics. A false report of the assassination of Egypt's Hosni Mubarak had triggered a mad celebration, with thousands of Kalashnikov rifles fired in short-lived jubilation.

On Tuesday night, Primakov met Saddam at a government guest house in central Baghdad awash in generator-powered lights. The Iraqi leader strolled into the house accompanied by his senior lieutenants, removed his trench coat, and unbuckled his gun belt. Primakov was startled by Saddam's appearance: he looked gaunt, as if he had lost thirty or forty pounds since their last meeting, four months earlier. Saddam, continuing a tirade begun earlier in the day by his foreign minister, Tariq Aziz, rebuked the Soviets for giving the "green light" to a United Nations war against Iraq. Baghdad would never capitulate, he vowed. Iraq's will to resist was "unshakable."

Even to a man as well versed in Iraqi wiles as Yevgeni Primakov, Saddam's war plan was puzzling. Baghdad appeared paralyzed by fatalism. In their session at the presidential palace the previous October, Primakov had reminded Saddam of the ancient fortress of Masada, where hundreds of Jews killed themselves in A.D. 73 rather than surrender to Roman soldiers. "Don't you think that, like Israel, you have developed a Masada complex?" Primakov had asked. Saddam nodded. "Then," Primakov continued, "your actions will to a great extent be determined by the logic of a doomed man." In early January, Saddam told a French diplomat: "I know I am going to lose. At least I will have the death of a hero."

To the extent that Saddam was guided by reason rather than a yen for martyrdom or willful obstinance, his strategy was similar to that of the Japanese during World War II: maintain the wan hope that the Americans would see the futility of trying to dislodge their opponents and eventually accept a negotiated peace brokered by a third party. Like the imperial Japanese, Saddam seems to have concluded that self-indulgent Americans lacked the discipline and taste for blood necessary to destroy its enemies.

Without question — as Aziz would admit after the gulf war — the Iraqi leadership underestimated the intensity of modern combat and the American willingness to demolish much of Iraq's infrastructure. Insular, naïve, plagued with delusions of glory that led him to liken himself to Nebuchadnezzar, the Babylonian warlord and conqueror of Jerusalem in 586 B.C., Saddam had an understanding of military matters that was unsophisticated to the point of juvenile. "They tell you that the Americans have advanced missiles and warplanes," he had declared in a speech shortly before the war began, "but they ought to rely on their soldiers armed with rifles and grenades."

Primakov, according to accounts he subsequently wrote for *Pravda* and *Time,* tried to jolt some sense into Nebuchadnezzar. "The Americans absolutely favor a broad-scale land operation, as a result of which the Iraqi task force in Kuwait will be completely destroyed. Do you understand? Destroyed." Primakov then sketched a Soviet plan for ending the war, beginning with Baghdad's announcement of a total and unconditional withdrawal from Kuwait as quickly as possible. Saddam nibbled at the proposal. How, he asked, can we be sure our soldiers will not be shot in the back as they withdraw? Would the air strikes end after the pullout? Would United Nations sanctions be lifted? He proposed sending Aziz to Moscow for further discussion, but Primakov chafed at the delay. "There is no time left," the Soviet warned. "You must act immediately."

The meeting ended with Saddam's promise of quick response. At 2 A.M. on February 13, Aziz delivered a written statement to Primakov: "The Iraqi leadership is seriously studying the ideas outlined by the representative of the Soviet president and will give its reply in the immediate future." Though it was hardly a passionate embrace of the Soviet overture, Primakov considered the note a critical first step. Before leaving Baghdad for the circuitous return trip through Iran, he sent a message to Moscow through the Soviet embassy. "There are," he cabled, "certain promising signs."

## Amariyah, Iraq

At 4:30 A.M. on February 13, precisely as planned, a pair of F-117 fighters arrived on station above Baghdad's southwestern suburbs. The two pilots identified the Al Firdos bunker through their infrared scopes and centered the roof in their laser cross hairs, scrupulously avoiding the nearby school and mosque. The first GBU-27 sliced through ten feet of reinforced concrete before the bomb's delay fuze detonated in a cataclysm of masonry, steel, and flame, jamming shut the shelter's heavy metal doors. The second pilot also laid his bomb dead center through the roof. Smoodge — the dust and smoke extruded from an underground target — rolled from the ventilation shafts.

The lucky ones died instantly. Screams ripped through the darkness, muffled by tons of shattered concrete and the roaring inferno that enveloped the shelter's upper floor. Sheets of fire melted triple-decker bunk beds, light fixtures, eyeballs. One survivor, Omar Adnan, a seventeen-year-old whose parents and three younger sisters perished, later described the conflagration: "I was sleeping and suddenly I felt heat and the blanket was burning. Moments later, I felt I was suffocating. I turned to try and touch my mother who was next to me, but grabbed nothing but a piece of flesh."

Black smoke boiled from the building as fire trucks wheeled into the compound. Rescue workers hammered futilely at the steel door; finally they cut it open and groped their way down the ramp into the hellhole below. Bodies lay in grotesque piles, fused together by the heat. Limbs and torsos were strewn across the floor. Eighteen inches of water flooded one corridor, the surface covered with a skim of melted human tallow.

For hours rescuers lugged victims out into the morning light, sometimes vomiting from the stench or collapsing in anguish beneath their unbearable loads. Lingering heat in the basement melted the plastic gloves on their hands. Smoke still curled from some of the bodies, so charred that only their size identified them as children. In the courtyard, the decapitated corpse of a woman lay next to the limbless, headless torso of a small girl. A Daihatsu pickup truck served as a makeshift hearse; its bed was stacked with bodies wrapped in soiled sheets. Ambulances raced away with several dozen survivors who had been pulled from the nether corners of the crematorium. Most suffered from burns, lacerations, and shock. At Yarmuk Hospital, four miles east of the shelter, the chief of surgery counted fifty-two burn cases. Without adequate electricity, water, or bandages, the hospital laid the naked patients uncovered on beds and, in many cases, waited for them to die.

Baghdad promptly claimed that only civilians had been inside Shelter Number 25. Tariq Aziz initially put the number at four hundred; later, the Iraqi government would tell the United Nations Human Rights Commission that 204 had perished. Residents of Amariyah insisted that the shelter had recently been opened to ordinary Iraqis, as well as to Jordanians and Palestinians living in the neighborhood. Children, they reported, had enjoyed the generator-powered video cassette player used to show Clint Eastwood and Bruce Lee movies at night after their families trooped down the ramp at dusk with blankets and sandwiches.

By midmorning five thousand Iraqis stood outside the shelter, watching the grisly cortège emerging from the basement. Sunlight poured through the yawning shafts where the bombs had burrowed. For the first time in the war, Western television cameras filmed without censorship. "Why did this happen?" Iraqis shouted at the correspondents. "Is this the way to win back Kuwait?" One man flung himself to the ground in grief, wailing incoherently. Others wept silently or beat their chests as each new victim was laid on the ground. *Allah akhbar!* they shouted again and again. God is great! God is great! God is great!

## Riyadh

Buster Glosson, who normally slept from dawn until noon, was awakened at 10 A.M. by a phone call from Dave Deptula. U.S. eavesdroppers had picked up emergency radio traffic in Baghdad. "There are indications that there were quite a few civilians in the Al Firdos bunker," Deptula reported. "It's going to be on CNN shortly." Glosson showered and walked across the Saudi air force compound to the Black Hole, where he found his staff staring at the television.

"I can't believe this," Deptula said bleakly. "Boy, did we fuck up." Glosson shook his head. "Listen, on the basis of the information we had available there was absolutely no reason for us not to target that bunker." Glosson called Colonel Al Whitley, the F-117 wing commander in Khamis Mushait. "Tell your pilots that it's important for them not to feel guilty about this," Glosson said. "They were given a target and told to strike it. They did what they were told to do."

The U.S. government immediately tried to shift the blame to Saddam, even suggesting that the Iraqis had deliberately sacrificed the civilians in Shelter Number 25 to turn world opinion against the Americans. In Riyadh, CENTCOM officers insisted that some of the dead were Mukhabarat members operating on the lower level of the bunker; the civilians, they postulated, were families of intelligence officers, secretly admitted as a perquisite granted to the elite. "I'm here to tell you," Butch Neal declared in a press briefing, "it was a military bunker."

The precise truth would remain uncertain, although without doubt U.S. intelligence erred, grievously, in failing to detect the presence of so many civilians. At the Pentagon, officers found themselves explaining why a war intended to eject the Iraqi army from Kuwait seemed to focus so intensely on Baghdad. "That's the head. That's the brain. That's where the missions come from," Joint Chiefs' spokesman Tom Kelly told reporters. "If the Iraqi army is going to mount a multidivisional attack and come out of their fortifications to head south into Saudi Arabia, that's where those orders would come from." Even so, Kelly added, "we are going to examine our consciences very closely to determine if we can't do something in the future to preclude this sort of thing."

Powell and Mike McConnell, carrying satellite photos and other evidence used to select the target, drove to the White House to face a drawn, pensive George Bush. Clearly, the images of charred bodies had spoiled the administration's effort to portray the war as bloodless. McConnell, who had felt nauseated after hearing the first report on the radio before dawn while driving to the Pentagon, reviewed the evidence in excruciating detail. "Mr. President," Powell added, "we don't have a case that you could take to the Supreme Court, yet we passed the commonsense test. We are convinced that this was a military target. We don't understand why those civilians were in there. But I assure you we have examined it and we stand as strongly behind the decision after the fact as we did before the fact."

In Riyadh, Glosson assembled a similar packet of evidence into three file folders — one for Schwarzkopf, one for Horner, one for himself — and drove to the MODA building. "I'm sure you will be asked as many or more questions about this as I will," he told the CINC. "Here's a chronology of all the information I had and the thought process behind why I elected to strike that target." Schwarzkopf was somber but calm. "Was that a proper target?" he asked Jack Leide. "Yes, sir," the intelligence chief answered, "it was."

CENTCOM staff officers began compiling schematic drawings of the shelter and other data to be given to the media in Riyadh — although information from the Baghdad spy remained top secret. Glosson and Schwarzkopf discussed tactics to minimize the risk of another disaster in Baghdad, such as limiting attacks to bridges and smaller targets that could not be stuffed with civilians. (Even "safe" targets carried risks, however. Fifteen hours after the Al Firdos strike, four British Tornados from Dhahran darted up the Euphrates to attack a highway bridge in Fallujah. A laser-guided bomb, apparently equipped with defective fins, veered sideways from the river and killed an estimated 130 people in a crowded marketplace.)

After a catastrophe like Al Firdos, Glosson knew, things would never again be the same. The horrific scenes from Amariyah, televised around the world, provided Saddam with an immense propaganda victory. But raids on the Iraqi capital, Glosson believed, had to continue. Already sorties north of the Euphrates had dwindled to less than half the number he considered necessary; no Tomahawks had been launched in nearly two weeks. A further curtailing of the strategic air campaign would hand the Iraqis a military victory by giving Baghdad an undeserved respite from the war and permitting Saddam to strengthen his grip. Surely, Glosson thought, the allies would not be so foolish.

## Washington, D.C.

In Colin Powell's view, however, the air campaign against Baghdad had nearly run its course. Even before Al Firdos, Powell had begun to suspect that American bombers were getting diminishing returns for their efforts in the Iraqi capital — "making the rubble bounce," as the chairman put it. How much more could be gained by further attacks, he asked himself, particularly with the public and press scrutinizing every dropped bomb? Was it in American interests to further terrify and devastate Iraqi civilians?

Powell believed the time had come for air attacks to focus almost exclusively on the Kuwaiti theater, "isolating and pounding the hell out of the Iraqi army" in preparation for the ground war. Ever atuned to public opinion and the geopolitics underlying any military venture, he sensed a growing unease over the incessant pounding of Baghdad. The effort to kill Saddam or his lieutenants — antiseptically classified as "leadership targets" — was something the country could accept, even endorse; but killing women and children, however inadvertently, was quite a different matter, particularly when television so vividly displayed the carnage. Another massacre like Al Firdos would destroy the allies' moral standing, Powell felt, and anyone who doubted that failed to understand war in the age of modern telecommunications.

The chairman may have underestimated the American tolerance for bloodshed. Al Firdos horrified, but it also hardened. For many, the initial shock soon yielded to a complex mixture of emotions: revulsion and disgust, certainly, but also anger at the enemy and a flinty determination to see the war through to victory. Americans had at least tacitly endorsed the demolition of dozens of Axis cities during World War II in the cause of military necessity. The public turned against the Vietnam War not as a consequence of the six million tons of explosives dropped, but because American soldiers were dying by the tens of thousands.

A national poll taken immediately after Al Firdos found that eight in ten Americans blamed the Iraqi government for the tragedy. Only 13 percent asserted that the United States should take greater pains to avoid hitting civilian areas in Iraq. Nearly 80 percent still backed Bush's decision to go to war, a level of support that had not diminished in weeks.

Without question, though, Powell had anticipated the qualms of the administration's senior civilians. Cheney, Wolfowitz, and others believed a threshhold had been crossed with the bombing of Al Firdos. At the White House, Brent Scowcroft and his deputy, Robert Gates, worried not only about American public opinion but also about sentiment abroad. Solidarity within the coalition — as well as support from the Soviets, the Chinese, and the Arab world — would be important if sanctions were to be maintained against Iraq after the shooting stopped. Powell took the initiative before he was pressed to act.

"We have got to review things to make sure we're not bombing just for the sake of indiscriminate bombing," he told his staff. "Let's take a hard look and determine whether a target's destruction is really required for prosecuting the war or whether it's just somebody's favorite target. If there's a target in Baghdad that we need to hit, then by God take it out. But don't target indiscriminately."

Powell himself would evaluate sorties proposed against the Iraqi capital, not to second-guess every mission but as a safeguard against imprudence. "The amount of additional damage we might inflict compared to the consequences of that damage takes on more policy and political overtones," he told Schwarzkopf on the telephone. "It has to be looked at differently . . . We don't want to take a chance on something like this happening again."

Although Schwarzkopf did not dissent, the chairman anticipated that Riyadh would not be wholly pleased with the new arrangement. With few exceptions, such as the recent curtailment of cruise missile launches, theater commanders had been given a free hand in waging the air war. But it was vital to remember, believed Powell, that the national interest ultimately lay in the overall success of the war; victory was a matter of perception as well as battlefield position. Military campaigns were not an end unto themselves.

To those who took issue, Powell had an answer waiting on a bookshelf in his Pentagon office, a passage he occasionally cited from Thomas Jefferson's First Inaugural:

> I shall often go wrong through defect of judgment. When right, I shall often be thought wrong by those whose positions will not command a view of the whole ground. I ask your indulgence for my own errors, which will never be intentional, and your support against the errors of others, who may condemn what they would not if seen in all its parts.

## Riyadh

On February 14, Schwarzkopf summoned Glosson to the MODA building. "I need to go over every target in Baghdad each day so that I can explain exactly why we're striking it and what we expect to gain — where it's located, what's in the surrounding area, everything," the CINC said. Glosson previously had given Schwarzkopf only generic descriptions of the strategic menu: "Sir, tonight we're going to focus on the Iraqi security apparatus and hit twelve targets in Baghdad." Now he and Deptula had to justify every mission beforehand, orally at first, then in writing. Schwarzkopf seemed unperturbed by the new arrangement; Glosson surmised that the CINC was preoccupied with preparations for the ground offensive.

Despite grumbling in the Black Hole about "Colin Powell's presidential aspirations," Glosson recognized the chairman's responsibility for taking account of political and strategic ramifications beyond Riyadh's ken. Yet if Powell commanded "a view of the whole ground," in Jefferson's phrase, Glosson felt he commanded a comparable panorama of the air. He believed the Ba'athist regime was tottering. The Air Force could not guarantee that continual bombing would topple Saddam, but it certainly heightened the probability of his demise and weakened his control over Iraq. Neither he nor Deptula could understand Washington's apparent failure to recognize the potential political advantage of continued, *discriminating* attacks on the enemy's levers of power.

To prove the strategic campaign had not yet run its course, Glosson gave Schwarzkopf a chart showing that, because of bad weather and sundry diversions, on day eighteen of the war strategic sorties had reached only the number originally planned by day seven. Glosson and Deptula had faith in the psychological value of rubble. By repeatedly smashing the central symbols of Saddam's rule — such as Ba'ath Party headquarters and the presidential palaces — air power could strip the Iraqi leader of his invincible aura. And contrary to the chairman's appraisal, the two airmen were convinced that the rubble in Baghdad had not yet begun to bounce.

Long before November 1, 1911, when an Italian pilot flipped an oversized grenade from his cockpit at a Libyan oasis, thus becoming the first aviator to drop a bomb from an airplane during combat, military strategists had debated the psychological impact of "celestial assault."

"It undoubtedly produces a depressing effect to have things dropped on one from above," a German theorist wrote of military ballooning in 1886. After World War I, the aviation writer R. P. Hearne added, "It is particularly humiliating to allow an enemy to come over your capital city and hurl bombs upon it . . . The moral effect is wholly undesirable." A British analyst estimated the psychological "yield" of RAF attacks on Rhine towns in the Great War to be twenty times the material damage. Among other consequences, air attacks kept the population in a perpetual state of unease; one French town endured only seven air raids but three hundred alerts, each of which emptied factories, robbed residents of their sleep, and nurtured discontent.

The most intriguing question, debated without resolution for nearly a century, was whether bombing could strip the enemy population of its will to fight or even incite an aggrieved citizenry to overthrow a regime that failed to protect them. Marshal Foch warned that massive air attacks might have such "a crushing moral effect on a nation" that its government would find itself disarmed. Adolf Hitler in 1939 envisioned an air offensive "against the heart of the British will-to-resist." Britain's Bomber Command subsequently decreed that "the primary objective of [RAF] operations should now be focused on the morale of the enemy civil population."

In the terrible laboratory of World War II, theory played out in practice with checkered results. Hitler's blitz on London was a manifest failure, ratifying J. M. Spaight's warning in 1930 that air attacks "may not smash the will to war [but] only harden it, intensify it." The Allied devastation of Berlin — and Hamburg and Essen and Frankfurt and Nuremburg and Cologne and fifty other cities — was not what brought the Nazis to heel. Arthur (Bomber) Harris, who became head of Britain's Bomber Command, concluded that in a police state even a demoralized people could be kept working and politically neutered. (After the RAF's incineration of Dresden, on February 13, 1945 — forty-six years to the day before Al Firdos — even the bellicose Winston Churchill questioned the "bombing of German cities simply for the sake of increasing terror.")

On the other hand, Allied strategists had hoped that air strikes on Rome would drive a wedge between Mussolini and his Italian subjects. Six days after the Italian capital was bombed, in July 1943, Mussolini fell from power. As Lee Kennett has pointed out in *A History of Strategic Bombing*, the attacks may have been more a contributing factor than the primary cause of Il Duce's undoing, but the results heartened those who believed that air power alone could heave over the enemy leadership.

In Japan, where eight million civilians were left homeless in the sort of methodical campaign Churchill called "dehousing," the postwar Strategic Bombing Survey concluded: "Even without the atomic bombing attacks, air supremacy over Japan could have exerted sufficient pressure to bring about unconditional surrender and obviate the need for invasion." A generation later, the devastating Linebacker II campaign — an eleven-day assault on North Vietnam in December 1972 that included more than seven hundred B-52 sorties — helped push Hanoi into an armistice. "It was apparent," the Air Force historian William W. Momyer wrote, "that airpower was the decisive factor leading to the peace agreement of 15 January 1973."

In the gulf war, the weight of this historical baggage pressed on Glosson, Deptula, and other denizens of the Black Hole. For more than seventy years, pilots had envisioned a day when air attacks would be precise and potent enough to bring an enemy to his knees, when strategic air power would indeed be "a war-winning weapon in its own right," as General Hap Arnold had once asserted. Michael Dugan, the Air Force chief sacked by Cheney in September, had spoken almost mystically of air power's "special kind of psychological impact." Dugan relegated American ground forces to a constabulary role — perhaps necessary for reoccupying Kuwait, but only after air power had so shattered enemy resistance that soldiers could "walk in and not have to fight."

Yet the faith that bombing alone could win the war was just that: faith. For the notion that a few more attacks in central Baghdad might create the critical mass leading to Saddam's overthrow or Iraq's surrender, Pentagon policymakers like Paul Wolfowitz found no credible evidence. Wolfowitz, who on January 17 would have been delighted with a victory wrought by air power, now insisted that crushing the

Iraqi army with ground forces was a vital U.S. objective, a necessary means to strip Saddam of the pretense that he had been thwarted only by high-technology air attacks. "If this war ends with the Iraqis able to say, 'We were never beaten on the ground,' that's going to be a strategic victory of sorts for Saddam," the under secretary argued.

Schwarzkopf concurred. His war plan had never embraced a strategic campaign that bombed the Iraqis into submission. Once Saddam and the Ba'athists had survived the first ten days or so of air attacks, the CINC concluded that the chances of the regime collapsing from within were nil. Anyone who thought otherwise failed to understand how stubborn Arabs could be. He also suspected that the Air Force was excessively sensitive to any guidance from Washington because it smacked of the straitjacket restrictions imposed during Vietnam.

Certainly the air planners were not immune to service parochialism and a yen for a victory wrapped in Air Force blue. All wars also develop an inexorable momentum, in which destruction can become its own rationale. In the main, however, Glosson and his lieutenants were driven by the honorable impulse to spare American soldiers the presumed bloodbath of a ground war. Their faith in air power — whether embodied in Tomahawks, Navy A-6s, or Air Force bombers — transcended service loyalty. But they ardently believed that the rush toward a ground attack reflected both ignorance of air power's lethality and the Army's determination to play a major role in the war. The war against Iraq would probably shape the structure and budgets of the American military for years to come; bringing several hundred thousand soldiers into the theater without allowing them a significant part in the victory, Glosson and Deptula concluded, was a difficult prospect for ground commanders to accept.

To some degree, the Air Force fell victim to its own precision. The reduction of Baghdad was wholly different from the flattening of Berlin or the firebombing of Tokyo, where a single raid in March 1945 killed more than eighty thousand Japanese and left sixteen square miles of cityscape in ashes. Although the allies would drop 88,500 tons of bombs in the gulf war — more than was dropped on Japan in the final six weeks of World War II — by far the largest number fell outside Iraqi cities. Eighteen thousand strategic sorties would be flown, including six thousand by "shooters," but only forty or so targets had been hit in Baghdad at the time of Al Firdos. Nearly a third of the strategic missions, for example, attacked Republican Guard units; three thousand hit airfields; and hundreds more dropped or damaged fifty bridges.

As President Bush himself had publicly stated, the allies had no quarrel with "the Iraqi people." Targets in Baghdad were chosen to weaken the regime, not to terrorize the citizenry. Except for the seats of government power, most of the capital remained unscathed. Iraqis might be without electricity, clean water, and dignity, but the prospect of an infuriated, frightened populace rising in rebellion seemed as unlikely in Iraq as in Hitler's police state.

Glosson and Deptula also suspected that Iraqi citizens — not unlike many Americans — were baffled about why the war to liberate Kuwait was being waged in downtown Baghdad. In his original Instant Thunder proposal, John Warden conceived of a strategic psychological operations plan; it was to convince the Iraqis that Saddam was the source of their misery and that the American-led alliance stood ready to help reconstruct Iraq if the demon was exorcised. Leaflets printed with similar messages had been dropped by the tens of millions in previous wars, albeit with limited results. ("Adolf Hitler has led you into this hurricane," read one message written by Bomber Harris and dropped after an attack on Frankfurt in 1943. "What you experienced this past night was like first raindrops which announce a coming storm . . . Take heed!")

But in the gulf war, the strategic "psyops" campaign never took shape, in part because of protracted disputes in Washington among the Pentagon, CIA, U.S. Information Agency, and others over who was in charge. (By contrast, the tactical psyops operation against Iraqi troops later was a brilliant success.) Not until February 26 would F-16s sprinkle half a million leaflets over Baghdad — "Saddam's First Line of Defense: Innocent Civilians," the fliers proclaimed — but by then the war was in its final hours.

As required, Glosson began presenting Schwarzkopf with written justifications for targets in Baghdad. Few suggestions were flatly rejected; many, however, fell into what the Black Hole soon dubbed the "no, not yet" category. When it came to strategic targeting, Schwarzkopf no longer displayed his usual peremptory command style. Decisiveness yielded to procrastination as the CINC awaited Powell's vetting. During a five-night stretch following Al Firdos, no bombs fell on downtown Baghdad except for two telephone exchanges hit immediately after the bunker disaster. On the night of February 18–19, stealth pilots again attacked Iraqi air force headquarters and the adjacent Mutheena Airfield downtown, but several days of bad weather aborted other sorties. Not until February 23 would another leadership target in Baghdad be bombed. The net effect was to give the capital a ten-day reprieve. In the two weeks before the Al Firdos attack, twenty-five targets had been struck by F-117s in downtown Baghdad; in the two weeks following the attack, only five would be hit.

Like their Army brethren, Air Force officers in Riyadh found themselves nominating targets with little assurance that they would be struck. Officers in Checkmate and the Black Hole complained to one another about a "Vietnam-style political mistake" similar to the bombing pauses that had given Hanoi time to regroup. Beyond those subterranean cubicles, however, scant grumbling could be heard. Chuck Horner at times expressed dismay at the new restrictions and periodically offered suggestions on prospective targets that Schwarzkopf and Powell might find acceptable. Yet, in Glosson's view at least, Horner's resentment never reached the level of outrage that Glosson himself felt. Even Mike McConnell, often Buster Glosson's staunchest ally on the Joint Staff, concurred with Powell's reasoning.

Among the targets to elude destruction were Baghdad's bridges. Nine bridges spanned the Tigris in the Iraqi capital; none had been targeted as of January 17. Then U.S. intelligence concluded that fiber-optic cables led from communications centers in the subbasements of the Al Rashid and Babylon hotels, snaking beneath one of the bridges — no one seemed certain which one — before running south along an oil pipeline to Basrah. (Western correspondents who later toured the bowels of the Al Rashid found nothing more sinister than a laundry.)

With blessings from Horner and Glosson, officers in the Black Hole had drafted a memo in January proposing that the Al Rashid and Babylon be bombed after fair warning was issued in a leaflet barrage. The proposal was denied by the CENTCOM staff, which recognized the political hazard of demolishing hotels. As an alternative, Glosson and Deptula ordered four bridges destroyed in late January and early February. After some difficulties because of bad weather and errant marksmanship, F-117 pilots struck them. But before the others could be dropped into the Tigris, Powell's restrictions took hold. The remaining bridges were spared.

Glosson, now thoroughly irritated, complained directly to McPeak, the Air Force chief. Those communications cables, he insisted, helped Saddam control his army in the south. The bridges were as vital to Baghdad's commerce, both military and civilian, as the Fourteenth Street, Memorial, and Key bridges spanning the Potomac were to Washington's well-being.

McPeak took up the cudgels in a private session with Powell. "We really ought to be knocking these bridges over," he urged. "Shit, I want every taxi in Baghdad to have to drive the maximum distance to get to the other side of the river. We're trying to get the Iraqis to consume their resources. Why reduce the expense of their operation?" But the chairman remained adamant. Many of the 126 highway bridges and nine railroad bridges south of Baghdad had been struck, he pointed out, and damage to the capital's infrastructure now affected civilians more than the military. Moreover, the political risk outweighed the military gain.

McPeak was convinced by Powell's logic. Glosson, stubborn and strong-willed, was not. Typically, he obeyed while looking for loopholes. One bit of ingenuity involved redefining Baghdad's city limits. If proposed targets within the capital had to be justified, then perhaps the capital could shrink. On one map in the Black Hole, Baghdad's central business district was shaded with dark ink. The perimeter of the shaded swath became, for targeting purposes, Baghdad's boundary. Targets outside it were considered beyond the capital, and thus exempt from written justification.

Even so, the number of "no, not yets" climbed. The Black Hole suggested, for example, bombing the Iraqi Ministry of Strategic Industry, which, U.S. intelligence concluded late in the war, helped oversee Saddam's nuclear bomb program. "Destruction will severely hamper regime's capacity to construct weapons of mass destruction," Deptula wrote in the proposal presented by Glosson to Schwarzkopf.

Of an attack proposed on Republican Guard headquarters, Deptula reasoned, "Destruction will disrupt ability to control forces in [Kuwaiti theater] and degrade the enemy's ability to reestablish, organize, and train the Republican Guard after the war."

In proposing to destroy a large statue of Saddam in central Baghdad, he wrote: "Removal will emphasize vulnerability of Saddam regime to allied attack and underline allied war focus on [Saddam] and not the people of Iraq. Will strengthen position of elements within Iraq opposed to Saddam's war. One of the largest monuments to the Iraqi dictator."

In each case the answer came back: not yet. Al Firdos had hastened the shift in focus to the primary objective of liberating Kuwait — by direct attack, not by indirect assaults on the enemy's more remote centers of gravity.

"It's as if someone changed the rules," Glosson complained. But Horner offered the shrewder observation: "We'll probably never appreciate just how much freedom we had."

# 11

# Misadventure

## Washington, D.C.

Curiously, the historical antecedent that most frequently haunted Washington policymakers during the Persian Gulf crisis was not Vietnam — although that war provided ghosts aplenty — but Korea. If Inchon offered a beguiling model of bold generalship, the "forgotten war" itself loomed as a monument to good intentions gone sour. Two aspects seemed particularly cautionary. First, in 1950 the White House and Pentagon had broadened American war aims in midfight, with disastrous consequences. Second, the war had dribbled to an inconclusive ending, leaving two hostile armies — and a large force of Americans — glaring across a treacherous no man's land for nearly forty years.

The Korean analogy arose often in the musings of the Small Group, the sextet of deputies responsible for much of the administration's policy before and during the war. Meeting one to three times a day, either in the White House Situation Room or by teleconference, the deputies served as alter egos for their principals: Paul Wolfowitz from Defense; Robert M. Kimmitt from State; Admiral David Jeremiah, vice chairman of the Joint Chiefs; Richard J. Kerr, deputy CIA director; Robert Gates, deputy national security adviser; and Richard Haass, the resident White House Middle East expert who also served as the deputies' scribe. The Small Group framed issues and drafted recommendations for Bush's war cabinet on subjects ranging from congressional tactics and Security Council strategy to the phrasing of the letter laid in front of Tariq Aziz on January 9. Almost never did deputies and principals disagree on fundamental policy — a symptom of pellucid harmony or, less charitably, Potomac inbreeding.

The paramount question of war aims had occupied the Small Group since September. What did the United States hope to achieve in fighting Iraq? How should victory be defined? Clausewitz had sorted wars into two types: those aimed at destroying the enemy's "existence as a state," and those directed toward limited conquests on the enemy's frontier. The gulf war clearly fell into the second category. But American wars traditionally tended to grow unlimited, a reflection, as the political scientist Robert E. Osgood noted in the 1950s, of "our profound distaste for the very notion of containment and limited war."

Korea was a case in point. Although most Americans tended to blame Douglas MacArthur for the widening of that conflict, certainly Harry Truman, Dean Acheson, and Omar Bradley shared the culpability. The war began with the American objective of ejecting North Korean invaders from South Korea and restoring the status quo ante bellum. Several months later, emboldened by battlefield success, MacArthur sent his troops across the 38th Parallel with the larger intent of destroying the enemy army, deposing the communist regime, and unifying Korea under a single, democratic government. China subsequently intervened, and the war dragged on for three years at a cost of 54,000 dead Americans.

In the gulf crisis, President Bush had explicitly laid out American policy objectives within a week of Iraq's invasion: the "unconditional withdrawal of all Iraqi forces from Kuwait"; the "restoration of Kuwait's legitimate government"; a guarantee of the "safety and protection of the lives of American citizens abroad"; and the enhancement of "security and stability of Saudi Arabia and the Persian Gulf." For those in the Small Group attempting to specify these guidelines, the final objective proved particularly hard to define. What combination of diplomatic and military action would suffice to maintain "security and stability" in the region? Could peace endure with the Ba'athist regime still entrenched in Baghdad, or did long-term stability require more ambitious action than that proposed in the current allied war plan?

As the triumph of the air campaign against Iraq became evident, the Small Group — stalked by the Korean poltergeist — resisted the urge to reach for a larger prize. By mid-February, Wolfowitz and others suspected that an armored spearhead to Baghdad would be an easier enterprise than anyone could have dreamed on January 17; yet all agreed that liberating Kuwait and demolishing the enemy army of occupation should remain the cornerstones of U.S. policy. Unless Saddam used chemical or biological weapons, the toppling of his regime would not be a formal war aim, however devoutly it might be wished.

Beyond the fear of ignoring history and thus being condemned to repeat it, several other factors obtained. Most of America's European allies, notably the French, adamantly opposed broadening the war. The

Saudis, Syrians, and other Arabs wanted Saddam removed but preferred divine intercession — or one well-placed bomb — to the messy uncertainty of a wider conflict. The UN Security Council resolution of November 29 had authorized the use of force to eject Iraqi troops from Kuwait, not to overthrow the Ba'athists in Baghdad. Even the U.S. Senate had barely given Bush endorsement of his limited war aims. Neither the Security Council nor Congress, the Small Group concluded, would countenance a war waged to conquer Iraq.

For the American military, the overarching lesson from Vietnam and Korea was the need for clear, immutable combat objectives. The Joint Chiefs — through Powell's oracular voice — insisted on a mission that was both finite and within the means of the forces at hand. As translated into Schwarzkopf's war plan, Bush's policy goal of regional security and stability meant eradicating Iraq's capability to build weapons of mass destruction — chemical, biological, and nuclear — and the destruction of Republican Guard forces in the Kuwaiti theater. The latter, in military parlance, did not entail obliterating every last Republican Guard platoon, but rather disabling the divisions as an effective fighting force. This would strip Iraq of its armored spearhead, thus blunting Saddam's most dangerous offensive capability and contributing to regional peace.

A drive north of the Euphrates River and on to Baghdad would fundamentally alter the war. At best, only the British would join such an attack, permitting Saddam to portray the invasion as an imperialist Anglo-American expedition. (Even in Schwarzkopf's current war plan, Saudi, Egyptian, and other Arab troops would not set foot in Iraq lest they violate the territory of a fellow Arab.) Although no Chinese hordes waited to spill across an Iraqi Yalu, the nature of the combat would likely change if Iraqi soldiers and civilians found themselves fighting for hearth and homeland north of the Euphrates. The river offered a convenient demarcation line between southern and northern Iraq. By remaining in the south, Schwarzkopf's legions could defend a mostly barren wasteland that provided a buffer zone against a counterattack into Kuwait. Should it drive farther north, the army of liberation would become an army of occupation, with incalculable logistical, political, legal, and military complications.

The U.S. government also worried, perhaps excessively, that a power vacuum in Baghdad might "Lebanonize" Iraq by dicing the country into warring duchies under the sway of Iran, Turkey, or Syria. Moreover, if the prospect of installing an American viceroy in Baghdad seemed quaint, no strong, pro-Western Iraqi alternative to Saddam waited in the wings to assume power. (Unlike Panama, where Guillermo Endara had stood ready to replace Noriega after the U.S. invasion, American planners searching for a successor to Saddam wistfully referred to a mythical creature dubbed "Teddy Tikrit.") Even MacArthur had warned that "nothing is gained by military occupation. All occupations are failures."

But how to justify leaving in power the tyrant whom George Bush had so often compared to Hitler? The Small Group — at times grimacing at the president's hyperbole, which inflated Saddam and diminished Hitler — coalesced around two rationales. The first was that Saddam resembled the poltroon of Munich and the Rhineland, not the monster of Auschwitz and Buchenwald; checking his ambitions now meant the former would never metamorphose into the latter.

The second rationale was a belief that Saddam would succumb to a coup once the defeated Iraqi army returned home to see the catastrophic consequences of his leadership. With luck — and Small Group members like Jeremiah and Haass thought the odds better than even — the Iraqi military would do the dirty work themselves.

As in Vietnam and eventually in Korea, the Persian Gulf War would be fought for relatively modest objectives. Unlike those earlier wars, the objectives in the gulf were plainly stated and rigidly maintained. Bush and his men concluded that the excessive price of total victory would be indefinite responsibility for rebuilding a hostile nation with no tradition of democracy but with immensely complex internal politics. This was — and remains in retrospect — a sensible strategic calculation.

What they failed to see as clearly, however, was that limited victory also exacted a price: eternal vigilance. Threats to the great oil fields in the south would persist, from Saddam or his successors or Islamic fundamentalists or militant Arab nationalists. Security and stability in the Middle East had never been, nor would it be, more than a fleeting dream. The enemy, in whatever guise, would have to be watched, contained, and periodically rebuffed by force of arms, perhaps for years, perhaps for decades.

In pondering how to conclude the war after a presumably successful ground campaign, the Small Group hoped for a crisp, resolute finale — something akin to a surrender ceremony on the deck of the *Missouri* — while fearing that a more likely outcome was what came to be known as "the ragged ending."

Since the nineteenth century, wars more often than not had been characterized by a pernicious inconclusiveness. Again, the Korean experience seemed ominous. Even if American war aims were achieved, with Kuwait liberated and the Iraqi army routed, the war would not necessarily be over. What if the Iraqis refused to capitulate? What if some units surrendered, but others kept fighting? Saddam appeared obstinate enough to transform his seven-hundred-mile southern border into a new 38th Parallel. Would forty thousand American troops be needed here, too, as a trip wire to forestall another invasion?

The prospect was chilling and wholly at odds with the first principle underlying the Small Group's mandate: win the war with minimum loss of American life and avoid a southwest Asian quagmire. Both American and British planners became so preoccupied with this vision of the ragged ending — "getting an arm caught in the mangle," in Margaret Thatcher's metaphor — that no clairvoyant foresaw an equally ragged alternative: twin rebellions by Kurds and Shi'ites, suppressed with sanguinary zeal by an Iraqi army that would throw in its lot with the Ba'athist regime. Moreover, the clarity of allied war aims may have worked in Saddam's favor. The Iraqi leader was able to gamble that even if he lost his bid to seize Kuwait, he would retain the sovereignty of his country.

Another Small Group obsession was the so-called nightmare scenario, in which Iraqi forces either relinquished most of Kuwait, keeping perhaps Bubiyan Island and a sliver of the Rumaylah oil patch, or withdrew completely to loiter just north of the border. Even more than Israeli intervention, such a gambit was likely to fray the American-led coalition; few allies seemed inclined to rally round a ground attack launched to liberate the Bubiyan mud flats. The greater the allied success in the air war, the greater the chance that Baghdad would try to avoid utter humiliation by retrenching to the north before Schwarzkopf could deliver the coup de grâce. This added to the pressure to launch the ground offensive sooner rather than later.

The Small Group spent weeks drafting a list of harsh measures to be imposed should Saddam announce a withdrawal. After first resolving that no Iraqi armor be permitted to leave Kuwait, the group concluded that a more realistic response would be to demand Iraq's abandonment of Kuwait City and Bubiyan Island within forty-eight hours of a cease-fire, as well as the immediate withdrawal from all border fortifications and a general migration northward by the entire Iraqi army. Intricate timetables specified the types of weaponry to be removed or surrendered and the geographic phase lines that Iraqi forces had to reach within a certain hour after announcement of a cease-fire. Failure to comply would result in continued allied attacks.

Such strictures, the Small Group believed, would quickly disclose whether the avowed retreat was genuine or duplicitous, while also forcing the Iraqis to abandon much of their heavy equipment. Yet the Republican Guard divisions, which had already repositioned in southeastern Iraq, would remain intact — and with them the core of Iraq's military power. More than ever, the enveloping sweep of the allied Left Hook seemed necessary to ensure that Saddam did not escape with a partial victory.

But a sudden burst of Soviet diplomacy revived Washington's fears of both the nightmare scenario and the ragged ending. On Thursday, February 14, Gorbachev cabled Bush to report that Primakov detected a change in "attitude" in Baghdad. Saddam no longer seemed resistant to an unconditional withdrawal. With Tariq Aziz coming to Moscow for further talks, the Soviet president suggested, "it would not be desirable to conduct any massive ground operations" until the discussions had run their course.

Publicly, the White House maintained a cool propriety, professing to welcome any peacemaking efforts that ended the war as long as Saddam complied with United Nations resolutions. Privately, Gorbachev's intrusion provoked exasperated grumbling. Scowcroft had long recognized Moscow's eagerness to forestall a ground war, to avoid the humiliation of its former client and a slugfest between American and Soviet weaponry. At the State Department, Baker clung to his conviction that Gorbachev would not betray the basic strategic commitment he had made to the allied cause in August. Yet the Soviet overture threatened, as Gates told the Small Group, to give Iraq "half a loaf." On the issue of delaying the ground attack, Bush made his position clear in a comment to Baker: "No way, José!"

On February 15, Radio Baghdad broadcast a long communiqué from Saddam's Revolutionary Command Council that for the first time suggested Iraq's willingness to relinquish Kuwait. The announcement sparked a celebration in Baghdad — with the usual automatic weapons fusilade in the city's streets — and an air of anticipation elsewhere in the world. Financial markets rallied, telephone circuits overloaded, Arabs in Riyadh and Cairo greeted one another with exclamations of *mabrouk!* — congratulations.

Bush, alerted to the communiqué by early morning television reports, arrived in the Oval Office at 7:10 A.M. and immediately repaired to his private study with Scowcroft, Baker, and Gates. For the next hour they studied the translations of the announcement that came fluttering page by page into the White House fax machine from the government's Foreign Broadcast Information Service. Within minutes, however, the Iraqi offer was seen to be a charade. Any withdrawal was contingent on an end to coalition air attacks, the removal of allied forces from the region within a month, Israel's surrender of the West Bank and Golan Heights, a new government in Kuwait, and the repeal of all Security Council resolutions.

The broadcast ended with a rhetorical blast at "the perfidious, the treacherous, and their imperialist masters." A demand that American taxpayers finance the rebuilding of Iraq particularly angered Bush, who immediately called John Major, François Mitterrand, and Turgut Özal to close ranks.

At 10 A.M., in an appearance before a group of scientists gathered in the Old Executive Office Building next to the White House, the president denounced the communiqué as "a cruel hoax, dashing the hopes of the people in Iraq and, indeed, around the world." In his most explicit appeal for an Iraqi rebellion since the August 2 invasion, Bush added, "There's another way for the bloodshed to stop, and that is for the Iraqi military and the Iraqi people to take matters into their own hands, to force Saddam Hussein, the dictator, to step aside." Reports of the jubilation in Baghdad had convinced the president that the war-weary Iraqis were ripe for insurrection. Wolfowitz and others suspected Saddam's "shimmy" was a direct consequence of the Al Firdos bombing and a belief in Baghdad that allied attacks had intensified in virulence.

Later in the day, Bush flew to Massachusetts for a tour of Raytheon's Patriot missile plant, gaily decorated with yellow ribbons, patriotic bunting, and hundreds of American flags. As the president and Barbara Bush strolled through the factory, workers broke into applause and darted up to request the First Lady's autograph. In the cavernous room where the missile radar was assembled, Bush stepped to a microphone, grinning and waving at daredevils scrambling on the steel girders overhead. "I am going to stay with it," the president vowed. "We are going to prevail, and our soldiers are going to come home with their heads high."

Israel's private complaints earlier in the week about Patriot deficiencies were defiantly brushed aside. "Thank God for the Patriot missile," Bush declared. "Forty-two Scuds engaged, forty-one intercepted." The adoring crowd roared its approval. "U-S-A! U-S-A!" they chanted, and the incantation echoed from the high ceiling with the deep, ferocious timbre of a war cry.

## Eskan Village, Saudi Arabia

Schwarzkopf's target date of February 21 for the ground offensive soon proved overly optimistic. As Wolfowitz had sensed during his trip to Riyadh with Cheney and Powell, the Marines could not get ready in time. Walt Boomer's abrupt realignment of his two divisions and the concomitant logistical complications caused the Marine commander on February 14 to ask Schwarzkopf for a three-day postponement. The CINC concurred in part to allow heavier air attacks against Iraqi positions in the western portion of the Kuwaiti bootheel, where the Marines would attack. Long-range meteorological forecasts also predicted better weather later in the month. Schwarzkopf passed the request to Powell.

The chairman was not pleased. Given Moscow's diplomatic ventures and the growing anxiety that Saddam might deprive the allies of a clear victory by withdrawing, any delay seemed dangerous. Powell also worried that postponement was a "slippery slope," with one delay begetting another. "The president," he told the CINC, "wants to get on with this."

For six months Schwarzkopf had been unsure exactly where the pressure points lay in Washington. Except for occasionally referring to "those hawks over in the White House" or suggesting that "my secretary wants this," Powell in his phone calls to Riyadh tended to keep the decision making at home veiled and anonymous. Rarely certain whether Powell's guidance reflected his own conviction or that of the civilian leadership, Schwarzkopf had difficulty divining what he called "the personality lash-up." If pressed by the Pentagon to attack prematurely, Schwarzkopf had told de la Billière and others in Riyadh, he would appeal directly to George Bush.

But in this instance a challenge of Powell's authority was unnecessary. Schwarzkopf had what he believed were solid tactical reasons to delay regardless of strategic concerns in Washington. After his initial hesitations, the chairman agreed. He had worried about the Marines and the heavy casualties they might suffer more than any aspect of the war. If an additional three days of bombing would help to minimize Boomer's losses, then a postponement was warranted. Bush and Cheney reluctantly concurred: the president found it much harder to say "no way, José" to his military commanders than to the Soviets. Powell called Schwarzkopf to say the request had been approved. The attack was rescheduled for February 24.

*

Intended to accommodate the Marines, the grace period also bought some extra time for the U.S. Army. Lieutenant General John Yeosock was trying to shift two large Army corps — comprising more than a quarter million troops and 73,000 vehicles — halfway across Saudi Arabia and into their attack positions without the Iraqis detecting the movement. As commander of Army Component Central Command, or ARCENT, Yeosock had been in Saudi Arabia since August 6. In those uncertain days, he later said, the most lethal U.S. Army weapon standing between Iraqi tanks and the Saudi oil fields was his pocketknife. Tall and craggy, with a gravelly voice deepened by the bark of a thousand commands in his long military career, Yeosock in the subsequent six months had worked doggedly to get his forces into the theater and ready to fight. Taciturn by nature, he seemed to his closest deputies to have become even more withdrawn under the lash of Schwarzkopf's displeasure, as if retreating into a shell to protect himself from the CINC's tirades.

Exhaustion and stress finally took their toll. Weakened by an attack of pneumonia and wracked with mysterious intestinal pains, Yeosock checked into a Riyadh hospital on February 14. He pleaded with Schwarzkopf not to relieve him of command, but when his condition deteriorated, he found himself on a medical evacuation plane bound for Germany, where surgeons promptly removed his gall bladder.

With Yeosock gone, probably for the duration of the war, Schwarzkopf asked Calvin Waller to serve as his new Army commander. The deputy CINC leaped at the chance. Although he was fond of Yeosock, Waller found his command style wanting. In Waller's view, ARCENT was afflicted with a bunker mentality. Instead of fusing his two corps into a single force, Yeosock let them operate almost as disparate armies preparing to fight two distinct wars. Waller had no intention of lingering in the ARCENT headquarters at Eskan Village, shuttling into nearby Riyadh for daily meetings with Schwarzkopf. Instead, he would lead from the front as a fighting commander. Now is the time, Waller told himself, to go out there and shake the living shit out of Freddy and Gary.

Fred Franks and Gary Luck, commanders respectively of VII Corps and XVIII Airborne Corps, welcomed the change. Luck in particular had chafed at Yeosock's temporizing and apparent reluctance to confront Schwarzkopf; in private comments to his staff, Luck expressed his annoyance at being subordinate to a man whom the Army had never entrusted with command of a corps.

Within hours of Yeosock's departure, both corps found Waller in their midst, prodding, questioning, exhorting. As he shuttled among the brigades and divisions, he put the same query to every commander: "How do you intend to fight this battle?" Vigorous and animated, his bulky figure at once imposing and reassuring, he breathed new fire into staff officers grown weary of reviewing their war plans. After Waller conducted one lively war game over a large map of southeastern Iraq, Stan Cherrie, the VII Corps operations officer, walked from the tent and crowed, "Today I finally feel like I'm playing on the 1960 Green Bay Packers." No longer did corps staff officers refer to ARCENT as "Amateur Hour."

Waller also demanded common sense. When VII Corps requested 35,000 TOW-2 antitank missiles and half a million hand grenades, Waller balked. "The enemy only has seventeen hundred or so tanks left in the entire Kuwaiti theater," he told the corps staff. "Not all of them are in the sector where you'll be fighting. Why do you need 35,000 TOW rounds? And hand grenades? This is a mobile armored corps. Where are you going to close with the enemy so that every man in the corps needs five hand grenades? Who in hell are you going to throw them at?"

Rare was the commander, from lieutenant to lieutenant general, who did not covet more trucks, more artillery, more intelligence, more ammunition. Rarer still was the commander who did not suspect that someone else was getting his share, and officers in XVIII and VII Corps kept a wary eye on one another. The two corps reminded Waller of fractious brothers whose blood ties and affection routinely yielded to petulant rivalry. ("Like a pair of ragamuffin boys," in Cherrie's phrase.) XVIII Corps had arrived first in the theater as the Army's premier quick-reaction force; though a relative latecomer, VII Corps was to conduct Schwarzkopf's main attack. Complaints multiplied as the ground offensive drew closer.

XVIII Corps resented Riyadh's order to surrender an artillery brigade, the 1st Cavalry Division, and a fleet of remotely piloted drones to VII Corps; VII Corps in turn suspected XVIII Corps of hoarding the Army's best maps and of inventing enemy troops in its battlefield sector to win back some of its lost firepower. Beginning in late January, XVIII Corps had pressed Riyadh for permission to "go deep" with reconnaissance, artillery, and attack helicopter missions that would pinpoint the enemy, reduce his forces, and acclimate American soldiers to combat. VII Corps, meanwhile, worried that overzealous deep operations would betray the Left Hook battle plan and draw Iraqi armored forces into the western desert. The two corps also haggled over communications equipment, four-wheel-drive vehicles, and boundary lines.

Similar disputes, Waller knew, occurred in all wars. With lives at stake, with honor and glory hanging in the balance, even petty issues assumed a fell urgency. The stress of imminent combat amplified every emotion, affirming, as every battlefield bloodletting affirmed, the eighteenth-century truism that "the human heart is the starting point of all matters pertaining to war."

The great westward migration of American troops had begun nearly a month earlier. XVIII Corps, facing a three-hundred-mile trek to anchor the allies' outside flank, moved first. Within a week of the initial air strikes, ten engineering battalions had begun carving roads, landing strips, ammunition dumps, and bivouac sites near the town of Rafha, roughly four hundred miles west of the Persian Gulf. Through the end of January and into February, 130,000 troops and 28,000 vehicles streamed across northern Saudi Arabia a brigade at a time, moving mostly at night and under radio silence.

Much of the traffic rumbled along Tapline Road, the asphalt two-lane running parallel to the border straight as a gunshot toward Jordan. At the dust-choked crossroads of Hafr al Batin, a military policeman in a fluorescent orange vest whistled through convoy after convoy of troops flashing victory signs and chanting, "No slack to Iraq." In August, when the corps first trickled into the country, some soldiers had ridden to their encampments aboard red Saudi double-deckers, in a scene reminiscent of the taxi fleet that rushed French troops to the front in August 1914. Now the Americans rode with their tanks and armored personnel carriers and howitzers, fully armed and ready to fight, toughened by long months in the desert.

The corps was an odd hybrid: a heavily armored division, the 24th Mechanized Infantry; a light paratrooper division, the 82nd Airborne; an "air mobile" helicopter division, the 101st; a cavalry regiment, the 3rd Armored; and the French Daguet division, a creature neither fish nor fowl that comprised a fleet of small tanks and a colorful assortment of hussars, dragoons, and Foreign Legionnaires. The 1st Cavalry Division had been peeled away by Schwarzkopf to serve as his theater reserve near the Wadi al Batin, ready to reinforce either VII Corps or Arab forces farther east.

Each American division was freighted with unique psychological baggage. Barry McCaffrey's 24th had not seen heavy combat since Korea and was eager to prove itself. James H. Johnson, Jr.'s 82nd carried the conviction that Schwarzkopf nursed a grudge against the Army's most celebrated unit. (To avoid further antagonizing the CINC, Johnson even refused to let his paratroopers wear their distinctive maroon berets.)

J. H. Binford Peay III's 101st, the Screaming Eagles of Normandy fame, faced recurrent questions about whether the helicopter had become a battlefield anachronism.

The corps commander was a wry, laconic Kansan with brawny forearms, a fondness for Skoal cherry tobacco, and a folksy drawl. "When you start to take yourself too seriously," Lieutenant General Gary Luck once warned, "that's when you get in trouble." Disdaining pomp and convention, Luck reportedly had transformed the dining room of the commander's mansion at Fort Bragg into a pool parlor; his dog, a black Labrador retriever named Bud, trotted about the corps headquarters with a nonchalance that perfectly reflected Luck's informality.

Luck was a Schwarzkopf favorite, but, like so many other senior commanders in Saudi Arabia, he had run afoul of the CINC. In late October he sent Schwarzkopf a casual but secret message suggesting possible alternatives to the One Corps Concept. A courier brought the CINC's reply in a sealed envelope: "You command the corps, I'll command the theater, and if you object, there are others who would like to command XVIII Airborne." Luck, showing the letter to Peay and Colonel Frank Akers, the corps operations officer, observed, "We're about to get our asses fired." He wrote a note of contrition. Schwarzkopf answered with a conciliatory message passed through an aide, who added, "That's about as much of an apology from the CINC as you're going to get." The incident passed, but XVIII Corps became more circumspect in forwarding suggestions to Riyadh.

Like Fred Franks, Luck had seen severe, bloody fighting in Vietnam as a Special Forces and cavalry troop commander. But there the similarities ended. Franks was precise, attentive to detail, and distressed by uncertainty. Luck made an art form of a "decentralized" style that gave his division commanders the power of feudal barons. Inevitably, personalities clashed, comrades fell in and fell out, grievances and rivalries festered and healed, only to fester again. Binnie Peay and Jim Johnson both admired Barry McCaffrey's brilliance and sang-froid, yet suspected him of manipulating the corps headquarters and placing the glory of his division above other concerns. (Peay, a courtly, earnest Virginian, at one point took his complaints about McCaffrey to Luck and Akers.) To their credit, when the time came to fight, the generals put aside their spats and mustered three superbly trained, superbly led divisions.

Luck maintained his tranquillity, aware that the fate of his corps was about to shift into the hands of subordinate commanders who would oversee the "cash transaction" of combat. After listening to officers from the 101st Airborne explain the intricacies of their planned helicopter operations, Luck replied with typical brevity: "Well, you're making air-assault history. Don't fuck it up." As the general turned away, a sergeant murmured admiringly, "Now *that's* guidance I can understand."

While XVIII Corps finished marshaling near Rafha, VII Corps prepared to make its move. From laager sites around Al Qaysumah, the corps would shift northwest 135 miles toward the border. No American commander then in uniform had ever marched an entire division, much less a corps. For nearly half a century, VII Corps had sat in West Germany waiting for the Warsaw Pact attack that never came, rarely maneuvering more than a battalion or a single brigade across the cramped Bavarian landscape.

To avoid the congestion of nearly 150,000 soldiers and 40,000 vehicles simultaneously crossing the desert — a force larger than the Third Army commanded by George Patton in World War II — planners intended to shift the units northward a single division at a time. But Fred Franks had a different idea. During one of his reveries before the map board, Franks realized that in both distance and angle of march the move west exactly replicated the corps plan of attack into the Republican Guard flank. On his office wall, Franks had framed a brief exchange between Robert E. Lee and Stonewall Jackson before the battle of Chancellorsville. "What do you propose to do, General?" Lee had inquired. "Go around," Jackson replied. "With my whole corps." The Left Hook in many ways resembled that Confederate attack, Franks believed; VII Corps now had a priceless opportunity to rehearse "going around."

His staff and division commanders resisted mightily. Maneuvering an armored division with 15,000 men and 350 tanks in a narrow zone was akin to turning an ocean liner in a tight channel: the potential for havoc was enormous. Among other obstacles, the corps would have to squeeze past a royal camel farm without leaving dromedary carcasses scattered across Saudi Arabia. Unlike Europe, where troops tended to bunch up on constricted roads, the open desert encouraged wide dispersal. "Shit," complained Ron Griffith, the 1st Armored Division commander, "we're really going to end up in each other's way."

Franks was adamant. After several days of his dogged campaigning — he reminded one officer of a ward heeler wheedling votes — the commanders capitulated. At dawn on February 16, with clear skies overhead and a stiff breeze blowing from the west, the corps surged forward.

Not since El Alamein had such a desert spectacle unfolded. On the left, the 2nd Armored Cavalry Regiment led two armored divisions abreast, the 1st and the 3rd. On the right, the Big Red One — the 1st Infantry Division — spearheaded Britain's 1st Armoured. From horizon to horizon the immense host pressed across the pan on a fifty-mile front: tankers and infantrymen, gunners and loaders, engineers and missile crews, military policemen, radio technicians, cooks, accountants, mechanics, electricians, chaplains, doctors, dentists, gravediggers.

Brown streams of dust boiled from beneath each tank and truck — hundreds of tanks and trucks, then thousands, then tens of thousands. Wave followed wave in an inexorable flood of steel, roiling the desert calm with the shrill creak of armored tracks and the whirr of turrets swinging toward imaginary targets, as crews raised their tubes and fixed their sights and in their mind's eye killed and killed again.

Northwest toward Mesopotamia the Army pressed, terrible and magnificent. A deep rumble rolled over the land, as from an avalanche or a stampede. The stench of diesel fuel hung in the air from the three million gallons burned each day to keep the corps moving. Helicopters scissored overhead, their pilots hooting in disbelief as they vainly searched for an end to the endless cavalcade.

Franks and Stan Cherrie watched from the ground for several hours, gawking like a pair of nineteen-year-old corporals. "Jesus," Cherrie murmured, "it's like being in the middle of the Spanish Armada." At midmorning, Franks took to the air in his Blackhawk. Except for maddening communication problems — "Damn it," the corps commander exploded at one point, flinging his radio microphone to the floor, "how can I command if I can't talk?" — the march proved surprisingly free of snarls. Battalions, brigades, and divisions avoided mingling. Logistics trains kept pace. Few vehicles broke down or got lost. The royal camel herd survived. "This is going to work," Franks told Cherrie, smiling broadly. "It's going to work."

## King Fahd International Airport, Saudi Arabia

Buster Glosson, as every officer in the Black Hole knew only too well, could be tyrannical, arrogant, and infinitely stubborn. Unless countermanded by Schwarzkopf or Horner, he assumed — and often proclaimed — complete suzerainty over the 2600 allied aircraft in Southwest Asia. Yet his blustery temperament obscured a shrewd, analytical intelligence and a gift for innovation. Bold enough to risk failure, Glosson infused the strategic campaign with élan and contributed several tactical triumphs, such as tank plinking. He served well the cause.

He also made mistakes. His audacity at times proved to be temerity dressed up in a cocked hat, leaving him in a scramble to correct his own errors. In the first three days of the air campaign, for example, Glosson had ordered 180 sorties against Iraqi bridges by aircraft dropping unguided munitions. His targeteers and intelligence officers had warned him that dumb bombs rarely proved accurate enough to destroy a bridge — a lesson painfully learned in Vietnam. But Glosson hoped that modern guidance systems would overcome such historical shortcomings. By the end of the third day, however, not a single Iraqi span had been destroyed. Frustrated and chagrinned, Glosson promptly restructured the bridge campaign. He ordered F-117s with laser-guided bombs into the fray and telephoned a civil engineering professor at his alma mater, North Carolina State University, to determine the most vulnerable aim points. (For most spans, he learned, bombs should strike just above the stanchions closest to shore.) On day four, Iraqi bridges began to fall, tumbling at a rate of seven to ten each week for the rest of the war.

Glosson also tinkered with the barrier combat air patrols — the bar CAPs — flown to prevent Iraqi planes from fleeing to Iran. By January 31, two days after F-15s began orbiting near the Iranian border, the Iraqi exodus had stopped completely. For four days not a single Iraqi jet left the ground. Glosson, moreover, was aware that many Air Force pilots had become reliant on go and no-go pills to cope with the fatigue of the long bar CAP missions. On February 4 he abolished the bar CAPs and shifted his fighters elsewhere.

More than thirty enemy planes immediately bolted for Iran. Glosson phoned the F-15 squadrons. "That's it," he declared. "I don't care if your guys stay on uppers and downers for the duration. I want these flights stopped. We'll straighten the pilots out with the doctors after the war." On February 6 the bar CAPs resumed, again stanching the exodus.

In mid-February Glosson's audacity led to another mistake, one that cost a pair of fighter planes and the life of a pilot. As the ground war approached, he had begun encouraging his airmen to be more aggressive in attacking Iraqi armor and artillery. Altitude restrictions, imposed during the first week of the war to keep pilots flying above most Iraqi antiaircraft fire, were dropped. Enemy SAM launches had diminished to the point that Glosson, with Horner's approval, urged his A-10 Warthog squadrons to "get down amongst them," dropping as low as four thousand feet so that pilots could use binoculars and rifle scopes to distinguish tanks from decoys.

"American soldiers are about to cross the border," Glosson advised in a message to all wings. "Now there is something worth dying for in Iraq. All restrictions are removed . . . Flight leaders will have authority to decide whether to drop bombs from fifty feet or five hundred feet or five thousand feet."

At the same time he began pushing the A-10s farther north. Sluggish and slow, the A-10 had been designed for close air support of ground troops in combat. Equipped with a seven-barreled rotary cannon that fired rounds the size of milk bottles, the aircraft was supposed to fly no more than a few miles beyond friendly lines on the kind of "trench strafing" sorties first developed in World War I. Yet the seven Warthog squadrons based at King Fahd Airport, northeast of Dhahran, had demonstrated unexpected success in hitting radar sites, antiaircraft batteries, and other targets deep in enemy territory. Torn between their desire for a larger role in the war and a sense of vulnerability, A-10 pilots soon found themselves attacking Republican Guard formations sixty miles above the Saudi border.

Glosson notified Horner that he planned to order A-10s on their deepest strike yet, against the Madinah Republican Guard division entrenched just north of the Iraq-Kuwait border. "Buster," Horner warned, "I'm not sure they can survive that far up over the Republican Guard."

"We've beat the enemy down to the point that I believe they can," Glosson said. "I feel obliged to try it because they've been so successful down in the first echelon. I'm going to do it unless you tell me I can't."

Horner replied, "It's your decision, but I'm doubtful it's the right thing to do."

Shortly after 3 P.M. on February 15 the first pair of A-10s crossed Kuwait's northern border to attack the Madinah. Flying in the lead Warthog was Captain Stephen R. Phillis, a thirty-year-old Air Force Academy and Fighter Weapons School graduate, widely considered his squadron's best pilot; 1st Lieutenant Robert Sweet flew on his wing. The mission made both men uneasy. Earlier in the day, a Warthog flown by one of the two A-10 wing commanders, Colonel David A. Sawyer, had limped home with three hundred holes in its tail from an SA-13 fired by Tawalkana gunners southwest of the Madinah.

After dropping cluster bombs on several tanks and trucks, Sweet and Phillis rolled in for a final strafing pass with their 30mm Gatling guns when an Iraqi SAM blew the aileron and flap from Sweet's right wing. "Oh, man, I'm hit!" he radioed. "Heading south." Within ninety seconds the aircraft began spiraling out of control. "Try to roll your wings level," Phillis urged. "I can't," Sweet answered. "I'm out. I'm out." At six thousand feet he ejected.

Ten seconds after his parachute opened, Sweet saw the fireball of his Warthog as it smashed into the desert below. For several minutes he drifted to the ground, Iraqi bullets streaking past him. He landed thirty yards from a T-72 tank. Staggering to his feet, Sweet raised his hands, but when the Iraqis kept shooting he bowed his head and covered his face with his arms. Several dozen soldiers rushed forward, tearing at his parachute harness and clubbing him with rifle butts. "God, don't hit me again," he pleaded. The savage beating continued until Iraqi officers pulled him away from the mob. He would spend the rest of the war in prison.

Phillis, flying aircraft number 37, courageously remained overhead, calling for search-and-rescue help and A-10 reinforcements until he too was hit. Pilots listening to his transmissions heard a final radio call — "Three-seven is bagged as well" — then silence. Not until March would his body be recovered.

Phillis's death and Sweet's capture infuriated the Warthog pilots. Why, they asked, was the A-10 being used interchangeably with attack aircraft capable of flying twice as fast? Wing commander Sawyer wrote a long letter to Horner on February 16. Reviewing the Warthogs' role in the first month of the war, Sawyer noted that in the first two weeks of February

> we took hits on seven aircraft, losing one of them. We expected to take a few more hits . . . because we were taking on the Republican Guard and flying sixty to seventy miles beyond friendly lines. Up until yesterday, I didn't think the [loss] rate was excessive, that is, serious enough that we needed to change guidance.
>
> But that was before we returned to the Tawalkana Division and visited the Madinah Division for the first time. After weeks of holding their fire, the Iraqis launched eight [infrared] SAMs at us yesterday, bagging two A-10s and damaging one, which I happened to be flying . . . Believe it or not, on the way home I flew over a flight of F-16s working a target approximately fifteen miles north of the Saudi border. A-10s over the Republican Guard and F-16s in the southern KTO [Kuwaiti Theater of Operations] does not compute.

Glosson, on hearing reports of the downed planes, knew immediately that he had erred. The sparse antiaircraft fire of the preceding weeks had deceived him into believing that the Republican Guard was battered to the point of cracking. Instead, the divisions had been digging in deeper, waiting for a chance to fight back.

He phoned the A-10 wings at King Fahd and canceled all missions deeper than thirty miles north of the Saudi border. Until the ground war began, no Warthogs would fly north of 29'30", a latitudinal line bisecting Kuwait Bay. The 30mm Gatling gun, which required attacks at lower altitudes for accuracy, would be used only to support allied troops locked in battle.

An hour after Glosson's call to King Fahd Airport, Horner walked into the Black Hole. "Pull the A-10s back, Buster," he ordered. "They're only allowed to hit the front-line divisions now."

"Yes, sir," Glosson answered. "That's already been taken care of."

## Phase Line Minnesota, Iraq

As allied legions massed along the northern Saudi border to deliver a killing blow to Iraq's army, officials in the United States hunted for something magical to prevent friendly forces from also killing one another. With essentially four corps prepared to attack — two Army, one Arab, and the Marines — the prospect of soldiers shooting the wrong targets in a fast-moving battle at close quarters loomed large.

Spurred by the Marine deaths at OP-4 during the battle of Khafji, the Joint Chiefs ordered a review of on-the-shelf technologies in hopes of finding a device that would enable gunners and pilots to distinguish friend from foe. On February 6 the Defense Advanced Research Projects Agency (DARPA) began to evaluate sixty proposals sorted into five categories: thermal imagery, infrared imagery, lasers, special radio frequencies, and visual devices. At the Yuma Proving Ground in Arizona on February 15, only six days before the scheduled ground offensive, DARPA started testing the most promising techniques.

The crash effort was belated and futile, attempting in a week's time to solve a mystery that had plagued warriors for thousands of years. Since the age of Homer men had inadvertently slain their comrades, although modern warfare, with its massed armies and lethal weaponry, had raised the potential for such killings from the snuffing out of individual foot soldiers with spears and arrows to the slaughter of battalions with errant cluster bombs and artillery salvos. Whether called fratricide, amicicide, blue on blue, friendly fire, or — as in official U.S. casualty reports from Vietnam — "misadventure," the phenomenon had become all too commonplace on twentieth-century battlefields.

French casualties from friendly artillery in World War I have been estimated at 75,000; across the trenches, Germans dubbed their 49th

Field Artillery Regiment the "48 1/2" because its gunners persistently lobbed rounds short of the intended target. With the rise of air power came new forms of misadventure. The French in World War I stopped using large metal canisters crammed with steel flechettes after an aviator accidentally dumped several on a detachment of Zouaves. At Monte Cassino and Venafro, Italy, in March 1944, misguided bombs struck a corps headquarters and killed nearly a hundred Allied troops. In the worst case of fratricide in World War II, Allied bombers in July 1944 decimated the 30th Infantry Division during the breakout from Saint Lô in France, inflicting 814 casualties and killing Lieutenant General Lesley J. McNair. At the battle of Hill 875 in Vietnam, a confused bomber pilot slew forty-two paratroopers and wounded forty-five as they fought a North Vietnamese regiment in November 1967.

Extrapolating from a review of 269 fratricide incidents in four modern wars, a 1982 study for the U.S. Army's Combat Studies Institute estimated that friendly fire accounted for "something less than 2 percent of all battlefield casualties," including roughly 15,000 of the 774,000 Americans killed and wounded in World War II. (In southeast Asia from 1961 to 1975, 1326 deaths were classified under "misadventure," about 3 percent of the total combat losses.)

In the gulf war, by contrast, 24 percent of the Americans who would be killed in action — thirty-five of 146 — and 15 percent of those wounded — seventy-two of 467 — fell victim to friendly fire. Most of the twenty-eight fratricide incidents had root causes similar to those which caused a Confederate sentry to shoot Stonewall Jackson at Chancellorsville in 1863: poor visibility, battlefield jitters, misidentification of the target, and the omnipresent "fog of war."

Several other variables account for the higher proportion in Desert Storm. Unlike earlier wars, nearly every fatality was investigated and documented; had the 290,000 U.S. battle deaths in World War II been subjected to similar scrutiny, the number of those attributed to fratricide would certainly have been much higher. Because the gulf war lasted only six weeks, few soldiers accumulated the battle experience that often prevents troops from firing on their comrades. Iraqi incompetence kept the overall American death toll low, concomitantly inflating the fratricide percentages.

Moreover, the factors that would give American forces superiority — desert warfare, modern tanks, a battle doctrine that stressed speed and violence — also spawned fratricide. Confusion and uncertainty thrive on a fluid, nonlinear battlefield. Simply gauging north from south in the open desert at night can be baffling; distinguishing comrade from enemy nearly impossible. The sights and computers in modern fire-control systems permitted tank, helicopter, and aircraft crews to spot and kill a presumed "enemy" at ranges measured in thousands of yards. Tank gunnery in particular stressed the importance of firing the first shot; crews were trained to detect and destroy a target in six to ten seconds. An Abrams tank crew using thermal sights could spot another vehicle more than three kilometers away, yet the blurry thermal image was so indistinct that positive identification was difficult beyond seven hundred meters.

"The impact of amicicide on combat power is geometric, not linear," Charles R. Shrader observed in the Army's 1982 study. "Each amicicide incident that results in friendly troops killed or wounded has an adverse effect on morale and confidence, disrupts the continuity of friendly operations, and represents one bomb, shell, or bullet that should have fallen on the enemy to reduce his combat power rather than our own."

All of this was known to American officers in Saudi Arabia. Fear of fratricide, though seldom discussed publicly, was never far from the mind of any competent commander. Rare was the colonel or general who could not cite personal knowledge of blue on blue in Vietnam. (A young corporal killed by U.S. artillery fire in Lieutenant Colonel Norman Schwarzkopf's battalion became the subject of the best-selling book *Friendly Fire*.) Commanders adopted various control measures and recognition symbols: battlefield phase lines, fluorescent orange panels, the inverted V daubed on every friendly vehicle. They distributed five thousand global positioning systems — compact electronic gadgets that triangulated satellite signals to give the user his exact grid location.

Yet somehow they were not truly ready, not for twenty-eight incidents, not for thirty-five dead and seventy-two wounded, not for the shame and mortification of having to confess: we killed our own. The agony was compounded by a reluctance, verging on cruelty, to promptly and fully disclose the details of fratricide deaths to widows and parents. In thirty-three of thirty-five cases the Army and Marines would know within a month after the war that the deaths had been caused by friendly fire. But in most instances the services waited four more months to notify the survivors how their sons and husbands really died. Several families continued to receive official accounts wholly at odds with the truth, including disgraceful, false assurances that their fallen soldiers had received medical care or last rites.

The United States government had spent $3 trillion rebuilding the American military in the 1980s. Yet the search for a simple safeguard to avoid unleashing that new-bought firepower on our own soldiers came down to a desperate week of testing in the Arizona desert when the gulf war was nearly over. DARPA scientists, poring over their sixty proposals, cobbled together a battery-powered beacon that could be seen through night-vision goggles five miles away. A protective collar around the front of the light prevented the enemy from seeing the flashing signal — unless the enemy lurked on the flanks with night goggles. A few "DARPA lights" arrived in the theater on February 26, by which time the friendly fire toll had tripled. It was too little and it was too late.

No misadventure seemed more pointless and pathetic than that which befell the 1st Infantry Division shortly after midnight on February 17. Having moved into attack positions as part of VII Corps's westward migration, the Big Red One slashed twenty holes in the sand berm marking the border, known on division maps as Phase Line Vermont. Artillery pounded enemy positions ten miles to the north, and more than a thousand soldiers, Task Force Iron, pushed three miles into Iraq to establish a screen line on a twenty-mile front along Phase Line Minnesota.

The task force was drawn largely from a brigade of the 2nd Armored Division, a Germany-based unit attached to the Big Red One for the war. At 9:45 P.M. on the 16th, troops commanded by Lieutenant Colonel James L. Hillman detected six enemy armored vehicles several miles north of the screen line. Half an hour later, they spied another Iraqi patrol creeping southeast. Darting in and out of view, firing flares and then disappearing into the desert's subtle folds, the Iraqis appeared both professional and daring. Several small hunter-killer teams crept to within half a mile of the Americans, drawing machine gun, mortar, and artillery fire.

Hillman had anchored his right flank with some scouts, including a Bradley Fighting Vehicle and an M-113, an armored box containing a special radar system used to detect movement on the ground. At 11:30, American gunners reported hitting an Iraqi vehicle with a TOW missile. With his forces thinly spread across a broad front, Hillman, fearful of being outflanked, requested Apache helicopter gunship reinforcements.

The Apache battalion commander was a forty-two-year-old lieutenant colonel named Ralph Hayles. A tall, aggressive Texan who had been flying helicopters for sixteen years, Hayles infused his unit with fighting spirit. "Gunfighters, sir!" his men barked whenever they saluted him. Division staff officers considered Hayles a bit of a cowboy, but knew him to be competent and experienced. In training sessions at Fort Riley and in Saudi Arabia, he had repeatedly stressed the importance of identifying enemy targets before firing. "I have high confidence," he told reporters during an interview three weeks earlier, "that we won't shoot any coalition forces."

The division commander, Major General Thomas G. Rhame, though something of a firebrand himself, had ordered his subordinate commanders to avoid direct participation in the fighting, the better to control their forces. Hayles chose to disobey. As he later explained to Army investigators, this would be the first night mission flown by his Apaches in support of U.S. troops in combat, and he regarded himself as the battalion's most qualified pilot. Moreover, flying conditions were wretched, with twenty-five-knot southerly winds and no moonlight. Hayles's driver found himself momentarily lost in a sandstorm during the hundred-yard trip from the battalion commander's tent to the Apache helipad.

Five minutes after midnight, three Apaches lifted off and headed north. Hillman's scouts shut off the ground radar, because its electronic emissions resembled those of Iraqi antiaircraft batteries. At 12:30, Hayles, flying under radio call sign Gunfighter Six, spotted the American screen line on his cockpit infrared scope. He raised Task Force Iron on the radio. Hillman described the disposition of his troops, noting that no friendly forces were forward of the map grid line known as 25 east-west.

Hayles's wingman, Captain Daniel L. Garvey, flying four hundred yards behind and to the right of Gunfighter Six, then noticed two rectangular vehicles that seemed to be sitting more than a mile north of Hillman's screen. Garvey pointed them out to Hayles, who again radioed Hillman. "Gunfighter Six has two big APC [armored personnel carrier] sort of vehicles . . . They do not appear to be part of your screen line. They're stationary. Let me look at them a little bit here." Hayles saw several small figures moving around. He concluded that Iraqi soldiers were shifting equipment from one vehicle, perhaps damaged by the earlier TOW shot, to the other.

"Yeah, those are enemy," Hillman reassured him. "Go ahead and take them out."

But Hayles was confused. Instead of flying due north, his formation had drifted northeast, so he was flying nearly parallel to Hillman's line rather than perpendicular. He also misread the Apache's navigational data on his fire control computer, which correctly informed him that the two targets were at a different grid location from the one Hillman's troops had reported. Garvey noticed the discrepancy on his own computer. "Hey, uh, Gun Six, this is Blue Six," he radioed. "I'll tell you, I'm getting a range of about four thousand meters, but it's not coming out right. We better go take a look. You think so?"

The Apaches crept closer, noses tilted downward by the wind. At 3800 meters, or roughly two miles from the targets, Hayles decided to rely on his own battlefield judgment instead of the Apache computers. "Can you still engage those two vehicles?" Hillman asked. "Roger," Hayles reported. "I could shoot those easy."

Still he hesitated. "Boy, I'm going to tell you, it's hard to pull this trigger." He tried to fire the Apache's 30mm cannon, but after three rounds the gun jammed. Aligning the larger vehicle in his sights, he readied a Hellfire missile and radioed, "Okay. I'll be firing in about ten seconds."

The first missile leaped from its rail in a halo of white flame. "I hope they're enemy," Hayles said, "because here it comes." The missile exploded dead on. "I guess you could say that hit it," Hayles said. "I'm gonna go ahead and shoot the second vehicle. It's still intact, but it's fixing to go away." A second missile struck home as the other Apaches opened with cannon fire.

Moments later an unidentified voice came over the radio. "Friendly vehicles may have been hit. Over."

"Roger. I was afraid of that," Hayles answered. "I was really afraid of that."

"Cease fire. Cease fire."

"Cease fire," he repeated. "I hope it's not friendlies I just blew up, because they're all dead."

Not all. One missile hit the Bradley anchoring the right flank, wounding three soldiers and killing Specialist Jeffrey T. Middleton and Private Robert D. Talley; the second destroyed the M-113, wounding three more. For a few panicky moments other soldiers in the screen line assumed that Iraqi sappers with rocket-propelled grenades had slipped behind them. A platoon leader nearly compounded the tragedy by firing on figures silhouetted against the flames; at the last instant he noticed that they wore American helmets. One Bradley crewman leaped from the burning vehicle then dashed back to save his badly burned platoon sergeant as ammunition detonated around them. The wounded were wrapped in flak jackets for warmth and bundled into another Bradley, which raced off to find a medic. The bodies of Middleton and Talley would not be recovered until dawn.

Hayles flew up and down the screen line, trying to fathom the unfathomable. At 2 A.M. he led the Apaches back to the helipad in Saudi Arabia, where he sat alone in the darkened cockpit for thirty minutes after landing.

Early the next morning, as ordered, he reported to Rhame's command post. There, the division commander slipped Hayles's gun camera cassette into a video player. The room fell silent but for the taped voices of the pilots. Hayles watched from a folding chair as the two doomed vehicles flickered into view on the screen, then exploded in a white flash. "I made a mistake," he told Rhame. "I made a mistake." Rhame played the tape again, then again. After a fourth viewing he turned to Hayles. "Okay," the division commander said brusquely. "That's it. No question about it."

Hayles was relieved of command for violating orders by leading the mission. Later in the day Fred Franks flew to Rhame's headquarters. The Big Red One had planned an Apache raid into Iraq that night, but the corps commander, after talking to Cal Waller in Riyadh, canceled the mission and ordered Task Force Iron to retreat below the border. Franks feared that Hillman's troops would be drawn into a major battle, upsetting Schwarzkopf's timetable and betraying the presence of the entire corps. "Don't show yourself in strength," Franks told Rhame. "And don't get yourself in a pitched ground fight, or we'll find ourselves with another Khafji on our hands."

By late afternoon on the 17th, Task Force Iron had slipped back through the twenty cuts in the berm, relinquishing its fingerhold on Iraq but leaving behind two black smudges as proof that the Americans had been there.

A few hours later, on the morning of February 18, the trail units from VII Corps closed on their assembly areas near the border to complete Franks's great march across the desert. The dust clouds subsided and the creak of tank tracks gave way to the sound of shovels and bulldozers. Company by company and battalion by battalion, the divisions dug in and draped themselves with camouflage netting. Across a four-hundred-mile front, from Gary Luck's corps in the west to Arab brigades along the Persian Gulf near Khafji, the coalition army began its final preparations: stockpiling fuel and ammunition, replacing shredded tires and worn-out engines, and all the while waiting, as so many armies before had waited, for the order to attack.

12

# Echo to Foxtrot

## U.S.S. *Tripoli,* Persian Gulf

General Schwarzkopf's decision to abandon the amphibious landing at Ash Shuaybah in favor of Operation Slash — a more modest assault against Faylaka Island — had sent the U.S. Navy back to its planning tables. To soften Faylaka with gunfire from *Wisconsin* or *Missouri,* while simultaneously preserving the threat of a full-blown invasion, battleships had to be able to move along a mine-free path within range of the island, which lay near the mouth of Kuwait Bay. Bereft of hard intelligence regarding Iraq's minefields, officers from the Navy's Mine Countermeasures Group (MCmG) aboard U.S.S. *Tripoli* fashioned what they believed was a prudent plan for clearing that path and easing into enemy waters.

Sweeping would begin at Point Echo, a spot more than thirty-five miles east of the Kuwaiti coast and presumably beyond the cast of Iraqi mine layers. An alley one thousand yards wide would be cleared from Echo to Point Foxtrot, fifteen miles to the northwest. At Foxtrot, minesweepers would then clear a three-by-ten-mile gunfire box south of Faylaka from which the battleships could unlimber in time to support the ground offensive, now set to begin on February 24.

This caution reflected the Navy's nagging disquiet about mines. "Damn the torpedos!" David Farragut had cried at Mobile Bay in 1864, steaming through mine-infested waters and into naval legend. But legend seldom noted that a ship in Farragut's squadron, the ironclad *Tecumseh,* promptly struck one of the damnable torpedos and sank like a stone in three minutes. Mines so bedeviled the American fleet in the Korean War that one admiral lamented, "We have lost control of the seas to a nation without a navy." A more contemporary misfortune befell the U.S.S. *Samuel B. Roberts* not far from Point Echo, where in 1988 an Iranian mine almost ripped the ship in half. If the spirit of

Mobile Bay in 1864 seemed far removed from the Persian Gulf in 1991, officers on *Tripoli* had only to recall the axiom that "every ship can be a minesweeper once" to keep their intrepidity sheathed. MCmG's thirty-one ships forsook Farragut's full speed ahead in favor of a more circumspect five knots.

On February 16 the American fleet, stiffened with five British mine hunters and a half-dozen helicopters flying from *Tripoli*, began trolling the Echo-to-Foxtrot channel. They finished shortly after sunset on the 17th. Finding nary a mine, the allied captains inferred that the Iraqi fields lay closer to shore.

This deduction was rational, cogent, and wholly wrong. Afloat on a sea of delusions, the Americans had miscalculated the position of Iraqi minefields by many miles. Six fields lay *east* — seaward — of Point Echo, forming a 150-mile crescent that bracketed nearly the entire coast of Kuwait. Four additional mine lines formed a picket just inside those fields to cover the deep-water approaches to Kuwaiti ports and the Iraqi naval base at Umm Qasr. All told, the Iraqis had dumped more than twelve hundred mines into the gulf.

The MCmG commanders had unwittingly steamed through the minefields on the way to Point Echo. That no ships were struck reflected both blind luck and Iraqi ineptitude. Many moored mines had broken free of their tethers and drifted away; others were "birdcaged," snarled in their mooring cables and thus too deep to be hit by a passing hull; still others sank to the bottom, apparently stuffed with too much explosive to float. After the war, ordnance experts estimated that 95 percent of the acoustically triggered influence mines were improperly primed and therefore inoperable.

Iraqi incompetence was matched only by allied ignorance. Although naval intelligence suspected that Iraq had begun laying mines in the fall, Schwarzkopf had refused Navy requests during Desert Shield to reconnoiter or sink the mine layers because of concern that such attacks could trigger all-out war before January 17. Even in mid-February the CINC had administered another tongue lashing to Vice Admiral Stan Arthur, accusing him of having moved into position for the Faylaka attack without proper orders. Thus blinded and bound, CENTCOM and MCmG were reduced to guesswork. As the senior British naval commander in the gulf, Commodore Christopher Craig, later observed, "Intelligence is perhaps a little like first love: one's ideals and ambitions are never quite matched by reality."

At 7 P.M. on February 17, the Americans detected radar emissions from a Silkworm missile battery believed to be operating behind a schoolhouse on the Kuwaiti coast. The missile cruiser U.S.S. *Princeton* promptly interposed herself between the fleet and the Silkworm site, while the MCmG ships weighed anchor and steamed east out of range, zigzagging to complicate Iraqi targeting. *Tripoli* led a procession of seven ships, including three minesweepers spaced behind her at two-thousand-yard intervals.

Six hundred feet long, displacing eighteen thousand tons, *Tripoli* seemed an unlikely choice to take the point. Designed for amphibious operations and named to honor the early-nineteenth-century campaign against the capital of the Barbary States, the ship looked like a small aircraft carrier. But the skipper, Captain G. Bruce McEwen, and the MCmG commanders had legitimate reasons for placing the ship first in line. With a crew of fifty-six officers and six hundred sailors, *Tripoli* could muster more lookouts than the smaller vessels behind; she had more night-vision goggles aboard than the minesweepers; and the towering flight deck afforded a better vantage point to watch for drifting contact mines, believed to be the only threat in the area.

At 4:20 A.M. *Tripoli* and her retinue turned back toward the coast. Now eight miles east of Point Echo, McEwen intended to reach Foxtrot by dawn so that his helicopters could begin sweeping the battleship firing box below Faylaka. Among nine lookouts posted on the carrier, three dangled over the water in metal cages suspended from the fo'c'sle. Three others stood on the bow, scanning the dark sea with their goggles and Big Eye, a tripod-mounted night-vision telescope, while also taking turns watching the more entertaining spectacle of allied aircraft passing far overhead.

None saw the LUGM-145 contact mine tethered fifteen feet below the surface dead ahead. Olive green, fanged with a trio of five-inch detonation prongs, the mine held 320 pounds of explosives. At 4:36 A.M., mine met ship.

The blast tore a jagged hole, twenty by thirty feet, through the one-inch hull steel of the starboard bow. *Tripoli* heaved up, then heaved again as the shock wave ricocheted from the gulf floor twenty fathoms below and struck the ship a second blow, cracking the keel. The anchor chain, welded from hundreds of ninety-pound links, rippled through the fo'c'sle like a garden hose. A bank of flight lights tore loose from the mast, smashing onto the signal deck. The explosion vaporized hundreds of gallons of paint and thinner stored in a hull locker, smothering the ship in a gray, malodorous miasma. Sea water gushed into the stricken vessel, tugging her down at the bow.

On the flight deck the concussion flung the lookouts back from the bow gunwales. Bosun's Mate Onzie Level lay on the deck, clutching a metal cleat used to tie down aircraft. "God," he wailed, "they're trying to kill us!" Level crawled to the phone. "Bridge, this is bow. We've been hit. We see no aircraft in the area. It must be a mine." A staccato whooping from the collision alarm clamored through the ship — "Brace for shock! Brace for shock!" a voice warned — followed by the shrill din of General Quarters. On the third deck, the ship's master at arms, Edwin Alvarez, was asleep in the vacant brig when the explosion threw him from his bunk to the deck, now rent with a six-foot gash. Spattered with paint from the demolished storeroom below, Alvarez wondered briefly if he were dead before concluding that a plane had smashed into the side of the ship. Grabbing a chain dangling from the second deck — the brig ladder was twisted beyond use — he pulled himself to safety.

Watchstanders on the bridge spied a gray nimbus of smoke and paint vapor rolling starboard along the hull. "All engines stop!" the conning officer ordered. At five knots, *Tripoli* would take five hundred yards to coast to a halt. Sailors in the fireroom far below, mistaking the mine for a boiler explosion, immediately shut down the ship's propulsion plant. Captain McEwen, who had been resting on the leather couch in his cabin, tugged on a pair of coveralls and reached the bridge in thirty seconds. With two crewmen reported overboard — an erroneous report soon corrected — he radioed a mine warning to the ships in trace and asked them to launch a search. "Everyone needs to go to general quarters stations in an orderly fashion," McEwen told the crew over the 1MC public address system. "We're not in danger of sinking. Please remain cool and professional." Then the 1MC went dead.

Sailors tumbling from their bunks into a fogbank of paint vapor assumed the ship had been struck with a nerve gas warhead. Several tugged on gas masks only to collapse from the fumes; one blue-hued seaman was revived with cardiopulmonary resuscitation. Aware that any spark would engulf the bow in a fireball, damage teams quickly rigged several large ventilation blowers to begin pushing the mist from the lower decks.

Despite the captain's assurances, *Tripoli*'s seaworthiness remained at issue. The crew, heeding a command to assume Condition Zebra, latched all watertight doors in five minutes, nearly half the standard time. But a quarter million gallons of water flushed into the ship's hull, flooding the forward pump room with fifteen feet of water. Water in number 1 diesel room was waist-deep and rising, a certain symptom of "free communication with the sea." The ship's engineer, Lieutenant Commander Steve Senk, who would win a Silver Star for his heroics, reported the ship down four feet in the bow. Crewmen attempting to plug the cracked bulkhead in number 1 diesel with baled rags soon abandoned the effort. Water completely swamped an adjacent storeroom and poured into the ship's magazine, lapping at fifteen hundred artillery shells stacked on the deck. JP5 fuel also began seeping from *Tripoli*'s bunkers into the magazine. Efforts to reclaim several flooded compartments with a portable pump proved futile. "It's like trying to pump out the Persian Gulf," Senk commented.

While McEwen supervised operations from the bridge, Senk and other officers pulled out the ship's blueprints and damage-control books. Like most large vessels, *Tripoli* was built with a succession of frames, progressively numbered from fore to aft, that served as a rib cage to strengthen and compartmentalize the ship. Thirteen "spaces," most of them forward of frame 19, had been flooded. Senk and McEwen, pondering how best to keep their ship from sinking, quickly sorted through what could be saved and what had to be surrendered to the sea. The damage-control documents made clear that if flooding extended aft through frame 26, *Tripoli* would capsize. There the line would be drawn.

Since before Trafalgar, shoring had been the accepted method of strengthening bulkheads separating one compartment from another. Damage teams now dragged out hundreds of metal shoring beams and six-by-six-inch timbers of Douglas fir, wedging them against *Tripoli*'s half-inch steel bulkheads. Others wedged rags into seeping cracks or jammed wooden plugs into the "stuff tubes" that carried wires — and now water — between compartments. Senk directed the shoring on two decks at frames 19, 23, 26, and 31, in a frenzy of hammering that McEwen likened to "an Amish barn raising." Helicopters hauled additional shoring lumber from the U.S.S. *Jason* and British ships waiting nearby. Sailors bracing the bulkheads in *Tripoli*'s freezer compartments first had to hack away eighteen inches of concrete and fiber-glass insulation with axes and picks.

An hour after the mine struck, Senk reported to the captain on the bridge: "We have the flooding contained. We're certainly not going to sink in our present condition." The engineer paused, then added, "Unless we hit another mine."

## U.S.S. *Princeton,* Persian Gulf

Like a mother hen shielding her brood, *Princeton* remained ten miles northwest of *Tripoli*, watching for Silkworms under Air Warning Red — "attack imminent." Commissioned just two years earlier, the cruiser was one of the Navy's great prides. Nearly six hundred feet long, with a crew of 350 and a price tag of $1 billion, she boasted an Aegis radar system capable of tracking aircraft or missiles hundreds of miles distant and shooting them down with an impressive arsenal. Her skipper, Captain E. B. Hontz, and most of her sailors were "plank owners," the first men to crew a new ship, and they beheld her with parental affection. (An earlier *Princeton* had been sunk in a mercy killing by American torpedos at Leyte Gulf in September 1944, eight hours after a 500-pound Japanese bomb turned the ship into an inferno.) In the gulf war the cruiser had provided air cover for Navy carriers, carefully counting all friendly planes returning from their bombing missions to be certain no Iraqis infiltrated the formation. After a personal plea from Hontz to Gulf Papa on *Wisconsin,* she had also been allowed to fire the last Tomahawks launched from the gulf — three TLAMs flung at an Iraqi communications station on January 29.

Hontz was awakened with news of *Tripoli*'s plight shortly before 5 A.M. The irony of the mishap did not elude him. On Sunday morning, eighteen hours before the mine strike, he had flown by helicopter to *Tripoli* at the request of MCmG officers for a meeting with Christopher Craig, the British commodore. Articulate, measured, and immaculately uniformed, Craig appeared slightly exasperated by his American comrades. For forty-five minutes he raised pointed questions about the safety of his mine hunters from Iraqi attack; he also wondered aloud whether "anyone should feel entirely comfortable about the minesweeping that has been done in this area." The MCmG officers offered assurances, increased the firepower devoted to protecting the British squadron, and adjourned to finish the Echo-to-Foxtrot sweep. Hontz, now responsible under the new arrangement for shielding H.M.S. *Manchester* from attack, flew back to his ship.

At 7 A.M. on the 18th, Hontz joined the watchstanders on *Princeton*'s bridge. At three knots — "bare steerage way" — the cruiser hardly rippled the calm surface, her creeping pace dictated in part by depths of only fifty feet, of which *Princeton* drew thirty-five. A dozen lookouts stood fore and aft, eager to claim Hontz's bounty of a steak dinner and three-day liberty to any swab spotting a mine. At 7:16 Hontz addressed his crew over the 1MC: "A couple of hours ago, the U.S.S. *Tripoli* struck a mine about ten miles from us. We're going to have to be extremely vigilant. The mine remains our primary threat."

As if on cue, an influence mine — probably an acoustically triggered, Italian-made Manta — detonated on the seabed directly beneath *Princeton*'s screws. An immense gas bubble spread beneath the fantail, nearly lifting the nine-thousand-ton ship from the water with a sound similar to a speeding automobile crashing into a brick wall. Hontz dropped the microphone and grabbed the chart table with both hands. Others on the bridge fell to their knees, clawing at the green linoleum. The shock wave traveled from stern to bow, flexing the cruiser like a fly rod so that the ship heaved and fell half a dozen times in as many seconds. The blast also triggered another mine three hundred yards off starboard amidships, and a second sequence of shocks buffeted *Princeton* laterally.

As the whipping subsided, the ship shuddered violently. Hontz, clutching the table, concluded she was disintegrating. He regretted his stubborn refusal to practice abandon-ship drills, which he had denounced as an exercise in "negative leadership." Now, he feared, the crew would pay for his recalcitrance.

Stumbling to the port side of the bridge, Hontz lifted the red cover from the General Quarters button, sounded the alarm, and yelled "All stop!" to the engine room. He then grabbed a walkie-talkie and radioed for help. A passing helicopter pilot answered immediately. "This is *Princeton,*" Hontz told him. "We've just struck a mine. The damage is unknown."

As the captain soon learned, the damage was devastating. At frame 472, forty feet from the stern, the shock had snapped steel I-beams like twigs, heaving the deck upward twenty degrees and nearly severing the fantail from the rest of the ship. At frame 260, a six-inch crack opened in *Princeton*'s aluminum superstructure, running from the doorway to the Aegis radar room on the main deck up through the radio room and down the other side of the ship. Steel teeth snapped from the elevation drive on the aft gunmount. Restraining bolts broke from several missile launchers on the fantail and four Harpoons burst through their membrane coverings before sliding back into the launcher tubes. In the crypto vault, where the ship's classified documents were stored, the shock sheared away twenty-two bolts fastening the door frame to a bulkhead, tossing the frame and its thick steel door twenty feet down a corridor.

The whipping action, most violent at the extremities of the ship, tossed dozens of sailors into the air. Sitting on the bow bullnose watching for mines, Seaman Jeselito Alino was flung three times into a canvas sunscreen overhead. Cut and bruised, he finally crawled beneath his chair and hung on. Petty Officer James Ford, standing behind Alino, executed a languid jackknife ten feet above the bow, then smashed into the deck, crushing his knee and suffering critical head injuries. In the aft gunmount, Michael Padilla had just relieved the watch when the blast heaved him repeatedly into the I-beams six feet overhead.

While medics attended the wounded, the rest of the crew scrambled to general quarters. Several sailors caught in the shower ran naked to their posts, clothes bundled under their arms. Unlike *Tripoli*, *Princeton* suffered no puncture of her hull, but she began flooding nonetheless. The shock severed a six-inch fire main at frame 472, spraying tons of sea water into the stern and swamping number 3 electrical switchboard. (Electrician's Mate Scott Smith, ignoring the red-lettered signs warning of 450 volts, waded through the room until the power was routed to an auxiliary switchboard.) The ship's chill-water pipes, used to cool radar and other equipment, also ruptured. Within minutes the overheated Aegis system shut down, blinding the ship and leaving her without the protection of either her fore or aft missiles.

Lieutenant Erich Roeder, who had just crawled from his bunk when the mine hit, grabbed his helmet and life vest, darted into the corridor, ran back to find his glasses, then sprinted toward his battle station in the fantail. *Princeton* had been turning to starboard, and the blast jammed the left rudder, leaving the ship as difficult to control as a car with one front wheel askew. Roeder crawled into the after steering room, a cramped compartment in the bottom of the ship where two steel control posts led down to the twenty-foot-long rudders. Drenched with fluid from a ruptured hydraulic line, the lieutenant and several seamen tried without success to manually crank the balky rudder back to center line. Their efforts were disrupted by an urgent order to abandon the after steering room because of fears that the fantail was about to tear away from the ship. As they scrambled to safety, the order was belayed. A few minutes later Hontz called down to check their progress. "So how you boys doing?" the skipper asked. "Well, Captain," Roeder replied, "we're scared as hell, but we're doing what we have to do."

The same could be said of the entire crew. Despite hands numbed by the icy spray from the chill-water lines, repairmen cut through a steel deck with torches, patched the leaks, and restored the Aegis system less than two hours after the mine strike. Other damage teams clamped the gushing fire mains and began pumping out flooded compartments. Roeder and his seamen, blending ingenuity and brute force, wrapped a long chain around the rudder post and by early afternoon managed to yank the port rudder into proper alignment.

*Princeton* sat dead in the water, ready to fight but unable to move. Hontz concluded that further stress on her fragile hull and fractured superstructure, either from another mine or simply the vibration of her engines, could be catastrophic. Eighty percent of the structural strength in the fantail had been destroyed, raising the specter of the stern suddenly filling with water and dragging the cruiser to the bottom. By midafternoon the captain had reached a painful but inevitable decision: the ship would be towed to port. *Princeton*'s war was over.

## U.S.S. *Tripoli*

Defiant despite her wounds, *Tripoli* launched half a dozen helicopters on minesweeping operations within two hours of the mine strike. That business-as-usual attitude notwithstanding, the carrier's plight remained precarious. Throughout the morning of the 18th, McEwen attempted to creep eastward in search of clear water. The captain issued engine and rudder orders that he repeatedly countermanded as lookouts spotted additional mines. For several hours the ship lurched forward and backward in an ungainly dance across the minefield. At 11:30 A.M. McEwen dropped the port anchor to consider his options, gambling that the battered windlass would still operate to haul the anchor back up again.

Flooding had ceased four hours after the explosion. Shoring continued apace, with two frames completely reinforced from starboard to port. Pumping — "dewatering," in naval parlance — also continued, though with limited success. Divers reported that cracks from the mine blast radiated fore and aft from the gaping hole. Even so, *Tripoli* appeared seaworthy. At 7 P.M. the carrier weighed anchor — the windlass worked — and again headed east, led by two sweepers. Shortly after midnight she cleared the minefield and anchored thirty miles east of Point Echo.

*Tripoli* would remain on station for five days to help finish sweeping the gunfire support box. She suffered a final scare when seas rose to six feet, causing a phenomenon known as "free surface effect," in which water surging through the starboard hole battered the ship with such force that terrified sailors could see the bulkheads "pant" in and out. Each surge pushed air from the flooded compartments through the ruptured bulkhead seals, sending an eerie howl — promptly dubbed the Monster — through the lower decks. But at length the sea calmed, the Monster returned to its lair, and *Tripoli,* battered but unbowed, retired from the field only when her bunkers ran low of fuel.

*Princeton*'s fate was less dignified than *Tripoli*'s. At 5 P.M. on the 18th, the salvage ship U.S.S. *Beaufort* snagged the cruiser with a towline and began hauling her toward Bahrain. *Princeton* skidded from the minefield at two knots, her exhausted crew bedded down with blankets and pillows at their battle stations. The sweeper *Adroit* led the procession, marking suspected mines with flares. *Adroit* quickly exhausted her flare supply and was reduced to flinging bundles of green chemical "sticks" overboard, where they glowed like goblins in the dark water.

Sixty tons of new steel awaited *Princeton* in Bahrain, enough to patch her for the transoceanic trip to an American drydock. As she cleared the minefield in the early morning, a Canadian warship sent over seventeen cases of beer in sympathy. Captain Hontz, his duty done, his ship saved to fight another day, retired to his cabin for private consolation. He wept.

Notwithstanding the embarrassing setbacks, the Navy managed to finish clearing the gunfire box in the eastern gulf. No more ships struck mines. *Missouri* and *Wisconsin* steamed north toward Point Foxtrot, and now Faylaka Island and Kuwait City lay within easy range of the battleships' big guns.

## Base Weasel, Saudi Arabia

The duty of a commander, Stonewall Jackson once decreed, is to "mystify, mislead, and surprise." To dupe the enemy was to enlist him in the cause of his own defeat, and deception in warfare was at least as old as the Greeks' wooden horse at Troy. Deceit could be as facile as a soldier camouflaging himself with tree leaves and grease paint, or as intricate as the creation of entire armies from whole cloth. The laws of modern war, codified by various international tribunals, distinguished between legal ruses — "those measures designed to mislead the enemy by manipulation, distortion, or falsification of evidence" — and illegal perfidy, the sort of trickery exemplified by soldiers feigning surrender.

Military dupery achieved the status of an art form in World War II. American commanders in the southwest Pacific deliberately created clouds of dust to delude the Japanese into believing two airfields were being built on New Guinea while a real field was constructed much closer to the enemy base at Wewak. Before the Japanese realized their mistake, a surprise attack in August 1943 destroyed two hundred enemy aircraft. Elsewhere in the Pacific, Claire Chennault periodically repainted his small fighter force in China to fool the Japanese into believing the American planes flew from different units and thus posed a larger threat.

In the European theater, British commanders showed a particular gift for deception by creating twenty-six imaginary divisions, of which the

Germans believed twenty-one to be real. The invasion of Sicily succeeded in part because of the celebrated "man who never was," a corpse dressed in a Royal Marines uniform left floating off the coast of Spain with a briefcase chained to his wrist; the false documents and credentials it contained led the Nazis to disperse their forces in Italy and the Balkans. By May 1944, German intelligence counted seventy-one Allied divisions in the Mediterranean; in truth there were thirty-eight.

Before Normandy, the Allies hinted at an attack through Scandinavia by a fictional Fourth Army, reportedly exercising in Scotland. Another pre-invasion ruse, Operation Fortitude, involved a feint toward Pas de Calais, two hundred miles from the actual landing site. Eisenhower concocted an imaginary army group, supposedly commanded by George Patton, bivouacked in dummy camps throughout East Anglia and communicating via phony radio traffic. The Germans positioned fifteen divisions at Pas de Calais, waiting for the assault that never came. As the philosopher Thomas Hobbes had observed, "Force and fraud are in war the two cardinal virtues."

American commanders in the gulf war drew inspiration from such fictions and studied historical ruses assiduously, for deceit was the fulcrum of Schwarzkopf's attack plan. The most potent lies are usually simple in concept, and CENTCOM's scheme was simple indeed. It sought to convince Iraq that the main allied thrust would come through Kuwait when in fact an army of nearly 300,000 was poised to envelop from the west — the latter-day equivalents, respectively, of Pas de Calais and Normandy. The Iraqis proved vulnerable to subterfuge, and Schwarzkopf exploited their gullibility with zeal.

The threat of an amphibious landing, to which *Princeton* and *Tripoli* had been sacrificed, was only part of the plan. As American forces secretly moved west and marshaled below the Iraqi border, they left behind "deception cells" to simulate their continued presence south of Kuwait. XVIII Airborne Corps's operation was typical. A few dozen soldiers set up a phony network of camps — the corps "headquarters," Forward Operating Base Weasel, lay thirty miles west of the Kuwaiti bootheel — that created the impression of a force of 130,000 scattered across a hundred square kilometers.

Most of this Potemkin village was built electronically. Portable VHF and UHF "emulators" scattered across the desert imitated radio transmission bursts, all orchestrated by a computer that cued the devices to concoct messages passed from corps to division and division to brigade. Enemy eavesdroppers heard only a static hiss, since the bursts — like real American radio calls — were securely scrambled, but the computer simulated sixteen different scenarios of various headquarters hissing back and forth. Other messages were broadcast "in the clear" — unscrambled — to exploit the notoriously poor security of Egyptian forces laagered north of Base Weasel; the Americans taped seven messages in Egyptian-accented Arabic, including one that began, "Soldiers from the 24th U.S. Division came by wanting to know where they could find water."

A deception platoon dashed about the desert cranking out "visual signatures" with smoke generators, and loudspeaker teams along the border broadcast tape recordings of tanks and trucks repositioning at night. Daily convoys shuttled from Base Weasel to Logistics Base Alpha, churning up great banks of dust; the drivers, donning the distinctive maroon berets of paratroopers, deliberately strutted about the truck stops on Tapline Road in full view of any lurking spies. North of Weasel, engineers carved hundreds of tank berms and built bunkers outfitted with generator-powered lights.

A fleet of decoys enhanced the illusion. There were inflatable dummies of fuel bladders, fifty-five-gallon drums, Humvees, and helicopters — complete with fiber-glass rotor blades. Two-dimensional mockups, made of plastic tubing and printed nylon fabric, showed tank turrets pointing north. More elaborate tank decoys, costing $4000 each, used metal frames, camouflage nets, and heat strips powered with small Honda generators to emit the infrared signature of an M1A1 Abrams. Radio traffic intercepted from an enemy corps in southern Kuwait confirmed that Iraqi intelligence believed XVIII Corps was still near the Kuwaiti bootheel as late as February 21.

The VII Corps deception cell practiced similar ruses, hoping to reinforce Iraqi suspicions that an allied attack into Kuwait would probably angle up the natural invasion route of the Wadi al Batin. To carry the illusion one step further, Fred Franks ordered a feint into the wadi by the 1st Cavalry Division, which had been noisily demonstrating along the Saudi border ever since the second week of February. Beneath overcast skies on February 20, the 1st battalion of the 5th Cavalry Regiment pushed through two cuts in the border berm, swung clear of a minefield, then maneuvered toward the flat western edge of the dry riverbed. Under orders to probe the Iraqi defenses without sustaining heavy casualties, the battalion pushed north for six miles in a diamond formation, detecting no sign of the enemy.

At noon the first burst of gunfire erupted. "Receiving fire," a scout platoon leader reported calmly, "returning same." Seven Iraqi soldiers dropped their rifles in surrender. But as American reinforcements moved forward to collect the prisoners, the wadi exploded with artillery, mortar, and antitank fire. Concealed behind a slight undulation in the terrain, the Iraqis had so cleverly hidden their guns in bunkers and beneath camouflage nets that American reconnaissance flights had overlooked them. As two U.S. tank companies rumbled forward to form a skirmish line, a 100mm antitank round destroyed a Vulcan air defense gun, killing the gunner. Another round hit the sight atop a Bradley and deflected into the gunner's seat, eviscerating a sergeant, breaking the legs of a soldier in the rear bay, and burning the vehicle commander, who was blown through the top hatch.

Bullets pinged off armor plating and chewed into the sand. The wadi boiled with shouts and screams, explosions and smoke. "Oh, my God, they're dead! They're dead!" a hysterical gunner shrieked over the radio. "There's blood everywhere." Private First Class Ardon B. Cooper, a strapping redhead, had moved forward to shield a wounded soldier with his body when a mortar fragment hit him in the throat. He stood, blood gushing from his mouth, just as an antitank round hit a second Bradley, blowing off the TOW launcher, spraying Cooper with more shrapnel, and severing an artery in his leg. He would die later that night despite eighteen pints of transfused blood.

For more than an hour the fight raged. Abrams rounds and Bradley cannon fire ripped up the wadi, smashing the enemy bunkers. American artillery shells howled overhead. A-10 pilots maneuvered behind the Iraqis, dropping bombs and stitching the ground with cannon fire. As the shooting ebbed, the battalion requested and received permission to withdraw. When an Abrams hit a mine while pulling back, the tank was demolished but the crew was uninjured except for bruises.

By nightfall the Americans had crossed the berm back into Saudi Arabia, bearing three dead and nine wounded. Some felt that bulling into the Iraqi defenses at midday had been foolhardy, but Fred Franks disagreed. Four enemy divisions guarded the wadi, convinced more than ever that they were blocking the American avenue of invasion. A dozen casualties, the corps commander concluded, was not an unreasonable price to perpetuate the ruse.

Schwarzkopf's effort to convince Iraq that the allies intended to attack through Kuwait posed a dilemma for the I Marine Expeditionary Force (MEF), which was preparing to do precisely that. Walt Boomer had won a three-day delay in the offensive, but his Marines still were struggling to divine the size, shape, and condition of Iraqi defenses in southern Kuwait. With regrettable mistiming, the Corps had retired its primary reconnaissance aircraft, the RF4C, the previous August. I MEF was forced to rely on remotely piloted drones, which had limited range and difficulty peering through oil smoke. The Marines accused CENTCOM of failing to deliver timely satellite and U-2 imagery; CENTCOM in turn accused I MEF of mishandling the thick sheaves of photographs that were provided. Recriminations whistled back and forth until the Marines finally sent several officers to the Defense Intelligence Agency to gather a planeload of new photographs.

The two Marine divisions also had pressed Boomer for permission to send teams into Kuwait on surveillance and commando missions. The MEF commander rejected virtually all proposals until mid-February. "You're bringing me ideas that are just going to get the entire patrol killed," he complained. "That's not even worth talking about." (Two proposals approved by Boomer proved fruitless: a February 10 attempt to capture Iraqi soldiers near OP-4, and the infiltration of a sniper team to shoot enemy artillery officers.) Boomer eventually agreed to broaden the reconnaissance and on February 17 the first of eleven small patrols slipped across the border, hiding during the day and prowling at night around the minefields ten to fifteen miles into Kuwait. Despite the reputation of Iraqi combat engineers as among the world's best, the Marines found barbed wire poorly strung and ill-maintained, few antitank ditches, and land mines that — like those strewn at sea — had been planted haphazardly.

While the two Marine divisions probed the actual attack zones, a two-hundred-man "ambiguity force" sought to convince the enemy that the MEF's intentions were directed elsewhere. The Army's deception cells wanted to portray an imminent attack up the wadi in westernmost Kuwait; their Marine counterparts tried to suggest an invasion through the bottom of the bootheel. (The actual Marine axis would angle in between the two.) At noon on February 18, Task Force Troy — symbolized by a wooden camel on wheels — began operating along a twenty-five-mile sector south of Al Wafrah in a zone recently vacated by the 2nd Marine Division. "We don't want the deception plan to be hokey," warned Brigadier General Thomas V. Draude. "We want to deceive the enemy, not amuse him."

Like the Army, the Marines used armor and artillery decoys, as well as taped broadcasts of tanks repositioning at night. The tapes even produced the diesel growl of Marine M-60 tanks instead of the high-pitched noise of the turbine engines in the Army's M1A1 Abrams. Task Force Troy added a few innovations, including "drive-by shootings" by TOW and cannon gunners speeding along the berm, a staged inspection of the border by a Marine officer wearing Boomer's three-star uniform, and several air strikes with napalm and fuel-air explosives — all intended to suggest the heightened activity preceding an attack.

The *pièce de résistance* was code-named Operation Flail. Still mourning the slaughter of their comrades in Beirut when a truck bomb blew up the barracks in 1983, the Marines elected to respond in kind. Before dawn one morning, two teams drove a mile north of the Marine lines in a five-ton truck and Chevrolet Blazer packed with 3400 pounds of C4 plastic explosives. Covered by artillery and mortar fire, the Marines centered the wheels, armed the detonators, locked the steering column, and jammed the accelerator.

As the drivers leaped clear, the trucks — obscured by smoke pots smoldering on their front bumpers — rolled toward the Iraqi lines. The Blazer traveled three hundred yards and veered into a ditch, but the five-ton continued another kilometer to the edge of the Iraqi fortifications. Both blew up in a pair of stunning fireballs visible for miles. The blasts inflicted little if any damage, but Beirut, if not avenged, was remembered.

## King Fahd International Airport, Saudi Arabia

Complementing Schwarzkopf's stratagem to deceive the Iraqi army was an equally elaborate effort to unhinge the enemy psychologically. Early in Desert Shield, with help from the Army's 4th Psychological Operations Group, the CINC had drafted a detailed plan for "the psychological preparation of the battlefield." The proposal languished in Washington for many weeks, victim of bureaucratic wrangling, legal skirmishes over the propriety of urging Saddam's ouster, and the scheme's reliance on covert action by the CIA. After Schwarzkopf delivered himself of a table-pounding tirade, Cheney approved a modified plan in December. Further delays followed as the Saudis considered the wisdom of provoking Saddam before the war began. On January 12, an MC-130 Combat Talon aircraft, flying along the Kuwaiti border, sprinkled more than a million leaflets advocating Arab brotherhood, and the most intense psyops campaign in military history was under way.

An army shorn of its will to fight is no longer an army but a rabble. "In war," Napoleon famously observed, "the moral is to the material as three is to one." Schwarzkopf, who retained personal oversight of allied psychological operations throughout the gulf war, hoped to do nothing less than convince Iraqi soldiers that their defeat was inevitable and their cause morally corrupt.

Psyops provoked deep skepticism through much of the U.S. military, which associated its practitioners with such stunts as playing loud rock music outside the papal nuncio's residence in Panama in a fatuous attempt to flush out Manuel Noriega. In Vietnam the modest success of the Chieu Hoi (Open Arms) program in enticing defectors was undermined by more dubious enterprises, such as Operation Tintinnabulation — a pulsating noisemaker used to harass the enemy — and Wandering Souls, a campaign meant to exploit Viet Cong fears of burial in an unmarked grave by playing eerie tape recordings supposedly representing the spirits of dead guerrillas.

Schwarzkopf brushed aside skepticism and plunged ahead, certain that the "killing of the enemy's courage" — in Clausewitz's phrase — was as vital as killing his troops. Showering enemy troops in the trenches with propaganda had been practiced since 1914, but never — ticker-tape parades excepted — had so much paper fluttered over such a small area in such a brief span of time.

By war's end nearly thirty million leaflets would be dropped. Like Madison Avenue executives, psyops experts spoke of "market penetration" and "target audiences." The 4th Psyops Group used six Army printing presses, as well as those of the Saudi military and a private printer, working around the clock to produce four-color fliers measuring precisely three by six inches — a size deemed most conducive to "fluttering" and thus the maximum flight time. Soldiers practiced their drops from a fifty-foot tower, measuring the time of descent to predict dispersion patterns. Laptop computers, programmed with the estimated wind speed for each thousand-foot increment of altitude, calculated the "footprint" of every leaflet mission. Fliers fell from the open bays of MC-130s and in special "bullshit bombs" — programmed to burst open at the optimum altitude — dropped from B-52s, F/A-18s, and F-16s.

Operating out of King Fahd Airport, the Army developed fifty different leaflets sorted into seven themes. An "intimidation" leaflet depicted an Iraqi coffin lashed to the roof of a taxi — a common sight during the Iran-Iraq war — with the Arabic caption, "Time Is Not in Your Favor." At Schwarzkopf's unorthodox suggestion, leaflets showing bombs raining from the belly of an airplane were sprinkled over units *before* a B-52 strike: "This is your first and last warning. Tomorrow, the 20th Infantry Division will be bombed! Flee this location now!" After the raid, another leaflet warned, "Yesterday we demonstrated the power of the multinational forces. Once again, we offer you survivors the chance to live." A day later the bombers struck again, followed by another paper barrage. Similar messages followed attacks with 15,000-pound "daisy cutters": "You have just experienced the most powerful conventional bomb dropped in the war. It has more explosive power than twenty Scud missiles . . . Flee south and you will be treated fairly."

Other messages urged soldiers to abandon their equipment, or stressed the hardships of war inflicted on families at home. One leaflet originally showed an Iraqi surrendering to an Arab soldier. Schwarzkopf, assessing the drawing with a connoisseur's critical squint, sent it back for more work. "We've got to reassure the enemy that it's okay to surrender to Americans," the CINC decreed. "They've got to know that we won't hurt them." An artist transformed the Arab soldier into an affable GI Joe by giving him a stronger chin, a shave, Popeye forearms, and an American Kevlar helmet.

Psyops teams also hired a smuggler to drop twelve thousand bottles off the Kuwaiti coast, each stuffed with a flier showing Marines sweeping ashore atop a towering wave. Some fifty thousand audio cassettes containing popular Arabic music interspersed with anti-Saddam messages were smuggled into Iraq and Kuwait. In addition, the Americans operated a covert radio station — the Voice of the Gulf — to broadcast news and propaganda on six frequencies from ground stations in Saudi Arabia and a pair of EC-130 Volant Solo aircraft flying near the border. The station beamed taped surrender appeals from Iraqi defectors, whose voices were electronically altered to prevent retaliation against their families. Sixty loudspeaker teams farmed out to Army and Marine units at the front broadcast similar appeals.

Whether any of this was having the desired effect on enemy morale remained an open question. CENTCOM and the Pentagon expected a flood of deserters; barely a trickle dribbled south, despite the millions of safe conduct passes rained on Iraqi divisions. Most of those who did desert were low-ranking conscripts whose tales of mass defections to the north, battered morale, and Iraqi execution squads may have been useful for American propaganda but were considered to be of dubious intelligence value. (After the war a survey of Iraqi prisoners indicated that nearly all had seen leaflets; seven in ten said the fliers had been a factor in their decision to surrender.)

Then on February 20 an incident in the west suggested that at least some Iraqi units were close to cracking. Apache pilots from the 101st Airborne spotted a network of bunkers at Thaqb al Hajj, forty miles north of the border and astride the division's planned invasion route. Helicopter gunships and A-10s attacked the elaborate complex for four hours, inflicting little damage. An airborne company landed nearby with a three-man psyops team broadcasting surrender appeals over a loudspeaker. White flags fluttered. Hundreds of Iraqis from the 45th Division poured from their holes. By early evening 435 enemy soldiers had been packed into Chinook helicopters for the short flight south.

Schwarzkopf's pleasure was soon soured by the press accounts, which were accompanied by photographs showing Screaming Eagle patches on the shoulders of the 101st troopers. "Goddam it," he fumed in the war room. "Any intelligence specialist worth his salt now knows I'm moving forces out there. If the Iraqis have any smarts at all, they'll know what I'm fixing to do to them."

## Riyadh

Allied success in deceiving and demoralizing the Iraqi army did little to assuage bickering within the American camp as the ground war loomed closer. Among other disputes, the struggle over who would control allied air attacks against enemy forces in the Kuwaiti theater continued with internecine fury, particularly after Cal Waller — who had been appointed to arbitrate the issue in early February — left the war room to replace Yeosock as the senior Army commander. Horner and Glosson vigilantly protected their targeting authority. When one of Schwarzkopf's colonels, Clint Williams, proposed distributing more sorties to the Army corps, Horner snapped, "That's not your business. That's mine."

As they had since January, both the Army and Marines accused the Air Force of giving short shrift to targets nominated by ground commanders. Boomer warned of potentially "disastrous consequences," since the Marines believed they were about to attack 40 percent of the enemy in the theater with only two divisions. "Who's running the goddam war?" he complained to his staff on February 16. "Is it the Air Force or the CINC? You've got to wonder." Boomer also grumbled to Schwarzkopf about the lack of B-52 strikes against Iraqi entrenchments in southern Kuwait. By the third week of February, the Marines had pulled virtually all of their warplanes from the strategic air campaign to concentrate on targets they deemed insufficiently attended to by the Air Force.

The Army lacked recourse to its own bomber fleet and complained with even greater pique. An XVIII Airborne Corps message voiced "grave concerns" at the dearth of air support. Fred Franks peppered Riyadh with complaints from VII Corps and at one point flew down to confer with the Air Force brass. Few ground commanders shared Horner's enthusiasm for tank plinking; they believed that enemy artillery posed the greater threat and that their own tanks could take care of most Iraqi armor. Not least among the Army's vexations was that Schwarzkopf himself — through decisions made during his nightly seven o'clock strategy sessions — was responsible for some of the last-minute changes that pulled bombers away from targets the Army wanted hit.

"Be patient," Horner counseled his Army and Marine compatriots. "We'll take care of it." Glosson believed, accurately, that many nominated armor and artillery targets had been destroyed without the ground commanders' knowledge, because pilots often struck targets of opportunity that did not appear on the formal air-tasking order. Every night allied pilots reported killing another hundred or more enemy tanks and armored personnel carriers. Glosson and others in the Black Hole were particularly mystified by the Army's stress on hammering conscript divisions near the border, given Schwarzkopf's fixation with the Republican Guard as *the* Iraqi center of gravity. Bombers already were dropping thousands of tons of explosives on border units at the expense of heavier attacks against the Guard.

The issue boiled over in a February 18 "situation report" written by Steve Arnold, the Army's frustrated operations officer, who likened Glosson and his utterances to "someone speaking in tongues." Stamped secret, Arnold's message rocketed around the theater, as well as to the Joint Staff in Washington and American commanders in Europe and Asia:

> Air support–related issues continue to plague final preparations for offensive operations and raise doubts concerning our ability to effectively shape the battlefield prior to initiation of the ground campaign. Too few sorties are made available to VII and XVIII Corps and, while air support missions are being flown against first-echelon enemy divisions, Army-nominated targets are not being serviced. Efforts must be taken now to align the objectives of the air and ground campaigns, and ensure the success of our future operations.

This complaint sent the Air Force scrambling to the barricades. Horner was livid. "Boomer comes to me with a back-channel message and it's just between Boomer and me. You've told the whole goddam world," he charged in a phone call to Arnold.

"You're right on that count," Arnold replied. "I hadn't thought about that and I apologize for it. Let me —"

"You don't understand," Horner interrupted. "This is not a discussion. I'm talking and you're listening."

After a ten-minute harangue, Horner hung up and sought out Waller before a meeting at the MODA building. "What's this all about?" he demanded, thrusting the message at the Army general. "How could you allow something like this to go out?" Waller scanned the offending paragraph and tried to smooth Horner's feathers. "I assure you, Chuck, that we're happy with what you're doing thus far," he told Horner.

Back at ARCENT headquarters, Waller gently rebuked Arnold: "Jesus Christ, you've got the Air Force all exercised. They're demanding an apology. Don't do that again." Arnold refused to apologize, and Waller, preoccupied with final preparations for the ground offensive, pressed the issue no further.

Like so many contretemps in the gulf war — or any war — this one proved to be a teapot tempest. On the surface it reflected a wrangle over battle doctrine and the eternal struggle between creatures aerial and terrestrial.

But the roots reached deeper, tapping old fears about sending forth soldiers to meet a formidable foe. Had the true size and fighting prowess of the Iraqi army been known, the issue of who controlled the bombs dumped on the 45th Infantry Division or the 10th Armored would have seemed less momentous. Still, having endowed the enemy with legions that did not exist and martial virtues they did not possess, the Americans were obligated to treat the coming battle as one akin to Armageddon.

The precise strength of Iraqi troops in the Kuwaiti theater in late February may never be known, but almost certainly the number was tens of thousands fewer than presumed by allied intelligence. When the Iran-Iraq war ended in 1988, the Americans held detailed "order of battle" data on the Iraqi army (some of them provided by Baghdad through a covert intelligence pipeline). Saddam's ground forces were believed to have grown from twelve divisions of 350,000 men in 1982 to fifty-six divisions of 1.1 million men shortly after the war ended with Iran. In the two years following that cease-fire, CIA and DIA analysts turned to more pressing issues elsewhere in the world. The Iraqi data grew stale. But for two months after the invasion of Kuwait, the old order of battle was used to estimate the size and configuration of Iraqi units.

Saddam, however, had restructured his army. Divisions that once mustered twenty thousand soldiers for the static war against Iran now numbered roughly half that number. The Iraqis further confused the Americans by replacing Republican Guard divisions, which spearheaded the invasion, with units from regular army corps as the Guard retrenched in southeastern Iraq.

U.S. intelligence experts, after examining satellite photographs and occasional radio intercepts, began to redefine the enemy. They realized that instead of eight artillery tubes in a battery, the Iraqis now had six; the seven tanks in an armored platoon had been pared to four. But satellites offered only limited information, making it difficult for the allies to track units as they moved around or mingled. Moreover, the same satellites were needed for mapmaking and, eventually, battle-damage assessments; these heavy demands meant that U.S. analysts occasionally lost track of entire Iraqi divisions. An intelligence system that for forty years had concentrated on defending Western Europe against the Warsaw Pact, proved inadequate when asked quickly to change focus in support of an offensive campaign in the Middle East.

Reluctant to trigger the war prematurely, Washington also forbade reconnaissance aircraft from flying over the Kuwaiti theater before January 17, further restricting efforts to obtain an accurate count. Even after the war began, intelligence gathering was hampered by the untimely retirement a year earlier of the SR-71 Blackbird, a spy plane capable of photographing a thirty-mile corridor while flying at two thousand miles per hour.

In mid-January the Pentagon estimated that the Iraqi occupation force included approximately forty-three divisions with 540,000 troops, a number predicated on an average strength of twelve thousand soldiers per division, plus assorted headquarters and auxiliary units. Another twenty-four divisions, comprising mostly ill-trained infantry conscripts, remained north of the Euphrates.

Estimates in Riyadh differed only slightly. As the allied ground offensive grew closer, Jack Leide and John Stewart, respectively the CENTCOM and ARCENT intelligence chiefs, believed the number of enemy troops in the KTO to be roughly between 450,000 and half a milion. CENTCOM's tally of Iraqi divisions had climbed from thirty-five on January 17 to forty in early February to forty-two in mid-February. This reflected both an expansion of the theater boundaries to include some enemy units heretofore excluded, and a continuing struggle to understand how Iraqi corps and divisions were organized — a process Leide called "by guess and by golly." The configuration of units near Basrah and the Euphrates Valley — including several Republican Guard infantry divisions — was particularly baffling.

This uncertainty over the Iraqi order of battle continued well into the war. Two infantry divisions believed to be facing VII Corps — the 31st and 47th — would not be encountered during the ground attack, apparently because they were not in the Kuwaiti theater. Analysts also discovered late in the war that they had inadvertently confused the identities of four armored divisions. Schwarzkopf's overestimation of the enemy's size and capability persisted to the last shot and beyond: as the war entered its final hours he would tell reporters that the allies were outnumbered at least "three to two . . . [and] as far as fighting troops we were really outnumbered two to one." Even after the war, Schwarzkopf would continue to assert that Iraq had 623,000 soldiers in the Kuwaiti theater.

Scant evidence supports the CINC's claim. Without a network of spies in the enemy camp, the Americans were slow to realize that many units were badly under strength; some divisions may have had as few as four thousand soldiers. Five weeks of bombing badly attenuated Iraqi divisions through casualties and desertion. ("We are not 'preparing' the battlefield," Dave Deptula had written on a sign posted in the Black Hole. "We are destroying it.")

Constitutionally wary of underestimating the enemy — another Vietnam legacy — the U.S. military was especially reluctant to estimate how many Iraqis had fled north. Satellites and U-2s provided a fair catalogue of Iraqi equipment, particularly armor and artillery, although this also would prove to be overestimated by 20 to 25 percent; in terms of gauging potential enemy resistance, the tally of Iraq's weaponry was more important than the raw number of troops in the theater. But a photograph that showed an artillery battery with one tube destroyed, one possibly damaged, and four intact could not reveal that the crews had deserted en masse, effectively spiking all six guns.

Erring on the side of caution, the Americans also endowed the enemy with an esprit and battlefield savvy that would prove wanting in the last days of the war. If the Iraqis misjudged the difference between fighting poorly trained, ill-equipped Iranians and a U.S. military bred to battle the Russians, so, ironically, did the Americans. One statistic proved particularly telling: the Americans shipped nearly 220,000 rounds of tank ammunition into the theater, of which less than 2 percent would be fired. Only as the war entered its final weeks did the intelligence picture begin to clear; notwithstanding confusion about enemy numbers, Leide and Stewart in particular came to recognize that the Iraqis were in very bad shape indeed.

How many able-bodied Iraqis remained in their trenches and revetments as G-Day approached? One postwar Army estimate put the number at roughly 300,000. Others considered half to two-thirds that figure more likely. A study by the House Armed Services Committee, extrapolating from estimates offered by captured enemy commanders, set the number at 183,000 of an original 362,000. (Presumed losses included 153,000 deserters and 26,000 killed or wounded in air attacks.)

Against this tattered remnant the allies marshaled 700,000 troops. The ease with which they routed the enemy from the theater would eventually mute public acclamation of the achievement, as if glory could be won only in a sea of blood. Certainly the intelligence shortcomings in gauging Iraqi mettle were serious and embarrassing; but to err in the other direction by misprizing the enemy — an unfortunate American proclivity on battlefields stretching from Bull Run to the Mekong Delta — would have been unpardonable if not calamitous.

"After Vietnam," Leide would observe following the gulf war, "I vowed to myself that I would never underestimate my enemy again." Whatever the sins of commission and omission committed by Riyadh and Washington in this matter, they were venial and not mortal. Whether the next war would be so forgiving remained to be seen.

## Washington, D.C.

Colin Powell had worried that allowing the start of the ground offensive to slip from February 21 to the 24th would encourage requests for further delays. That concern now proved well founded. As the third week of February slipped past, Schwarzkopf appeared reluctant to press ahead with the attack as scheduled. Phone conversations between chairman and CINC regarding G-Day became ever more intense. Schwarzkopf not only wanted time for the Marines and the Army divisions to position themselves, but he also had promised Walt Boomer seventy-two hours of fair skies in order to guarantee allied air cover.

Powell found himself caught in the middle. Bush, Scowcroft, and others in the civilian leadership were eager to bring the war to a quick finish. If Saddam were to feign a withdrawal or actually begin to pull out of Kuwait, such a move, the White House feared, would crack the coalition. Powell wanted Schwarzkopf to be comfortable with his timetable, but the chairman saw little reason not to launch the attack on the 24th or even earlier.

Yet when he pressed Schwarzkopf to consider advancing the schedule by a day or two, the CINC dug in his heels. Citing meteorological predictions of foul weather on the 24th and 25th, Schwarzkopf proposed instead that the attack be postponed a couple of days more.

Powell relayed the request to Cheney. "Norm's under a lot of pressure," the chairman explained. "He knows he's going to go. He wants to go, but he's trying to be considerate of what his commanders want. He's worried about the weather, and he's a little high-strung right now."

"Colin, you're making it hard," Cheney said, barely concealing his exasperation. "There's a limit. What reason do I give the president?"

For virtually the first time in the war, Cheney intervened directly rather than working through Powell. In a conference call to Schwarzkopf, with the chairman also on the line, the secretary asked the CINC to make the case for postponement. The air campaign had lasted more than a month, with remarkable success. Schwarzkopf had been given everything he asked for — more, in fact, than he had asked for. An enormous army was poised to strike. Norm, Cheney said pointedly, tell me your reasons for wanting an additional delay.

Schwarzkopf believed — as he later acknowledged in his memoirs — that an Iraqi withdrawal would mean an allied victory without further allied casualties. He still feared that the coalition could lose five thousand dead and wounded in just the first two days of the ground war. "Time is on our side," he told the secretary and chairman.

But in Cheney's view, allowing the enemy army to escape at this point would be a strategic blunder. The purpose of the war was not only to liberate Kuwait, but also to destroy Iraq's offensive capability. Hearing no persuasive reasons for delay, Cheney urged Schwarzkopf to "wrap it up."

Powell subsequently placed another phone call to Riyadh from his office. "Norm," he began, "you've got to give me some good stuff on this one because I don't really understand it."

To Powell's surprise, Schwarzkopf exploded. "What if we attack on the twenty-fourth and the Iraqis counterattack and we take a lot of casualties because we don't have adequate air support?" In a bellowing rage, the CINC accused Powell of political expediency and timidity in refusing to confront Bush and Cheney. "My responsibility is the lives of my soldiers," he added. "This is all political."

Powell had prided himself in ignoring Schwarzkopf's occasional effrontery, but this went beyond the pale. The suggestion that only the theater commander was concerned about American lives infuriated the chairman. With a roar that filled his office, he lashed out. "Wait a minute, buddy! Don't you patronize me! Don't pull that on me, that *we* don't care about soldiers."

As abruptly as it began, the storm subsided. "Sometimes I feel like I'm in a vise, like my head is being squeezed in a vise," Schwarzkopf pleaded, his voice unsteady. "Maybe I'm losing it."

Powell paused for a moment. It was important to re-establish some equilibrium, to recapture what he thought of as "the usual Colin-Norm routine." We need to take a little time on this, the chairman told himself. It will work out. Let's cool it.

"Norm," Powell said, "you know that at the end of the day we will do what you want. I will make any recommendations necessary for you to do what you want. But on this one, you've got to give me some help as to why you think we should change the date. But we'll support you. We always have. I always have."

A few hours later, Schwarzkopf called back. Good news, he told Powell. The weather forecast looked better. Boomer and the Marines would be ready. Four A.M. on the 24th looked like a good time to launch the attack after all.

But one last issue required resolution before the ground offensive began. The Central Intelligence Agency and CENTCOM had been unable to find common ground on the bugaboo issue of battle-damage assessment, BDA. CENTCOM's estimates of Iraqi tanks destroyed continued to climb, from 476 on February 1, to 862 on February 11, to 1439 on February 16. Now the number approached seventeen hundred, or nearly 40 percent of Iraq's armored force in the theater. CIA analysts, examining satellite photos for blown turrets or shattered hulls, could confirm only about a third.

Schwarzkopf growled at the agency in the cloister of his war room while publicly maintaining the mien of a warrior closing in for the kill. In a newspaper interview on February 19, he had described an Iraqi army "on the verge of collapse" from the daily loss of two tank battalions to allied air strikes. The war, he added, resembled "a beagle chasing a rabbit."

Powell, perhaps more owl than beagle, agreed with the CINC's assessment but thought it prudent to refrain from gloating. "If we start making claims and we can't deliver," he told Tom Kelly, "we're in real trouble." With Powell's blessing, Kelly disavowed Schwarzkopf's comments in a briefing to the Pentagon press corps and warned, "There's still a lot of fighting to be done."

Although a CIA analysis concluded that Iraqi forces had been "heavily degraded," the discord over dead tanks was nettlesome enough for the agency's director, William Webster, to alert Bush. The president asked Scowcroft to investigate, much to the annoyance of Cheney and Powell. Schwarzkopf was incredulous. "For Christ's sake, Colin," the CINC pleaded. "From the standpoint of military judgment you know what these guys are saying is wrong. Look at the number of bombs we've dropped. I can't have somebody second-guessing us back in Washington and doubting my judgment as theater commander."

CENTCOM's final estimates before G-Day portrayed an enemy army that was badly battered. Assessments compiled by the Army's John Stewart placed Iraq's 4th Corps — entrenched in southwestern Kuwait — at only 58 percent of its original combat strength. The enemy's

7th Corps, located west of the Wadi al Batin, was in even worse shape, at 42 percent. Even the three heavy Republican Guard divisions — the Tawalkana, Madinah, and Hammurabi — were rated between 57 and 72 percent.

CIA analysis, however, tended to place the enemy units at 75 to 85 percent intact. The agency sidestepped the issue of whether the ground attack should be postponed, but both CENTCOM and the Pentagon presumed that was precisely where the agency's argument led. Harry Soyster, the DIA director, attempting to mediate, met in downtown Washington with Webster and his lieutenants. When a CIA official insisted that "these figures from CENTCOM are inflated," Soyster brought him up short. "Whoa! That's the wrong word. 'Inflated' suggests that they've been deliberately pushed up when in fact it just reflects a different way of counting." The mediation ended.

On February 21 at 4 P.M. Scowcroft gathered the disputants in his West Wing office with the intention of either effecting a reconciliation or at least "getting somebody to shut up," since the press had got wind of the dispute. Powell, Cheney, and Mike McConnell represented CENTCOM; Webster brought one of his national intelligence officers, David Armstrong. Working from a briefing paper cobbled together at Langley during the previous two days, Armstrong reviewed the BDA issue from the beginning of the war. He argued that CENTCOM's methodology had changed twice since January 17; that pilot reports had been initially accepted without allowing for exaggeration or error; that "double and triple kills" of the same target were significant. He closed by disavowing any CIA interest in usurping Schwarzkopf's command prerogatives and reiterating that, regardless of the tank tally, Iraq's army was "highly degraded."

McConnell, armed with charts and graphs, quickly rebutted. Before leaving for the White House, he called Colonel Richard Atchison, Schwarzkopf's deputy intelligence officer. "How comfortable are you?" McConnell asked. "Right now I'm extremely comfortable," Atchison replied. "Can I tell you the morale of this or that division? No. Can I tell you they're on their fannies? Yes."

The CIA, McConnell told Scowcroft, was largely limited to satellite images of the battlefield, a view often obscured by bad weather and other limitations. Riyadh benefited from tons of other intelligence not readily available to the CIA, including U-2 and RF-4 photography, pilot reports, and radio intercepts. Moreover, the agency focused almost exclusively on the Republican Guard, whereas some of the heaviest bombing had fallen on Iraq's regular army divisions (in part because of the U.S. Army's success in diverting attacks to enemy divisions closer to the border).

Scowcroft, looking like a man with a toothache, listened as the arguments flew back and forth and then cleared his office of all but the senior policymakers. The schism disturbed him, as it disturbed the president. But Powell and Cheney seemed to have robust confidence in CENTCOM's evaluation. To Cheney, the dispute was yet another Vietnam legacy: the CIA remained wary of "light at the end of the tunnel" optimism from field commanders. Cheney believed that Webster either failed to understand the limitations of his agency's intelligence collection in this war, or that it was "cover your ass time" at the CIA.

Like the defense secretary, Scowcroft believed the moment of truth had come for the allied ground attack. Rejecting CENTCOM's BDA judgment, he realized, would signal a devastating loss of confidence. And any delay could give Saddam time to wriggle from the noose. Scowcroft saw no alternative except to side with the military. Declaring the matter resolved, he dismissed the meeting with a sardonic "Go back to work."

Thirty minutes later Powell called McConnell, who had returned to his Pentagon office. "The issue is settled," the chairman announced. "The CIA will not report on it anymore."

## Moscow

If the CIA had been chased from the field, the Soviets had not. Kremlin diplomacy, endured by Washington as one endures a benign but faintly malodorous cousin, now showed alarming signs of progress. On February 18, after making the circuitous trek through Iran, Tariq Aziz had arrived in Moscow. Gorbachev immediately offered him a peace plan on behalf of the allies, with whom the Soviet president had consulted not at all. In exchange for Iraq's immediate withdrawal from Kuwait, Gorbachev guaranteed that Saddam and his regime would survive the war; that no reparations would be exacted; and that other regional issues — including Arab-Israeli disputes — would eventually be addressed. "The timing is crucial," Gorbachev warned. "If you cherish the lives of your countrymen and the fate of Iraq, you must act without delay."

Aziz immediately repaired to Baghdad. Gorbachev cabled a three-page summary of his pitch to the White House, where Bush had just returned from a weekend at his summer home in Maine. Among other shortcomings, the plan gave Saddam six weeks to vacate Kuwait. "The clearer this gets, the worse it gets," Scowcroft told Bush. "It's designed to make things as easy as possible for Saddam."

The president replied with a private message to Gorbachev on Tuesday morning, the 19th. Any armistice, Bush insisted, required Iraq's evacuation of Kuwait within four days (thus guaranteeing the abandonment of most armor and heavy weaponry), the immediate release of allied war prisoners, and a detailed disclosure of all minefields, like those which had crippled *Princeton* and *Tripoli* the day before. Searching for the proper wooden stake to drive through the heart of the Soviet proposal, the president had settled for a curt public snub: "It falls well short of what would be required."

Bush's rebuff sat badly in Moscow. "That plan was addressed to the Iraqi leadership," Bessmertnykh, the foreign minister, observed acidly, "so he rejected a plan which did not belong to him." With a humaneness that was in scant evidence during the long Soviet war in Afghanistan, the ubiquitous Middle East envoy Yevgeni Primakov declared, "The slaughter must be stopped . . . A people is perishing."

Bearing new marching orders from Saddam, Aziz returned to Moscow. After touching down in a light snowfall at Vnukovo Airport shortly before midnight on February 21, he was immediately whisked in a black limousine to the Kremlin, where he and Gorbachev dickered across a lacquered table for two hours and twenty minutes. Aziz agreed to an unconditional withdrawal from Kuwait and the immediate release of prisoners of war. He balked, however, at Soviet attempts to impose the withdrawal timetable and insisted that all United Nations sanctions be lifted after two thirds of Iraq's forces left Kuwait.

Vaguely alluding to "technical reasons," Aziz continued to suggest that six weeks might be required to dismantle the Iraqi occupation. Primakov, slouched with his arms crossed at the end of the table, pointed out that the August invasion had taken but a few hours. Aziz countered that only two divisions initially invested Kuwait; several hundred thousand troops had entered the emirate since then. Gorbachev leaned across the table, his patience clearly fraying. "The proposed deadline," he warned, "can and must be reduced to a minimum."

At 2:40 A.M. the Kremlin spokesman, Vitaly Ignatenko, rushed into the Soviet Foreign Ministry building on Smolensk Square. "Please excuse me for having kept you so long, but I do think that the effort was worthwhile," he told a clutch of sleepy journalists. "They finished just ten minutes ago . . . The response is positive." Ignatenko recited an eight-point agreement, conspicuous for its lack of a specific withdrawal timetable. "I can say [it is] a very good morning," Ignatenko gushed. "I do think we can give some applause here."

## Washington, D.C.

No one was clapping in the White House. At 6:47 P.M. Washington time, just moments after Ignatenko's announcement, Bush took a call from Gorbachev in the president's upstairs study. (The Soviet leader had phoned earlier but had been forced to wait twenty minutes while the White House hunted down a translator.) "Let us work at it over the weekend," Gorbachev urged. "Don't put a deadline on it. We need time. We can work this out." Aziz had promised that Saddam was ready to leave Kuwait unconditionally, but more time was needed to work out the details.

Bush, bracketed by Baker and Robert Gates, listened as Gorbachev ticked off the points of agreement. Aziz's concurrence seemed wholly at odds with Saddam's defiant rhetoric during a radio broadcast earlier in the day, delivered with his usual flair for invective and apocalyptic imagery. "There is no other course than the one we have chosen," the Iraqi leader had declared, "except the course of humiliation and darkness, after which there will be no bright sign in the sky or brilliant light on earth."

Baker and Gates sifted through Gorbachev's words for signals. The secretary of state saw no sign that the Kremlin was abandoning its strategic commitment to Washington; Baker recalled how often Eduard Shevardnadze, the former foreign minister, had warned that the only language Saddam understood was that of force. Gates interpreted this gambit as a final effort by the Soviet bureaucracy to save a client state from obliteration. Gorbachev, though earnest, seemed detached. Gates, straining to hear tones of desperation, detected none.

The proposal, Bush told Gorbachev, still fell short of the allied conditions. Not only did the Iraqis seem determined to amble out of Kuwait at their own pace, but there was no mention of assistance in locating mines, restoring the antebellum government of Kuwait, or payment of war reparations. A withdrawal would satisfy only one of the Security Council's twelve resolutions. Gorbachev promised to keep negotiating.

Bush cradled the receiver. "We've got to get this thing back under control," the president told his aides. "It's slipping away from us." He hurried downstairs to rejoin a small dinner party that included the Air Force chief, Tony McPeak, and Supreme Court Justice David Souter. When he finished his meal, the president bundled his guests into an armored limousine and raced six blocks east to Ford's Theater for a performance of Leslie Lee's *Black Eagles*, a play about the Tuskegee airmen who formed the first black bomber squadrons in World War II. McPeak, settling into an orchestra seat, eyed the infamous presidential box where Lincoln had been shot and his theater guest wounded.

"I don't mind sitting near you, Mr. President," McPeak murmured, leaning toward Bush. "But I don't know if I want to sit right next to you." The curtain rose, the drama unfolded. "Maybe we're fighting on the wrong damn side," a Black Eagle named Nolan declared as the second act drew to a close. "No, we're on the right side," Black Eagle Leon replied, "but the right side has got itself a hell of a lot of problems." The curtain fell and at 10:20 P.M. the motorcade rolled back through the White House gates. Bush bade his guests good night and strode upstairs to his study.

Eight men awaited him: Baker, Scowcroft, Quayle, Gates, Cheney, Powell, Chief of Staff John Sununu, and the press secretary, Marlin Fitzwater. With the ground attack scheduled to begin in less than forty-eight hours, none favored a postponement for further Soviet mediation. Scowcroft and Gates believed that the Iraqis were manipulating Moscow in hopes of fracturing the allied coalition. Saddam must not emerge with what the West Wing called "a Nasser victory," an allusion to the former Egyptian president, whose stature in the Arab world soared after he defied the Western powers. Cheney, still dressed in a tuxedo from a reception hosted by the queen of Denmark, pensively chewed on the earpiece of his glasses. The Soviet gambit vexed him. Moscow had no troops committed, nothing at risk. Cheney felt the impulse to snap, "Buzz off, Gorby. What do we need you for?" He sensed that others in the room felt equally inclined, in Powell's phrase, to "stiff" the Soviet leader.

Yet Bush was not among them. "We're not going to sunder our relationship with Gorbachev," said the president, sitting behind his oak desk. Soviet support in the UN Security Council had been critical and would remain so in any postwar armistice arrangements.

"But," the president added, "we are going to go forward with the campaign. We're going to get this thing resolved." Bush looked over a typed sheet listing objections to the Gorbachev-Aziz agreement. "It's not enough to just say we don't accept the Soviet plan. I went through this with Gorbachev on the phone, and he knows it's unacceptable and he knows the specific reasons why. We ought to lay them out to the whole world."

Powell, casually dressed in a green turtleneck and sport jacket, had scribbled some notes to himself earlier in the evening, including the cryptic "Can't kick them out. Can't let it dangle." Having just wrestled both Schwarzkopf and the CIA to the ground in setting G-Day for the 24th, the chairman was reluctant to open the issue again. Saddam's delaying tactics were clearly beyond compromise.

Before the war began, the very process of setting the January 15 deadline for Iraqi compliance had drawn the country and coalition together. It had rallied support and forced a dénouement. Perhaps, Powell suggested, a similar ultimatum would work now. "If you get them out by noon Saturday, Mr. Gorbachev, you get the Nobel Prize. If you don't, we kick Saddam's ass."

"I think that's a good idea," Bush said. "Has some merit."

A public deadline also would prime the Army and Marine divisions preparing to attack from Saudi Arabia, the chairman added. Allied troops would "know exactly what to be looking for and when." Cheney and Scowcroft concurred. Baker also liked the idea, while noting that enough time had to be allowed to notify more than three dozen allied governments. For a few minutes the men debated whether to give Saddam until Sunday or Monday to begin his withdrawal, which would require postponing Schwarzkopf's offensive. In the end, the war cabinet agreed on Powell's suggestion of noon Saturday, Washington time, eight hours before the scheduled attack.

The meeting adjourned at midnight. Baker and Scowcroft began calling the allies. Bush walked down the hall to bed.

On Friday morning, part of the administration brain trust gathered in Scowcroft's office to polish the ultimatum. Should Saddam capitulate by noon Saturday, the Americans favored giving the Iraqi army but four days to evacuate Kuwait; the British and French thought that unrealistic and successfully advocated expanding the timetable to one week. (The Pentagon estimated that in a week's time Saddam could salvage less than half his surviving combat equipment and only a fifth of his ammunition stocks.)

On other demands the Americans held firm: the abandonment of Kuwait City and release of all prisoners within two days; the immediate withdrawal from defensive belts along the Saudi border and the gulf islands; the disclosure of all booby trap and mine locations. Once Saddam agreed to pull out, allied pilots would suspend their attacks for two hours to give the Iraqis time to begin their retreat. The moratorium would last as long as the withdrawal continued in good order.

Paul Wolfowitz, sitting across the room from Scowcroft's desk, wondered whether even the Saturday deadline was too generous. "Not that I'm bloody-minded, but I really think if Saddam's able to pull his army out of Kuwait and pretend that it was never defeated," Wolfowitz said, "it's more likely that he or somebody else in Baghdad can make a comeback someday with an aura of invincibility."

Dennis Ross, one of Baker's State Department deputies, reminded Wolfowitz that just a few weeks earlier he would have been elated had Iraq withdrawn, obviating the need for a ground war. "Aren't you nervous about the outcome?" Ross asked.

"No," Wolfowitz replied. "I think it's going to be over very quickly."

In the small study adjacent to the Oval Office, Bush took a phone call from France's Mitterrand, who soon made clear that he shared none of Wolfowitz's bellicosity. The French president dug in his heels, arguing that Saddam should be given more than a day to acquiesce. Shouldn't we consider allowing them at least seventy-two hours?

As Bush fumbled for a reply to this unexpected burst of Gallic *raison*, Gates hurried in from the West Wing with a new intelligence report, which he handed to the president. Satellites had detected many new pillars of smoke in the Kuwaiti oil patch. Plumes curled from the processing centers at Ash Shuaybah and Mina al Ahmadi, and from the vast Burqan field. Several dozen wells had been destroyed in the previous five weeks, but this appeared to be a scorched-earth campaign. More than 150 wells now blazed. Other wellheads, sabotaged but unignited, sprayed geysers of oil that pooled in thick petroleum lakes.

Bush pounced on the report. "Look what's happening," he told Mitterrand. "We can't wait. We can't give him seventy-two hours."

Gates walked back to the West Wing, grinning. "Once again," he told Wolfowitz, "Saddam Hussein has pulled our chestnuts out of the fire."

At 10:40 A.M., the president stepped behind a small podium in the Rose Garden, his face deeply lined. For three minutes, in a finger-jabbing excoriation that invoked Saddam's name eight times, Bush laid down his marker.

> We learned this morning that Saddam has now launched a scorched-earth policy against Kuwait, anticipating perhaps that he now will be forced to leave. He is wantonly setting fires to and destroying the oil wells, the oil tanks, the export terminals, and other installations of that small country. Indeed, they are destroying the entire oil-production system of Kuwait.
>
> I have decided that the time has come to make public with specificity just exactly what is required of Iraq if a ground war is to be avoided. Most important, the coalition will give Saddam Hussein until noon Saturday to do what he must do, begin his immediate and unconditional

withdrawal from Kuwait. We must hear publicly and authoritatively his acceptance of these terms.

Bush finished his statement and turned on his heel. The die was cast.

As he had promised, Schwarzkopf began phoning his division commanders a few hours later. In a terse, emotional call to the field headquarters of the 24th Infantry Division, which Schwarzkopf had once commanded, he told Barry McCaffrey: "It's a go. They've set Kuwait on fire. We also think they're torturing our pilots."

He hesitated for a moment to gather himself, then added, "I know I can count on you to do what is required. Godspeed."

# PART III
# LAST WEEK

# 13

# The Biltmore

## Baghdad

Day by day the crude calendar scratched on the wall of David Eberly's cell had filled with hatchmarks, each slash representing another small triumph over starvation, torture, and madness. With only a week left in February, Eberly could not yet bring himself to etch a new page for March. Instead, he began listing hymns alphabetically, scratching titles on the red brick with his tiny drainplate screw — "Amazing Grace," "Blessed Assurance," "Count Your Many Blessings," "Do Lord," "Everlasting Arms" — until the wall was covered with the careful runes of a man of faith.

Each morning he tried to do twenty or thirty pushups, an ever more difficult ordeal for a body now reduced to barely more than a hundred pounds of bone and sagging flesh. Then he paced: two steps by three steps by two steps by three steps, ten laps clockwise followed by ten laps counterclockwise. Calculating two hundred laps to a mile, he tried to cover five miles each day, periodically resting to work on his hymnal or to daydream about his family. He re-created, in minute detail, vacations that he had taken with Barbara and Timm: packing the car; driving south on Interstate 95 to Florida; strolling into the hotel at Disney World; staring at the turquoise water of the swimming pool. Sometimes he walked the streets of Brazil, Indiana, retracing his boyhood paper route, house by clapboard house; or he relived camping expeditions on the James River, feathering the paddle of his canoe as he scanned the leafy shoreline for a perfect glade in which to pitch his tent.

His olfactory sense had dwindled, then vanished, overpowered by the stench of a body unwashed since January 19. In its stead his hearing sharpened, and he strained to catch every creak and whisper that drifted through the Biltmore cell block. For hours he listened to the guards' radio, imagining that he was beginning to comprehend snatches of Arabic. The most cherished sound was the clang of the gurney trundling down the corridor with the afternoon bucket of broth. Like gunshots, the small steel portals in the cell doors opened and slammed shut as each prisoner was fed in turn. Eberly always stood ready, small orange bowl in hand, eager for the brief glimpse of the hallway when the shutter flew open. "Soup again?" he invariably asked, gesturing for a bit more in the bowl. Sometimes the guards replied with spittle and snarled epithets — "America! George Bush! Dog!" — but twice he cadged the bounty of an extra half ladle.

Breakfast typically was a few sips of water and a half slice of bread hoarded from the previous day. Late one night in mid-February, a stranger in a red-and-white kaffiyeh had appeared like an apparition at the cell door and handed over five dates. Eberly devoured one and hid the rest in his sock. For three successive mornings he allowed himself a date, which he savored as he stalked in circles around the cell; each lasted at least 250 laps. The fifth he saved for Sunday, and took the entire morning to nibble it, molecule by sweet molecule.

The interrogations, though less frequent now, had degenerated into sessions of terror. Eberly sat, blindfolded and helpless, while his captors clubbed him in the head or whacked at his legs, demanding details of the allied ground attack. Several times he felt the blunt jab of a pistol barrel against his skull. Hearing the snick of a cocking trigger, he wondered dimly whether he would also detect the sound of hammer striking cartridge before his brains splattered across the room. A finger tightened on the trigger, the muzzle pressed tighter into his scalp, and with a metallic snap the hammer fell. For the briefest instant he dangled between being and nothingness until the world reasserted itself, drawing him back from oblivion with symptoms of this life: the thump of his heart, sweat beading beneath the blindfold, a sharp slap. The clip had been removed from the gun.

Only sleep offered respite from fear and the constant cold, which seeped into his bones through the walls and the floor. Using his fist as a pillow, he gave himself up to vivid dreams, of home, of flying, of bowls brimming with dates. But waking before dawn, when the cell was dark and silent as a sarcophagus, he sometimes felt his mind slipping away, felt the bright, sharp edges of reality — always so crisp in his pilot's world — begin to dissolve. He wondered whether sanity was a finite commodity that could run short, as one ran short of sugar or flour in the larder. Lying there on the cold slab, he learned to reach out and lay his palm on the icy concrete. That simple act of contact with the floor, stark and irrefutable, seemed to draw him back from the abyss.

On the evening of the 21st, as the glimmer of outside light faded from gold to gray, sounds of sobbing carried down the corridor. The cries grew louder, turning to the shrieks of a man gone mad. Eberly heard heavy footsteps and the clink of keys as guards approached and opened a cell. Now the sounds jumbled together: the bark of commands, the thuds of a body being beaten, screams, grunts, more screams. "La! La! La!" No, no, no! Eberly listened in horror, praying for the victim, until at last the cries died to a whimper.

The next evening the shrieks began again, and again the guards answered with a savage beating. But this time, as the howling continued, the madman was dragged into the corridor and chained to Eberly's door. Pacing furiously, Eberly could hear the rain of punches and kicks, punctuated by the clear sound of a pistol cocking. Oh, God, he thought, not here. Suddenly the door swung open and a guard ordered him out. He caught a glimpse of the poor wretch — a shirtless, howling Arab — bleeding on the floor as the guard led Eberly to a small room at the end of the corridor.

"Lower your trousers," an Iraqi ordered. "We are checking for sexually transmitted diseases." Eberly slowly untied the drawstring, anticipating unspeakable tortures. Then he realized that the Iraqis were hunting for Israeli pilots and other Jews by checking to see which prisoners had been circumcised. He stood impassively until the examination was completed, then fastened the yellow pants and shuffled back to his cell amid a barrage of kicks and slaps.

One afternoon, shortly before supper, he was led down the rear stairs to the interrogation room. Blindfolded as always, listening for the slightest rustle that warned of a blow to the head, he sensed several people in the room. This time the questions turned to taunts, to fantastic and clumsy claims evidently designed to fracture his will. Seventy thousand Americans had died, his captors asserted, and two thousand prisoners were rotting in a camp south of Baghdad. "Did you know that Mubarak has been killed? We have shown that he is part of the American CIA." Bush, too, was dead. "Do you think Quayle will use nuclear weapons?"

Eberly pondered the question, momentarily dropping his guise of befuddlement. "Yes," he answered firmly. "He's not going to let us lose. We will win this war."

"What is the plan for the ground attack?" someone demanded yet again.

Eberly canted an ear toward the voice. "You mean the Army still hasn't attacked?" he asked. There at least was a bit of news. He shook his head. "I don't know."

"We take you downtown," the interrogator warned, "and show you some of the damage your pilots have made, maybe let our people take out some of their anger on you."

Eberly shrugged, his voice trailing off. "Well, if you must do that . . ."

"If you don't start cooperating, we will throw you back in that hole," a voice threatened.

Eberly thought of Brer Rabbit. Oh, yes, please, the briar patch. A rough hand grabbed him by the shirt and dragged him back to the cell.

The session, he estimated, had lasted three hours. Seventy thousand dead? Two thousand prisoners? Surely that wasn't possible. On the floor of his cell he found his soup bowl, a thick skim of grease congealed on the cold broth. Eberly wolfed it down. He hoped that in some small way he had helped to mislead and frustrate the enemy.

Although in recent weeks he had spent little time imagining an escape, now he began to plot how a rescue team might attack the Biltmore to free him and his fellow prisoners. He envisioned special forces commandos stealing up the narrow rear stairwell in the dead of night; darting through the prison with silencer-equipped weapons; killing the unsuspecting guards one by one before opening the cell doors and leading the men to safety.

He carefully popped the stitches on the edge of his blanket, tore away the trim, and used it to fashion shoelaces, a belt to hold up his baggy prison trousers, and bandages to support his weak ankles. Maybe he would have to run, to make a desperate sprint toward a waiting helicopter as bullets pinged at his feet. Maybe not. Whatever his fate, he wanted to be ready.

## Riyadh

For three weeks, almost from the moment they arrived at Ar Ar to begin Scud hunting in Mesopotamia, Wayne Downing and Delta Force had been working on plans to free the American prisoners of war. As in Pacific Wind — the prewar scheme to assault the U.S. embassy in Kuwait — Downing and his men spent endless hours contemplating avenues of attack, the proper balance of firepower and mobility, phases of the moon, and a hundred other variables. Downing often discussed the plans with Schwarzkopf, who weighed the odds of success against the prospect of negotiating the prisoners' release.

As the CINC had confided to Barry McCaffrey, U.S. intelligence suspected that the Iraqis were mistreating the POWs, which provided further incentive to free them as soon as possible. The survival rate of soldiers and airmen in captivity was not heartening. Thirteen thousand American prisoners had died of injuries, maltreatment, and disease in German and Japanese camps in World War II; of 766 confirmed U.S. prisoners in Vietnam, 106 perished in captivity. Yet the record of past efforts to liberate American captives was poor. In November 1970, a brilliant and daring raid on the Son Tay prison camp near Hanoi failed when commandos belatedly discovered that all the prisoners had been moved three months earlier. Operation Eagle Claw, the attempted rescue ten years later of U.S. hostages in Tehran, had been a debacle.

By the third week of February, Delta and American intelligence had identified three possible detention sites, among them the Abu Ghuraib Prison outside Baghdad. (None of the three was the Biltmore.) Only one appeared reasonably accessible to a raiding party. Downing never felt comfortable with the rescue plans, nor did Schwarzkopf, who sensibly chose to hold Delta in check at least until the ground offensive unfolded.

Buster Glosson, Dave Deptula, and others in the Black Hole also pondered the POWs' whereabouts because of fear that they would be bombed inadvertently. Glosson believed the most likely site to be a secret police compound at the north end of Saddam's palace in Baghdad. The compound remained off limits to U.S. bombers.

Frustrations in the Black Hole over Colin Powell's bombing restrictions had not abated. Deptula continued to draft written justifications for targets in the Iraqi capital only to have them slide into the now familiar "not yet" category. These included the People's Army headquarters; the "Baghdad presidential bunker," a suspected alternate national military command post that had been bombed but not destroyed early in the war; and "Baghdad barracks southwest," identified as "facilities associated with [the] personal security of Saddam Hussein." Glosson still chafed at the restrictions, filling his diary with irate scribbles. Other Air Force planners also were convinced that the Army leadership was determined to launch the ground campaign lest air power reap all the credit for winning the war. One senior Air Force officer wrote in his journal: "CINC and CJCS [Powell] and Army refuse to acknowledge what air is doing to destroy the ground army in Iraq. This coverup has been orchestrated by the CINC and chairman in a manner which they can deny."

Yet not all targets in Baghdad were proscribed. As the shock of the Al Firdos catastrophe receded, Glosson extracted permission to hit several buildings occupied by the security police, as well as hangar facilities at a downtown airfield (in case Saddam tried to flee the country in an executive jet).

After several unsuccessful efforts to target Ba'ath Party headquarters — only two communications buildings in the sprawling complex had been bombed — Glosson waited until he could slip into the chair next to Schwarzkopf's when no one else was within earshot. "Look, sir, I've just got to talk to you about this target," he said. "It's the pillar of power for that regime and it's still standing. What a symbol that has to be for the people of Iraq. I don't know how we can *not* destroy it." Schwarzkopf studied Glosson's face for a moment before commanding, "Take it out." Glosson hurried back to the Black Hole before there was any change of heart; a few hours later bombs rained down on the party headquarters.

Another target in the "leadership" category also worked its way onto the air-tasking order. East of the Tigris in downtown Baghdad stood a three-story white building with rows of small windows on either side, almost like portholes on an ocean liner. In Baghdad the sleek structure was known as the White Ship. On the Air Force documents that scheduled the target for destruction by four F-117s on the night of February 23, it was identified as Regional Headquarters, Iraqi Intelligence Service. ("Organization responsible for monitoring political dissidents and suspected subversive activity," Deptula had written in his justification.) Some of those unfortunate enough to live in the building knew it by yet another name. They called it the Biltmore.

## Baghdad

In the Biltmore cell block, Saturday the 23rd had been notable only for its monotony. David Eberly received his daily ration of broth around 2 P.M. Shortly after sunset, a guard he had never seen before opened the cell door and peered inside. "May I have some water, please?" Eberly asked, holding out his cup. To his surprise, the guard took the cup and returned with a few ounces of water. "I thank you for being so kind," Eberly said slowly. The guard gave him a blank look and locked the door. Eberly wrapped himself in his blanket and again sank to the floor.

Two hours later he heard the abrupt scream of an aircraft engine and the slap of footsteps in the corridor. He scooted against the wall and draped the blanket over his head. The plane sounded low, so low that he assumed it was an F-111 at five hundred feet. As the jet swooped past, an electrical crackle filled the air.

The first bomb struck with a stupendous roar near the generator outside, perhaps fifty feet away. Shards from the narrow glass window sprayed the cell like shrapnel as the blast wave rolled across the cell block; the building swayed against Eberly's back. Too stunned to move, he waited for the floor to collapse or the ceiling to cave in. Was he sitting against the proper wall? Should he be braced against the door for the overhead protection of the steel frame, or would the door blow in on him? Muffled voices pierced the block. "Get us out of here! Can anybody hear? Let us out!"

Thirty seconds later he heard the shriek of another jet and the crackle that he now recognized as the last rattle of a laser-guided bomb. The second explosion seemed to detonate almost on top of the cell. With a crash, the ceiling in the corridor collapsed. He heard pipes bursting and the gush of water below. The cries for help turned to terrified screams. Eberly tucked his arms and legs into a ball as a third bomb slammed into the rear of the building, which wobbled under the impact. Dear God, he prayed, it's up to you. Thy will be done.

Still as death he waited, barely breathing. He assumed the attackers had come in a four-ship formation, but a minute passed, then another, without a fourth bomb. He stood slowly and shuffled toward the door. Smoke and dust choked the air. Voices echoed through the darkness, American voices shouting American names.

Above the cries he again heard the sounds: an aircraft engine and then the fatal crackle. He threw himself against the wall and bellowed "Incoming!" The last bomb smashed into the compound with unspeakable violence. How, he wondered, could the floor and ceiling remain intact?

Voices filled the corridor outside his cell. Several walls had collapsed, freeing some of the prisoners. Eberly's mind raced. They'd need transportation, maybe a bus. And weapons. And someone who spoke Arabic. At least a map. He imagined them clattering south through the darkened city.

Men called out their names and units. "Grif?" he yelled. "Are you okay?" No answer. Panic gave way to excitement. "Where are you from?" someone asked. "CBS News," a voice answered. Eberly heard the crunch of boots on broken glass and shouts in Arabic. "Here they come!" he warned. "Knock it off." He scooped up the scrap of bread he had been saving, tucked it inside his yellow shirt, and sat down.

Keys rattled in the lock. Grunting and straining, the guards pried open the warped door. A hand grabbed Eberly and yanked him from the cell; he wrenched free from another hand that tried to tug the blanket from his shoulders. "Hurry! Hurry!" a guard demanded, dragging him by the collar. By lantern light he saw that the corridor was choked with debris, much of it the fallen ceiling. Finding the path blocked by a smashed stairwell — most of the steps had turned to rubble — the guard dashed back into the corridor and down a different flight of stairs, hauling Eberly behind. They waded through water gurgling across the basement, a scene that reminded Eberly of an old submarine movie. "It's okay, it's okay," he assured the frantic guard, watching for a chance to break and run.

They climbed another flight of stairs and emerged into a courtyard awash with moonlight. Eberly felt overwhelmed, stricken by this sudden sweep of open sky after so many weeks of darkness. Through the rubble they stumbled to a bus across from a row of white stone houses. Guards were everywhere, screaming and herding their charges onto the seatless vehicle with kicks and blows.

Eberly sank to the floor on the left side, drawing up his legs and hooding his head with the blanket. Iraqis swaggered down the center aisle, ripping the shirts of prisoners who failed to keep their faces covered. Leaning against the man on his left, he whispered, "I'm Eberly. What's your name?" A vaguely familiar voice replied, "You've gotta be shittin' me." Eberly waited for a guard to pass, then repeated his name. "I know who you are," the man said. "I'm Roberto Alvarez from CBS. I interviewed you twice before the war." The cameraman was among a four-man CBS crew captured in southwestern Kuwait on January 21; farther back in the bus guards savagely pummeled the crew's correspondent, Bob Simon.

Another prisoner, screaming maniacally, was flung aboard the bus. Peering beneath the edge of his blanket, Eberly recognized the barefoot and shirtless Arab madman. Blood streaked his face and chest as two guards fell on him with truncheons. When the screams continued, the Iraqis dragged him through the door and chained him to the rear bumper. The bus lurched forward for a block and stopped long enough for the madman to be unchained and heaved aboard again.

For nearly an hour they rode through the shadowy streets of Baghdad. Three guards stalked up and down the aisle, clubbing those who dared whisper. Eberly estimated that thirty prisoners were aboard, most of them American. From beneath his blanket, he caught several glimpses of a huge, waxing moon silvering the palm fronds. How beautiful, he thought; how incredibly beautiful.

The bus stopped. For a few minutes the guards bickered among themselves, either lost or uncertain of where to take the prisoners. The bus started up again and they rumbled past a stone tower. After a massive steel gate rolled aside, they entered what looked like a deserted junkyard. In fact, it was Abu Ghuraib Prison, soon to be dubbed Joliet Prison after a scene from *The Blues Brothers,* a popular Hollywood movie. The prisoners were herded into a two-story cell block.

Eberly found himself in a cramped cell with ten other men, all American fliers, packed so tightly that they had to cross-weave their legs in order to sit. Although disoriented and weak from starvation, Eberly was elated to see that one of the men was Tom Griffith. In hoarse whispers the airmen took roll, memorizing one another's names to bear witness if any man was left behind when freedom came.

Eberly studied the gaunt, stubbled faces. After weeks of solitary confinement, with only his thoughts and imagination for company, he was bewildered by this sudden multitude, so comforting and yet so jarring. Until dawn the men swapped stories of SAMs and tracers, of flaming airplanes and barbarous Iraqi guards. The optimists speculated that their release was imminent; the pessimists fretted that a year might pass before Saddam set them loose. Eberly found a scrap of cardboard in the corner of the cell; overcome with exhaustion, he centered it beneath his hips and curled into a fetal ball. The voices droned on — the soft Southern drawls and pinched Midwestern vowels of his countrymen. His eyelids grew heavy. "Too bad for the colonel," he heard someone whisper sympathetically. "What a way to finish a career." Too bad? But he was alive, among friends. Someday, soon perhaps, he would be home again with Barbara and Timm. Too bad? He had endured. He had kept faith. And now he slept.

## Task Force Grizzly, Kuwait

In the cramped confines of the Riyadh war room, Schwarzkopf and his staff busied themselves with a thousand final details: reassuring the Egyptians and other Arab forces, monitoring fuel and ammunition stocks, examining the latest intelligence for signs of Iraqi repositioning.

The CINC's timetable for the ground offensive called for an exquisitely sequenced attack. At 4 A.M. on Sunday morning, February 24, the flanks of the vast allied front would strike first, with XVIII Airborne Corps lunging into Iraq in the far west while in the east two Marine divisions and the Arab troops of Joint Forces Command East (JFC-E) plunged into the Kuwaiti bootheel. Yet these thrusts would be but supporting attacks — deceptions, in effect — with which Schwarzkopf hoped to confuse the Iraqis and draw enemy reserves toward the allied wings.

The main attack would not begin until twenty-six hours later, at dawn on Monday. Then the allied center would surge forward — with VII Corps heading north (and ultimately east) toward the Republican Guard in the so-called Left Hook, followed an hour later by the Egyptians and other Arab units of Joint Forces Command North (JFC-N) angling into western Kuwait. Here lay the heart of the campaign: VII Corps was charged with Schwarzkopf's primary military objective, destroying the Republican Guard, and JFC-N bore responsibility for the political objective of liberating Kuwait City.

In the days and hours before the formal launch of the ground campaign, however, skirmishing intensified along the entire length of the allied line. Divisions inched forward, in some cases crossing the border into Kuwait or Iraq to better position themselves for the attack. Artillery barrages and air strikes hammered enemy defenses from the Persian Gulf coastline in the east to As Salman, several hundred miles inland. Although the full fury of allied power would remain in check until Sunday and Monday, the attack crescendo already had begun to build.

Certainly the most formidable Iraqi defenses lay along the Saddam Line in southern Kuwait, and there the allied air armada pounded the enemy relentlessly. A week before the ground war was to begin, fourteen F-117s had demolished a dozen pumping stations and feeder lines supplying the enemy fire trenches — long ditches that could be filled with flaming oil. B-52s stepped up their attacks on troop formations and artillery batteries. (Most Marine and Army engineers preferred that the B-52s avoid striking minefields for fear that deep bomb craters would complicate breaching operations; they originally had hoped that air-burst fuses — designed to avoid cratering by exploding the bombs a few feet above the ground — would solve the problem, but tests in California and the Kuwaiti elbow showed that the fuses tore away barbed wire and other obstacles yet failed to create enough overpressure to detonate most mines.)

Marine bombers also dropped napalm and fuel-air explosives (FAE) on Iraqi artillery and infantrymen as part of a no-quarter campaign "to kill things that kill Marines." Invented during World War II, napalm had achieved such notoriety in Vietnam — where the jellied gasoline was formally known as Incendergel — that the Joint Staff in Washington briefly debated whether to risk renewed public outcries by using the weapon in the gulf war. But fuel-air explosives were as nasty as napalm. The bombs enveloped a target with an aerosol of ethylene oxide, which was then ignited to create an explosive overpressure that swept across the ground at Mach six with a force comparable to a nuclear blast. "The effect of an FAE explosion within confined spaces is immense," a CIA document warned. "Those near the ignition point are obliterated. Those at the fringe are likely to suffer many internal, and thus invisible injuries, including burst eardrums and crushed inner organs, severe concussions, ruptured lungs . . . and possible blindness." FAE and napalm — like the 15,000-pound BLU-82 bombs dropped on the enemy — were intended to wreak psychological as well as physical havoc.

To steal a march on the Iraqis even before G-Day, Walt Boomer had asked Schwarzkopf for permission to slip several thousand Marines from his 1st Division into Kuwait at least two days early. Four battalions — two from Task Force Taro and two from Task Force Grizzly — would infiltrate across the border to mark lanes through the Iraqi mine belts and provide spotters for Marine artillerymen, thereby expediting the February 24 attack by Task Forces Ripper and Papa Bear. Schwarzkopf agreed to Boomer's infiltration proposal, but with a warning: "If we suddenly declare peace, make goddam sure you haven't started the ground war and that you can pull your men back out."

The CINC nearly proved prescient. As the four battalions slipped across the berm into southern Kuwait, Gorbachev's peacemaking gambit and Bush's ultimatum deadline played hob in the war zone. Commanders found themselves uncertain how aggressive they could be without scuttling the Soviet efforts to negotiate an armistice. On 1st Division's eastern flank, Task Force Taro infiltrated Kuwait in two columns on the night of February 22. By sunrise on the 23rd the two battalions had penetrated eight miles beyond the border, nearly reaching the first mine belt south of Al Wafrah. The Marines hid in foxholes throughout the day on the 23rd, undetected by Iraqi observers.

Boomer, heeding a warning from Riyadh, cautioned the 1st Divison commander, Major General Mike Myatt, to "avoid doing anything irreversible" before Bush's Saturday deadline expired at 8 P.M. (noon in Washington). Myatt relayed the order to the Taro commander, Colonel John Admire. Deep in enemy territory with two thousand lightly armed Marines burrowed like crabs in the desert pan, Admire mulled over the precise military definition of "irreversible" and concluded that it meant avoiding a pitched battle. After sundown on the 23rd, Taro crept forward on hands and knees to begin carving three paths through the minefield, still undetected as they probed the sand with bayonets, listening for the telltale clink of metal striking metal.

On the division's western flank, though, the infiltration ran into snags. By early morning on February 22, 2200 Marines from Task Force Grizzly had penetrated twelve miles into Kuwait to reach the mine belt. Like Taro, Grizzly burrowed into the sand and draped itself with camouflage netting. But at noon on the 22nd, Iraqi gunners spied the westernmost battalion and began lobbing artillery shells. F/A-18 and Harrier jets attacked the enemy positions, and Marine batteries along the Saudi border also returned fire. When the shooting subsided, engineers moved into the minefield on the night of the 22nd, only to be driven back at 9 P.M. when Marine bombers struck targets less than eight hundred meters from Grizzly's lead scouts.

This unexpected bombardment infuriated the task force commander, a tall, outspoken colonel named James A. Fulks, who radioed the division headquarters demanding to know "what the hell's going on." By the time the confusion was sorted out, night was spent and Grizzly had under twenty-four hours to find a path through the minefield. When the task force regrouped on the 23rd, scores of Iraqi deserters began streaming across the sand, including several who were shot in the back by their comrades as they tried to surrender. Artillery and mortar fire erupted again, lightly wounding two Marines. The supposedly surreptitious infiltration threatened to escalate into full-scale combat.

Fulks began planning a "firepower breach" that would blast a channel through the mines with plastic explosives and allow him to storm across with an entire battalion. But at 2 P.M., Myatt called Grizzly with the warning against taking any "irreversible action" before Bush's deadline six hours hence. "You have to wait for approval," Myatt cautioned. "You can't put any significant forces on the other side of the obstacle belt yet." Fulks again lost his temper, this time at being given "an impossible mission" with "my hands tied."

As the sporadic shelling continued, a group of Iraqi defectors offered to show the Americans a passage through the mines. Fulks sent forward a team of fourteen Marines to suppress sniper fire while the Iraqis marked a lane through the antitank mines with chemical lights. Myatt now authorized Grizzly to put "security elements" across the minefield to protect the breach; Fulks, seizing the opportunity, shoved an entire battalion through.

But eight hundred yards beyond the antitank belt, the Marines discovered an unanticipated field of antipersonnel mines eighty yards across. A combat engineer, Staff Sergeant Charles Restifo, led a squad into the second field on hands and knees to carve out a trail ten feet wide with bayonets. At 9 P.M., an hour past the presidential deadline, Myatt ordered Grizzly to push north "in force." By early morning on the 24th, both battalions had threaded their way across Restifo's footpath and were preparing to blast a wider channel for the armored forces that would soon follow.

## King Khalid Military City, Saudi Arabia

Schwarzkopf's iron grip on the 9400 special operations forces (SOF) in Saudi Arabia loosened slightly as G-Day approached. Except for Delta's secret wanderings through Mesopotamia and five unsuccessful Scud-hunting missions by Green Beret teams in eastern Iraq, few "cross-border" ground operations had been approved before mid-February. To the dismay of his more impetuous lieutenants, the CINC used his special operations units conservatively: Navy SEALs hunted for sea mines and scouted Kuwaiti beaches from the shallows; Army helicopter crews waited for search-and-rescue opportunities that rarely came; Air Force special operations planes, with Army psyops squads, dropped leaflets and the occasional 15,000-pound bomb.

The 5th Special Forces Group farmed out more than a hundred teams to allied units in response to the CINC's request for "ground truth." (After some initial puzzlement — "What the fuck is 'ground truth?' " muttered the group commander, Colonel James W. Kraus — the Green Berets concluded that Schwarzkopf sought a candid assessment of Arab combat prowess and fighting spirit.) Among other services, the teams tried to convince Saudi troops that their flimsy sand fortifications — "kill-me berms," in the American vernacular — would not stop enemy tank rounds. They also implored commanders in the six Kuwaiti brigades to prevent vengeance killings during the liberation of their homeland. "Make sure they understand," Kraus told his soldiers, "that he who commits the last atrocity is the one who is remembered."

But as the hours ticked down toward the ground offensive, SOF operations grew bolder. Several teams flew by helicopter deep into Iraq on "trafficability" missions, touching down long enough to snap photographs and collect soil samples that confirmed the desert's strength to support the heavy tread of American armored columns. To reinforce the threat of a Marine amphibious assault, two speedboats carrying a dozen SEALs raced north from the Saudi port of Al Mishab on February 23. Fifteen miles off the Kuwaiti coast at Mina Saud, the SEALs slipped into Zodiac rubber raiding boats and closed to within five hundred yards of the beach. From there the men swam to shore, stringing a line of buoys — identical with those used to mark amphibious landing zones — and implanting six satchels, each filled with twenty pounds of plastic explosive, near the waterline. When they returned to their speedboats, the SEALs peppered the coastal bunkers with machine gun fire. Automatic timers detonated the satchels along the beach, and air strikes added to the cacophony.

While SEALs blew up the beach in Kuwait, ten Special Forces reconnaissance teams prepared to infiltrate deep behind enemy lines to watch for repositioning tank forces and other signs of Iraqi counterattacks. Five teams from 5th Special Forces Group would be inserted north of the Euphrates, supplemented by a sixth team in the far west. (Corps and division commanders also infiltrated several long-range surveillance detachments on similar missions closer to the Saudi border.) Four teams from 3rd Special Forces Group would be positioned in the VII Corps's attack corridor to scout the Republican Guard.

Deep reconnaissance required patience, courage, and a willingness to live like a mole for up to a week in the claustrophobic confines of a desert burrow. Each "A-team" consisted of six to ten men, led by a captain or a warrant officer, and usually included one or two 18 Bravos (weapons experts), 18 Echoes (communications specialists), 18 Charlies (engineers), and 18 Deltas (combat medics able to perform battlefield surgery if necessary). Most were also crack shots, capable of hitting a skull-sized target at five hundred yards with an M-16.

For weeks the teams had trained in isolation outside King Fahd Airport. Each man carried a rucksack weighing 150 to 200 pounds. A team's kit might include silencer-equipped weapons, claymore mines, grenade launchers, food, five-gallon water bladders, more than a thousand rounds of ammunition, night-vision goggles, camouflage material, periscope, laser range finder, sniper scopes, and, most important, radios. Because any sudden gesture could betray them in the open desert, the men practiced moving with exaggerated deliberation, almost as though operating under water.

The construction of spider holes, or hide sites, required special artistry, and each team developed its individual touches. A typical hole for three men and their rucks measured four by eight feet — accommodations likened to "puppies in a litter" — and was built with a subterranean frame of plastic pipe and plywood. The frame was covered with chicken wire and burlap — some teams favored a particular British weave called Hessian — that was then spray-painted a desert tan and covered with sand. (Particular attention was paid to the hue of the sand, which changed as it dried in the sun.) Teams avoided shrubbery lest it attract grazing camels; urine and feces were deposited in five-gallon bladders and stored in the hole to avoid luring dogs. The soldiers took elaborate pains to "sterilize" their hide sites before climbing in, sweeping away boot prints with whisk brooms and even using their fingers to resculpt any tire tracks in the area.

On the late afternoon of February 23 eight of the teams gathered in hangars at King Khalid Military City, north of Riyadh. (Two others staged from Ar Ar.) Plans called for the infiltration helicopters to cross the border shortly before 8 P.M., giving the teams at least six hours on the ground to dig their hide sites before dawn. Some men made last-minute videotapes for their families or huddled with the chaplain. Those who were superstitious tucked away their good luck charms or saw to it that their boot laces were lashed just so. One nervous team leader showed signs of balking until his battalion commander, Lieutenant Colonel Frank Toney, snapped, "Captain, you get paid for one thing: to put your ass on the line."

Two missions were scrubbed, one because of uncertainty over the boundary line between VII and XVIII Corps attack sectors, another because of enemy movement on the proposed infiltration site. Further delays resulted from confusion over whether CENTCOM had officially authorized the missions, given Bush's 8 P.M. ultimatum; helicopters carrying several teams were temporarily held up as they flew toward the border, thus delaying the missions for a few hours.

By midnight, most of the reconnaissance teams were deep into Iraq. A few dug in without incident. Captain Edward J. McHale and his five men set down near Hill 499, southwest of the Tawalkana Division. After filling in the divots left by the Pave Low helicopter, the team built two hide sites, unable even to see the holes they were digging because of the thick oil smoke that blanketed the desert soon after their arrival. They would remain in place for sixty hours, seeing nothing but two Bedouin and a camel, and thus assuring VII Corps that no counterattack appeared imminent.

Others were less fortunate. One team landed in an area that had appeared, in satellite photos, to be strewn with boulders; the "rocks" proved to be Bedouin tents, and the team was immediately extracted. Another heard Arabic voices all around and aborted. A third found the desert flooded from a recent storm; unable to dig in, the team returned to Saudi Arabia.

For the remaining teams, including two inserted halfway to Baghdad, the ease with which they burrowed into enemy soil and seemed to vanish from sight would prove illusory. With dawn would come discovery, and a belated realization that even the most cunning camouflage was no match for a sharp-eyed child.

## Phase Line Becks, Saudi Arabia

The center of Schwarzkopf's planned attack lay nearly two hundred miles west of the Persian Gulf, where VII Corps slowly coiled itself with the agreeable conviction — soon to be shattered — that the corps had an extra day to prepare before striking. Fuel tankers and ammunition trucks lumbered forward, division commanders pored over their maps, and thousands of soldiers scribbled final letters home, groping for words to convey the fearful exhilaration that seizes an army on the eve of battle. Many settled for: I'm scared and I love you.

Like Walt Boomer, Fred Franks wanted to occupy a patch of enemy territory before launching his main attack on Monday morning, February 25. He had deliberately weighted the bulk of the corps's firepower on his left wing, beyond the barriers of the Saddam Line, which dribbled to a halt farther east. More than eight hundred M1A1 tanks, two thirds of the corps's armor strength waited below the border on the left in a column only twenty miles wide. By cutting gaps through the berm on February 23 and pushing a dozen miles into Iraq, Franks hoped to give his corps a flying start toward the Republican Guard divisions 130 miles to the northeast.

This penetration mission fell to the 2nd Armored Cavalry Regiment, which served as the spearhead for the 1st and 3rd Armored Divisions stacked abreast. Beginning at the border berm, known as Phase Line Becks, the regiment received orders to move to Phase Line Bud by late afternoon on the 23rd. (Other map demarcations named after beers were Coors, Corona, and Lite.) When the attack began in earnest at dawn on the 25th, the cavalry was to lead the corps to Phase Line Smash, seventy miles above the border. There Franks intended to determine how they would destroy the Tawalkana, Madinah, and Hammurabi Republican Guard divisions. On the corps's right wing, the 1st Infantry Division had orders to breach the remnant of the Saddam Line for the British 1st Armoured Division, which would veer to the east, freeing the Big Red One to join her sister divisions in the north.

The 2nd Armored Cav comprised eight thousand soldiers, 125 tanks, 116 Bradleys, and a helicopter fleet, all divided into four squadrons and further subdivided into troops of roughly 150 soldiers each. As the oldest cavalry regiment on continuous active service in the Army, the unit wanted for neither élan nor heritage. Organized in 1836 by Andrew Jackson to battle the Seminoles, the regiment subsequently fought at Antietam, Gettysburg, and beside Theodore Roosevelt's Rough Riders in Cuba. In World War I, it was the Army's only mounted combat unit, and it galloped into battle on French draft horses. A generation later, mounted on tanks, the regiment won a Presidential Unit Citation for relieving Bastogne during the Battle of the Bulge. For the past thirty years, the 2nd Cavalry had been headquartered in Nuremberg, whence it patrolled a four-hundred-mile stretch of the East German and Czech borders.

The current commander was a man considered one of the Army's ablest officers, Colonel L. Donald Holder, a laconic, lanky Texan whose troopers regarded him with a devotion unto adoration. Like many officers of his generation, Holder had returned from Vietnam disillusioned, disgusted, and prepared to resign. Instead, he remained in uniform and joined the informal cabal of loyalists determined to bring the Army back from the dead. Armed with a master's degree from Harvard, he taught history at West Point and in 1980 joined the department of tactics at Fort Leavenworth. There he wrote the final draft of Field Manual 100–5, consecrating AirLand Battle Doctrine as the Army's new scripture for waging war. Holder returned to Leavenworth in the mid-1980s and as director of the School of Advanced Military Studies — the Jedi Knights — helped revise and advance the doctrine in the 1986 edition of FM 100–5.

By temperament he was unflappable, a bit remote, tucking his emotions behind a mask of command that kept his subordinates at a respectful distance. In the back of a small notebook Holder jotted trenchant observations about the major generals commanding the corps's four divisions, as an understudy might review the acting of those at center stage. His influence on Franks — an old cavalryman with an ill-concealed affection for the regiment — was significant. No one more than Holder embraced the metaphor of desert combat as naval warfare, in which armored forces maneuvered like fleets across the bounding main.

Now, beyond the berm that rose like a golden wave frozen in midcurl, theory would play out in the mêlée of combat. Here the toil of the last fifteen years would be assayed, and the concepts enshrined at Leavenworth — agility, depth, synchronization, and initiative — undergo trial by fire.

Holder also had a demon to exorcise. His father had taken command of an armored cavalry regiment in Vietnam, only to be killed a month later in a helicopter crash. "I've got to be careful," he told his operations officer, Major Doug Lute, in a wry, rare allusion to the tragedy. "My family tradition in war isn't a good one."

At 1:10 P.M. on the 23rd, the corps radio networks opened after a week of strict silence. Twenty minutes later the 210th Field Artillery

Brigade, joined by three howitzer batteries and a multiple-launch rocket system (MLRS) battery began shelling Iraqi positions ten to fifteen miles across the border. American officers stood with computer printouts of the enemy targets, methodically checking each obliterated bunker and revetment like items on a grocery list. The MLRS barrage was particularly horrific. Each salvo of twelve rockets streaked down range in a tangle of white smoke trails; each round burst above the target in a "warhead event" that rained 644 submunitions the size of hand grenades. A dozen rockets sprayed nearly eight thousand grenades — each capable of cutting through two inches of steel — across thirty acres. The gunners also fired eight Copperhead artillery rounds, guided by laser beams from helicopter spotters hovering far forward. Six rounds struck home, destroying primarily T-55 command tanks.

Holder watched the guns boom and the rockets flash from their launchers a thousand yards to the east. Wild cheering broke from the crews and the cavalrymen poised at the border, like the guttural roar he had heard from his troopers when the first jets zoomed overhead the night the air war began. Twenty years had passed since he had last seen combat, and he had forgotten how ferocious young soldiers became when blood rose in the gorge. He felt the hair bristle on the back of his neck.

Two rounds of white phosphorus burst overhead, signaling the end of the bombardment. The regiment surged forward as psyops loudspeakers played "The Ride of the Valkyries." Helicopter scouts darted toward the northern horizon, shadowed by a stately procession of tanks and Bradleys, gliding across the desert like men-of-war running before a fair wind. Engineers slashed the berm with bulldozers for the divisions that would soon follow: twenty-five holes on the left for 1st Armored and eighteen on the right for 3rd Armored. Holder watched from the hatch of his swaying M-113 command track, his cavalry kerchief pulled up to keep the dust from his nose and mouth.

Before dusk the regiment halted on Phase Line Bud, fifteen miles into Iraq. Any enemy soldiers left alive from the artillery barrage had fled north, taking their dead and wounded. Here the regiment was to wait for thirty-six hours until the launch of VII Corps's main attack.

"We've trained hard for this," Holder told his assembled staff in a brief homily that no doubt echoed exhortations before Chancellorsville and San Juan Hill. "The regiment has always fought brilliantly. I expect the officers to do their duty. God bless the 2nd Cavalry."

The end had begun.

# 14

# G-Day

## Washington, D.C.

George Bush spent the final hours before the ground offensive on the telephone. Gorbachev called shortly before the noon deadline on Saturday and caught the president and James Baker in the Camp David gymnasium. Taking the call on a phone in the locker room, Bush complimented the Soviet leader for his efforts but turned down a final request to delay the attack.

Gorbachev halfheartedly argued that there was still room to negotiate without an "escalation of combat operations to a new, even more destructive stage." Nevertheless, he added, whatever happened in the next few days should not impair the vision Bush and he shared "of a new world." Without disclosing that the ground assault was scheduled to begin in a few hours, Bush reminded the Soviet president that Saddam at his peril had ignored the first allied deadline of January 15, a day before the air campaign began. Like that ultimatum, Bush implied, this one was no bluff. Gorbachev ended the twenty-minute conversation by signing off in English: "Okay, goodbye."

Although Tariq Aziz had formally endorsed Moscow's peacemaking plan — now altered to a six-point proposal — Baghdad's ruling Revolutionary Command Council denounced Bush's deadline as "an aggressive ultimatum to which we will pay no attention." Washington greeted this bluster with relief: by now, few wanted to see the curtain come down before the final act. Bush subsequently spoke with the leaders of Japan, Turkey, Egypt, Britain, Germany, Canada, and Australia, as well as congressional leaders and the four living American former presidents. Iraq had chosen to ignore the allied threat, Bush told one and all; the time had come to oust the invaders by force of arms.

At 9:30 P.M. Saturday — it was 5:30 A.M. Sunday in Riyadh — the presidential helicopter *Marine One* banked around the Washington

Monument and over the Reflecting Pool before settling on the South Lawn of the White House. Scowcroft, Quayle, Sununu, Marlin Fitzwater, and Richard Haass waited in the cold moonlight. Bush emerged from the passenger bay, saluted the military aide standing at the bottom of the helicopter steps, and walked quickly into the Oval Office.

Preliminary reports from the front were heartening. Schwarzkopf's offensive had begun as scheduled ninety minutes earlier, with supporting attacks by the Marines and JFC-E on the allied right flank and XVIII Airborne Corps on the far left. The main attack by VII Corps still was fixed for early Monday. No confirmed use of chemical weapons had been reported. In Kuwait City, however, the Iraqis apparently had imposed a reign of terror, with executions, rapes, torture, and the taking of hostages. Some of the accounts would prove to be exaggerated, but now they confirmed Bush's fundamental conception of the war as a struggle between good and evil.

At 10 P.M. the president strode into the White House press room. "The liberation of Kuwait has entered a final phase," he told a national television audience. "What we have seen is a redoubling of Saddam Hussein's efforts to destroy completely Kuwait and its people. I have therefore directed General Norman Schwarzkopf, in conjunction with coalition forces, to use all forces available, including ground forces, to eject the Iraqi army from Kuwait."

Bush's left arm trembled slightly as he finished reading from the notes spread on the podium before him. "Tonight, as this coalition seeks to do that which is right and just, I ask only that all of you stop what you are doing and say a prayer for all the coalition forces, and especially for our men and women in uniform, who at this very moment are risking their lives for their country and all of us."

## Umm Gudair, Kuwait

Schwarzkopf's earlier hesitations in launching the ground attack had reflected not only his desire to give the Marines more time to get ready, but also to assure them of fair skies. "We need good weather," Boomer had advised Schwarzkopf in mid-February. "We're still outnumbered, though not as much as we thought we were. But we depend on aviation to complement our ground forces." CENTCOM meteorologists, tracking the highs and lows sweeping east from Europe and the Mediterranean, had forecast a favorable front pushing across southwest Asia around the 24th. Like Eisenhower before Normandy, Schwarzkopf anxiously watched the elements and tried to divine the heavens. "The weather," Chuck Horner had assured the CINC, "is going to be okay." "All right," Schwarzkopf told Boomer, "we'll get you seventy-two hours of good weather."

In this he was mistaken. "We probably didn't get seventy-two minutes," Royal Moore, the Marine air chief, later lamented. A purple bank of oil smoke had rolled over the Marines late in the afternoon of Saturday, the 23rd, followed by steady rain and a thick fog that cut visibility in some areas to a hundred yards.

Unhappy but undaunted, the Marine regiments had begun moving into their attack positions around midnight on Sunday as Bush was making his telephone calls from Camp David. Psyops loudspeakers played "The Marine Hymn" at earsplitting volume in honor of the forty-sixth anniversary of the celebrated flag raising on Iwo Jima's Mount Suribachi. "Now we will attack into Kuwait, not to conquer but to drive out the invaders and restore the country to its citizens," Boomer proclaimed in a message to his I Marine Expeditionary Force. "Your children and grandchildren will read about your victory in the years to come and appreciate your sacrifice and courage."

Boomer's main attack lay with the 2nd Marine Division halfway up the western border of the bootheel. William Keys, the division commander, planned to breach both mine belts between the Umm Gudair and Al Manaqish oil fields with 6500 men from the 6th Marine Regiment; they would be followed by the Army's Tiger Brigade — equipped with M1A1 tanks and better night-fighting equipment than the Marines possessed — which would then swing north to secure I MEF's western flank before seizing Al Jahra and Mutlaa Ridge to block Iraqi escape routes from Kuwait City. Keys's operations officer, Colonel Ron Richard, estimated that the division would suffer a thousand casualties.

The 6th Marines broke camp at 4:30 A.M. and pushed across the line of departure an hour later, just as the first of three hundred enemy artillery and mortar rounds fell. Despite repeated radio warnings of "Oscar! Oscar!" — the code word for incoming shells — the Marines soon realized that Iraqi gunners, with no forward observers or airborne spotters, were firing blindly. Foul weather notwithstanding, Royal Moore pushed a fresh pair of Marine fighters over the battlefield every seven and a half minutes. More than 140 artillery tubes and nine rocket launchers also pummeled the Iraqi batteries with counterfire.

By 6 A.M. the regiment had rolled twelve miles into Kuwait to Phase Line Elk, forward edge of the first minefield. Breaching an obstacle belt, the Marines had realized during weeks of rehearsal, was strikingly similar to crossing a beach in an amphibious landing. Speed and precise timing were vital to avoid stalling in the enemy kill sack. Keys's plan called for cutting six lanes the width of a tank, evenly spaced across a three-mile front. From left to right, the breach lanes were labeled Red 1, Red 2, Blue 3, Blue 4, Green 5, Green 6.

Five hundred engineers, commanded by Major Gary Wines, fanned out along the barbed wire fence the Iraqis had erected to avoid blundering into their own minefield. As three M-60 tanks fitted with mine plows or steel rakes rumbled to the head of each prospective lane, demolition experts moved into position with line-charge launchers. The most sophisticated launcher, the Mk-154, carried a trio of two-inch cables, each studded with a ton of plastic explosives and lashed to a rocket. When fired, the rocket streaked thirty-five yards down range, unspooling the cable and draping the line across the minefield. From the safety of his armored cab, a Marine driver triggered the explosives, blowing up the mines with "sympathetic detonation." The tanks then rolled into the channel to "proof" the lane by pushing away any remaining mines with plows and rakes.

In Red 2 the breach went smoothly, although the Marines quickly discovered that half the line charges would not "command detonate," requiring the driver to scurry from his cab and manually prime the cable with a blasting cap. In Red 1, at 6:35 A.M., a Fox chemical weapons vehicle detected mustard gas and sarin nerve agent; a second detector appeared to confirm the discovery. Marines in the lane requested that Red 1 be shut down. The regimental commander, Colonel Larry Livingston, instead ordered the men to keep moving. The report soon proved to be a false alarm but not before hundreds of Marines had scrambled into MOPP 4, the complete ensemble of chemical protective gear that included mask, gloves, hood, and bulky overshoes. By 6:55 A.M., Red 2 had cleared the first mine belt, with Red 1 soon to follow.

On the right flank, however, breaching slowed to a near standstill. Confusion swept the battlefield: men bellowing, artillery rounds bursting, line charges and mines exploding. Gary Wines arrived at Blue 4 to discover two tanks disabled by British bar mines, rectangular devices apparently confiscated from Kuwaiti stocks after the August invasion. One M-60, its plow wrapped around the tracks, had to be towed from the lane. Wines managed to guide the other out with frantic hand signals.

Moving to Green 5 and 6, he found several unpleasant surprises. Not only was the field a hundred meters wider — nearly double the width at Red 1 — but the Iraqis had implanted a baffling number of mines, two to three times the expected density. Three tanks stood stranded in the lanes, with either their tracks or plows shattered. One line charge arced across an overhead power line; when it detonated, the blast failed to sever the line. Another charge snarled around the antenna of a tank at Green 5 and a ton of plastic explosive flopped across the turret. "Don't fire!" the frightened crew pleaded. "Don't fire!"

Slowly, the Marines bulled their way through. Iraqi artillery dwindled to inconsequence, as did small arms fire from enemy infantrymen. By midmorning, all six lanes had cleared the first belt. The Marines pushed toward the second minefield two miles beyond, and again the right flank proved troublesome, with Green 5 and 6 reduced to only two tanks and a pair of line-charge launchers of an original twenty-three.

As Larry Livingston's battalions on the left and center punched across the second belt, hundreds of Iraqis from the 7th and 14th Infantry divisions threw aside their rifles and surrendered. By early afternoon the regiment had reached Phase Line Red, six miles beyond the mines, and Tiger Brigade began its passage. Mines had claimed eleven vehicles, including nine tanks, from 6th Marines; fourteen men had been wounded, none fatally.

Keys and Ron Richard also crossed the barriers to Phase Line Red, where a captured Iraqi colonel was brought before them. Unlike the Americans, now grimy with oil smoke, the kneeling Iraqi was immaculate, his uniform crisp and tailored, his shoes polished, and, most improbably, his nails manicured. "He looks a helluva lot better than I do," Keys muttered.

The Americans pulled the enemy commander to his feet, removed his handcuffs and blindfold, and handed him a carton of juice. Richard, wondering how to effect a formal surrender, said through an interpreter, "This is Major General William Keys, call sign Pit Bull, commanding general of the 2nd Marine Division."

The Iraqi nodded. "I present myself to you this afternoon."

"We believe we hit a chemical mine in your area," Pit Bull growled.

The enemy colonel, clearly flustered, shook his head. "No! Absolutely not! We have no plans for chemicals."

"Are there chemicals?" Keys demanded.

"No, we would never use chemicals on the Americans."

As the Iraqi dropped to his hands and knees and began sketching the disposition of his forces, Richard interrogated the colonel's operations officer, who had been captured at the same time. "We're very disappointed today," the officer confided. "This is not going well for us. The shock of you coming through our flank disoriented me. I could not make good decisions." The Marines, he added, were known to the Iraqis as "the angels of death," a term Richard found immensely pleasing.

*

Fifteen miles south, at the bottom of the Kuwaiti bootheel, the 1st Marine Division found the Iraqi defenders equally dispirited. On February 23, Task Forces Grizzly and Taro had penetrated the first minefield to secure, respectively, the division's west and east flanks. Shortly before dawn on the 24th, Task Force Ripper began its breach in the center of the division line. Like the 6th Marines, Ripper soon learned that the finicky line charges often failed to detonate without manual priming. Yet enemy artillery proved no more accurate or enduring than it was against the Marines in the north; of forty-two Iraqi guns detected by the Marines' counterfire radar, twenty-six were hit immediately with artillery fire and sixteen were struck from the air.

At 6:44 A.M. Ripper's lead battalion fired green and white flares to signal the completion of a lane through the first belt. Another opened twenty minutes later, despite the loss of an M-60 tank to a mine blast. As Ripper finished its six lanes, Task Force Papa Bear — named for the call sign of Marine Commandant Al Gray — moved into the minefield on Ripper's right and began carving another eight channels. By 10:30, all fourteen lanes had been cleared.

Again casualties were light. Of the ten Marines from Papa Bear hit with mortar fire, two were badly wounded. A lance corporal's left leg was shattered when he stepped on a mine while clearing a bunker with three other Marines. He was evacuated to Fleet Hospital Number 5, where he died two days later.

On the left flank, six Humvees and fifteen trucks from Task Force Grizzly were pushing into the second mine belt amid a fusillade of enemy rockets when machine gun and tank rounds — mistakenly fired by Ripper — raked the convoy. "We're taking fire!" yelled Sergeant Gordon Gregory. "Get out of your vehicles!" Two dozen Marines bolted from their cabs and, to Gregory's dismay, plunged into the minefield. He ran forward, pulled them back, and steered the men toward a depression in the desert a hundred yards to the northeast.

Several tank sabot rounds skipped across the ground between the vehicles. Gregory now spotted another Marine, Lance Corporal Christian J. Porter, still sitting behind the wheel of his truck. "Get out!" Gregory shouted. "Your truck's a target!" Porter opened the door and put one foot on the ground just as an American tank round streaked through the passenger window and across the cab, blowing a large hole through his chest. Gregory shrouded the dead man with a poncho. Before Marine officers managed to impose a cease-fire, Ripper had fired fifty-five tank rounds at the convoy, destroying two trucks and an amphibious assault vehicle. Only poor gunnery prevented more friendly fire casualties.

Shortly after noon, despite more salvos of enemy artillery, Ripper reported its lead forces across the second belt. For a few precarious moments commanders feared that a deep rumbling on their right flank heralded an armored counterattack from the Al Burqan oil field; the Marines quickly realized, however, that the roar came not from enemy tanks but from the inferno of sabotaged wells. Encumbered now with prisoners, Ripper began wheeling west toward Al Jaber Air Base. Papa Bear angled northeast on a trajectory toward Kuwait City International Airport.

From his command post south of the border, Boomer studied the laconic battle reports from his field commanders. "Progressing smoothly — timely manner," 2nd Division noted. "Things going well," added 1st Division. Thousands of additional Marines poured through the breaches behind the combat spearheads. No enemy chemical weapons had been confirmed, nor had the dreaded onslaught of heavy artillery materialized. Marine intelligence had predicted that Iraqi troops would counterattack within four hours after the American assault began, but thus far the enemy seemed too stunned to react coherently.

"None of our fears materialized," Boomer told a reporter. Iraqi soldiers — at least the ragged conscripts consigned to defend the bootheel — were unwilling to die in the cause of Kuwait. "They never were that good," the Marine general added. "We made them into something they weren't."

## Objective Rochambeau, Iraq

More than two hundred miles west of the Marines, the left flank of Schwarzkopf's forces found the Iraqis even less prepared than their comrades in the Kuwaiti bootheel to contest the allied offensive. Amid solemn invocations of Lafayette, Normandy, and two centuries of Franco-American solidarity, the French had pressed forward early Sunday morning with a battle order of succinct élan: *"Attaquez!"* Scouts from the 6th Armored Division pushed through a light drizzle and sporadic Iraqi artillery fire, followed by a 13,000-man force of dragoons, marines, legionnaires, and American paratroopers. Swept up in the fraternal spirit, the 2nd Brigade of the 82nd Airborne Division dispatched a message to their Gallic brothers-in-arms: *"Côté à côté soldats français et américain nous écrirons une page d'histoire."* Side by side, French and American soldiers will write a page of history.

It would be written slowly. Under pressure from Paris to minimize combat casualties, the French plodded forward with a deliberation that soon irked their American *confrères* in XVIII Corps headquarters. Within hours the corps operations officer, Colonel Frank Akers, was on the radio to his French counterpart with a less convivial message: "François, get your ass moving. Why are you guys taking so long?"

As the anchor of Schwarzkopf's western wing, the French had been given two objectives. Objective Rochambeau, named for the commander of French forces in the American Revolution, was a patch of high ground crowned with a communications complex thirty-five miles beyond the border; thirty miles farther north lay Objective White, the town of As Salman, where the Iraqi 45th Infantry Division had established its headquarters near an airfield. (As Salman had also served as a launch pad for Scuds fired at Riyadh.) By capturing Main Supply Route Texas, a paved highway stretching from the border to As Salman, the French would block any Iraqi reinforcements from the west and open a logistics line for American forces pushing into the Euphrates Valley.

This battle plan had evolved after negotiations between Washington, Paris, and Riyadh, as the allies sought to balance Schwarzkopf's military needs, French military capabilities, and expeditionary politics. The French force, known as the Daguet (young deer) Division, had originally been under Saudi control. Before Christmas, however, the French commander, Brigadier General Jean-Charles Mouscardes, privately asked Schwarzkopf for a new arrangement. "For God's sake," Mouscardes pleaded, "you've got to get me out from under the command of the Saudis." Without disclosing Mouscardes's true concerns — Saudi military incompetence — Schwarzkopf deftly convinced General Khalid that the French provided precisely the proper force needed to protect the allied left wing. On January 20, the CINC formally shifted the Daguet to the control of Gary Luck's XVIII Corps.

Luck in turn assigned one brigade from the 82nd Airborne to accompany the Daguet *côté à côté*. Under the corps's attack scheme, the rest of the 82nd would follow behind the French attack to sweep up pockets of enemy resistance before pressing northeast beyond As Salman toward the Euphrates. Luck also ordered a sizable portion of his twenty-two artillery battalions to support the French. Instead of trailing several miles behind the infantry battalions, as dictated by normal gunnery doctrine, the American artillery would be tucked up within a kilometer of the advancing forces to be closer to enemy artillery positions and offset the longer range of Iraqi guns.

That the French would fight with competence and esprit the Americans had no doubt, even after Mouscardes took ill in early February and was evacuated to France, to be replaced on February 9 by Brigadier General Bernard Janvier. ("I cried with sadness and rage against this dirty carcass that prevented me from carrying out to the end of the marvelous adventure," Mouscardes later commented.) Many French soldiers had seen action in Chad or Lebanon. Carrying particular cachet were two regiments from the French Foreign Legion. Formed in 1831, the Legion brought to the desert a reputation for discipline, courage, and ornery independence embodied in the motto *Legio Patria Nostra*, "The Legion Is Our Country." According to one famous story, a new recruit, on being asked his civilian profession before joining the Legion, replied, "I was a general." Such panache impressed the American paratroopers.

Whether the French attack would move at the blitzkrieg pace favored by the Americans, however, had remained an open issue. In contrast to the American obsession with detailed planning, the French seemed alarmingly laissez-faire and disinclined to commit themselves to a battle plan. Major General Jim Johnson, commander of the 82nd Airborne, warned Luck that the Daguet would take two days to capture As Salman, exposing the 101st Airborne Division to a possible Iraqi counterattack near the Euphrates Valley. In mid-February, Johnson had offered to seize the town with his division in twenty-four hours. The proposal was rejected, in part because it would have left the Daguet without a role.

If the French knew of these misgivings, they maintained a diplomatic silence. Like other allied forces along the border, the Daguet Division had seized several miles of Iraqi territory before G-Day, pushing far enough north on February 23 to surmount a steep sand escarpment that had been abandoned by the Iraqis several days earlier. When the attack began on the 24th, Janvier's forces moved on two axes. On the left, across the open desert, the forces included both Legion regiments, the 1st Hussars Airborne Regiment, and a commando unit known by the unfortunate French acronym CRAP. On the right, bracketing MSR Texas, the units included the 2nd Brigade of the 82nd, the French 4th Dragoon Regiment, and sundry engineers and artillerymen.

"Everything," Janvier later explained, "worked against our movement: the spreading out of units, a dense night, violent sandstorms, and rain . . . The difficult terrain resulted in multiple flat tires." French tactics resembled those used by the Americans in Vietnam — a cautious advance preceded by intensive artillery shelling and air strikes. Iraqi artillery proved fitful and ineffective; American radar systems so swiftly pinpointed the enemy guns that counterbattery salvos often returned fire even before the incoming rounds landed. Contrary to the allies' worst fears, the Iraqis used no chemical weapons and had none stockpiled.

The French advance had traveled only six miles when enemy prisoners began swarming from their bunkers with a great fluttering of white flags. Many were emaciated and reported that they had been reduced to five tablespoons of rice and six ounces of water a day. Their numbers soon grew to more than two thousand, further miring the attack.

By midafternoon the French reported the capture of Rochambeau. In truth, as Luck and Akers soon discovered, they had reached the southern edge of the objective. To the Americans' consternation, French tents began popping up across the desert despite more than an hour of daylight left. Akers's pleas for greater speed were gently rebuffed. Janvier had decided to wait until morning before pressing toward As Salman. "To avoid mistakes," he told his staff, "it's better to delay."

Gary Luck, the XVIII Airborne commander, planned to sever the Euphrates with three sharp chops. First, a brigade from the 101st Airborne on G-Day would construct a forward logistics base at Objective Cobra, ninety miles into Iraq and twenty miles east of As Salman. Then, on G + 1, another brigade of Screaming Eagles would arrive by helicopter in the river valley between Samawah and Nasiriyah, "putting a cork in the bottle" by cutting Highway 8. Finally, the armored forces of Barry McCaffrey's 24th Division would sweep into the valley farther east, between the Talil and Jalibah air bases, before wheeling toward Basrah.

At 3 A.M. on Sunday, February 24, two Apaches and two Blackhawks had flown toward the border with the intention of planting radio beacons — like electronic bread crumbs — to guide the four hundred helicopters assigned to descend on Cobra a few hours later. Leading the small advance flock was Lieutenant Colonel Dick Cody, who had commanded the attack against the Iraqi early-warning sites more than five weeks before, on the opening night of the war. No sooner had Cody crossed into Iraq, however, than his four helicopters plowed into a dense fog bank that forced the pilots to slow to a crawl. A few minutes later he heard an urgent call from one of the OH-58D helicopters lingering near the border to relay radio messages back to division headquarters.

"No Mercy Six," reported the pilot, using Cody's call sign, "I've just lost my wingman and I'm IMC." IMC — inadvertent meteorological condition — meant the helicopter was lost in the fog. Cody immediately turned south again. Four miles from the last known position of the missing helicopter, he spotted the burning carcass of an OH-58. He set down near the wreckage and ran toward the flames, expecting to find two dead pilots. Both crewmen emerged from a foxhole, dazed but alive. They had clipped the ground and crashed while feeling their way through the fog, and had pulled each other to safety before the helicopter exploded.

Cody hurried back to his Apache and radioed the division headquarters. "My recommendation is that you're going to lose a lot of aircraft if you don't delay this thing." The division commander, Binford Peay, reluctantly agreed.

For more than two hours they waited — pacing, fidgeting, cursing the fog. Peay's 1st Brigade, commanded by Colonel Tom Hill, sat by the helicopters at fifteen separate assembly sites. At 7 A.M. Cody reported to Hill, "There's still fog midway out. It may cause you some problems." Hill called Peay for the third time that morning and recommended that they take the risk anyway. At 7:27 the first wave of sixty-six Blackhawks, ten Hueys, and thirty twin-rotored Chinooks headed north at 120 knots, flying only ten feet above the ground. As they crossed the berm, Hill — apparently concluding like so many senior officers that history demanded an epigram to mark the occasion — radioed back to his operations center: "Let it be known that I just crossed into Iraq. *Jihad*, motherfucker!"

Cody had moved forward to MSR Virginia, just north of Cobra, where an Iraqi battalion opened fire from a concealed bunker complex. Antiaircraft rounds the size of flaming tennis balls swept past the cockpit. He returned fire with his 30mm cannon and pulled south while his wingman whipped a Hellfire missile into an enemy pillbox. Hill's vanguard, Alpha Company of the 1st battalion, 327th Infantry Regiment, spilled from their helicopters shortly after 8 A.M. about a mile south of the road. As F-16s and A-10s hammered the Iraqi fortifications, Alpha Company wheeled their 105mm howitzers from the Chinooks and began lobbing artillery shells. Three hundred enemy soldiers soon surrendered, and all resistance ceased.

Wave upon wave of helicopters swept into Cobra, a flat, sere wasteland that resembled the moon more than any earthly landscape. Traffic controllers popped canisters of colored smoke to direct pilots to the proper landing zone. Humvees with TOW missiles rolled down the Chinook tail ramps and raced to defensive positions on the perimeter. The first consignment of more than a million gallons of fuel arrived in huge rubber sacks slung beneath the helicopter bellies. Pallets of ammunition, food, and water soon littered the desert. Though the fog had burned off, churning rotor blades soon raised a brown cloud so thick that Hill ordered all pilots to set down for an hour until the dust subsided. But the flag had been planted, the colony established. Beyond the northern horizon, the Euphrates Valley beckoned.

## Swayjghazi, Iraq

Far removed from the great sweep of armies in the south, two Special Forces A-Teams waited in their burrows above the Euphrates, watching for any sign of a counterattack from Baghdad. One eight-man squad from 5th Group's 1st Battalion had been inserted by helicopter the previous night near the tiny village of Swayjghazi. The team leader, a stocky, tobacco-chewing chief warrant officer named R. F. Balwanz, herded his team into a shallow drainage canal, where they built a pair of hide sites three hundred yards west of Highway 7, the two-lane blacktop running from the Iraqi capital to Nasiriyah.

With the farmland around Swayjghazi lying fallow for the winter, American intelligence had concluded that local peasants would have little reason to venture out of their village and into the fields. This assumption, as Balwanz immediately discovered, was absurd. No sooner had the team finished sterilizing its holes at sunrise than dozens of Iraqis began drifting across the countryside. Peering through a narrow slit in the camouflage, Balwanz counted at least fifty people: women gathering firewood, children playing, men herding goats and sheep.

Soon the Americans heard the singsong of young children capering along the canal bank. The voices suddenly hushed. Evidently sensing intruders in the ditch, the children ran shrieking toward the village. Several sharpshooters burst from the holes, training their weapons on the fleeing youngsters. For a terrible moment the Americans wondered whether to fire. "Do we shoot them, chief?" a soldier asked. "Naw," Balwanz replied, "we're not going to do that." Instead, the team slogged east in the muddy ditch for four hundred yards and set up a defensive perimeter.

A few minutes later the children returned, this time accompanied by a young man wearing a robe and sandals. With nowhere to hide, the Americans waited in a tight circle, weapons at the ready. The Iraqi's eyes widened at the sight of eight heavily armed soldiers crouched in his field. "Peace be with you," Balwanz called in Arabic. The man said nothing; he turned and hurried back toward the village. Perhaps, Balwanz told himself, the civilians would ignore them.

It was not to be. Thirty men carrying rifles soon emerged from Swayjghazi, fanning out across the fields. A moment later four trucks, a Land-Rover, and a bus rolled up from the south. A hundred and fifty Iraqi soldiers spilled onto Highway 7, apparently summoned from a communications compound three miles away. Balwanz radioed XVIII Airborne Corps headquarters 150 miles to the south. "Contact is imminent. We're going to have a fight here. We need to be picked up and we need close air support."

Two enemy platoons funneled into the canal; two others circled toward the American flanks. Curious women and children drifted from the village to watch. Balwanz ordered all rucksacks and classified communications equipment tossed into a pile, salvaging only the weapons and a single satellite radio. Setting a one-minute fuse in a block of C4 plastic explosive, the team again scurried east to a new fighting position where the ditch made a ninety-degree bend; the equipment blew up behind them just as Iraqi scouts reached the pile.

The crack of rifle fire swept across the field. Bullets sang overhead or chewed into the dirt embankment. The Americans fired back with M-16s and a pair of 203mm grenade launchers. "Shoot only the soldiers, no civilians," Balwanz commanded. "Conserve your ammunition." With a warbling war cry the Iraqi troops dashed forward, dived for cover among the furrows, then ran forward again. Enemy bodies soon littered the field, but the Americans were still outnumbered fifteen to one. A sense of doom stole over the team as the enemy noose tightened. Balwanz turned to see two of his men glumly waving to each other in a last, sad gesture of farewell.

Just when their predicament seemed desperate, the roar of Air Force F-16s washed over the battlefield. The first payload of cluster bombs shattered the highway, where fifty Iraqi reinforcements and a dozen vehicles had congregated. Using a small survival radio and a flashing mirror to signal his position, Balwanz directed other strikes to within two hundred yards of his flanks. The fields blossomed with fire around them as the team huddled in the ditch. "One shot, one target," Balwanz warned, glancing at the dwindling stockpile of ammunition. "Nobody fires on automatic. Take your time, pick out your targets. Let's keep them at bay."

With the Air Force overhead, the tide turned. Another Iraqi charge was repulsed with more cluster bombs and a bold counterattack back down the canal by Balwanz and one of his sergeants. Several more hours would pass before two search-and-rescue helicopters dared venture into the firefight. Dispirited and in disarray, an estimated 150 of their comrades now dead, the Iraqis failed to spot Balwanz and his men as they slipped in pairs from the ditch to an earthen berm three hundred yards away. Then, from the south, the Americans heard the throb of rotor blades. Two Blackhawks touched down almost on top of them. The men flung themselves into the open bays. Within fifteen seconds, they were gone.

Seventy miles to the west, another Special Forces A-Team had been inserted by helicopter shortly before midnight on the 23rd. Splitting into a pair of three-man squads, the team took up positions fifteen miles apart above the Euphrates River town of Samawah. The southernmost squad had been compromised almost immediately. Bolting across the desert, the men would travel southwest for three days before being picked up at a rendezvous point.

The other trio — Master Sergeant Jeffrey Sims, Sergeant First Class Ronald Torbett, and Staff Sergeant Roy Tabron — had hiked five miles from the helicopter drop-off point before building a hide site in a ditch three hundred yards north of the village of Oawam al Hamzah. Throughout the morning of the 24th the soldiers kept vigil over Highway 8, assuring XVIII Corps that no Iraqi armored units had yet rolled toward the French forces at Rochambeau or the Screaming Eagles at Cobra. Shortly after noon, however, Sims spotted a pair of slender figures wandering from the village toward the drainage ditch. Squatting in their lair, the Americans watched nervously as a young Iraqi girl and an old man strolled toward them across the furrowed fields.

Sims never knew precisely what betrayed him. Maybe it was the small satellite antenna concealed on the burrow roof; perhaps a glint of metal or subtle changes in the hue of the sand drying around the burrow. Whatever the reason, the girl stopped abruptly and pointed at the hide site. The old man shuffled forward, his eyes searching the ditch. Then he gathered his robes and sat on the ground only inches from the hole.

The Americans flung back the heavy roof and burst into the open with drawn pistols. Tabron grabbed the old man, who tugged open his coat to show he was unarmed. "We are your friends," Sims said in Arabic. The Iraqi, clearly skeptical, announced that Iraqi soldiers were garrisoned nearby. No more eager to kill civilians than Balwanz had been, Sims ordered them released. The terrified girl and the elderly man hurried back into town; the Americans moved five hundred yards down the ditch in search of a better fighting position.

Fifteen minutes later rifle fire erupted from houses on the outskirts of Oawam al Hamzah. A bus clattered to the edge of the field and fifty Iraqi soldiers spilled out the door. As the enemy leader flashed hand signals to his dispersing men, Sims shot him dead. To the north, Torbett saw a woman driving a tractor with soldiers in tow on a flatbed trailer.

Squinting through the sniper scope of his M-16, he squeezed off a single shot at nine hundred yards and, in an astonishing feat of marksmanship, sent an Iraqi flying from the trailer.

Sims radioed XVIII Corps with a request for immediate extraction. To protect his flanks, he ordered Torbett to move two hundred yards to the left; Tabron shifted a similar distance to the right. Two more buses pulled up and a hundred soldiers poured onto the battlefield, weapons blazing.

With only three hundred rounds of ammunition apiece, the three Americans fired on automatic only when the Iraqis attempted to overrun them. Sims felt himself dipping and soaring through wild mood swings, from depression as the enemy seemed to gain the upper hand, to elation whenever gunfire drove them back. He took stock of his arsenal: an M-16, five grenades, two claymore mines, fifteen pin flares. They would never be captured alive, Sims resolved; he preferred death to the prospect of being paraded like a circus beast on Iraqi television.

Ninety minutes into the firefight, a single F-16 streaked through the thick clouds overhead. Sims fired fourteen flares — one red dart after another flashing skyward — before the pilot saw them. A pair of cluster bombs boiled through the enemy on the right flank, five hundred yards from Tabron. The pilot made several more passes, walking his bombs ever closer, before breaking away to refuel. When he saw a sudden movement on his left, Sims whirled around and raised his rifle. "Ron? Is that you?" he barked. "You'd better tell me something or I'm going to blow you away." "It's me!" Torbett yelled, waving an arm. Sims lowered his weapon.

Sims's radio distress call had been relayed to the flight line at Rafha, 170 miles south, where a Blackhawk crew commanded by Chief Warrant Officer Jim Crisafulli had been resting after a long night spent inserting SF teams behind the lines. Within four minutes Crisafulli had cranked his engines; fifteen minutes later the helicopter lifted off with Crisafulli, his co-pilot, a pair of Special Forces soldiers, and two crewmen manning the Blackhawk's twin door guns.

Having dodged heavy gunfire on the previous night's mission, Crisafulli initially flew a zigzag course toward the Euphrates Valley at 160 miles an hour, keeping the helicopter five to ten feet above the desert. But Sims's pleas, audible over a high-frequency channel, grew ever more desperate. "If they can't get here in twenty minutes," he finally warned, "they may as well not come." Still forty-five minutes away, Crisafulli abandoned the evasive flight plan and pointed the Blackhawk's nose straight for Oawam al Hamzah.

The town soon loomed ahead, a clutter of boxy houses bracketed by

Highway 8 and the muddy ribbon of the Shatt al Hillah. Crisafulli believed Sims to be west and north of town, beyond a skein of electrical cables stretching across a field from the highway. "We're going to have to go under the power lines," he warned the crew. But as the helicopter swooped between two stanchions, another set of wires — much lower than the first — suddenly loomed two hundred yards ahead. Crisafulli yanked back on his control stick and the Blackhawk soared almost straight up, plastering the men in back against the cabin ceiling.

After looping over the wires, Crisafulli again dived for the ground, but the Iraqis had seen them. "We're taking fire!" the crew chief yelled. Hundreds of muzzle flashes winked from the irrigation ditches and narrow dikes. Above the screaming engines and throb of the rotor blades Crisafulli heard the hammer of his door guns while the helicopter danced amid the enemy tracers.

Jeff Sims, watching the Blackhawk buck and heave along the power lines, fired his last pin flare. From the corner of his eye Crisafulli spied the red phosphorous rocket streaking five hundred feet into the air behind the helicopter and assumed the Iraqis had launched a surface-to-air missile. He pitched the Blackhawk into a sixty-degree dive at 140 knots before the crew chief yelled, "No, no, it's a flare! I see them!" The trio was south of the power lines after all. Again Crisafulli lurched over a skein of wires — pinning his crew to the ceiling once more — and swooped beneath a second set.

"One o'clock!" the crew chief shouted, pointing at three figures sprinting across the desert. Crisafulli kicked the helicopter into a 180-degree turn and slammed into the ground less than fifty yards from Sims. The two SF soldiers leaped from the bay, firing at the Iraqis with their M-16s while the door gunners swiveled from side to side, machine guns roaring. Enemy rounds pinged against the fuselage and shattered the cockpit windows. Plexiglass sprayed across the flight controls. More bullets smashed the Blackhawk's electronic jamming pod and ricocheted off two of the rotors, ripping holes in the honeycombed blades.

Lugging the radios as bullets nipped at his heels, Sims waited until Tabron and Torbett threw themselves through the Blackhawk's door. Fifty yards beyond the tail rotor he spotted several enemy soldiers rushing toward the helicopter. "You've got three Iraqis right behind you!" he shouted. The door gunner wheeled around and fired a long savage burst, slicing all three men in half.

Sims leaped aboard. Crisafulli lifted off. Wind whistled through the shattered cockpit as a final burst of tracers whizzed past. The Blackhawk pelted across the desert. Oawam al Hamzah receded from view, then dropped below the horizon. Beneath the singing of the battered rotors, nine men shrieked with joy, astonished to be alive.

## Riyadh

In Schwarzkopf's war room, success on the allied flanks was greeted with a mixture of jubilation and disbelief as the tiny symbols representing the Daguet, 101st, 1st Marine, and 2nd Marine divisions inched up the map. Arab units also were moving north: on the coastal road east of the Marines, troops of Joint Forces Command East had opened six lanes into Kuwait at 8 A.M. against token opposition from Iraqi divisions that had been all but obliterated by weeks of air strikes and naval gunfire.

Three hours into the Marines' attack, CENTCOM realized that Boomer was approximately eight hours ahead of schedule. Casualty reports from the French and Americans seemed so improbably low that Colonel Tony Gain, the CINC's current operations officer, doubted their accuracy and wondered whether units in the field were withholding bad news. Schwarzkopf paced before the map boards, his elation tempered only by frustration at being in a bunker hundreds of miles from the fighting.

"When you are winning a war," Churchill once observed, "almost everything that happens can be claimed to be right and wise." No battle plan ever appeared righter or wiser than the one now unfolding. Yet in success lay hazard. Every man in the war room could see that the Marines were rapidly establishing a salient that exposed their left flank to counterattack, particularly from the Iraqi 10th and 6th Armored divisions positioned west of Kuwait Bay. In retrospect, the likelihood of Iraqi commanders summoning the tactical wherewithal to exploit such an opportunity would seem ludicrous; on the morning of February 24, however, the threat looked very real.

Schwarzkopf's plan called for the Marine flank to be shielded by Joint Forces Command North, spearheaded by 35,000 Egyptians on the left and a Saudi-Kuwaiti-Syrian force on the right. But the JFC-N attack was not scheduled to begin until the next day, G + 1. Moreover, the Egyptians could be notoriously slow. The Syrians, after first refusing to cross the border, had agreed to attack but only as a rear guard; and Saudi military power struck no one with terror. The CINC's deputy intelligence officer, Colonel Chuck Thomas, pulled together a team of analysts who concluded, on the basis of sketchy reports from the front, that the Iraqis "could not coordinate a multibrigade counterattack — that at best they can put a brigade together."

Boomer shared the CINC's concern about his flank. "We're moving faster than we thought we would, particularly this first day," he told Schwarzkopf on the phone.

"What about moving the main attack up?" the CINC asked, referring to VII Corps and JFC-N. "Would that help?"

"Yes," Boomer replied, "I think that would be a good move."

Schwarzkopf, swiveling in his leather chair, studied the attack timeline printed on a small scrap of paper. Both VII Corps and the heavy armor from XVIII Corps — McCaffrey's 24th and the 3rd Armored Cavalry Regiment — were scheduled to launch at G + 26 hours, 6 A.M. on February 25. Months of intricate planning had gone into this plan. The twenty-six-hour delay in unleashing the main attack was a carefully calculated effort to allow Iraqi corps commanders and Baghdad time to conclude — given the havoc wreaked on the enemy's communications system — that the allied thrust focused on Kuwait. A thousand logistical nuances also were predicated on a staggered attack sequence. Could the Army and the Arabs go early without hopelessly snarling their fuel and ammunition trains?

Sitting on Schwarzkopf's right was Calvin Waller, who only a day before had reverted to his role as deputy CINC. To Waller's chagrin a pale and feeble John Yeosock had suddenly reappeared in Riyadh, looking, in the words of one incredulous officer, "more like he belonged in the morgue than in a war room." His defective gall bladder gone, cigar clamped between his jaws, Yeosock had dragged himself from a hospital bed in Germany and pronounced himself again fit for duty as ARCENT commander. After failing to persuade Yeosock to assume some of the DCINC's responsibilities instead, Schwarzkopf agreed. "Cal," he told Waller, "I'm going to tell you what I've heard you tell other people: you've got to dance with the girl you brought to the ball." Waller nodded, blinking back his disappointment. "Sir," he replied, "not a problem."

Waller's eight days in command of ARCENT's Third Army had convinced him that the troops were ready to fight regardless of the timetable. A commander, he believed, must be prepared to exploit success. Allied forces driving across Europe during World War II had, in less than a hundred days, captured objectives that conservative war planners predicted wouldn't be reached for a year. "To hell with the plan," Waller urged. "We can't let the Marines get so far ahead of everybody else that they could be enveloped or attacked from the flank."

Schwarzkopf concurred. Even the fragmentary reports from the front, including radio intercepts from the Iraqi 3rd and 4th corps defending the Kuwaiti bootheel, suggested the wholesale collapse of the enemy. If they moved quickly, the allies could knock the enemy completely off balance by forcing them to fight while retreating.

The CINC phoned Yeosock at ARCENT headquarters at 9:15 A.M. "It looks as if the defenses are crumbling. Can you go early? If so, how much time do you need to get ready?" Steve Arnold, the ARCENT operations officer, immediately called both corps headquarters. Gary Luck declared himself ready to launch the 24th Division with two hours' warning. John Landry, the VII Corps deputy commander, relayed the query to Fred Franks, who had already moved toward the border. Franks then called Yeosock directly: "I believe we can do it, John."

"I got an answer from XVIII Corps that they can attack on two hours' notice," Yeosock explained. "Okay," Franks said. "Sounds about right to me, too. Go ahead, but let me check with Tom Rhame and the Brits and my other subordinate commanders." In his diary Steve Arnold wrote: "Iraqi defenses may be falling apart. [We] may go into pursuit mode . . . French have taken over 3500 prisoners. No casualties yet in Third Army. Great start!"

The enthusiasm expressed by Franks and Arnold masked several nagging worries. Both VII Corps and ARCENT had drafted a "go-early option" many weeks before in the event that Bush ordered a sudden attack. But as that prospect faded, planning had focused almost exclusively on G + 26 hours. Franks had anticipated a full twelve hours of daylight for Rhame's 1st Infantry Division to open lanes through the Saddam Line for the British 1st Armoured; a night breach was complex, dangerous, and unrehearsed. Elaborate synchronization tables had also been constructed to keep the corps's 146,000 troops massed together — thus giving Franks his armored "fist" against the Republican Guard. Accelerating the attack could imperil that scheme, particularly if the left wing led by the 2nd Armored Cavalry dashed across the desert and left behind the forces in the breach.

Franks called his commanders, one by one, and found each chafing for a fight. Ron Griffith of 1st Armored and Don Holder of the 2nd Cavalry pronounced themselves ready with two hours' notice. Paul Funk, commander of the 3rd Armored, said he could launch at noon. Rhame had already radioed corps headquarters, asking for permission to begin the breach at 1 P.M. The British also declared themselves game, though the accelerated attack meant their Challenger tanks would have to move sixty miles to the border under their own power, since there was no time to load them onto heavy-equipment transporters.

The Army awaited Schwarzkopf's decision while the CINC consulted with General Khalid, commander of the Arab forces, to be certain JFC-N also could launch early. Shortly before noon, the Kuwaiti resistance reported that Iraqi forces had blown up the desalination plant in Kuwait City, which suggested to Schwarzkopf that the occupiers intended to abandon the capital. At 12:30 P.M. — shortly after Task Force Ripper punched through the second minefield in Kuwait — the CINC passed word through the chain of command: the main attack would begin fifteen hours ahead of schedule, at 3 P.M.

Some in the war room harbored deep skepticism. Tony Gain recalled how difficult it had been to accelerate a simulated attack in training by just two or three hours; launching a huge corps into combat fifteen hours early seemed impossible, even daft. "My God," Gain confided to another officer, "either the CINC's brilliant, or he's about to throw the Army into a gross failure because they won't be able to execute his order."

If Schwarzkopf himself had any reservations, they remained hidden behind an air of confidence. Later in the afternoon he left the war room for a ten-minute session with reporters in the press room of the Hyatt Regency Hotel. A slight smile played across his face as he spoke of "remarkably light" casualties.

"The war is not over yet," the CINC cautioned, "but so far we're delighted with the progress of the campaign . . . We're going to go around, over, through, on top, underneath, and any other way it takes to beat them."

## Phase Line Colorado, Iraq

At 2:30 P.M. the prelude to VII Corps's attack began with an artillery barrage as devastating, in intensity if not duration, as any ever unleashed in combat. Thirteen tube battalions and ten MLRS rocket batteries, parked almost wheel to wheel, dumped more than eleven thousand rounds down range in thirty minutes. Six hundred thousand MLRS bomblets raked a target area measuring only twenty by forty kilometers. (By contrast, Napoleon's famed gunners fired twenty thousand rounds during the entire battle of Waterloo.) The bombardment, despite being scaled back after Schwarzkopf accelerated the attack, still achieved a tempo more than twice that of the British cannonade at the Somme in July 1916, when 1.5 million shells fell in seven days. The effect in the Iraqi trenches — where each 155mm round exploded with a lethal bursting radius of fifty meters — can scarcely be imagined.

While awaiting CENTCOM's final instructions on when to attack,

Fred Franks had toyed with a last-minute modification to his maneuver plan. Don Holder, already fifteen miles deep in Iraq with the 2nd Cavalry, received a message from corps headquarters suggesting that his regiment and the 3rd Armored Division veer sharply to the east to strike Iraqi reserve units in the flank — this instead of driving directly north toward the Republican Guard. (Whether the brainstorm had come from the corps commander or his staff was unclear.) Holder immediately flew back to Franks's headquarters below the Saudi berm, where he found the equally alarmed Paul Funk of the 3rd Armored.

"We shouldn't change the plan with everybody prepared to drive off to the north," Holder urged Franks. "We don't want to start splitting up the corps now." Beyond jeopardizing the corps commander's desire to mass his combat power against the Guard, the proposal appeared to violate one of the most sacred dictums of warfare: never take counsel of your fears. Franks swiftly agreed. "Yeah, we're not going to do that," he said. "Just stay focused on the enemy." Holder flew back to the regiment. "Well," he told his staff with a grin, "we just won the first battle."

At 3 P.M. the artillery lifted. On the left wing Holder's cavalrymen cantered forward; behind them the two armored divisions lumbered toward the border. On the far right, a brigade from the 1st Cavalry Division punched up the Wadi al Batin with yet another feint intended to reinforce Iraqi convictions that the main attack would come through the dry riverbed.

The key to the initial corps assault, however, lay in the center with Operation Scorpion Danger — the sundering of the Saddam Line. Not since Normandy had the Army undertaken a major breach across a fortified defensive barrier. Franks had selected for the task the same unit chosen to assault Omaha Beach, the 1st Infantry Division. It was a good choice, both in terms of the division — the Big Red One, boasting battle streamers fairly won from the Marne to Tunisia to the Tet counteroffensive, had also served as Patton's spearhead across Sicily — and its commander. Major General Thomas G. Rhame, the pride of Winnfield, Louisiana, was noisy, profane, and relentless; in aggressive drive he had few peers. Admiring officers from the 2nd Cavalry referred to the 1st Division as the Wild Bunch, to Rhame as the Wild Thing.

Rhame's lead brigades pushed forward on a nine-mile front. The breach site had been chosen where satellite photographs indicated a seam between two Iraqi divisions: trenches of the 48th Division in the east were carved in a "lazy W" pattern with interlocking fields of fire; those of the 26th Division in the west were laid in straight lines without imagination or tactical aptitude. Unlike many American commanders, Rhame was pleased with the intelligence provided him before G-Day. Detailed templates — drawn from U-2, satellite, and Pioneer drone photography — showed the exact position of every large-caliber Iraqi weapon on maps that were updated daily in messages code-named Orient Classic. He estimated that the two divisions facing him had been reduced to no more than 60 percent of their prewar strength, and that 90 percent of their artillery had been destroyed. American doctrine called for an attacking force to be at least three times stronger than an entrenched defender; his own strength, Rhame calculated, was twelve times that of the enemy at the point of attack.

"I didn't come here to fight fair," he would proclaim after the war. "I came to put maximum destruction on this son of a bitch with as few American casualties as possible." In this Rhame succeeded. On the division's left the 1st Brigade began its attack with a thunderous volley of tank fire into the trench line beyond the minefield. Four enemy tanks atop a low bluff were destroyed with TOW missile shots from a range of three thousand yards.

Using plows, rakes, and steel rollers, soldiers began cutting eight lanes across the mines, mirroring a similar number begun by 2nd Brigade on the right. (Plagued during rehearsals with a 60 percent misfire rate from line-charge launchers, the Army had little use for the contraptions during the breach.) Iraqi resistance was feeble; a few dozen poorly aimed artillery rounds fell around the 1st Brigade. Many enemy soldiers were inexplicably facing west or southwest, offering the Americans their flank. Those who survived the artillery barrage were hardly in fighting trim.

During the three months spent studying the art of the breach, Army tacticians had concluded that the conventional practice of clearing enemy trenches with dismounted infantrymen was ill conceived. That the trenches had to be "sterilized" was undisputed: a hidden Iraqi sniper armed with a rocket-propelled grenade could wreak havoc on ammunition and fuel trucks. But both computer war games at Fort Leavenworth and mock battles fought in the Mojave Desert showed that breaching slowed to a crawl — and friendly casualties soared — if soldiers left their armored vehicles to root through enemy fortifications on foot. The British had come to similar conclusions, codified by Major General Rupert Smith in a simple adage: "Don't dismount your blokes."

Consequently, the 1st Division had devised techniques for entombing any Iraqi soldiers who failed to surrender promptly. An M9 armored bulldozer rolled along the trenches, collapsing the walls as Bradley fighting vehicles sprayed the enemy positions with machine gun and

25mm cannon fire. No psyops surrender appeals were broadcast for fear that such a move would slow the attack and give the Iraqis time to fire chemical artillery rounds. In several instances the Americans found a use for the temperamental line charges by firing them down the trench lines. The shallow trenches — typically no more than three feet wide and three feet deep — simplified the burial tactics, as did the baffling Iraqi practice of piling the spoil from the fortifications on the south side, where it could be pushed back into the ditches. ("Those dumb shits," Rhame had commented while studying reconnaissance photographs of the trenches, "why are they doing that?")

A hue and cry would arise after the war when the division's breaching techniques were disclosed, as if burying the enemy was less humane than eviscerating them with tank fire or eleven thousand artillery rounds. In truth, similar tactics had been used since the advent of armored warfare in World War I; against the Japanese, beginning with the bloody fight for Tulagi in the South Pacific, U.S. Marines had buried the enemy in their caves and bunkers whenever possible rather than dig them out. The tactic had been reviewed by a United Nations conference on conventional weaponry during the late 1970s and left unregulated as a "common, longstanding tactic entirely consistent with the law of war."

Twenty-nine minutes after entering the minefield, 1st Brigade broke through the Saddam Line; 2nd Brigade on the right punched across about the same time. Franks was worried enough about the entombment of enemy troops to call his assistant chief of staff, Colonel Mike Hawk. "Mike, I want you to get with the JAG [judge advocate general] and tell him that when the 1st ID made their penetration they led with plows and rakes, and some Iraqis were buried. I want to know whether that's any breach of the rules of war." Hawk conferred with the corps lawyers and reported back to Franks: because the division had been under fire, the tactic was not considered an improper use of force.

The Big Red One had sustained two casualties during the breach; one was a soldier killed by a mine. More than five hundred Iraqis were captured by day's end. A subsequent afteraction report, signed by Rhame and dated 14 March 1991, estimated that "some 150 enemy soldiers who chose to resist were plowed under."

## Washington, D.C.

Only the sketchiest reports of allied progress had reached the White House on Sunday morning, when the president climbed into his limousine for the short drive across Lafayette Square to Saint John's Episcopal Church. The motorcade rolled past the rearing equestrian statue of Andrew Jackson, gaily waving his cocked hat in a gesture of perpetual triumph. At 7 A.M. — it was 3 P.M. in Riyadh, the precise moment of VII Corps's attack — Bush walked through the green double doors and down the center aisle to pew fifty-four, where every chief executive since James Madison had worshiped. The president nodded greetings to Baker, Cheney, Quayle, and others from his war cabinet gathered for a brief prayer service.

A soft blush of blue, orange, and green light filtered through the stained glass depiction of the Last Supper behind the altar. Saint John's had long offered refuge to commanders-in-chief seeking spiritual solace. Built in 1816 after much of Washington was burned by the British, the church had been designed by the architect Benjamin Latrobe, who boasted, "I have just completed a church that made many Washingtonians religious who had not been religious before." Lincoln had prayed here during the Civil War, as had Woodrow Wilson in World War I and Lyndon Johnson during Vietnam.

"Ever since the war in the Persian Gulf began in January, a hymn has been running through my mind," the rector, Dr. John C. Harper, told his distinguished flock. "The hymn is a prayer for peace, and it begins, 'O God of love, O King of peace, make wars throughout the world to cease.'

"It is comparatively easy, I think, to indict a nation such as Iraq and slight our own national responsibility, or to concern ourselves with the intransigence of Israel, while at the same time ignoring the failures of the Western democracies to provide a better example than has sometimes been the case . . . We hear a great deal these days about the allied coalition, but might we not extend that concept to a wider family of nations, to those who oppose us?"

Bush listened intently, his head canted in concentration. In the pew behind him, his defense secretary pulled out a pen and a scrap of paper. Moments before leaving his house in northern Virginia, Cheney had spoken by phone to Colin Powell. The chairman bore only glad tidings. The ground attack was unfolding not only on schedule but with unanticipated ease.

"Mr. President," Cheney scribbled, "things are going very well. Dick." He passed the note to Bush, who studied the message and smiled. From the pulpit, the Reverend Harper concluded his homily. "The cross is what binds us, mixed-up people that we are," the pastor said. "So that in the end we find ways of making peace in the world by making peace in ourselves."

After the service Cheney and his family joined Bush and the First Lady for coffee in the upstairs living room at the White House. Using a colorful map pulled from a copy of *Time* magazine lying on the coffee table, Cheney pointed out the allied attack corridors. Then he and the president repaired to the privacy of an adjacent sitting room, where Cheney recounted Powell's report in detail. Two Marine divisions had crossed the minefields and other barriers with little difficulty, he told Bush. Allied casualties appeared to be extremely low — perhaps four dead. Schwarzkopf was so confident of a rout that he had accelerated his main attack by nearly a day.

Their worst fears — of chemical warfare, of another Asian quagmire, of body bags by the thousands — seemed ever more remote. Rarely given to emotional outbursts, Cheney could scarcely contain his excitement. "It's hard to believe how well we're doing," he added.

"Thank God," the president said. "That's great news."

## Phase Line Dixie, Iraq

As dusk rolled over the battlefield in VII Corps's attack sector, the 1st Infantry Division's lead battalions reached Phase Line Colorado, about six miles beyond the obstacle belt. Engineers marked the breach lanes with plywood signs and beacons fashioned from plastic water bottles filled with fluorescent liquid. They also began increasing the number of lanes from sixteen to twenty-four to accommodate the 28,000 soldiers of the British 1st Armoured Division pressing from the south.

The British had been placed under Fred Franks's formal control on January 27. His plan called for them to pass through the Big Red One before turning east to protect the corps's right flank against Iraqi reserves positioned near the Wadi al Batin. Comprising two brigades — the 7th Armoured, or celebrated Desert Rats, and the 4th — the British division thus would have the shortest distance to travel of any VII Corps division, a consideration made necessary by the dubious reliability of the Challenger tank. Nearly every extra tank engine and transmission in the British Army had been sent to Saudi Arabia; the Army of the Rhine in Germany particularly had been picked clean of spare parts.

To the Americans, whose units rarely boasted a lineage more than two or three generations old, the British regiments seemed fantastically ancient. The Queen's Royal Irish Hussars was an amalgam of two regiments — one formed in 1685, the other in 1693. The 17th/21st Lancers, also a combined regiment, recruited largely from Lincolnshire and Nottinghamshire, wore insignia drawn from the death mask of James Wolfe, who was killed in the assault on Quebec in 1759. (Churchill had fought with the 21st Lancers against the Dervishes at Omdurman in "Britain's last great cavalry charge.") The Royal Scots, chartered in 1633, vied with France's Picardy Regiment for the honor of the world's oldest military unit in continuous service. When a Picard once boasted, "We guarded the tomb of Christ," a lad from the Royal Scots is said to have replied, "If we'd done the job, He wouldn't have got away so easily."

British unit terminology so bewildered the Americans — "Fusiliers," perhaps inevitably, was perverted to "fuselage" — that the British were given cards printed with their regimental names to present to U.S. military policemen directing traffic into the breach. However close the battlefield fellowship — and it became tight indeed — the Yanks and Brits would remain two peoples divided by a common language.

Obsessed with a strict military conformity, the Americans found their British cousins quirky and informal. Soldiers often wore long hair; officers called superiors and subordinates alike by their Christian names; no afternoon seemed complete without what Americans referred to as the "tea thing"; uniforms were anything but uniform, with no two soldiers dressed exactly alike and many donning *shermaghs*, a sort of Arab headdress strictly forbidden in the American ranks. Tales of British eccentricity were legion: the Staffordshire lad who, on being told to get rid of his pet sand viper, kissed the reptile goodbye and was badly bitten; the commander who mustered his regiment in the desert to have them search the sand for his lost hunting horn, carried by his grandfather in the Great War; the 7th Armoured soldier, endowed with more pride than spelling prowess, who emblazoned his chest with a tattoo of a helmeted rodent and the inscription DESSERT RAT.

Commanding this host of idiosyncratics was a character of appropriate whimsy and good fun, Major General Rupert Smith. Known as the Pin-up General for his good looks and raffish demeanor — the nickname rather embarrassed him — Smith brooked no stuffy formality. Sweater draped around his neck as though he were going to a rugby match, Smith returned the stiff American salutes with a casual wave and cheery "Hallo, there! Hallo!" On one occasion, on climbing with several of his officers into a Humvee, he was asked by the skeptical American driver, "Where you guys from?" "Where do you think we're from?" Smith asked. "You're sure as hell not in the U.S. Army," the driver replied, rolling his eyes. "You can't be Mexican. Are you Canadian?"

Notwithstanding his casual mien, Smith was an accomplished tactician whose military talents and formidable intellect were admired by Franks and the other division commanders. In contrast to the Americans — who churned out reams of plans, appendixes, annexes, contingencies, and timetables — the British were almost as relaxed as the French. "I refused to make a plan," Smith later declared. Even as the 1st Division was plunging into the breach, he had not decided whether they would be followed initially by his 7th Brigade or the 4th. The proper course, Smith believed, would manifest itself as the battle unfolded. "Don't worry," he assured Franks, "I'll make it up when I see what the enemy is doing."

Other than surrendering — or dying — the enemy was doing little. Darkness, however, proved a tougher foe for the allies. Night had fallen with only half of the Big Red One across the minefield and the British still gathering below the border berm. At eight o'clock Franks called Rhame and Smith by satellite phone from his forward headquarters, a cluster of four tracked vehicles parked in a circle with a canvas awning stretched between them.

Although the earlier rain had lifted, visibility in the breach was almost nil: the desert seemed to soak up any light like blotting paper absorbing water. For soldiers without thermal sights — and that included everyone except the tank and Bradley crews — the Pentagon's oft-repeated claim to own the night was balderdash. Franks had visions of Rhame's logistics trucks wandering into the minefield or rolling over unexploded artillery rounds.

"What's your assessment?" he asked. "Do you think we ought to continue the attack all night? Or should we stop and resume at first light?"

Rhame reported that his third brigade, with most of the division's armor, had yet to cross the breach. "I recommend we wait until daylight," he told Franks. "We'll continue to fire artillery tonight, and in the morning we'll blow through." By noon on February 25, he promised, the division could reach Phase Line New Jersey in the north and begin passing the British to the east. Smith concurred. Franks approved the delay.

But what to do about the left wing, with the 2nd Armored Cavalry and two armored divisions? Holder's cavalrymen were now nearly fifty miles deep, close to Phase Line Dixie. Funk and Griffith, though advancing steadily, had fallen behind. (The 3rd Armored band had stood on the berm and serenaded the tanks across the border with "Garry Owen" and "Pennsylvania 6-5000," while tuba players in cumbersome chemical warfare suits gaily swung their instruments in time to the music.) Most of Griffith's 1st Armored had been more than twenty miles south of Iraq when Schwarzkopf accelerated the attack; now the division was struggling in the dark across terrain strewn with boulders. Meanwhile, an artillery regiment and much of the logistics tail assigned to support the left wing against the Republican Guard were stacked up at the breach. The timetable, so carefully calculated by Franks to achieve "synchronization in time and space," was beginning to fray.

Franks studied the map and meditated. Earlier in the day he had offered a silent pledge to his troops: *I hope I fulfill my end of the bargain. I know you have.* Now the corps commander faced one of the most difficult decisions of his thirty-six years in uniform. On the one hand he had vowed for weeks that, once begun, the attack would admit no pause — the forbidden "P-word." Franks had been so adamant about maintaining the momentum that his deputy, John Landry, remarked that he seemed "almost to have a fire burning inside him" whenever the issue arose. Any delay of the Left Hook could permit the enemy time to reposition or, worse yet, escape.

But circumstances had changed. In Riyadh, staff officers at ARCENT and CENTCOM had begun talking about pursuit, which suggested a headlong dash after a fleeing enemy. To Stan Cherrie, Franks's operations officer, such talk was recklessly premature; it could lead to "a latter-day Pickett's charge," Cherrie warned, with 140,000 soldiers galloping across the desert in a wild rush that would "send fratricide off the charts." And unless the British were given time to position in the east, the Iraqis could counterattack into the flank of the corps's left wing as it swung to the northeast.

Moreover, to minimize American casualties in the battle against the Republican Guard, Franks still believed he needed to mass his combat power. The three-division armored "fist" remained paramount. He knew what two of those divisions would be — the 1st and 3rd Armored; the third was uncertain. Schwarzkopf thus far had refused to cut loose his theater reserve, the 1st Cavalry, lest they be needed to reinforce the Egyptians. It appeared, then, that the final finger in the fist would have to be Rhame's Big Red One.

Around 10 P.M. on Sunday, Franks radioed his commanders for advice. "If we continue to move," Holder replied, "you're going to have me into the Republican Guard tomorrow and nobody behind me to mass against them. My recommendation is to hold and resume the attack at first light."

"That will allow us to mass sometime Tuesday," Franks said, "probably late Tuesday morning, to smack the fist into the Republican Guard." The two armored divisions behind him, Holder added, could continue through the night to push forward in an effort to close the gap with his cavalry. Both Funk and Griffith, however, thought it prudent to pause for refueling and to consolidate their forces. Darkness already had slowed all movement to a crawl. Griffith, whose front trace was now roughly eighteen miles into Iraq near Phase Line Apple, also warned that if his 1st Armored continued to advance through the night, he would hit the Iraqi 7th Corps rear at Al Busayyah, possibly alerting the Republican Guard to the American threat advancing from the southwest.

Franks spoke often of a commander's ability, when the need arose, to "call an audible," a football term referring to an abrupt change of play by the quarterback at the line of scrimmage. Now he called one. The corps log recorded his decision: "halt, refuel, ensure formation postured to achieve mass when attack recommences at [first light]." He radioed Yeosock in Riyadh and quickly sketched his predicament. The corps would stop until dawn. The British would then be prepared to blitz through the breach at noon on Monday, freeing the 1st Infantry to drive north and join the two armored divisions for the final fight. Yeosock offered no objection.

Yet the pause surprised the VII Corps staff, who debated their commander's decision in hushed tones. Landry was stunned. Certain that the decision had been foisted on Franks by Riyadh or that the attack had somehow foundered, Landry called Stan Cherrie from the corps main headquarters farther south in Saudi Arabia: "Stan, I don't understand this at all." How many times had Franks decreed that there be no respite in the attack?

Cherrie explained how nightfall had delayed the breach and how Franks was determined not to let his left wing outrun the right. "Okay," Landry replied, hardly concealing his skepticism, "I can understand that." As usual, Franks seemed to have carefully weighed the evidence before coming to a decision. No plan, the German field marshal Moltke had once warned, ever survives contact with the enemy.

Yet Landry was not reassured. It hardly seemed possible, he reflected, that this sudden change of heart could pass without incident.

15

# On the Euphrates

## Riyadh

Schwarzkopf had retired to his basement bivouac for a few hours of sleep late on the night of February 24. Around 4 A.M. on Monday, G + 1, he returned to the war room looking refreshed and confident, a commander who sensed that victory was within reach. His entire professional life had been spent in preparation for this moment, for an armored battle the likes of which had not been seen in half a century. The successes of XVIII Corps and the Marines, with very few casualties, seemed nothing short of miraculous. Always a man who allowed his emotions to ride close to the surface, the CINC took no pains to conceal his jubilation.

The big wall map across from his chair told the story. In the far west, the French straddled Objective Rochambeau and were preparing to push on toward Objective White at As Salman. If the Daguet Division had lost a little time in pushing north, that had been obviated by Binnie Peay's 101st Airborne, now entrenched at Cobra and preparing to spring into the Euphrates Valley. Also, XVIII Corps logisticians had discovered that the desert was firm enough to support some supply trucks, thus reducing the importance of MSR Texas to As Salman.

On Peay's right, Barry McCaffrey's 24th Division — attacking with 240 Abrams tanks along the combat trails X-Ray, Whiskey, and Yankee — had bolted across the desert toward the Euphrates, in the words of Butch Neal, "like a striped-ass ape." By the time Schwarzkopf returned to the war room this Monday morning, one of McCaffrey's brigades had reported reaching Phase Line Lion, eighty miles into Iraq.

In the east the Arab forces of JFC-E were moving smartly toward Kuwait City along the coast road, their progress slowed mostly by prisoners. The two Marine divisions had seized much of the Kuwaiti bootheel, overrunning three Iraqi divisions and capturing eight thousand prisoners. JFC-N, as expected, had been slow out of the blocks, in part because the Egyptians and Saudis received only enough breaching equipment to cut eleven lanes through the obstacle belts. The Egyptians, when ordered to attack early on Sunday, pushed three battalions to the edge of the first minefield, then wheeled around and headed south at nightfall to laager behind the safety of the Saudi border.

CENTCOM intelligence now portrayed the enemy's forward echelons as virtually annihilated. In the Iraqi 3rd Corps sector of eastern Kuwait, the 7th, 8th, 14th, 18th, and 29th divisions were all assessed as "combat ineffective." The 5th Mechanized and 3rd Armored divisions had been badly mauled; the latter was attempting to dig in between Kuwait City International Airport and Al Jahra. Several divisions and three special forces brigades in Kuwait City remained near full strength, but facing east to thwart an amphibious landing. After the *Tripoli* and *Princeton* mine strikes, Schwarzkopf had scaled back the planned assault on Faylaka Island before canceling the mission altogether. Instead, Marine helicopters conducted feints in the direction of the coast, racing toward the beach before peeling back out to sea. Just the threat of an amphibious attack sufficed to keep a sizable portion of the enemy force turned in the wrong direction.

Only when Schwarzkopf focused his attention on VII Corps did a frown cross his face. The frown quickly became a scowl. Map stickers representing Franks's divisions had advanced not at all.

"What the hell's going on with VII Corps?" the CINC demanded. "This map is wrong."

"No, it's not wrong, sir," Neal replied. "We just updated it."

"Get me the position reports."

The war room bubbled with activity as staff officers culled through the latest reports relayed from VII Corps through ARCENT. Neal checked the grid coordinates for the two armored divisions, the 2nd Cavalry, the Big Red One, and the British. "Sir," he reported, "that's where it plots."

The CINC flushed with anger. Instead of swinging into the Republican Guard with the fury of a broadsword, VII Corps was teasing the Iraqis with rapier pricks. The contrast between the two armored divisions still near the border and Schwarzkopf's former unit, the 24th Infantry, now nearly midway to the Euphrates, was appalling. "Get me Yeosock!" he barked.

For the next thirty-six hours, Schwarzkopf's frustration with VII Corps periodically boiled over. Yeosock fielded one irate call in his own war room at Eskan Village, where he was bracketed by Steve Arnold and John Stewart, respectively the ARCENT operations and intelligence chiefs. "I've got these map plots and I want to know what's going on," Schwarzkopf began, his voice rising to a roar. The senior Army commander attempted to explain, but the CINC cut him off. "Goddam it, I don't care what the problems are. I told you I wanted you to keep pace and you're not doing it. If Franks can't handle the job, I'll get someone who can. I've got Waller champing at the bit."

Arnold and Stewart, who could hear Schwarzkopf's rage crackling through the phone, exchanged anxious glances. When Yeosock hung up, looking like a man who had just been physically pummeled, he reported with understatement, "The CINC's not happy." Schwarzkopf, he added, had threatened to relieve both him and Franks.

In the CENTCOM war room — four hundred miles from the front and forty feet underground — it appeared that VII Corps was ceding the initiative to the Republican Guard, allowing the enemy to choose between counterattacking or fleeing. Schwarzkopf's fury reached new intensity after a call from Colin Powell. The chairman, who earlier had urged his theater commander to "make sure the Republican Guard doesn't escape," now demanded to know why VII Corps was being sluggish. Powell told Schwarzkopf that, while he was reluctant to second-guess his battlefield commanders, he found it difficult "to justify [the performance of] VII Corps when you see what the 24th is doing. What are the 1st and 3rd Armored divisions doing?" Powell ordered the CINC to call Yeosock and "tell him the chairman is on the ceiling about the entire matter of VII Corps."

In private conversations with senior officers from the Pentagon and XVIII Corps, the CINC would later complain that VII Corps was hidebound and inflexible, reluctant to amend its original plan and slow to exploit Iraqi vulnerabilities. He condemned Franks as a pedant whose preachings on tactics and corps operations masked battlefield timidity. In his memoirs, Schwarzkopf would further suggest that VII Corps's dawdling let some enemy forces elude destruction.

Such criticisms were unfair and unwarranted. McCaffrey's division had encountered no enemy resistance at all (and his 2nd Brigade was about to be stalled by treacherous terrain at Phase Line Lion). Neither Riyadh nor Washington fully appreciated the burden imposed on VII Corps when the attack was advanced by fifteen hours. Nor did they seem to understand the rigors of breaching at night, or Franks's ambition to minimize American casualties by massing his combat power, a major precept of American military thought since the 1840s. ("There is no higher and simpler law in strategy," Clausewitz had warned, "than that of keeping one's forces concentrated.") While few officers doubted that Schwarzkopf was genuinely angry with VII Corps, some suspected he was exaggerating for effect: Cal Waller, for example, believed that at least part of the CINC's tirade was a calculated performance intended to light a fire under the Army.

Light a fire it did. "CINC upset with VII Corps not moving fast enough," Steve Arnold wrote in his journal. Arnold wondered whether Schwarzkopf understood precisely what Franks was attempting. The corps commander was attempting to orchestrate one of the most remarkable armored maneuvers in military history. Despite the pause, he was still running well ahead of ARCENT's "synchronization matrix" — a chart printed with the anticipated battle schedule — which called for VII Corps to be in position to destroy the Republican Guard at H + 74 hours, or dawn on Wednesday.

To his credit, Yeosock withstood Schwarzkopf's outbursts. Rather than bludgeon his subordinates in turn, Yeosock conveyed the CINC's wrath through a call from Arnold to John Landry at VII Corps's main headquarters. "We're really getting beat up by the CINC over the slowness of the attack," Arnold reported. "What's going on? You've got to get it moving quicker."

Landry immediately called Franks at his forward command post and found the corps commander momentarily at a loss for words. All units had again begun moving forward in good order; ARCENT had been informed of the reasons for the pause; friendly casualties were almost nonexistent. Why was there such turmoil in Riyadh? "Christ," Franks murmured, more to himself than to Landry, "I don't understand."

## Al Khidr, Iraq

The assault into the Euphrates Valley by the 101st Airborne Division was intended not only to sever Baghdad from its army of occupation, but also to shock Saddam with the sudden appearance of American paratroopers on his southern doorstep, only 140 miles from the Iraqi capital. The chosen site lay midway between Samawah and Nasiriyah, south of the town of Al Khidr. Here the river curled in an oxbow close to Highway 8, forming a natural chokepoint for traffic between Baghdad and Basrah.

At eleven o'clock on Monday morning, after an hour's delay to await reports from scouts prowling the river valley, the lead four-ship formation of Chinook helicopters lifted off from the division assembly area in Saudi Arabia for the 150-mile flight into Iraq. Task Force Rakkasan — the term, a legacy from the paratroopers' occupation of Japan after World War II, meant "falling umbrellas from the sky" — would haul the 3rd Brigade's artillery and most of its TOW missiles to Landing Zone Sand, twenty-five miles below Highway 8. From there the heavy equipment would be pushed overland to rendezvous with two thousand infantrymen set to storm into the valley on Monday evening. (The Chinooks lacked the fuel capacity to fly all the way to Highway 8 and make it back to the forward logistics base at Cobra; moreover, the big, lumbering helicopters would have offered an easy target for any Iraqi gunners near Al Khidr.)

As happened so often in this war, the elements bedeviled the best-laid American plans. Late that morning Binnie Peay called his 3rd Brigade commander, Colonel Robert Clark: "The weather forecast for tonight is bad. If we wait until this evening to put in the infantry, we may have to scrub the mission. I want you to consider going early." After briefly weighing his options, Clark issued new orders: rather than assault the valley at night, the infantry would fly up later that afternoon.

At LZ Sand the first of sixty Chinooks staggered in, each carrying two TOW missile launchers mounted on Humvees. A light rain spattered the desert as the pilots found the proper landing sites — marked with red, orange, and blue panels — and lingered only long enough to dump their loads before peeling south for Cobra. The Rakkasan commander, Lieutenant Colonel Tom Greco, marshaled sixty vehicles and a rifle company, and shoved them north toward Highway 8. The convoy soon found the muddy track nearly impassable. Forced to muscle the cannons through axle-deep muck like exhausted doughboys manhandling their caissons across the Ardennes, the Rakkasans plodded north. Division planners had hoped that the twenty-five-mile trek could be made in several hours, thus putting the brigade's heavy weaponry into the hands of the infantry shortly after their arrival in the river valley. Instead, the march would take all night and part of the next morning.

Sixty-six Blackhawks, each crammed with fifteen to seventeen paratroopers, began lifting off from Saudi Arabia shortly after 3 P.M. for the seventy-five-minute flight to Highway 8. A chaplain had distributed prayer cards among the frightened soldiers, but as the helicopters crossed into Iraq one trooper tore the brown wrapper from a *Playboy* magazine. "Gentlemen, this is what we're fighting for!" he crowed, holding the centerfold high overhead and then flipping through the other nude photographs. "And this! And this! And this!"

This first wave of a thousand soldiers roared into the Euphrates Valley, touching down at three landing zones south of Highway 8 near Al Khidr. Leaping from the Blackhawk bays into a sea of black mud, they sank up to their knees. An Air Force sergeant, topheavy with radios, slipped and fell face down; a trio of paratroopers was needed to pull him to his feet. The throb of retreating helicopters faded in the south, leaving only the sounds of yapping dogs and the slurp of boots being wrenched from the mud at each slow step.

Clark, the brigade commander, scanned the wretched terrain. As far as he could see in the fading light, soldiers struggled through the bog, dragging their mortars and TOWs, scraping mud from their legs with bayonets, and listening for the expected rush of enemy artillery shells. Rain fell harder, drumming on helmets and rucksacks. Ponchos flapped in the wind. Several hundred yards to the north, Clark saw the yellow stab of headlights on Highway 8. An Iraqi civilian stopped his sedan and leaped from the car. He gawked in disbelief, then fled toward the river, slipping and falling three times before he disappeared from view.

Clark found a muddy dike angling south from the highway and within ten minutes had set up his satellite radio and called division headquarters. "The Screaming Eagles," he reported, "have landed on the Euphrates." (Gary Luck, always sensitive to grandiloquence, later complained, "Who sent that tacky message? That does not do credit to the 101st.")

The Screaming Eagles were indeed on the Euphrates, but as prisoners of the ground they held, immobilized by mud, their heavy weapons mired far to the south. Reluctant to bring in more soldiers only to have them exposed to artillery or tank fire, Clark canceled the second lift of Blackhawks, leaving a thousand paratrooper reinforcements to await further orders.

Less than a kilometer to the northwest of Clark's position, three hundred soldiers from the 3rd Battalion anchored the brigade's left flank under the command of Major Jerry R. Bolzak, Jr. (The battalion commander, Tom Greco, was with the Rakkasans at LZ Sand.) Bolzak sheltered in a narrow irrigation ditch, watching his soldiers erect several large signs across the highway with warnings in English and Arabic: "U.S. Military Operations — Keep Out." At dusk he heard the crackle of small arms fire from Alpha Company, entrenched along his left perimeter. Fifteen Iraqi infantrymen had crept to within two hundred yards; the Americans drove them off with a barrage of 60mm mortar fire.

Bolzak, who had recently been reading *War and Peace*, thought of Tolstoy's portrait of the Russian commanders and how little influence they had had on the events swirling around them on the battlefield. Even in this inconsequential skirmish on an Iraqi mud flat, the observation seemed apt: except for marginal decisions — such as whether to use the 60mm or 81mm mortars — Bolzak knew that much of what happened followed a course beyond his control.

Mud had ruined two of the battalion's four TOW missile launchers, but Bolzak moved the remaining pair and his machine guns into ambush positions overlooking Highway 8. To avoid shooting farmers or other civilians who wandered unawares into the kill sack, the gunners allowed solitary vehicles to pass, but truck convoys were fair game. As Bolzak watched a single pair of headlights moving southeast from Samawah shortly after dark, a burst of red tracers ripped across the roadbed. He radioed, "What are you doing? I only see one set of headlights." "Sir, there are four vehicles traveling together," the company commander answered, "but only the lead one has his lights on."

Bullets tore through the first truck, which swerved across the highway and crashed into the guardrail. The rest of the convoy screeched to a halt, and the drivers scampered into the night. (One would be found dead the next morning, laid out with hands folded across his chest and his head pointed toward Mecca.) Bolzak's scouts moved forward to search the wreckage. "What have you got?" Bolzak radioed eagerly, envisioning a cargo of Scud missiles. "Sir," came the reply, "they were carrying onions."

The highway was cut, the main corridor connecting Baghdad with Basrah and Kuwait now under American sovereignty. Rain and wind worsened by the hour, grounding allied warplanes and attack helicopters. But the additional forces would not be needed: Clark and his brigade awaited the counterattack that never came, the artillery salvos that never fell. An armored column with two hundred vehicles, reported by the Air Force to be moving from Nasiriyah, reversed course and headed east — to Clark's immeasurable relief.

Throughout the night other convoys — some laden with booty from Kuwait City — attempted to run the roadblocks only to be destroyed with machine gun and mortar fire. When one military truck tried to ram through a barricade made of a pair of captured Honda Preludes, a sergeant emptied two full clips from his M-16 at a range of twenty yards and killed thirteen Iraqis. On the brigade's eastern flank, a paratrooper battalion swept through a large oil-pumping complex and routed a platoon of defenders after a brief firefight.

With daylight came the bedraggled Rakkasans and their heavy weapons, including explosives for blowing craters in the highway pavement. Villagers put aside their terror — "Please don't eat us!" some screamed when the paratroopers first appeared — to plead for food. Bedouin women bared their breasts to show that their children were hungry.

The mayor of Al Khidr presented himself and asked permission to loot the shattered trucks now littering Highway 8. "My people," he explained, "have no water, no food, no power." Flush with the victor's generosity, the Americans agreed, and Iraqis swarmed over the carcasses like crows before staggering across the mud with flour sacks and loaves of bread and hundreds of onions cradled in their arms.

## Haql Naft Al Burqan, Kuwait

Heavy fog, mixed with purple oil smoke, shrouded the Kuwaiti bootheel as the Marines prepared to push north on the second day of their offensive. The wind, which in late winter usually blew from the northwest, had shifted from the southwest, pushing most of the petroleum fumes back into the faces of the Iraqi defenders. On the western lip of the Al Burqan oil field, however, where four thousand Marines from Task Force Papa Bear held the 1st Division's right flank, visibility was less than a hundred yards. Here, at last, the enemy counterattacked.

It began with a surrender. Around 9 A.M. on Monday, the Papa Bear commander, Colonel Richard W. Hodory, was leaning over a map spread across the hood of his Humvee when a tank and two armored personnel carriers loomed out of the fog fifty yards away. Hodory had positioned his 1st Tank Battalion on a forward screen line; he wondered why one of the M-60s should now wander back to the command post. Then he recognized the distinctive round turret of an Iraqi T-55. A burst of Marine gunfire raked across the enemy vehicles. "What the shit's going on?" Hodory yelled, amazed that they had got so close despite an entire tank battalion to his front.

A white flag appeared. The enemy crews were disarmed and handcuffed. Their leader, an English-speaking officer in his late thirties who identified himself as Major Adai, claimed to be the commander of the 22nd Brigade of the 5th Mechanized Division. (He later was found to be only a battalion commander.) Adai told Hodory that he and several company commanders wanted to surrender, but that most of the regiment was prepared to fight. He handed over his map with an Arabic legend that translated as "Secret/Confidential. 3rd Corps division distribution/Al Burqan Province Called Kuwait." In a steady voice, Adai told Hodory he was married and the father of three children; he had been wounded twice in the war with Iran and had arrived in Kuwait the previous summer under the impression that he was liberating the country from American invaders. The Iraqi counterattack against the Marines, he added, would begin soon.

The news was not entirely unexpected. Although the Marines had fretted over their exposed eastern flank — the Arab E-PAC forces were still twelve miles southeast along the coastal highway — the dense thatch of pipelines in the Al Burqan and Al Ahmadi fields seemed to form a natural barrier on I MEF's right flank. Any armored force would have to negotiate aboveground pipes every few hundred yards, in addition to the petroleum lakes and flaming wellheads left by Iraqi saboteurs.

But the previous night, while searching a command bunker near Al Jaber Air Base, Marines from Task Force Ripper had discovered Iraqi documents detailing plans for an attack "out of the flames." At 3:30 A.M. signals intelligence further confirmed that two enemy brigades were marshaling in the Al Burqan. Armed with a "target quality solution" — intelligence jargon for the precise position of enemy forces — 1st Division commander Mike Myatt ordered four artillery battalions to fire at two grid coordinates simultaneously. Shortly after 8 A.M., more than three hundred rounds fell on the two locations; the gunners then raised their tubes and fired again a thousand yards farther east. "Watch this, General," Myatt's operations officer, Lieutenant Colonel Jerry Humble, warned. "It's going to be like rabbits coming out of the briar patch."

The first rabbits flushed moments after the informative Major Adai offered his map to Hodory. Undetected in the fog, Iraqi tanks had pushed through the American screen line from the southeast. Suddenly, tank rounds and machine gun tracers ripped through the command post, sending Hodory and his officers diving for the ground. For ten minutes pandemonium swept the Marine encampment. Confused shouts were drowned out by booming tank fire and the chatter of automatic weapons. Muzzle flashes licked through the mist. The stench of cordite and diesel smoke rolled across the battleground. Radios screamed with orders and counterorders.

Had the enemy attacked on line, hitting the Marines with the mass of concentrated armor, they might have overrun Papa Bear's command post. But the assault was piecemeal. Iraqi marksmanship was atrocious, consistently six to ten feet high. Firing Dragon missiles and AT-4 antitank weapons, the Marines quickly blunted the attack and forced the enemy back in disarray. One corporal courageously grabbed an AT-4, dashed through the tracer fire, and destroyed a T-55 with a shot into the tank's right flank. Hodory ordered the 1st Tank Battalion to pivot east and then south. As the Marine M-60s creaked to the crest of a low hill, the fog suddenly began to lift. There before them lay the enemy regiment, milling about in hapless confusion like sheep in a slaughtering pen.

Eight miles to the north, Mike Myatt had set up the 1st Division's forward command post on the western fringe of As Subayhiyah oasis, an improbable grove of evergreen trees known to the Marines as the Emir's Ranch. At 9:30 A.M. Myatt and his deputy, Brigadier General Tom Draude, were on the radio monitoring Papa Bear's battle when several explosions shook the tent. "What the hell is that?" Myatt demanded.

Staff officers quickly rolled up the canvas flaps and saw, through the smoky haze several hundred yards to the east, Marine pickets locked in combat with another Iraqi brigade at the tree line. Artillery rounds burst nearby; Myatt surmised that the enemy was using radio direction-finding equipment to home in on the division's "antennae farm," which had been erected near the command post. After calling Task Force Ripper for reinforcements armed with TOW missiles, Jerry Humble radioed I MEF headquarters: "We need some help. Send all the Cobras you can."

"Jerry," a MEF staff officer replied, "everybody's in a fight."

"No, we're in a *real* fight at division forward," Humble insisted. "Listen." Using an old Vietnam technique, he held the radio handset aloft so that the explosions were clearly audible.

"Oh, shit," the other officer replied. "I hear."

Forewarned by the enemy documents captured at Al Jaber, Myatt had wisely chosen to reinforce the lone rifle platoon guarding his command post. The task had fallen to Captain Eddie Stephen Ray, a strapping company commander from south-central Los Angeles who had once played offensive guard for the University of Washington football team. Ray was furious at being pulled from Task Force Shepherd, where he had been scheduled to lead the regimental attack Monday morning; for half an hour he argued futilely with the task force commander, pleading to be spared the ignominy of guard duty. Still seething, Ray had arrived at the command post shortly after 1 A.M. on Monday with forty Marines from his 2nd Platoon and seven LAV-25 armored personnel carriers. He positioned the LAVs on a screen line a quarter of a mile from Myatt's tent.

When the Iraqi attack began, Ray heard the boom of artillery to the northeast, then saw a round detonate a hundred yards away. The harsh clatter of gunfire drifted south from Myatt's rifle platoon. Wary of fratricide, Ray ordered his men to remain in place while he raced north in an LAV to a Marine machine gun nest behind a sand berm. Two hundred meters to the east the silhouettes of Iraqi BMPs — armored personnel carriers — glided among the trees. Marine infantrymen fired two light antitank rockets. Both missed, but the BMPs stopped, and Iraqi soldiers poured from the rear hatches and fanned out through the grove. For the first time Ray realized that Myatt, Draude, and the rest of the division staff were in immediate danger of being captured or killed.

The firing intensified. Enemy bullets snipped at evergreen branches and spattered around the machine gun nest. Ray's gunner pumped a volley of 25mm cannon fire into one BMP, then another. Standing exposed in the LAV's upper hatch, trying to make out the shadowy shapes darting among the trees, Ray saw a yellow flash explode from the right. A Sagger missile streaked overhead. "Sagger! Back up!" he yelled to the driver. The LAV rumbled in reverse as a rocket-propelled grenade exploded a few yards short. Crimson 25mm rounds tore into a third BMP; the vehicle exploded, flinging bodies into the air.

Ray radioed his platoon leader. "I'm five hundred meters north of you on the screen line. Take a due north azimuth. I want three vehicles on each side of me when you get here, hundred-meter intervals. Do not veer east. We're firing to the east." He could not call in artillery; the Iraqis were too close. But a pair of Cobra gunships swooped in beneath the pewter overcast, and while the LAVs moved on line, Ray directed cannon and rocket fire from the Cobras by using tracer rounds to mark enemy infantry positions.

Uncertain of his flanks and alarmed by frantic reports of more Iraqi vehicles, Ray decided to counterattack. He dashed first to the southern edge of the trees, then to the north. Seeing no evidence that the enemy was trying to circle behind him, he ordered the LAVs into the forest as the Cobras rolled in and out above the screen line.

Again the Iraqis showed poor marksmanship and an inability to organize. Enemy dead carpeted the forest floor. Wounded soldiers dragged themselves through the trees, leaving scarlet blood trails. By now at least two dozen vehicles were ablaze, singeing the tree branches and erupting in fountains of fire as ammunition cooked off inside. Ray pushed forward for a thousand yards, then paused to round up prisoners and rearm the LAVs, each of which carried six hundred rounds of 25mm ammunition. The Marines searched a solitary building hidden in the glade, where they discovered an arsenal with two thousand weapons and a large cache of rice.

More Cobras arrived overhead; the LAVs again moved forward. The trees ahead boiled with cannon and rocket fire. A group of prisoners broke free and scampered through the woods. As his gunners trained their sights on the backs of the fleeing enemy, Ray told the men to hold their fire. "Just let them go," he ordered.

After covering another thousand yards, the Marines reached the eastern fringe of the Emir's Ranch. Beyond lay the wasteland of the Al Burqan, now swept with the express train roar of burning oil wells. A few surviving BMPs vanished into the smoke. Thirty-eight other vehicles had been destroyed and three hundred Iraqis captured. Ed Ray, who would win the Navy Cross — second only to the Medal of Honor among awards for heroism — wheeled around and headed back toward Myatt's command post to await further orders.

In the south, Papa Bear found itself locked in what the 1st Division would claim as the biggest tank battle in Marine Corps history. However historic, the fight soon turned into a rout. For three hours, while Marine bombers struck deep to the north and east, Hodory's tanks and LAVs shot everything that moved without a white flag flapping.

Particularly devastating was a flight of four Cobras led by Captain Randy Hammond out of Lonesome Dove, the Marines' makeshift airstrip just west of the bootheel. During their first sortie the Cobras — each of which carried four TOWs, fourteen rockets, four Hellfires, and seven hundred 20mm rounds — shot their entire munitions load in ten minutes, leaving a swath of burning enemy equipment.

After rearming and returning to the battlefield, Hammond and his pilots formed a skirmish line and moved northeast at twenty knots, destroying trucks, tanks, and personnel carriers. The 20mm cannon, which slaved automatically whenever the Cobra pilot swiveled his head, blew enemy infantrymen to pieces. Some Iraqis waved white flags, then fired from the rear as the Cobras passed. Others lay on their weapons, feigning death. The enemy returned fire with machine guns, surface-to-air missiles, 85mm antiaircraft fire, and even rocket-propelled grenades — hitting nothing. On their third and final sortie the Cobras demolished a two-story police post with seven Hellfires and three TOWs before dumping hundreds of cannon rounds and four more Hellfires into enemy troops clustered on Hill 371, the highest promontory in the central bootheel.

By noon the enemy was spent. The 1st Marine Division would count nearly a hundred armored vehicles destroyed and several hundred prisoners captured. However disastrous and ineffectual, the counterattack had one salutary consequence for the Iraqis: it temporarily halted the division's drive toward Kuwait City. At a cost of two brigades, the enemy had bought itself twenty-four hours.

## Al Khobar, Saudi Arabia

After nearly six weeks of war, the Iraqi Scud provoked more curiosity and contempt than fear among its intended victims. Except for a volley of six missiles on February 21, enemy launches since midmonth had dwindled to an average of one per day. Attacks against Israel had been confined to occasional firings from an area near Al Qaim, along the Syrian border. In Saudi Arabia the Scud was the butt of innumerable jokes. (Question: "How many Iraqis does it take to fire a Scud?" Answer: "Three. One to aim, one to fire, and one to watch CNN to see where it lands.") Air Force counter-Scud bombing and the perceived prowess of the Patriot missile had lulled many into a false sense of safety. Warning sirens, which once triggered a frantic fumbling for gas masks and a rush to the bomb shelters, now seemed a screaming inconvenience. Rather than seek cover, many stepped outside or repaired to the roof in hopes of glimpsing the fell beauty of missile dueling missile.

On Monday evening, February 25, all that would change. At 8:32 an early-warning satellite spotted the infrared plume of a Scud launch from the "Qurnah basket," in the marshy bottoms fifty miles northeast of Basrah; two minutes later an alert flashed from Cheyenne Mountain in Colorado. Within sixty seconds the warning reached Foxtrot battery of the 2nd Battalion, 7th Air Defense Artillery Brigade, positioned north of Dhahran. At 8:38, Foxtrot detected the missile streaking high over Kuwait more than sixty miles away.

Because the Scud remained well outside Foxtrot's "engagement zone" — the range and altitude in which the Patriot system automatically fired — Foxtrot's crew watched the Iraqi missile on their radar scope for thirty seconds without launching. Likewise, Delta and then Echo batteries tracked the Scud but did not shoot, because the target was also beyond their respective zones.

The southern batteries of 2nd Battalion, Alpha and Bravo, were positioned on Dhahran Air Base, long designated by the Army as the most precious military asset in eastern Saudi Arabia. Bravo, however, had shut down four hours earlier for maintenance work on its radar system. At Alpha battery 1st Lieutenant Gerald M. Dailey and his assistant, Specialist Samuel M. Luse, sat in the engagement control van, listening to radio reports from Echo and Delta and waiting for the Scud to show up on their scope. On beginning his shift at 6 P.M., Dailey had run through the usual checklist: missile inventory, status panels, radar operations. All appeared normal. He and Luse double-checked the system after hearing the Scud alert. Again the Patriot seemed to be "100 percent go."

They were deceived. An abnormality had crept into Alpha's computer software that blinded it to the Scud, now plummeting toward Dhahran at four thousand miles per hour. When a Patriot system was running, its computer accumulated an infinitesimal "timing error" of one microsecond for every second of operation. (A million microseconds equal one second.) Normally the error was too small to affect the missile's performance; moreover, whenever the system was "rebooted" — turned off and then back on — the timing error was erased and reset to zero. Had the Americans been fighting a more sophisticated foe, like the Soviets in Europe, the Patriot crews would have turned their equipment off and moved at least once a day to prevent the enemy from homing in on radar emissions.

Against Iraq, however, the batteries remained relatively stationary. Alpha had operated continuously for a hundred hours — more than four days — and the accumulated timing error had reached 360,000 microseconds, roughly a third of a second. Investigators would later conclude that this "surveillance range gate error" caused the Patriot radar to miscalculate the projected path of the enemy missile. After initially detecting the Scud, the Patriot computer estimated the missile's path and directed the radar to search a particular section of sky in order to verify and track the target. But because of the timing error, this search was approximately seven hundred yards from the Scud's actual trajectory. When Alpha's radar saw nothing to confirm the initial glimpse, the battery's computer assumed that the skies were clear.

U.S. Army missile technicians had incorporated corrective measures in a new software package developed as a result of information provided by the Israelis on February 11. Using data-gathering equipment shunned by their American counterparts in Saudi Arabia — the Americans feared the equipment would interfere with missile operations — Israeli crews had detected a significant range gate error after only eight hours of continuous operation. The Army technicians assumed that most of their crews were rebooting their computers every few hours, so the timing problem was considered minor.

The new software, Version 35, was intended primarily to help the Patriot more easily distinguish a Scud warhead from other debris when Iraqi missiles disintegrated in flight; the timing modification was described as "a minor improvement which had no impact on system performance." Computer tapes containing Version 35 arrived at McGuire Air Force Base in New Jersey on February 22, where they sat for a day before being flown to Saudi Arabia with another shipment of PAC-2 missiles. Once they arrived in Riyadh, late on the night of February 24, the tapes were loaded onto cartridges and trucked to twenty-one Patriot batteries scattered across the Arabian Peninsula. The new software would not arrive in Dhahran until the morning of February 26 — twelve hours too late.

Inside the Alpha engagement van, Lieutenant Dailey and Specialist Luse strained at their scope for ninety seconds after Delta battery reported seeing the Scud head south. Perplexed by the blank screen, they again checked their equipment, unaware that the timing error had caused the system to disregard the incoming missile. Again the Patriot seemed to be working properly: radar plainly showed "air breathers" — aircraft — flying in the vicinity. As they watched and waited in the sealed van, a fiery streak flashed from the northwest, plunging unseen through the heavy overcast.

Three miles from the air base, in the Dhahran suburb of Al Khobar, sirens wailed at 8:42 P.M. Soldiers browsing through Souk's Supermarket instinctively glanced up at the ceiling. In the crowded toy store next door, Saudi parents herded their children together and hurried toward the exit. Behind the shopping center lay an immense metal warehouse, two hundred feet long and fifty-five feet wide, which had been converted into an Army barracks. Four neat rows of cots stretched across the cavernous building. In one corner, ropes had been strung and draped with ponchos to fashion a dressing room for female soldiers. Around the cots were strewn M-16s, duffel bags, stuffed animals, and sacks of cookies sent from home.

One hundred and twenty-seven soldiers had bivouacked in the warehouse, more than half from a single Army reserve unit: the 14th Quartermaster Detachment. In the wake of Vietnam the Pentagon had deliberately woven the reserves and National Guard into the country's military force structure so that never again could a president undertake a major war without drawing the nation at large into the conflict. Of the 245,000 reservists who had been summoned to active duty for the gulf conflict, many were now in Southwest Asia as doctors and dentists, truck drivers and dock workers, pilots and mail handlers. The reserves were at war, and thus the nation.

The 14th Quartermaster was America in microcosm. Drawn largely from western Pennsylvania, the unit had been activated on January 15, dispatched to Virginia for a month of training as water purification specialists, and flown to Dhahran on February 19. In civilian life the soldiers were clerks, secretaries, college students. Specialist Anthony E. Madison, the father of two, had studied bricklaying while hoping to be a professional boxer. Specialist Christine L. Mayes had served in Germany as an Army cook, then joined the reserves to save money for college. Sergeant John T. Boxler, a Vietnam veteran, had lost his job in the dying steel industry before finding new work with the U.S. Postal Service.

When the sirens sounded, someone opened a door at the end of the warehouse, provoking catcalls and demands that it be shut. For reasons never understood, the warning in Al Khobar came less than thirty seconds before the missile hit, even though the Scud launch had been detected ten minutes earlier. Some soldiers reached for their gas masks. Others, including those facing the graveyard shift on guard duty, dozed on their cots. Since no bomb shelters lay close to the warehouse, the troops had been told to remain inside during air raid alerts. A nickel poker game continued, as did a game of Trivial Pursuit — although the number of players was reduced from nine to eight when one soldier stood up and scurried away after spotting a mouse.

The move probably saved his life. A moment later, six of the remaining eight players were dead. The Scud — possibly the only Iraqi missile that failed to break apart in flight — smashed through the center of the metal roof. For an instant the force of the penetration caused sheet metal and I-beams to buckle inward, folding over the soldiers like closing petals on a tulip. Then the warhead detonated, gouging in the slab floor a crater four feet deep and twelve feet in diameter. The blast whipped heavy concrete and metal fragments forty yards in all directions; smaller shards traveled twice as far, raking the warehouse like grapeshot. Flaming debris — sleeping bags, boots, blankets — rained through the air like sparks from a Roman candle. Sheets of fire soon engulfed the warehouse.

Screams rent the night. Private First Class Anthony Drees had been fumbling for his helmet when the explosion knocked him to the floor. He grabbed his legs, feeling bloody holes where his thighs had been. Another soldier lay nearby, both legs gone, his face contorted in bewildered, mortal pain. Chased by flames, Drees dragged the dying man toward the wall, then crawled outside, fighting an overpowering urge to sleep.

Scores of firemen and military policemen swarmed around the blazing barracks. M-16 ammunition crackled in the flames. A smell of sulfur mingled with the stench of charred flesh. Soldiers wept, survivors embraced. Medics wrapped burn victims in foil sheets. Buses were converted into ambulances, their windows smashed so that the stretchers could be hoisted inside. Helicopters swooped in to pluck away the most critically wounded.

The inferno at last died out, leaving a smoldering skeleton of twisted girders. If the carnage was a consequence of blind chance and bad luck, it also was a reminder of the allies' extraordinary good fortune heretofore. Had the catastrophe occurred early in the war, certainly the psychological impact at home and in the field would have been far greater. Had the missile struck Israel with comparable devastation, it is unlikely that any amount of pleading from Bush or Cheney could have stayed the Israeli sword. As it was, the Al Khobar calamity cast but a brief shadow across the good news pouring from the battlefield, providing a sad footnote in the chronicle of triumph.

The Scud killed twenty-eight Americans and wounded ninety-eight. Of the dead, thirteen were members of the 14th Quartermaster Detachment, including Anthony Madison, Christine Mayes, and John Boxler. Thirty-nine others were wounded; the unit's casualty rate was 75 percent. At home in Greensburg, southeast of Pittsburgh, where the 14th Quartermaster had its headquarters, news of the tragedy raced through town nearly as fast as the flames that had swept the barracks.

As the dead were identified in Saudi Arabia, Army staff cars began crisscrossing the snow-clad Pennsylvania countryside. On hearing of the attack, Marlene Wolverton had called the Greensburg armory hourly for word of her husband; at dawn on Tuesday two sergeants appeared at her door. "We are here from the president of the United States and the secretary of the Army," said one man, fighting back tears, "to inform you that your husband, Richard, was killed in action on February 25 in Dhahran, Saudi Arabia." Married only eight months, widowed in an instant, she resisted. "Stop! I don't want to hear it! Don't tell me this!" In helpless rage she ripped the yellow ribbons and American flag from the front door.

For others the news arrived even more cruelly. In a large room on the armory's second floor, above the drill hall now packed with reporters, dozens of terrified mothers, fathers, husbands, and wives kept vigil. They peppered a major general with questions about Scuds and Patriots and casualties. So insistent were their queries, so desperate the hunger to know — yes or no? alive or dead? — that the Army decided to forgo its usual procedure of delivering death notifications at home. Instead, a chaplain was asked to give the fatal verdict in the armory.

As eighty people watched in dread, he began with the parents of

Specialist Beverly S. Clark, a twenty-three-year-old secretary. The chaplain led the dead woman's mother from the room. She thumped his chest with her fists, sobbing, "My baby! My baby! She was my baby!" Then a second family was summoned, and those left waiting heard shrieks from the corridor and wails of unspeakable anguish. The chaplain returned to call a third name; another family shambled out, and again cries of lamentation echoed in the hallway.

Men and women remaining in the room clung to one another like prisoners awaiting execution, praying and weeping, their eyes pinched shut in supplication. Finally, before a fourth name could be pronounced, someone confronted the armory's first sergeant: "You can't do this!" The sergeant agreed. The death march halted. Further notifications would be delivered at home, where at least the living could mourn their dead in privacy.

## Phase Line Smash, Iraq

Not until after the war would Fred Franks learn of Schwarzkopf's threat to dismiss him, but there was no mistaking CENTCOM's distress at the pace of VII Corps's attack. After his tirade in Riyadh, the CINC called the corps command post by satellite phone later on Monday. Franks had gone to confer with the 3rd Armored Division, so Stan Cherrie took the call and gave Schwarzkopf a quick status report on the corps's attack.

"You've got to keep pushing. I want you to keep pressuring the enemy," Schwarzkopf urged. "Do you know who Bobby Knight is?"

"Yes, sir, I sure do," Cherrie said. The Indiana University basketball coach was renowned for both his temper and a swarming, aggressive style of play.

"Good," the CINC said. "I want you to keep the Bobby Knight press on them. Keep pressing."

ARCENT also called, wanting to know why the corps seemed so sluggish compared to the 101st and 24th divisions farther west. Cherrie had already grown weary of hearing about XVIII Corps's exploits, and now he lost his temper. "Because they don't have any enemy out in front of them, that's why!" he snapped. "All they're doing is running over rats and lizards."

Franks responded to the pressure by urging his commanders forward while clinging to his *idée fixe* of striking the Republican Guard "with an almighty crack," in the phrase of Rupert Smith. Certain battlefield realities, however, could not be swept aside by exhortation or force of will. The passage of the British 1st Armoured Division, begun shortly before noon, was tedious and time-consuming — not unlike the process of pouring fluid through the narrow neck of a bottle. Plodding logistics trains could not be left in the dust without risking empty fuel tanks or ammunition racks farther north. Prisoners had become a serious encumbrance across the battlefield, a common phenomenon in desert warfare since a routed opponent had neither terrain nor vegetation to mask his withdrawal. (In December 1940, when the British in North Africa captured 130,000 Italians, prisoners were forced to organize their own unescorted shuttle service to POW camps.)

Late in the afternoon Franks and Cherrie flew by helicopter to the 2nd Cavalry Regiment, which had laagered along a dirt road — misnamed Phase Line Blacktop — approximately fifty miles from the northwest corner of Kuwait. The weather was worsening by the hour, with high winds and rain that would soon shut down air operations. Don Holder had ordered his cavalry squadrons to form a "hasty defense" in a horseshoe formation to await the two armored divisions driving from the southwest.

This decision pleased Franks not at all. "Who said anything about defense? No defense. No defense," he said heatedly, rapping Holder's map with his fist. "I want you to push on through. Find the enemy and fix him. Keep in contact, but don't become decisively engaged."

Franks had long intended that once the corps reached Phase Line Smash, which bisected the desert from the Wadi al Batin in the southeast on past Al Busayyah in the northwest, he would decide how to attack the Republican Guard. Now the decision was upon him: Holder's cavalrymen had crossed Smash en route to Blacktop.

Studying the map and intelligence reports, Franks tried to surmise Iraqi intentions. In the past twenty-four hours the Americans had assembled a clear picture of an enemy that was battered but still struggling to mount a defense. Among Iraq's forward echelons the war was all but over. Captured officers reported that neither the 25th nor the 26th Infantry Division had had any contact with corps headquarters for more than a week; morale was abysmal, desertions flagrant. Of eighteen artillery tubes in one Iraqi regiment, eighteen had been destroyed.

But three Republican Guard divisions remained entrenched just above the Kuwaiti border: the Tawalkana in the west, the Madinah in the center, the Hammurabi in the east. A fourth division, the Adnan, lay outside Basrah near the Euphrates. Two brigades from the enemy's 12th Armored Division had swung south toward the Wadi al Batin, apparently reacting to the 1st Cavalry's feint. Another armored brigade, the 50th, had moved northeast to serve as a screen in front of the Tawalkana Division.

Holder and his intelligence officer, Major Steve Campbell, believed the 50th Brigade and the Tawalkana Division intended to form a blocking force, behind which the Iraqi 3rd Corps could retreat from Kuwait. Campbell predicted that the Tawalkana would await the American attack close to the Kuwaiti border, with enemy scouts extended five miles farther west.

Thus warned, Franks again urged Holder to "reach out and find the enemy, but just touch them." He then flew back to his command post, now located about fifty miles north of the Saudi border with the 3rd Armored Division. The British, Franks saw, were making steady progress. Rupert Smith had elected to send his 7th Armoured Brigade through the breach first, followed by the 4th Brigade. He would then wheel his division in a hard right turn to the east. Each British brigade was assigned a series of objectives named after metals, all leading into Kuwait: Copper North, Zinc, Platinum, and Lead for the Desert Rats in the north; Bronze, Copper South, Brass, Steel, and Tungsten for 4th Brigade in the south.

A greater problem was the American 1st Armored. Slowed by prisoners and a firefight with remnants of the 26th Infantry, the division was finally nearing Objective Python, the crossroads town of Al Busayyah. Defended by two enemy battalions, Al Busayyah comprised approximately fifty buildings and a corps logistics base. Ronald Griffith, the division commander, had hoped to capture Python by 6 P.M.

"I can go ahead and take Al Busayyah tonight," Griffith told Franks in a call at dusk. "But there's a network of wadis south of town, and I'm afraid if I start pushing up those wadis in the dark I'm going to lose some tanks. I'll do whatever you tell me to do and I can probably be on Phase Line Smash by midnight or one o'clock. But I'd prefer to knock the crap out of them all night long with artillery and then go in at first light."

Franks agreed. Losing soldiers in a confused fight at night in broken terrain seemed pointless, no matter how eager Riyadh was to push the attack. "Okay," he told Griffith, "but I want you to press hard. I want you to be on Phase Line Texas [near Smash] by nine o'clock tomorrow morning. Not later than." "Right, sir," Griffith replied. "Got that."

The separate fingers of the corps had begun to clench. The fist was forming. "It's time, Stan," Franks told Cherrie. "This is it." He had made up his mind. Rather than continuing to arc toward Basrah, the corps would pivot due east to deliver the knockout blow from the Left Hook. Schwarzkopf had yet to authorize release of the 1st Cavalry to join the attack; therefore the Big Red One would leave the breach and pass through Holder's regiment on Tuesday, coming on line with 1st and 3rd Armored. McCaffrey's 24th Infantry in the north and the British 1st Armoured in the south would thus give the allies at least five heavy divisions in a steel wall from the Euphrates Valley to southern Kuwait. But to maintain momentum, supply lines had to remain open. To his chief logistician, Colonel Bill Rutherford, Franks said, "I need fuel. Don't let me run out. Keep pushing fuel. Brute force logistics. Push, push, push."

Among the stacks of contingency plans drafted in the preceding weeks lay a nine-page document entitled "Frag Plan 7" and signed "Franks, LTG [lieutenant general]." Frag Plan 7, copies of which had been distributed in advance throughout the corps, was predicated on the Republican Guard's remaining more or less stationary. It provided a blueprint for wheeling the corps to the east and hitting the Iraqis in the flank. With minor modifications, such as shifting the British sector slightly south and adding deep attacks by the 11th Aviation Brigade, the plan seemed to fit the battlefield circumstances almost perfectly.

Late Monday night Franks issued a one-sentence order, distributed to his commanders via a laptop computer message: "Frag Plan 7 placed in effect."

Only in retrospect could one recognize that two distinct visions of the battlefield had taken hold of the American command forty-eight hours into the ground war.

For Schwarzkopf and his subordinates in Riyadh, the enemy was all but beaten. The virtual collapse of Iraqi resistance in Kuwait, the stellar success of Boomer's Marines, and the ease with which the French and XVIII Corps pressed their attack in the west all suggested a routed foe for whom escape was the overriding impulse. The specter that for months had haunted Schwarzkopf and the nation — of ten thousand American casualties in a protracted slugfest — had vanished. Now the enemy's disarray should be exploited with audacity and speed, bringing the war to a swift, triumphant conclusion.

Yet for commanders at the front, particularly for Franks and his lieutenants, a dangerous, unpredictable enemy remained at large. The collapse of conscript divisions in the Kuwaiti bootheel had little bearing on whether the Republican Guard and other armored brigades would flee or fight. Five enemy divisions remained clustered around northeast

Kuwait; American intelligence indicated that these units were repositioning toward rather than away from the oncoming VII Corps. If allied victory now appeared certain, many lives still hung in the balance between good military judgment and bad. The final days of the attack would seem anticlimactic to everyone except those who had to carry the fight to its finish.

# 16

# Upcountry March

## King Khalid Military City, Saudi Arabia

On Tuesday morning, February 26, a military passenger jet banked around a cluster of towering thunderclouds and landed on the airfield at King Khalid Military City, two hundred miles north of Riyadh. Built by the Saudis as a frontier outpost — complete with an equestrian center, an Olympic-sized swimming pool, cascading fountains, and a graceful mosque trimmed in green stone — KKMC rose from the wasteland of the Ad Dhana Desert with Oz-like splendor. To the Americans, who used the immense compound as an air base and logistics center, it was known as the Emerald City, gateway to Iraq.

Brigadier Generals Steve Arnold and John Stewart, still queasy after their turbulent flight, stepped from the jet onto a tarmac slick with rain. On this third morning of the allied ground offensive the two ARCENT staff officers had decided to visit both XVIII Corps and VII Corps; they were eager to judge for themselves the progress of the Army's attack. Inside the terminal, however, they discovered that bad weather had temporarily grounded all connecting helicopter flights to the north. Arnold also found an urgent message awaiting from John Yeosock: call as soon as possible.

He and Stewart walked to an Army communications van near the terminal, where Arnold phoned ARCENT headquarters. "Things are not going well down here," Yeosock confided. "The CINC is very, very unhappy with the speed of the attack. He wants us to move things faster. He doesn't understand why things are so slow. We need to go brief him and explain why the hell it's taking so long. I want you to come back and help me work this out."

Promising to return as soon as the weather permitted, Arnold hung up, once again baffled by Schwarzkopf's grievances. The CINC himself had slowed XVIII Corps, asking Gary Luck to hold up McCaffrey's 24th

Division at Phase Line Viking near the Euphrates Valley "until VII Corps can catch up." Schwarzkopf also had temporarily restrained Walt Boomer's drive through southern Kuwait because of concern that the 2nd Marine Division's left flank was exposed by the slow-moving Egyptians. CENTCOM's order, combined with heavy rain, oil smoke, and the counterattack against Mike Myatt's 1st Division, had brought the Marines to a dead stop through the night of February 25.

Despite the frustration caused by the atrocious weather, every officer in Riyadh — Schwarzkopf included — could only be elated that the allied casualties were low. As for VII Corps and its Left Hook, Arnold thought Franks's current maneuverings and the massing of forces under Frag Plan 7 were competent, even brilliant. To charge headlong into the Republican Guard, on the assumption that the Iraqis would not fight, was to risk a bloody nose or worse. Arnold pulled out his journal and scribbled: "CG [Yeosock] very upset. CINC in a rage. Wants us to get VII Corps moving faster. CG wants me to come back."

Perhaps, Arnold told Stewart, Cal Waller could better explain precisely what Schwarzkopf wanted. Arnold placed a call to the MODA war room. Waller was unavailable, so Arnold left a message and wandered over to the airfield terminal.

A few minutes later Waller phoned back. Stewart, unable to spot Arnold outside the van, took the call. "Sir, we're very confused about what's going on," he told the DCINC. "We're just trying to get a feel for the environment there and what we can do to make sure it's put right."

"The attack just isn't going right," Waller said.

"All right, sir. Got that one."

"You guys already have the plan that we put into —"

Waller was standing by the front table in the war room with a cup of coffee in one hand and the phone receiver in the other; he stopped short when Schwarzkopf wheeled around to face him.

"Cal," the CINC snapped, "stay out of ARCENT's business."

Stunned, Waller began to protest. "Sir, I'm only trying to clarify —"

"Cal, they've got a commander now," the CINC interrupted. "You're no longer the ARCENT commander. Get off the phone."

Waller nodded, seething. To Stewart, who was unaware of this byplay in Riyadh, he said, "John, the environment here is not worth a shit. Talk to Yeosock and do what you think is best."

"Okay, sir," replied Stewart, now thoroughly baffled.

"I haven't got any more for you, John."

Waller hung up the phone and stalked from the war room. In all the years he had known Norman Schwarzkopf, not until this moment had the man ever left him speechless with anger.

At KKMC, Arnold returned to the van. While Stewart was recounting the peculiar conversation with Waller, Yeosock called. "John, you were just talking to General Waller."

"Yes, sir," Stewart agreed.

"What were you talking to him about?"

"Sir, we were trying to get a feel for the environment to see if Steve and I could support you better."

Yeosock sighed. "It's not going right today, John. I just got another call from the CINC chewing my ass out again because my G-2 [intelligence officer] called the DCINC."

"That's not exactly what happened, sir. But I apologize."

"It's okay, John," Yeosock said before hanging up. "Don't worry about it."

Watching the rain sweep across the runway, Stewart and Arnold settled in to wait for a return flight that would carry them back to Riyadh, their mood now as gloomy as the weather.

## Phase Line Bullet, Iraq

No sight is dearer to a soldier, Napoleon once observed, than the knapsack on the back of a retreating enemy. Iraqi knapsacks, crammed with loot from Kuwait City, were now in full view. Pressed by the Marines and JFC-E, thousands of 3rd Corps troops had fled into the Kuwaiti capital. Kuwaiti resistance fighters, working with the CIA, reported more signs that the city was to be abandoned. In the Black Hole late on Monday night, Buster Glosson had received word from the resistance — this time passed through the Royal Saudi Air Force — that enemy forces were forming convoys with both military trucks and confiscated civilian vehicles. Glosson directed a JSTARS aircraft — a modified Boeing 707 capable of tracking movement on the ground much as AWACs tracked planes in the air — to focus its radar on the highway leading north of Kuwait City.

At 1 A.M. on Tuesday, Glosson called Colonel Hal Hornburg, commander of the 4th Wing at Al Kharj. JSTARS now reported unmistakable evidence of an Iraqi exodus: the aircraft's radar scope was peppered with tiny green crosses, each signifying a moving truck, tank, or car. At least 150 vehicles had been detected near Al Jahra on the six-lane highway leading from the capital toward Basrah. "I know you've just finished a hard night's flying," Glosson told Hornburg, "but I've got a job for you. I want you to put your guys back in their aircraft, fly over to Kuwait, and stop a convoy. Stop it at all costs."

Hornburg rousted his F-15E squadron commanders. Before dawn, battling an overcast that in places closed to within two hundred feet of the ground, a dozen Strike Eagles hammered the fleeing procession. After first dropping cluster bombs near Mutlaa Pass on the lead vehicles, the pilots struck the rear of the convoy to prevent a retreat into Kuwait City. For the next forty-eight hours Air Force, Navy, and Marine bombers flew hundreds of sorties against what became known, in the hyperbole of the gulf war, as the Highway of Death.

Schwarzkopf's exasperation with VII Corps was fueled by anxiety that the enemy would slip from his grasp. Although aircraft could and did wreak havoc on the Iraqi withdrawal, only ground forces had the wherewithal completely to sever the avenue of retreat. By Tuesday afternoon the corps's leading edge was approximately thirty miles from the Kuwaiti border and another forty miles from the highway. Fred Franks continued to urge his commanders forward, despite perpetual worries about fuel that were aggravated after ninety logistics trucks bogged down in the desert on his left flank.

The great wheeling of the corps to the east under Frag Plan 7 was nearly complete. In the south the British pushed across the first of their metallic objectives — Copper and Bronze — and Rhame's Big Red One bolted from the breach in a hell-for-leather dash to catch up to Holder's cavalry. In the far north Griffith's 1st Armored dumped more than three hundred MLRS rockets and fourteen hundred 155mm shells on Al Busayyah before attacking at first light on Tuesday with armor and mechanized infantry. By 9 A.M. the bunkers surrounding the town had been overrun, but a stubborn Iraqi battalion still held the main street and many of Al Busayyah's fifty buildings. Ordering most of his division to continue eastward, Griffith left a battalion task force behind to lay siege with a 165mm demolition gun powerful enough to flatten a house with a single round. By noon Al Busayyah, reduced to rubble, had fallen.

Between 1st Armored and the 2nd Cavalry lay the thirty-two battalions of Major General Paul Funk's 3rd Armored. A bald, large-framed native of Roundup, Montana, possessor of a doctorate in education and the wings of a Cobra pilot, Funk had recently witnessed dozens of mock tank battles as commander of the Army's National Training Center in the Mojave Desert. The coming fight, he knew, was likely to be brief, violent, and confused — all the more so since the drenching rains of Monday night had given way to a blinding *shamal* with thirty-knot winds and visibility that rarely exceeded a few hundred yards.

With his helicopter scouts grounded by the weather, Funk was uncertain of the enemy's position. The turn to the east had squeezed 3rd Armored into a narrow ten-mile front; this gave the division the mass of extraordinary killing power but almost no room to maneuver. Funk positioned his 2nd Brigade on the left, to his north; 1st Brigade on the right, to the south; and held 3rd Brigade in reserve. But as the division crossed Phase Line Bullet, his greatest concern was for a unit in which he had once served as a junior officer, the 4th Squadron of the 7th Cavalry. Equipped with Bradleys but no tanks, the squadron was groping through the sandstorm on Funk's extreme southern flank. The 7th Cavalry had an illustrious heritage marred by misfortune: butchered by Sioux Indians at the Little Big Horn in 1876, it had again been decimated after blundering into an ambush in the Ia Drang Valley, where the unit lost 151 dead and nearly as many wounded on a single day in November 1965.

At 3 P.M. the squadron commander, Lieutenant Colonel Terry Lee Tucker, radioed Funk to report that "visibility down here is about three hundred meters." Funk briefly debated whether to pull the cavalry behind the shield of 1st Brigade's tanks. Reluctant to withdraw his most agile scouts, he chose to wait — a decision he soon regretted.

The narrowing of the division sector fell hardest on Tucker. From a zone five kilometers wide — three miles — the squadron had been progressively squeezed against the southern boundary with the 2nd Cavalry. Tucker radioed the 1st Brigade commander, Colonel William L. Nash, to ask for a three-kilometer sector. With three large battalions of his own wedged into a seven-kilometer front, Nash denied the request. "I'm going to pinch you again," Nash replied. "It's now down to one click [one kilometer]."

Tucker again began shifting his men to the south. Bradley commanders stood erect in their open hatches, squinting through the swirling sand. Leading the squadron were the twenty Bradleys of Alpha Troop, commanded by Captain Gerald S. Davie, Jr., a young West Pointer from the north shore of Massachusetts who had amused his soldiers a few days earlier by reading aloud the dialogue between Cassius and Brutus in Shakespeare's *Julius Caesar*.

Davie had placed his 1st Platoon as Alpha's vanguard; 2nd Platoon followed five hundred yards behind; the seven Bradleys of 3rd Platoon trailed in reserve. At 3:30 P.M. Tucker advised Davie that the division was in pursuit, leaving the troop commander with the impression that no substantial enemy forces lay between him and the Kuwaiti border. A few minutes later 1st Platoon crested a low ridge, and gunfire raked across the Bradleys.

"We've got crunchies up front," a Bradley commander reported, referring to dismounted enemy infantrymen. Having encountered pockets of Iraqis a half dozen times in the previous two days, Tucker saw little cause for alarm. Alpha edged forward. Gunners peering through their thermal sights spotted first the small red dots of foot soldiers and then, abruptly, the larger crimson rectangles of BMP armored personnel carriers and tanks. Tucker relayed the report to Funk's command post, adding, "I don't know what we've got."

What they had, as he and Davie discovered soon enough, was a screen line of six T-72 tanks and eighteen BMPs from the Tawalkana Division. The Americans, blinded by the *shamal*, had rambled into an Iraqi kill sack, closing to within six hundred yards of the enemy guns. The deep boom of tank fire erupted across the desert. Davie ordered his 2nd Platoon forward to join the 1st, reinforcing both flanks with three more Bradleys. Alpha, which had been squeezing into the prescribed one-kilometer front, now fanned out again.

Davie fired a red star cluster to signal his position, but the wind promptly whipped the flare into the ground. Several hundred yards to his left, on Alpha's extreme northern flank, Bradley number Alpha 2–4 had loaded a TOW missile and was preparing to fire. But neither Davie nor the Bradley crew realized that 2–4 had crossed in front of a tank battalion from 1st Brigade, which also was spreading into attack formation half a mile behind the cavalry.

Without warning an M1A1 sabot round sliced into the Bradley just below the turret ring on the left side. The shell ripped across the lap of the gunner, Staff Sergeant Kenneth B. Gentry, searing away his thighs, then tore through the lower left leg of the Bradley commander, Sergeant Raymond Egan, before blowing out the other side of the hull.

The squadron's command sergeant major, Ronald Sneed, sped toward Alpha 2–4. Short and thickset, with a rolling gait like that of a sailor on the quarterdeck, Sneed had spent five years in Vietnam with the 173rd Airborne Brigade, yet never had he passed a more intense hour of combat than the one unfolding near Phase Line Bullet. Green tracers threaded the air from Iraqi infantrymen only fifty yards beyond 2–4. Sneed's gunner sprayed the enemy trench with 25mm cannon shells. Milky smoke poured from 2–4, mixed with Halon, a gaseous fire retardant automatically released when the Bradley was hit. Egan lay writhing on the ground. As Sneed scrambled from his hatch, a T-72 fired from less than six hundred yards. The round landed thirty yards short. A second round just missed Sneed's Bradley, spattering the hull with dirt and knocking him flat. He scrambled to his feet and sprinted to Alpha 2–4, where he helped Egan into the back of another Bradley, then dragged Gentry from the gunner's seat. Small arms fire beat a tattoo against the Bradleys. Mortar rounds detonated along the ridge. Tucker, who had moved to within a hundred yards of 2–4, spotted an Iraqi RPG (rocket-propelled grenade) team maneuvering toward the stricken Bradley; his gunner killed them. Egan would survive his wounds, as would 2–4's driver, who had suffered burns to the head. Gentry quickly lost consciousness and died several minutes later.

TOW missile and cannon fire swept from the thirteen Bradleys. A BMP exploded in a pillar of fire, then another, and another. Radios screamed with calls for help. "We need a medic! We need a medic!" someone pleaded. Davie managed to get the soldier to identify himself — he was a crewman who had been wounded in Alpha 2–6 — then added, "Now shut the fuck up and get off the radio net."

In Alpha 3–3 an Iraqi machine gun round ricocheted off the radio, wounding the Bradley commander in the hip. His gunner continued to fight while another crewman tried to stanch the bleeding. Davie, hoping to prevent the Iraqis from zeroing in on stationary muzzle flashes, urged the troop to keep moving. Ammunition dwindled; several Bradleys reported themselves "black," or reduced to less than a quarter of their stocks. Coaxial machine guns jammed repeatedly, forcing the crews to use cannon fire against the Iraqi infantrymen darting in and out of foxholes.

Now trouble developed on Davie's right flank. An Iraqi RPG blew a hole in the transmission of Alpha 3–6. Immobilized, the Bradley commander continued to fire as Lieutenant Michael J. Vassalotti moved forward in Alpha 3–1. When Vassalotti arrived, the crew of 3–6 scrambled from the crippled Bradley just as a Sagger missile struck its left front slope with a blinding flash, wounding all four men with shrapnel. Bleeding and dazed, the men tumbled into the rear of 3–1.

But as Vassalotti sped west in retreat, two M1A1 sabot rounds from 1st Brigade struck the front slope just below the turret, triggering a spray of Halon and igniting several 25mm rounds. Miraculously, none of the eight men inside was killed. Vassolotti, temporarily blinded with flash burns, was thrown from his hatch. Draped across the machine gun barrel, unable to see or communicate with the terrified driver, who also had been wounded, the lieutenant clambered down the front of the careering Bradley to the hatch and ordered the driver to stop.

Davie did not realize that much of the fire raking the troop came from his own division, but Tucker had spotted muzzle flashes from the rear. He radioed Nash, the brigade commander: "I'm concerned about those guys behind me." He then called the armored battalion commander directly. After a sharp exchange — the commander, apparently unaware that his crews had opened fire, initially denied that his tanks were shooting — the battalion was ordered to cease firing.

Yet one final tragedy remained to be played out. As 3rd Platoon pulled back, a sabot round — investigators never determined whether it was fired from 1st Brigade or the 2nd Cavalry to the south — slammed into the turret of Alpha 2–2 from the left rear. The round blew through the back of the gunner, Sergeant Edwin B. Kutz, killing him instantly and wounding two other crewmen. Davie pulled his Bradley alongside 2–2. An enemy soldier dashed across the battlefield three hundred yards to the east. Davie's coaxial machine gun was among the few that had not jammed, and he fired more than a hundred rounds before hitting the Iraqi. The wounded men from 2–2 were rescued, but Kutz's body was so tightly wedged into the turret that it could not be retrieved until the next morning.

Davie fired a barrage of smoke grenades, masking his withdrawal behind a dirty brown cloud. The Iraqis had been savaged, most of their BMPs and tanks destroyed. All fourteen Bradleys involved in the fight had been hit with shrapnel, small arms, or RPG fire; three of them had been demolished by American sabot rounds. The nameless skirmish was hardly comparable — in magnitude, consequence, or infamy — to the Little Big Horn or the Ia Drang Valley, yet two men were dead and another dozen wounded, most at the hands of their comrades. The cavalrymen had fought bravely, unaware for the most part that they were caught in a crossfire. Three men would win Silver Stars for heroism. General Funk, chastened by the prospect of more friendly fire casualties, ordered his division to halt until daylight.

As he pulled the remnants of Alpha Troop to the west, Davie for the first time saw 1st Brigade's tanks behind him. Irritated and perplexed, he wondered why they had not joined the battle.

"Men at some time are masters of their fates," Cassius observes in Act I of *Julius Caesar*, the very lines Davie had read to his men. "The fault, dear Brutus, is not in our stars, but in ourselves."

## As Salman, Iraq

At last *le drapeau tricolore* snapped in the wind above the captured town of As Salman, code-named Cleves by the French occupiers. Objective White, the westernmost goal in the allied attack plan, had fallen.

Moving with the same ponderous deliberation that had so frustrated XVIII Corps early in the ground offensive, the Daguet Division had shoved north from Rochambeau on Monday, February 25. Surrendering

Iraqis poured from their fetid bunkers in such numbers that military policemen ran short of plastic handcuffs. By midafternoon the 4th Dragoon Regiment — fortified with another grand battle order: *"Attaquez et exploitez!"* — reached the southern outskirts of As Salman. Despite American eagerness to seize the town and adjacent airfield at once, the French again laagered for the night.

U.S. Army psyops teams broadcast surrender appeals; helicopters dropped leaflets; American officers paced impatiently. At dawn on Tuesday, the French 3rd Marine Infantry Regiment swept into what proved to be an abandoned town. Of a prewar population variously estimated at three to seven thousand, the French found a dozen civilians, including a few Bedouin and the mayor, as well as fifteen bedraggled soldiers.

Goats and stray dogs roamed the deserted streets. Half-eaten meals of rice and onions lay strewn on kitchen tables. Obsolete military maps adorned the police station walls, with green thumbtacks marking Iraqi lines no longer held and red tacks representing allied positions long since advanced. Mortar shells littered the courtyard of a one-story school. Iraqi military uniforms, labeled "Made in Romania," hung from a rack in the As Salman hospital. French soldiers entering one house from the rear found a rocket rigged to detonate when the front door opened. The booby trap was defused, but a short time later two soldiers were killed and several wounded when either a mine or an unexploded cluster bomb blew up during a search of the fort crowning a ridge north of town.

With its French allies ensconced in As Salman, the 82nd Airborne Division was now free to turn east in an effort to catch the rest of the corps pushing into the Euphrates Valley. Gary Luck, commander of XVIII Corps, wanted the division to open Main Supply Route Virginia from As Salman to Objective Grey on the corps's eastern flank so that ammunition and fuel trucks could resupply American tanks driving toward Jalibah Air Base. Thwarted by the Daguet's pedestrian pace, unhappy at playing a peripheral role in the attack, the 82nd now suffered further aggravation at the hands of its sister division, the 24th Infantry, commanded by Barry McCaffrey. Pushing east on MSR Virginia, the paratroopers quickly rolled through the 101st Airborne corridor near Objective Cobra only to be stopped by a roadblock as they entered the 24th's sector at Objective Brown.

"By order of General McCaffrey," a military policeman declared, "no one goes any farther." Jim Johnson, the 82nd commander, pushed to the head of the column. "Now wait a fucking minute," he snapped. "Our mission is to get to Objective Grey, and it's getting dark."

For several hours the standoff persisted. Johnson, livid, sent his deputy to negotiate with the 24th. In Johnson's view this was yet another example of the imperious McCaffrey refusing to acknowledge the legitimacy of any unit's mission other than his own.

The 24th, not surprisingly, had a different perspective: instead of opening the logistics line, the 82nd had pinched it off by clogging MSR Virginia. McCaffrey's staff expected Johnson's brigades to wait until they were summoned forward. Now ninety-five petrol tankers, 160 ammunition trailers, and nearly a hundred water trucks were stuck in a traffic jam behind the paratroopers while armored commanders in the north pleaded for fuel. The 24th would blame the 82nd for at least a seven-hour delay in pressing the attack.

As the furious Johnson prepared to run the blockade — "Move aside," he declared, "we're coming through" — the contretemps was finally resolved. The 82nd continued east toward Grey, the 24th north toward the river. Bad blood boiled in both camps. Months later the episode still rankled. Searching for the exact words to capture his vexation, Johnson would observe: "It was easier to work with the French than it was with Barry McCaffrey."

While the divisions squabbled in the east, fatal misfortune struck in the west. The airfield at As Salman was considered a critical way station for ferrying wounded soldiers back to Saudi Arabia from the Euphrates Valley. Consequently, Alpha Company of the 27th Engineering Battalion had been left behind with orders to "get that field up and running as quickly as possible." The biggest task was clearing the runway and airport access road of unexploded bombs left by five weeks of allied air strikes.

Ordnance disposal — "depolluting," the French called it — was a test of patience, competence, and nerve. Although the dud rate of both Air Force bombs and Army artillery shells allegedly had improved since Vietnam, thousands of unexploded munitions littered Kuwait and southeastern Iraq. Some experts believed that soft sand sometimes cushioned the impact, exacerbating the malfunction rate of fuses designed to explode on contact.

Standard Army procedure for bomb disposal called for "blowing in place." A one-pound block of C-4 plastic explosive was set next to the munition; an engineer then activated a three-minute fuse, scuttled to safety several hundred yards away, and waited for the explosion. The task bred both professional pride and a robust fatalism. Two weeks earlier the 27th Engineers had marched in formation across the Saudi desert, loudly chanting a jody call:

Oh hell, oh hell, oh engineers,
What the hell are we doing here?
Oh mama don't you cry,
Your little boy is gonna die.

Alpha Company spent Tuesday morning blowing up approximately eighty BLU-92 submunitions from cluster bombs that had split open in midflight to disgorge a stack of long, slender cylinders. Each cylinder contained enough explosive — if detonated — to dismember anyone attempting to repair the battered roads or runway. Another two dozen unexploded BLU-92s remained to be cleared from the airfield when work was suspended for three hours because of high winds whipping the submunitions around the tarmac.

When the wind ebbed in the afternoon, work resumed. For unknown reasons, Alpha Company abandoned standard safety procedures. Seven experienced engineers — the company commander, Captain Mario Fajardo, a platoon leader, a platoon sergeant, and four enlisted soldiers — began collecting the cylinders by hand and stacking them in a pile. Investigators would later speculate that they were trying to save time, conserve their supply of plastic explosives, or minimize damage to the runway.

When fifteen cylinders had been collected the engineers gathered around Sergeant First Class Russell G. Smith — thus violating another safety procedure by congregating in a group rather than working only in pairs. Smith sank to his knees and appeared to be priming a fuse next to the stack when, for reasons again unknown, the munitions detonated.

A tremendous roar shook the airfield. The blast instantly blew all seven soldiers to pieces. Uniforms and equipment vaporized; none of the men's weapons was ever found. A medical platoon spent the rest of the day picking up arms, legs, and heads, which were flown in nine body bags to a mortuary in Saudi Arabia. Pathologists would labor for weeks to positively identify the dead.

By late afternoon the survivors from Alpha had finished clearing the airfield. In a convoy that looked like a funeral cortège, with flags flying at half mast, the company pushed east to catch up with the rest of the 82nd. The catastrophe provoked sorrow, disbelief, and anger. There was even a proposal to deny the dead their Purple Hearts as a posthumous rebuke. Although quickly dismissed, the suggestion served as a reminder that however strong the ties that bind soldiers in combat, pettiness also lurked on the battlefield.

## Riyadh

The Iraqi army, having demonstrated little aptitude for the simpler requisites of warfare, now faced the most difficult of military maneuvers: a withdrawal under fire. No tougher test of generalship could confront a commander than the exigency of conducting a coherent retreat. Even great battle captains, of which Iraq was singularly bereft, often found themselves overmatched by simultaneously trying to organize an effective rear guard capable of slowing enemy pursuit; evacuate the wounded and, if possible, the dead; and rally dispirited and humiliated soldiers, whose every step in retrograde was a reminder of defeat.

Rare was the army that could retire gracefully from the field. Perhaps the most celebrated retreat was that of Xenophon, the Athenian warrior, who after the battle of Cunaxa in 401 B.C. — not far from present-day Baghdad — led the Ten Thousand on a five-month *anabasis* (upcountry march) along the Tigris, fighting off Persian cavalry, hostile tribesmen, and winter hardships to reach safety on the Black Sea. A successful withdrawal usually required not only pluck by the vanquished, but also miscalculation or ineptitude by the victor. Some 340,000 British and Allied soldiers escaped from Dunkirk in 1940 after Hitler halted his Fourth Army tanks at the Aa Canal, foolishly relying on Hermann Goering's boast that the Luftwaffe would destroy the trapped force. ("It is always good to let a broken army return home to show the civilian population what a beating they have had," the Führer later explained.)

Even orderly retreats often were costly. After the Allies breached the Gustav Line in May 1944, the Germans maintained reasonable discipline while pulling out of Italy but still sustained seventy thousand casualties. Six years later, when Chinese soldiers poured into North Korea from Manchuria, the U.S. Army's 2nd Division braved a murderous gauntlet during the withdrawal from Chongchon River at a cost of three thousand dead and wounded. During the same sad week, troops from X Corps and the 1st Marine Division endured an even worse ordeal while retreating from Chosin Reservoir in temperatures so bitterly cold that mortar tubes cracked and blood plasma froze solid.

When a commander failed to prevent retreat from becoming rout, the results could be even more horrific — witness the Prussians' loss of 140,000 men while being chased by Napoleon across half of Europe after Jena and Auerstädt in 1806; or Bonaparte's own headlong dash from Russia six years later; or the Union Army's flight from First Bull

Run in 1861, when a tidy withdrawal became a stampede "miles long and a hundred yards wide" all the way back to Washington.

Clearly the Iraqi retreat was no latter-day *anabasis*. In some respects it was grimmer than Bull Run or Chongchon River, since neither Irvin McDowell in northern Virginia nor MacArthur's commanders in Korea faced an enemy who owned the skies as well as the land. Lacking cover but for the fickle veil provided by bad weather, the Iraqis were forced to flee across open country patrolled by hundreds of allied bombers.

On Tuesday afternoon the Army's Tiger Brigade, leading the 2nd Marine Division, had rolled over Mutlaa Ridge to capture the high ground and highway interchanges bordering Al Jahra at the west end of Kuwait Bay. (Although CENTCOM's script had called for this key terrain to be captured by the Egyptians, Walt Boomer brushed the plan aside with a curt "Bullshit. *We* are going to take that junction.") At least fifty Iraqis were killed in bloody fighting around the Al Jahra police station. Allied armor now blocked the only overland escape route from Kuwait City, completing the encirclement of the capital begun earlier in the day with air strikes on the Highway of Death.

The Iraqis were not without a withdrawal plan, of sorts. Unfortunately for Baghdad, the scheme was known to the Americans.

For months Iraq had practiced admirable "emissions control," EMCON, the art of minimizing radio and other electronic signals to prevent the enemy from eavesdropping or homing in on the radiating source. Allied intelligence had concluded that willful violators of Iraqi COMSEC, communications security, might be executed. But the adroit use of military radio networks is a highly perishable skill unless it is routinely practiced. If EMCON helped preserve Iraqi security, it also stripped the army of proficiency in orchestrating military maneuvers — a self-inflicted wound known in the intelligence world as "EMCON suicide."

A vast array of allied antennae — aboard satellites, aircraft, and on the ground — strained to intercept any stray Iraqi signal. As the air war took its toll on the enemy command and communications network, some units had "come up" with occasional radio transmissions. The Americans hotly debated whether it was better to eavesdrop on such broadcasts or electronically jam them. (Among the U.S. jamming systems operating in northern Saudi Arabia was the so-called Sandcrab, a 5000-watt contraption with an antenna larger than a football field.) In some cases only the cryptographic data at the beginning of the message were jammed, confusing the Iraqis and causing them to broadcast uncoded, "in the clear."

The Republican Guard, however, had remained silent since September 15, communicating primarily by secure land lines. Using intelligence gathered by the National Security Agency and other sources, CENTCOM assembled a blueprint of the Iraqi communications architecture in the Kuwaiti theater. A day before the ground attack, the intelligence chief Jack Leide had decided that of fourteen "nodes" — suspected headquarters or antennae farms — ten would be destroyed by air strikes or long-range artillery. Four others, all in the northern sectors controlled by Republican Guard divisions, were left intact in the hope that the enemy would resort to radio communications.

The strategy worked. Once the allied attack forced the Iraqis to begin moving, stationary land lines became an impediment. Late on February 25, for the first time in five months, the Republican Guard came up on the air with an FM radio broadcast from the Tawalkana commander, directing his subordinates to begin forming a defensive line against the allied onslaught. As the division struggled to mount its defense, a transmission from Basrah early on the 26th — either from the Republican Guard commander or his chief of staff — warned the Tawalkana that they were violating COMSEC. The Tawalkana commander angrily replied that with the American attack well under way he had little security left to protect.

This squabble, monitored by the National Security Agency and relayed to Riyadh, was followed six hours later by other intercepted messages. Another Republican Guard division — subsequently recognized as the Hammurabi — was informed by Guard headquarters in Basrah that heavy equipment transporters (HETs) had been dispatched to haul their T-72 tanks northward; heavy-duty tractors also had been sent to pull out the division's artillery. The Madinah Division, in turn, was instructed to burn unneeded equipment and move west by southwest into blocking positions.

This knowledge was most helpful to the allies. Leide carried the translated intercepts to Schwarzkopf in the war room, where, in combination with JSTARS radar images of enemy vehicles being repositioned, they gave a clear picture of enemy intentions. The Tawalkana commander, not without courage, was trying to cobble together a defensive line with three of his own brigades plus a pair plucked from other divisions. His command-and-control appeared so shaky, however, that he was trying to direct individual battalions rather than work through his brigade subordinates.

"The Tawalkana is the sacrificial lamb at this point," Leide's deputy, Colonel Chuck Thomas, told Schwarzkopf. "They're going to delay us as long as they can." The Madinah would form another screen line, while the Hammurabi and any forces that had escaped from southern Kuwait pulled back to the Basrah pocket, a strategy Iraqi commanders had used in the war against Iran.

The Defense Intelligence Agency now assessed twenty-six of forty-three Iraqi divisions in the Kuwaiti theater as "combat ineffective." More than thirty thousand prisoners had been captured. Iraqi tactical reserves west of the Wadi al Batin were collapsing under pressure from the British. Colin Powell, who before G-Day had estimated that the ground attack would take two weeks at most, now believed the end was in sight after three days. "Norm," the chairman told Schwarzkopf, "this is going so well that it won't take much longer."

Schwarzkopf still had one armored division that had not been committed to the fray, the 1st Cavalry. Waller and Steve Arnold had pressed without success to get the division released in order, as Arnold put it, "to reinforce success, to make the attack on the Republican Guard overwhelming." For two days the CINC resisted. "What the hell do I do if the Iraqis focus their counterattack on the Egyptians?" he asked. "What if we're running off to the north, having no trouble, while they're beating up on the Egyptians? If that happens, I've got a problem."

Now, however, the threat of a counterattack had dissipated. The Egyptians had finally reached their objective — the Al Abraq military barracks in western Kuwait — and turned east toward Ali As Salem Air Base. John Tilelli, the 1st Cavalry commander, believed that Iraqi defenses were weak enough to permit an attack straight up the Wadi al Batin. Schwarzkopf again demurred, but agreed to let the division pull back from its feint up the wadi, swing west through the breach opened by the Big Red One, and race north to join VII Corps — a 200-mile trek. That would give Franks five divisions — including the British — and the cavalry regiment.

Franks again phoned Yeosock, who repeated Schwarzkopf's displeasure with VII Corps's pace. "Do you think I should talk to the CINC myself?" Franks asked. "Yeah," Yeosock agreed, "sounds like a good idea. Why don't you call him?"

At 4 P.M. on Tuesday, clutching a stubby grease pencil used to mark his rain-streaked map, Franks called the war room by satellite phone from his forward command post. Struggling to make himself heard over the wind and the boom of artillery fire, he explained to Schwarzkopf the disposition of his divisions. The 2nd Cavalry had already found the Tawalkana screen line; soon they would give way to Rhame's 1st Division. But Franks also worried about a bypassed brigade of enemy troops on his right flank. When he suggested sending some forces south to clean out the pockets — a proposal initially recommended by Yeosock — Schwarzkopf interjected, "Fred, for Christ's sake, don't turn south! Turn east. Go after 'em!"

Franks quickly agreed. "We know where the enemy is," he told the CINC. "We think we have a seam down there and we're going to push the Big Red One through it. Tomorrow I'll have three divisions massed against the Republican Guard. I figure it will take us forty-eight hours or so, with those three divisions intact, to get through them. I don't think they see us coming. We're going to surprise them."

Schwarzkopf, perhaps reluctant to rattle the corps commander in the middle of a fight, made no mention to Franks of his anger. "Fine. Press the fight, Fred," he said. "Keep the pressure on." Schwarzkopf told Franks of the radio intercepts, quickly recounting the squabble between the Tawalkana commander and Republican Guard headquarters. HETs, the CINC added, apparently had been sent to extract the Hammurabi. Heavy fog was predicted for Wednesday morning, so a night attack seemed advisable. "You'll have good shooting tonight," the CINC said. "Don't let them break contact. Keep 'em on the run."

Franks signed off, his exhaustion temporarily masked by a fresh surge of adrenaline. At 5:30 he flew back to the corps headquarters and briefed his staff. "We're going to drive the corps hard for the next twenty-four to thirty-six hours, day and night, to overcome all resistance and to prevent the enemy from withdrawing," Franks warned. "We'll synchronize our fight, as we always have, but we'll have to crank up the heat. The way home is through the Republican Guard."

## 73 Easting, Iraq

The eight thousand soldiers of the 2nd Cavalry, leading VII Corps across Iraq much as the regiment once led Patton's Third Army across Europe, had been on the move since shortly after 6 A.M. Colonel Don Holder for the first time brought all three maneuver squadrons abreast. He carefully controlled their progress by means of north-south grid lines called eastings. (The 81 Easting was one kilometer east of the 80, and so on.) Franks had ordered the regiment forward to the 60 Easting, approximately twenty-five miles from the Kuwaiti border. Hesitant to embroil his cavalry in a slugging match before Rhame arrived with the 1st Division's three hundred tanks, the corps commander advised Holder to probe only with his scouts. Holder interpreted this order broadly; rather than expose the lightly armed scouts, he instead moved forward the entire regiment — including his 125 M1A1s. "We're all scouts," he told his staff.

That the Tawalkana lurked near the 70 Easting had been suspected since Monday afternoon. The suspicion was confirmed at 7:15 A.M. Tuesday when an OH-58 helicopter crew spotted an Iraqi T-72 tank, a sure sign of the Republican Guard. An hour later cavalrymen destroyed two enemy personnel carriers with TOW and mortar fire; a captured Iraqi captain, schooled at Fort Benning and fluent in English, provided further intelligence. Several other skirmishes erupted across the regimental front as the cavalry bumped into pickets deployed five to seven miles west of the main enemy line. An Iron Troop platoon routed five Bedouin with camels after noticing army boots beneath the men's robes and grenade launchers protruding from the animals' packs. "The regiment," Holder reported to Franks, "has found the security zone of the Tawalkana Division."

Franks ordered the cavalry to edge forward to the 70 Easting. The intensifying *shamal,* which was bedeviling other American units across southeastern Iraq and Kuwait, now grounded Holder's helicopters. Instead of being able to look at least ten miles deep, the regiment found visibility limited to the reach of their thermal sights. As the afternoon slipped by, Holder inched his squadrons forward past the 62 Easting, then 63, 64, 65.

Sitting in the rear of his command track, Holder turned to the regimental operations officer, Major Doug Lute. "If the enemy is getting out of Kuwait," he asked, "why die for it?" Franks had demanded caution; the cavalry had accomplished its mission by fixing the enemy's location. Artillery and Air Force bombers were ready to pounce. Rhame would soon arrive with the Big Red One. "We don't want to bloody ourselves if we don't have to," Holder added. "I'm not going to impale this regiment for the sake of body counts or glory."

By 4 P.M. — about the time Franks called Schwarzkopf to clarify his plan — the howling westerly wind had lessened slightly, although visibility with the naked eye was still less than half a mile. Flat and barren, the terrain in many places rippled just enough to hide enemy armor until the Americans had closed within a thousand yards or so — less than half the distance at which U.S. tank and Bradley gunners preferred to shoot.

The three cavalry troops on Holder's left flank — Ghost, Eagle, and Iron — each mustered approximately 140 soldiers mounting twelve Bradleys and nine M1A1s. By chance, all three were commanded by West Point classmates of Gerald Davie, whose unfortunate troop in the 7th Cavalry was fighting battalions from both the Tawalkana and his own 3rd Armored Division farther north. Eagle Troop, under Captain H. R. McMaster, Jr., had just passed the 67 Easting when machine gun fire poured from a cluster of buildings behind a sand berm. (The compound proved to be the headquarters for the enemy's armor training area, the Iraqi equivalent of the U.S. Army's National Training Center in California.) As several Bradleys returned the fire with their 25mm cannons, McMaster wheeled his nine M1A1s on line and ordered a simultaneous volley of tank rounds and TOW missiles. Cinderblock walls exploded, roofs collapsed, flames danced from the shattered buildings.

Taking the point with his tanks in a wedge and the Bradleys tucked behind to protect the flanks and rear, McMaster swung north past the burning outpost. At 4:18 P.M. the troop crested a low hill. "Contact!" his gunner yelled. "Tanks, direct front!" McMaster dropped down inside the hatch to peer through his own sights. At least eight T-72s lay behind shallow dirt fortifications in a "reverse slope" ambush, hoping to surprise the Americans as they came over the rise with their turrets silhouetted against the sky. "Fire, fire sabot!" McMaster ordered.

With a deep roar the round burst from the muzzle at one mile per second. The barrel jerked back in violent recoil and the Abrams's front wheels lifted three inches off the ground. The first enemy tank, seen as a small rectangle in the thermal sights, blew up as the round struck just above the turret ring. McMaster's loader slammed another shell into the breech. "Up!" he barked. Again flame licked from the muzzle and a second T-72 exploded, the turret ripping free of the hull in a fountain of orange fire. A pair of enemy sabot rounds burrowed harmlessly into the sand on either side of McMaster's Abrams. Ricocheting bullets struck sparks off the hull. McMaster's gunner fired for the third time; a third tank, only four hundred meters away, blew up.

Eagle Troop now creaked down the slope at twenty miles an hour. Game but overwhelmed, the Iraqi tankers fired wildly, then sat paralyzed while their automatic loaders — which required ten seconds between shots, an eternity in armored combat — slowly pushed home another round. (The tubes of reloading T-72s raised up slightly before dipping toward the ground, a sure sign that the enemy crew was momentarily helpless.) Iraqi commanders strained desperately to see through the smoke and haze. American intelligence had expected at least some of the enemy tanks to be outfitted with Belgian thermal sights. In this brigade of the Tawalkana, however, gunners possessed only ordinary telescopic lenses, a fatal disadvantage. Here the blind fought the sighted.

Within four minutes all of the T-72s and several BMPs in the first line of defense were in flames. Bradley gunners "scratched the backs" of the M1A1s, killing enemy infantrymen who hid or feigned death until the American tanks rolled past and then tried to shoot them in the rear grill with RPGs. Smoke boiled from the shattered Iraqi hulls, where the charred bodies of crewmen lay draped over their hatch rims. Ammunition cooked off in spectacular orange pinwheels. The stench of burning hair and flesh drifted across the desert.

A mile beyond the first line lay a second tier of seventeen T-72s parked across a three-mile front. McMaster, reminded on the radio by his executive officer that the 70 Easting was the regimental limit of advance, plunged ahead. "I can't stop," he replied. "We're still in contact. Tell them I'm sorry." Once again the Iraqis stood their ground and fought as the Americans charged forward; once again the enemy line disintegrated.

At 4:40 P.M. McMaster halted Eagle Troop on the 74 Easting, forming a defensive circle not unlike a wagon train. A hundred crackling fires cut through the afternoon gloom. Mortar crews set up their tubes and dumped shells — set to detonate twenty feet above the ground — on enemy infantrymen fleeing among the dunes to the east. Eagle would be credited with the destruction of twenty-eight tanks, sixteen personnel carriers, and thirty-nine trucks — with no American losses. The entire fight had lasted twenty-three minutes.

On Eagle's right flank Iron Troop paused at the 68 Easting as small arms fire poured from the same buildings McMaster had just passed on the north. The cinder-block compound again shuddered under a volley of American tank and cannon fire, this time from the south. Iron's scouts now spotted several vaguely rectangular shapes two miles down range. Edging forward, Captain Daniel B. Miller saw the unmistakable silhouette of Iraqi tanks. Positioned behind a network of L-shaped berms, the enemy turrets slowly swung toward the oncoming Americans. "Action front, action front," Miller warned. "Follow me." With two Bradley platoons positioned on each flank, Iron's tanks attacked in a flying wedge.

The enemy line quickly crumpled and collapsed. Miller pushed his scouts forward five hundred yards, just beyond the 70 Easting. Soon a second formation of enemy tanks — apparently counterattacking from the southeast — moved into range. Gunfire again erupted. One TOW appeared to skip across the hull of a T-72, hitting another in the turret. Iron Troop's 4th Platoon swung to the south and caught the enemy in a crossfire.

As the second Iraqi echelon dissolved, Iron took its only casualties of the fight. A 1st Platoon Bradley, commanded by Sergeant First Class

Ron Mullinix and temporarily disarmed by an electrical malfunction, headed for the protection of an empty tank revetment. Mullinix spotted a small globe of fire streaking across the desert from the south: an errant TOW from Killer Troop, which had mistaken the Bradley for an Iraqi armored vehicle. The missile struck the Bradley turret, ricocheted, and ripped off the driver's hatch. Shrapnel spattered Mullinix in the legs and peppered the back of his driver, Private First Class Gregory Scott.

Fire swept through the Bradley, melting the radio and igniting the machine gun ammunition box. Coaxial rounds sprayed the turret. The gunner, Sergeant Kirk Alcorn, had been flung from his hatch by the blast. His face burned and eyelids seared shut, Alcorn crawled across the TOW launcher and leaped to the ground. Scott rolled screaming across the desert, his back and arms afire. Mullinix hobbled over to help beat out the flames and rip off the driver's burning flak jacket. The three wounded soldiers, all of whom would survive, were bundled into the rear of another Bradley to await a medic.

The third and final act in the Battle of 73 Easting unfolded to the north, where Ghost Troop anchored the regimental left flank. Unlike the relatively brief fights in the Eagle and Iron sectors, this one lasted several hours.

On learning of Eagle's contact on his right, Captain Joseph F. Sartiano, Jr., had pushed his two Bradley platoons to the flanks and moved Ghost's eight tanks forward in the center. "Troop on line," he ordered. "Tanks lead." Having requested and received permission to move past the 70 Easting in order to see beyond a rise in the desert pan, Ghost crept forward on a three-mile front.

Sartiano's gunner, a twenty-year-old corporal from Arizona named Frank Wood, pressed his forehead against the vinyl headrest on his thermal sight. He could see a curious lump directly ahead, but it emitted no telltale heat signature. When Wood depressed the button that activated his laser range finder, crimson numerals flashed in the viewpiece: 587 meters. Switching to his day scope — a ten-power telescopic lens — Wood found the same amorphous mass through the veil of swirling sand. "Hey, sir, what's this?" he called. "Take a look." Sartiano, wearing goggles and a bandana knotted across his mouth and nose, dropped down from the open hatch and squinted into his own eyepiece on the right side of the tank.

At that moment Wood saw movement. Eight or nine soldiers swam into view, gathered near an armored personnel carrier. "Holy shit! Troops!" he yelled. "These guys don't have their hands up!" Wood fired a machine gun burst as the Iraqis scrambled toward the rear ramp of the BMP. Bodies jerked grotesquely when the bullets struck home. Men slumped to the ground, struggled to their knees, and pitched forward again. One slug caught an Iraqi in the back of the skull, blowing his head open. As Sartiano turned his attention to the rest of the troop, Wood destroyed the BMP with a tank round.

The Iraqi tanks and other vehicles, Sartiano realized, had been sitting still with their engines off, reducing their heat signature. Almost the entire troop had begun firing at the same time. For the first fifteen minutes the radio burst with excited chatter: "Engaged and destroyed one enemy tank!" "Engaging two vehicles." "Enemy tank destroyed."

As in the south, the Tawalkana had established a defensive line. Tanks and personnel carriers were parked behind sand mounds — "kill-me berms" — that obscured the Iraqis' vision, restricted the free play of their turrets, and offered little protection against TOWs or tank rounds. Rather than push through the line as Eagle had done, Ghost remained along the 73 Easting — firing, repositioning laterally, firing again.

Using tactics developed in countless gunnery sessions and exercises, the American crews methodically selected their targets, shooting those at closest range first before working deeper into the enemy ranks. Gunners on the flanks fired at targets "from the outside in"; gunners in the middle fired "from the inside out." Bradley sights were more powerful than those on the M1A1, and the turret — or doghouse — was a foot higher than the Abrams turret, giving the Bradley crew slightly better visibility and allowing them to use tracer rounds to pinpoint targets for the tankers.

The Tawalkana fought back, though ineffectually. Bullets pinged off the American tanks, and a few mortar rounds detonated overhead, shredding the duffel bags and water cans lashed to the Abrams and Bradley hulls. Sartiano spotted several beautiful blue bursts as Sagger missiles streaked harmlessly overhead.

A few minutes into the fight, Iraqi gunners found their mark for the first and only time. On Sartiano's left flank, where 1st Platoon was firing at enemy infantry and personnel carriers, a 73mm round from a BMP grazed the front slope of Ghost 1–6. "What was that?" shouted the Bradley gunner, Sergeant Nels A. Moller. Those were his last words. A second round struck beneath the TOW launcher, killing Moller and wounding a soldier in the back before blowing through the rear ramp. The driver and Bradley commander leaped from the burning vehicle. Other cavalrymen pried at the jammed ramp with a crowbar before hoisting the wounded crewman out through the cargo hatch on top.

(Hours later, when Hawk Troop moved forward to relieve Ghost, the lead tank mistook 1–6 for an enemy vehicle; a 120mm round destroyed what was left of the smoldering Bradley, where Moller's body still lay.)

The Tawalkana defensive line was no more. For the next three hours, however, enemy tanks, personnel carriers, and trucks poured into Ghost's kill sack. At 5:40 P.M. an Iraqi company counterattacked from the northeast. Bradleys from 1st Platoon destroyed four enemy vehicles with TOWs at twelve hundred meters, blunting the charge. Iraqi infantry scrambled from the rear of their personnel carriers only to be cut down with machine gun and cannon fire. Eleven more tanks and BMPs blew up as the Americans called in artillery strikes and lobbed more than two hundred mortar rounds at the enemy.

Sartiano, pleading for air support, shifted several tanks to reinforce 1st Platoon on the left. Together they shattered another company-sized counterattack from the northeast. The American gunners watched and waited in ambush, chain-smoking cigarettes between firefights. Shortly after 8 P.M. a third enemy force of sixteen vehicles appeared from the southeast. With Iraqi engines running hot and the weather improving, the enemy was now plainly visible to Sartiano's gunners at ranges of four thousand yards or more. Some were destroyed with long-range tank or TOW shots. A few ambitious Bradley gunners fired at targets beyond the 3700-meter TOW range, causing the missiles to "squelch" by snapping their control wires.

Most of the killing along the 73 Easting was the handiwork of direct fire, but not all. The 210th Field Artillery Brigade had fully expected to be left behind as the cavalry squadrons thundered across Iraq. Instead, the deliberate advance ordered by Franks and effected by Holder and other commanders permitted the guns to keep pace. Now the corps commander's determination to concentrate his combat power paid dividends. From 5:30 to 9 P.M. two thousand howitzer rounds and a dozen MLRS rockets dumped 130,000 bomblets in front of Ghost, Eagle, and Iron, terrorizing the enemy and ripping up Iraqi supply trains.

Adding to the mayhem was a procession of Air Force bombers. While an OA-10 spotter plane circled overhead to mark targets with white phosphorous rockets, F-16s and A-10s attacked with bombs and cannon fire. Sartiano would long believe the Air Force had abandoned him in his hour of need, but in fact ten Falcons and six Warthogs decimated Iraqi artillery and armored forces east of the front line. Several Apaches, airborne again as the *shamal* waned, destroyed an artillery battery with Hellfires and ripped up a convoy fleeing northeast on a two-lane highway near the Kuwaiti border.

At 10 P.M. the battlefield fell silent but for the crackle of flames and the periodic boom of burning ammunition. In four hours Holder's three cavalry troops had fired three hundred TOWs and tank rounds, plus seven thousand cannon and machine gun rounds. More than two hundred Iraqi armored and wheeled vehicles were destroyed. Nearly all of one Tawalkana brigade and elements of the 12th Armored Division had been obliterated. Because troops on the regimental right had encountered no T-72s, Holder was able to assure Franks that the cavalry had found a gap in the enemy defenses to the south. Into that breach the corps commander could now steer the Big Red One, which began filtering through the 2nd Cavalry before midnight.

Other battles would be more destructive than 73 Easting. Other units would fight with the same proficiency demonstrated by Holder's dragoons. Yet in this first major engagement against the Republican Guard, the U.S. Army demonstrated in a few hours the consequences of twenty years' toil since Vietnam. Here could be seen, with almost flawless precision, the lethality of modern American weapons; the hegemony afforded by AirLand Battle doctrine, with its brutal ballet of armor, artillery, and air power; and, not least, the élan of the American soldier, who fought with a competence worthy of his forefathers on more celebrated battlefields in more celebrated wars.

Here the terrible truth of this war was wholly revealed: the enemy never had a chance.

17

# Liberation

## Washington, D.C.

Hardly had the rout begun when Colin Powell started searching for an appropriate moment to stop it. As early as Monday, G + 1 of the ground offensive, Powell advised Schwarzkopf that a quick end to the fighting appeared likely. On Tuesday he repeated the assessment, suggesting Thursday as a possible cease-fire date; to Bush and others in the Gang of Eight, the chairman declared that "it's not going to be long," since the allies were "close to breaking the Iraqi army." In another call to Schwarzkopf, on Tuesday evening, Powell again suggested that a "window of ending" had opened. Very soon, the chairman believed, the allies could legitimately proclaim that "this is over; all we're doing is killing people."

Schwarzkopf would assert in a postwar interview that he had qualms about ending the fighting prematurely; in his memoirs he subsequently recanted. Certainly neither Powell nor CENTCOM staff officers detected any doubts at the time. The chairman hung up Tuesday night believing that he and the CINC had arrived at "a shared conclusion," with "no divergent assessment." Schwarzkopf asked Chuck Horner how much advance warning the Air Force would need to halt the bombing campaign. Horner replied that three or four hours' notification would suffice, observing that if necessary "I could continue to bomb Iraq until they're down to two stone axes and a pushcart." He too agreed that there was little profit in pressing the war to the point of slaughter.

"I knew that we were going to clean their clock," Powell told the Joint Chiefs intelligence officer, Mike McConnell. "I knew we were going to win this and I knew we were going to have relatively small loss of life on our side. But I had no idea it would happen with this speed."

*

In the railroad metaphor occasionally heard in Washington during the Persian Gulf crisis, George Bush was depicted as the locomotive engineer who drove the war train. Firm hand on the throttle, clear eye on the track, the president by force of will had plowed through all obstacles — a hesitant Congress, an anxious citizenry, a reluctant Pentagon — to bring the confrontation with Baghdad to a military dénouement.

If Bush was the engineer, then Colin Powell was the brakeman. Clearly the initiative to end the war came from the American military, and, given the chairman's dominance over the Joint Chiefs, that meant Powell. Having relied on Powell's expertise in constructing and executing the battle plan, Bush was disinclined to reject his counsel now that the campaign was nearly complete. This reflected both the president's faith in the chairman's judgment and his reluctance to be a commander-in-chief who intruded in the military's province. The contrast to Lyndon Johnson's ham-fisted meddling in Vietnam was surely to Bush's credit; in the main, the congruence between chief executive and chairman during Desert Storm was striking. With one notable exception — the president's insistence on diverting whatever military assets were necessary to keep Israel out of the war — Bush made no major decision that contradicted counsel of his uniformed advisers or commanders. (Franklin Roosevelt during World War II had made nearly two dozen, though he did so over four years as compared with the six weeks of the Persian Gulf War.)

As G + 3 dawned on Wednesday, Powell's immediate worry was the carnage being inflicted by Horner's pilots on the fleeing Iraqis. Tuesday had been the biggest day of the air war to date, with 3159 sorties, many of them flown against forces retreating north of Kuwait City. Vivid reports from the press pools in Saudi Arabia began to depict the onslaught as had never been done before. "There's nothing like . . . seeing the bullets come out and hammer the target," boasted one A-10 pilot, who with his wingman claimed twenty-three kills in just three sorties. The front page of Wednesday's *Washington Post* described how allied airmen "swarmed over Iraqi armor and truck columns, slaughtering the scattering vehicles by the score in a combat frenzy variously described as a 'turkey shoot' and 'shooting fish in a barrel.' "

No doubt the reports were accurate. B-52s, Marine Harriers, F-15Es — just about anything with wings and a bomb rack — were participating in the attacks. Some of the most intense sorties were flown by A-6s from U.S.S. *Ranger,* one of four carriers in the gulf. *Ranger* had steamed far enough north so that her attack planes, each laden with six thousand pounds of ordnance, could strike without wasting time refueling in midair; the carrier on Wednesday shifted to "flex deck," which meant that bomb loaders slapped on whatever munitions happened to be at hand as the planes landed, refueled, and immediately took off again to the strains of "The William Tell Overture" pealing from the public address system. The carrier battle group commander, Rear Admiral R. J. Zlatoper, reminded his pilots to avoid nonmilitary targets. "Our job," Zlatoper warned, "is to take out Iraqi armor and armored personnel carriers, and not buses."

Yet such surgical precision was impossible, given the denseness of the traffic pressing north. Mines and cluster bombs had blocked the Highway of Death above Al Jahra, but scores of vehicles veered around the obstruction or struck out across the open desert. Other convoys followed a second highway that swerved to the northeast along the coast past Bubiyan Island. (The latter was hit especially hard by Navy pilots. After the war approximately fifteen hundred destroyed vehicles would be counted on the Al Jahra highway, and another four hundred on the coastal road. Approximately 2 percent were tanks or armored personnel carriers.) Little talk of overkill could be heard in the bustling ready rooms aboard ship or at air bases ashore; that would come later. Implacability now seized many pilots, who were eager to avenge comrades killed in combat or paraded as prisoners on Iraqi television broadcasts. In a few squadrons these final sorties of the war became known as "sport bombing."

A growing number of commanders, however, were disturbed by the lopsided nature of the fight. That the marauding allied air attacks also presented a potential public relations problem was evident in Riyadh and Washington. On Tuesday morning, several hours after F-15s first bombed the exodus at Al Jahra, a CENTCOM officer had assured reporters, "There's no significant Iraqi movement to the north." When Saddam announced on Tuesday that he was withdrawing his army from Kuwait because of "special circumstances," Bush responded with scorn.

"He is not withdrawing," the president declared in the Rose Garden on Tuesday afternoon. "His defeated forces are retreating. He is trying to claim victory in the midst of a rout." Saddam, Bush asserted, intended to "regroup and fight another day." The president vowed that the allied campaign would continue with "undiminished intensity."

The fear that Iraq could somehow spin a political victory from military defeat haunted Bush and his war counselors. To deny Saddam the status of an Arab martyr, the allies had to subject him to public humiliation. No quarter would be given, the White House had decreed Monday night, unless the Iraqi leader "personally and publicly" accepted all allied conditions. If Saddam were unmistakably diminished, other Iraqis might be emboldened to kill or depose him — a fantasy by then widely indulged throughout the U.S. government.

In Powell's view, though, the allied onslaught had reached a point of diminishing returns, much as the bombing of Baghdad had by the time the Al Firdos bunker was destroyed two weeks earlier. Legally, he knew, the allies were on solid ground. The law of war — the orders signed by Abraham Lincoln before Gettysburg were an example — permitted an attack on enemy combatants, whether advancing, retreating, or standing still. The Geneva Convention of 1949 forbade the killing of an enemy clearly trying to surrender, a prohibition observed with admirable discipline in the gulf war. Those Iraqis who abandoned their vehicles and fled afoot — and they numbered in the thousands — were unmolested, but those who fired back or remained with retreating convoys were fair game.

Yet politically and morally the chairman had qualms. Powell's voice had been among the loudest arguing for limited — and militarily expedient — war aims. He wanted no part of a war that required an extended American occupation or a protracted hunt for Saddam. He ridiculed suggestions that the U.S. Army press on to Baghdad to impose democracy on Iraq, as if "lots of little Jeffersonian democrats would have popped up to run for office." Since early August he had argued that it ran counter to American interests to eviscerate Iraq and leave a power vacuum that strengthened Iranian or Syrian influence in the Middle East. (Some Pentagon civilians wondered whether Powell's stance reflected geopolitical conviction or was a convenient rationale for curbing further military entanglement.)

For Powell, these political judgments had clear military consequences. The prevalent American military philosophy since the Civil War had embraced a "strategy of annihilation," the relentless bludgeoning of an enemy to destroy his armed forces and ability to wage war. But the gulf war, as Powell repeatedly insisted, "wasn't going to be a battle of annihilation." Rather, the chairman saw the conflict as "a very careful application of power in a measured way." It was a limited war with limited objectives, not unlike — strategically at least — European wars of the eighteenth century. In Powell's mind, when those objectives were achieved, the war should stop.

Beneath this rationality lay certain emotional and intuitive coefficients. The prospect that U.S. soldiers and airmen would be remembered primarily for the ruthless destruction of an enemy in retreat weighed heavily on the chairman. By Thursday, McCaffrey's 24th Division and the VII Corps would be at the gates of Basrah. That entailed the clear risk, long avoided by Powell, of aggressive young troops in proximity with potentially hostile civilians. (Powell and Schwarzkopf had both served in Vietnam with the Americal Division, the unit responsible for the massacre at My Lai; more than two decades later that catastrophe still stained the honor of the U.S. Army.) The chairman anticipated that Americans and allies alike would soon see televised images of the carnage in northern Kuwait, and react with outrage. To blight the dazzling performance of the U.S. military with images of a "turkey shoot" was both unnecessary and foolish.

Without question the allied attacks could have continued. Schwarzkopf's avowed intention was to "drive to the sea and totally destroy everything in our path." But, Powell wondered, to what end? To eradicate a few more Iraqi battalions? The impact on Saddam's postwar military would likely be minimal. North of the Euphrates Iraq still had at least twenty divisions, which would undoubtedly survive the war. Even in combat, the chairman believed, chivalry should reign; compassion should be extended to a prostrate foe.

Powell's sentiments were defensible and cogent, even magnanimous. But some of his premises, as he surely knew, were open to question. The belief that Americans would recoil from the Highway of Death may have misread a culture steeped in violence; the nation had endured one of the bloodiest civil wars in history and two world wars in this century. The national contempt for the military that developed during the Vietnam War had taken years to form — not a day at Mutlaa Pass — and was a consequence more of American casualties than of the enemy's. And the issue was still theoretical: no pictures of the Highway of Death had yet appeared on American television networks, nor would they until after the war was over.

Moreover, if a large number of Iraqis escaped, second-guessing would be inevitable. One of Lincoln's bitterest disappointments was George Meade's failure to pursue Lee's army across the Potomac after the Union victory at Gettysburg: Lincoln likened his general to "an old woman trying to shoo her geese across a creek." In 1944, a hundred thousand German soldiers had been enveloped in the Falaise pocket west of the Seine; Omar Bradley failed to close the gap, permitting roughly forty thousand to get away and triggering a debate that persisted for decades.

Restraint in war, born of nobility or of baser military motives, often carried its own hazards. To Powell's archive of maxims regarding the judicious use of power could have been added a pronouncement from Marshal Foch: "The will to conquer is the first condition of victory."

*

Early Thursday morning Powell, vice chairman David Jeremiah, and nearly a dozen other senior Pentagon officials filed into the Crisis Management Room, a small vault across the corridor from Tom Kelly's office. A guard at the door checked the names of those entering against an authorized roster. Here the daily war briefings were conducted, with staff officers providing a quick, secret update of intelligence and operations from the theater.

As the men slid into their chairs around a long rectangular table, several slides flashed from a projector onto the screen at one end of the room. In the reconnaissance photographs snapped above the highways north of Kuwait City, the carnage could be seen at a glance: hundreds of wrecked tanks, trucks, and automobiles jammed bumper to bumper on the roadbed, many afire, a few bolting across the desert in flight.

The faces of those at the table showed no satisfaction. A few of the men seemed to wince, to shift uncomfortably in their chairs. The pictures conjured up Sherman's description of Shiloh, where "the scene on this field would have cured anybody of war." For many sitting in the Crisis Management Room, the implication of the photographs was clear: the enemy had been vanquished. The war was all but over.

## Jalibah Air Base, Iraq

In Kuwait and southern Iraq, of course, it was not over. That the war had entered its last full day probably occurred to no one on the battlefield, although many sensed the surge toward its finish. Few days in American military history would be more remarkable for the sheer sweep of armies and the rout of a reeling enemy. It was to be a day of conquest and controversy, of brilliance and occasional blunder, of individual acts of mercy amid wholesale killing. And then, with the abruptness of a blown fuse or a thrown switch, the war would stop.

The pressure on the collapsing Iraqis from the Marines in the south and VII Corps in the west was now intensified by the pressure of Barry McCaffrey's 24th Division in the north: thirty-four battalions, 26,000 troops, eighteen hundred tanks and other armored vehicles. Much as the French had considered the Ardennes a barrier against a German attack in 1940, the Iraqis apparently believed the broken terrain south of the Euphrates Valley to be impenetrable by American armor. But at 1 A.M. on February 27 the division's 1st Brigade reached Battle Position 102 overlooking Highway 8, only seventy-five miles from Basrah.

Soldiers blockaded the highway, captured twelve hundred prisoners, and by dawn destroyed a hundred enemy vehicles with tank and TOW fire. Bedouin squatting along a ridge watched the booming guns, politely clapping when a round struck its target. Iraqi farmers draped their huts with white flags and directed the American tanks around their tomato fields. Below the highway, at Objective Gold, the brigade discovered an immense munitions dump, seventy-odd square miles covered with more than a thousand storage bunkers — enough ammunition to supply an army for many months.

Although the brigade and cavalry squadron on his left flank had been slowed by boggy terrain and fuel shortages, McCaffrey ordered the division's other two brigades to turn east as directed by XVIII Corps headquarters under Contingency Plan Ridgway. At 6:30 A.M. five artillery battalions began shelling Objective Orange — Jalibah Air Base. The confused Iraqi defenders, believing themselves to be under another air attack, opened fire with eighty antiaircraft guns. As the enemy laced the empty sky with tracers, two battalions from the 2nd Brigade crashed through the perimeter fence from the south and a third attacked from the northwest. Several dozen Bradleys and M1A1s raced down the main runway, shooting up hangars, fuel tanks, an enemy tank battalion, helicopters, and twenty parked fighter planes.

By 10 A.M. Jalibah had fallen. McCaffrey flew over the airfield in his Blackhawk, nodding with approval at the billowing columns of smoke. Landing amid the wreckage, he wrapped the 2nd Brigade commander, Colonel Paul Kern, in a bear hug. Gary Luck had estimated that the 24th Division would need four days after the beginning of the ground attack to capture Objective Orange; the mission had taken sixty-seven hours.

While most of the division refueled for the final drive toward Basrah, McCaffrey persuaded Luck to authorize an armored raid on Talil Air Base, forty-five miles west of Jalibah. Although Luck's operations officer, Colonel Frank Akers, argued against the attack, calling it a sideshow, McCaffrey countered that Iraqi forces in the west jeopardized his plan to put a million-gallon fuel depot at Battle Position 102. With the corps commander's consent, a tank battalion raced up the Euphrates Valley, skirted the thirty-foot-high sand berm surrounding the airfield on three sides, and blew into Talil from the north with guns blazing. The raiders shot up half a dozen parked MiGs, two helicopters, a cargo plane, and infantry bunkers before hurrying east to rejoin the division. They also left behind the smoldering hulks of two M1A1s and two American personnel carriers — deliberately scuttled with tank rounds after they became mired in a network of irrigation ditches.

The Talil foray would remain a sore point for months after the war.

Akers believed the raid was a colossal mistake that slowed the division as much as twenty-four hours and permitted some of the Republican Guard to escape. "If you hadn't turned toward Talil," he later told McCaffrey, "you would have caught them all. You got distracted." McCaffrey and his battle staff adamantly disagreed; the need to wait for more fuel and provisions had caused whatever delay there was.

In truth, Talil appears to have stalled part of the division, not the main attack force. By early Wednesday afternoon, McCaffrey had ordered two brigades and the 3rd Armored Cavalry Regiment — which Luck had placed under his control — to push toward Basrah. Twenty thousand soldiers rumbled east, smashing everything in their path and capturing hundreds of prisoners. By dusk the division would reach Phase Line Axe, twenty miles from the Hawr al Hammar causeway, to begin a massive artillery barrage against Iraqi forces scattered across the western lip of the Basrah pocket.

The division scheme for capturing the city envisioned one brigade maneuvering to the north, another seizing the canal in the west, and a third hooking into the suburbs from the south. Once encircled, McCaffrey believed, Basrah would fall in twelve hours.

Early Wednesday morning Luck had called Binnie Peay with new orders for the 101st Airborne: "Get east as fast as you can. The enemy's trying to get out through Basrah." Peay directed his 2nd Brigade, reinforced with the 12th Combat Aviation Brigade, to fly ninety-five miles from Cobra to Forward Operating Base Viper, southwest of Jalibah. By 1 P.M. a new logistics compound had blossomed in the empty desert. Dozens of helicopters queued up for fuel, hopping forward like kangaroos as the line inched toward the gas pumps.

Throughout the afternoon four Apache battalions attacked heavy traffic on the two-lane Basrah causeway across the Hawr al Hammar and in the marshlands stretching north along Highway 6 toward Qurna. Oil smoke cut visibility to less than a thousand yards, forcing all pilots to fly with thermal sights. Waves of attack helicopters flew down the Hawr — a wide, shallow lake — to unleash Hellfire and rocket volleys on the fleeing convoys. Burning trucks careered into the water or plowed through the marshes. Hundreds of soldiers fled on foot into the thick reeds around the lake, many waving white flags, many others firing automatic weapons.

For the Apache crews, this chaos posed an ethical dilemma. If they honored the white flags, the pilots put themselves at risk; if they returned fire they might hit some of those attempting to surrender. One flight leader radioed his battalion commander, Lieutenant Colonel

Dick Cody, who was monitoring the fight from FOB Viper: "Sir, we're taking fire from everywhere. It's getting crazy up here." "Roger," Cody replied. "Do not lose an aircraft." This guidance, Cody believed, was plain enough: if the enemy resisted, shoot everyone. "Roger that," the pilot answered. Rocket and cannon fire ripped through the marshes along what soon would be known as Damnation Alley.

South of the Hawr causeway, near the Rumaylah oil field, enemy gunners on Wednesday afternoon brought down an Air Force F-16 — one of eight allied planes lost during the war's final week. As the pilot ejected and floated to earth, the Apache crews listened to his radio calls for help: "My God, they're shooting at me! They're shooting at my canopy!"

Blackhawk 214, a search-and-rescue helicopter attached to the 101st, flew headlong into the fray but was raked with gunfire. Two accompanying Apaches also were hit before managing to get away. The Blackhawk pilot veered left in a desperate attempt to escape, but the crippled helicopter smashed nose first into the sand at 130 knots, snapping off the tail boom. The fuselage cartwheeled across the desert. The pilot, co-pilot, and three other crewmen were killed; three survived to be taken prisoner. Among the survivors was the female flight surgeon, Major Rhonda L. Cornum, who suffered two broken arms, a bullet wound, and assorted other injuries. The F-16 pilot, Captain William F. Andrews, also was captured.

In contrast to the pandemonium east of Basrah, in the sector north of the city known to the Americans as Engagement Area Thomas, the Apaches found fewer targets than General Peay had anticipated. Not a single tank was spotted; most fleeing Iraqi armor apparently had holed up in Basrah and its suburbs. With his pilots exhausted, fuel and ammunition stocks running low, and persistent difficulties in coordinating attacks with the Air Force, Peay late on Wednesday suspended further raids into EA Thomas. Instead, he drafted a plan to insert a brigade along Highway 6 on Thursday morning. The assault, quickly approved by corps commander Luck, would seal off Basrah from the north.

## Kuwait City

For the U.S. Marine Corps, the war had nearly run its course. Boomer reported that unless Iraqi defenders offered unexpected resistance around MEF Objective C — the international airport south of Kuwait City — "our part of the mission is complete."

In the west, Tiger Brigade straddled the sole land route out of the Kuwaiti capital. The stink of rotting bodies hung over the Al Jahra highway, drawing packs of wild dogs. Among the smoldering enemy convoys lay an emporium of goods pilfered during the Iraqis' final kleptomaniacal binge: furniture, television sets, new shirts in plastic wrappers, carpets, a wedding dress, children's books, cutlery, artificial flowers, scuba gear, jewelry, hair spray, red nail polish. By early Wednesday William Keys's 2nd Division had halted along Phase Line Bear, the Six Ring Road south of Kuwait Bay. Thousands of captured Iraqis, some with bare feet wrapped in sandbags, shuffled into prison compounds. The five hundred men of one enemy battalion encircled themselves with concertina wire and waited patiently for American scouts to take them into custody.

In the east, Tank Force Grizzly swept through Al Jaber Air Base, where the fleeing Iraqis had booby-trapped doors, stairwells, and a mosque with Italian antipersonnel mines. (The mosque, desecrated with graffiti and human feces, also had been used as an enemy latrine.) The rest of Mike Myatt's 1st Division had begun enveloping the international airport early Tuesday evening, using an impromptu battle plan sketched on a cardboard rations box. By 11 P.M. a battalion from Papa Bear had breached the airport's perimeter fence.

At 6 A.M. Wednesday, as a gray dawn filtered through the oil smoke, Task Force Shepherd cautiously pushed toward the runways. Except for three bewildered enemy stragglers, the airport lay deserted. Battleship gunfire and air strikes had battered the hangars. Debris littered the tarmac. The passenger terminal and an adjacent hotel had been vandalized. The Marines hauled down an Iraqi flag, raised the American and Kuwaiti colors, and then — after contemplating proper political etiquette — lowered the Stars and Stripes. MEF Objective C had been captured.

Should glory be found in the gulf war, nowhere did it seem more likely than in the liberation of Kuwait City. Army Special Forces and the Marines — neither known for an aversion to fame — had assiduously studied street maps of the capital. Now the prize beckoned.

But Schwarzkopf had different ideas. With his own sense of political etiquette and a prudent reluctance to clear urban buildings with American troops, the CINC had decreed months before that Arab forces would recapture the city. As early as August, General Khalid, the Saudi commander, had confided to Schwarzkopf his "dream" of personally liberating the capital. "I will be standing in downtown Kuwait City when the emir arrives," Khalid had vowed. "I will hand him the Kuwaiti flag and tell him Kuwait is free." Although Schwarzkopf considered Khalid's vision somewhat ridiculous, he retained his conviction that Arab forces should play the prominent role in capturing the city. On learning that a small Marine patrol had reached the U.S. Embassy Tuesday night, Schwarzkopf was furious. Butch Neal called Boomer's headquarters on the CINC's behalf to remind the Marines that "the Arabs have to liberate Kuwait City. We don't want Americans in there pre-empting them."

Yet with the capital now in reach, squabbling broke out among the Arab allies. The Kuwaitis, much to Khalid's irritation, insisted on overseeing the liberation of their own capital. While trying to accelerate JFC-N's glacial pace, Schwarzkopf settled the bickering with a combination of diplomacy and fiat: the Kuwaitis would take the lead, with other Arab forces — notably Egyptian armored units — providing additional firepower as needed. Except for Marine reconnaissance teams and Green Beret advisers, most American forces would remain in the suburbs. The capital was carved into six sectors, each assigned to a Kuwaiti unit: the Liberation, Immortality, and Martyr brigades from JFC-N, and the Truth, Full Moon, and Victory brigades from JFC-E.

The American high command had long been more concerned about infamy than about glory in the liberation. As early as December, Dick Cheney had voiced fears of reprisals against Iraqi prisoners or the thousands of Palestinians living in Kuwait City. Such qualms were hardly assuaged when the Kuwaiti crown prince, living in exile in the Saudi city of Taif, warned, "Some of the Palestinians collaborated with Iraqi troops . . . When we go home, we shall check on them." For more than a month the 5th Special Forces Group advisers had lectured their Kuwaiti protégés about the importance of avoiding atrocities. The Americans also worried about fighting among rival resistance groups, or between those Kuwaitis who had remained under Iraqi occupation — roughly a third of the country's 600,000 citizens — and those who had fled.

At 9 A.M. Wednesday morning the first columns of JFC-N troops crossed in front of the 2nd Marine Division to enter the capital from the west. (The liberation was delayed two hours while the Egyptian commander, to the Americans' consternation, awaited President Mubarak's authorization to advance.) JFC-E units also pushed up the coastal road from the south.

By now few Iraqis remained in the city. Kuwaiti troops, driving pickup trucks with 50-caliber machine guns mounted in the beds, roamed the rubbled streets, setting up roadblocks and dealing with suspected collaborators. Precisely how many vengeance killings took place will never be known, but dozens of Palestinians were beaten severely enough to require hospitalization and some were summarily executed. Enough blood-curdling tales of retribution reached the American Army that Colonel Jesse Johnson, Schwarzkopf's special operations commander, eventually sent soldiers to sixteen police stations in the capital to demand that the Kuwaitis "knock that shit off." Middle East Watch, an American human rights organization, would report that by early March six thousand Palestinians had been placed in detention.

However iniquitous, the Kuwaiti thirst for revenge was not without provocation. The Iraqis had caused more than a thousand Kuwaiti civilian deaths, according to a war crimes report compiled by the U.S. Army and submitted to the UN Security Council in March 1993. Victims were tortured with electric drills, electric prods, and acid baths; others were shot, dismembered, or beaten to death. Rape was commonplace.

The once pristine capital had been sacked with a vengeance worthy of Hulagu Khan, the Mongol conqueror who pillaged Baghdad in 1258. Beginning in mid-August, Iraqi troops had begun loading booty into containers at Ash Shuwaikh port, where they were shipped to Umm Qasr and then sent north by barge or truck. Convoys had hauled away more than a million ounces of gold from the Central Bank of Kuwait, as well as jewels from the city's gem market. Iraqi looters had ransacked 170,000 houses and apartments, pilfering porcelain sinks, toilets, light fixtures, rugs, drapes, and even light bulbs.

They took seven marine ferries and twenty shrimp trawlers from Kuwaiti ports; baggage-handling equipment, runway lights, and fifteen airliners from the international airport; experimental sheep from the Kuwait Livestock Company; beef carcasses from storage freezers; granite facing from downtown skyscrapers; twenty thousand plastic seats from Kuwait University stadium; hearses and grave-digging backhoes from city cemeteries. Most of the country's half million cars, buses, and trucks had been stolen or stripped. Eight movie theaters and seventeen sports clubs had been vandalized; at least nineteen libraries were systematically stripped of their collections.

Most of Kuwait's 1330 oil wells and twenty-six gathering stations had been sabotaged. An estimated eleven million barrels gushed from the fractured wells every day, with roughly half of the crude burning up and the other half forming immense petroleum lakes. Parliament, government ministries, hotels, the emir's palace, department stores, telephone exchanges, and most other public buildings had been plundered; many had been torched.

Scuttled ships blocked harbor channels; sabotaged utilities left the capital with no electricity and little water. At the city zoo, elands and buffalo had been eaten, all 326 birds and most mammals had died of neglect, and a U.S. Army veterinarian, wielding a metal detector, found a bullet embedded in the shoulder of Dalal, the resident Indian elephant. Kuwaiti citizens — the estimates ranged from a few dozen to several thousand — had been seized as hostages earlier in the week and transported north in a final, contemptible act of barbarity.

Beneath a sooty pall likened by one British journalist to "a winter's day in Victorian England," Walt Boomer led a small Marine convoy into downtown Kuwait City on Wednesday afternoon. Boomer had expected the capital to be deserted, but thousands of jubilant Kuwaitis jammed the streets, showering the Americans with candy and cigarettes. "Thank you, thank you, U.S.A.!" they cheered. "God bless Bush!" Resistance fighters fired rifles and waved flags from the rooftops. Stone-throwing children pecked the eyes from wall murals of Saddam; sobbing women in black *abayas* blew kisses to the grinning Marines or held up babies to be hugged. Young men built bonfires of Iraqi dinars — for seven months the only legal tender in occupied Kuwait — and flashed photographs of the exiled emir.

Someone tossed a red, green, and black Kuwaiti flag to Boomer as he stood atop the amphibious landing vehicle that served as his mobile command post. Eyeing the abandoned Iraqi fortifications that seemed to crown every overpass and street corner, the Marine commander could sense how repressive the enemy occupation must have been. The war of liberation had cost the lives of twenty-two Marines; eighty-eight others had been wounded. At this moment, however, Boomer had no doubt that the sacrifices had been worthwhile. As he surveyed the delirious mob, the hardships and doubts of the past seven months receded, leaving him with the sweet sensation of having fought the good fight, and won.

## Objective Norfolk, Iraq

No cheering throngs awaited VII Corps during its final drive through the remnants of Iraq's defenses. Thousands of enemy soldiers fled or capitulated, but thousands more held their ground with a desperation rarely displayed in earlier fighting. As the corps gathered momentum in its sweep toward the Basrah–Kuwait City highway, Fred Franks jockeyed his forces with the intent of bringing five heavy divisions on line by late Wednesday. The 1st Cavalry, after pulling back from its feint up the Wadi al Batin, had swung west through the breach and was galloping north to plug Franks's extreme left flank, where it would tie into the 24th Infantry.

Pending the 1st Cav's arrival, the corps commander already boasted, from north to south, the 1st Armored, 3rd Armored, 1st Infantry, and British 1st Armoured divisions — a wall of tanks and armored vehicles eighty miles wide, moving with the inexorable power of an avalanche. For the first time Franks began contemplating how best to encircle all remaining Iraqi forces south of Basrah, seining the enemy army like a school of trapped fish.

The prospect that a substantial number of Iraqis could get away was clear to Franks by early Wednesday morning. JSTARS radar detected hundreds of vehicles still moving north through Kuwait and southeastern Iraq despite the allied air strikes along the main highways. Iraqi tanks, armored personnel carriers, and some trucks were able to leave the paved roads to flee cross country. Two Apache raids by the 11th Aviation Brigade — one before midnight on Tuesday and the other at three on Wednesday morning — found more targets in northwest Kuwait than they could destroy.

But to avoid tangling with Air Force and Navy aircraft, Army pilots had been prohibited from flying beyond the 20 Easting, parallel to and roughly fifteen miles west of the Basrah–Kuwait City highway. Franks urged ARCENT headquarters at Eskan Village to have CENTCOM adjust this limit of advance in order to unleash the Apache fleet on the fleeing enemy. But ARCENT, CENTCOM, and Air Force staff officers in Riyadh failed to agree on the necessary modifications, and the issue apparently was never taken to Schwarzkopf for adjudication. As the hours ticked by, more and more Iraqi forces reached the safety of the Basrah pocket.

Although he was frustrated, Franks was preoccupied with the immediate fight facing his divisions. Tom Rhame's Big Red One had begun filtering through the 2nd Cavalry shortly after the shooting stopped along the 73 Easting late Tuesday night. For six hours the division bulled its way toward Objective Norfolk near the Kuwaiti border, destroying Iraqi tanks, trucks, and, regrettably, each other. In three separate episodes, confused and exhausted 1st Division troops fired on their comrades. By the time Rhame halted to refuel at Phase Line Milford, six soldiers were dead and thirty wounded, with five M1A1s and five Bradleys destroyed. All of the deaths and most of the damage was self-inflicted, in yet another reminder that the war was rarely more lethal than when the Americans turned their guns on themselves.

Undaunted, Rhame regrouped for the march into western Kuwait while his troops rounded up hundreds of prisoners. "We're facing a broken enemy," Rhame radioed Franks late Wednesday morning. "Contact is light."

"What do you recommend?" the corps commander asked.

"I'd like to press east to Objective Denver and cut the Basrah–Kuwait City highway."

"Can you get there by dark?"

The division commander studied his map. The road lay less than forty miles ahead, and helicopter pilots now saw no Iraqi force larger than a company. Rhame very much wanted to reach the highway, to smash whatever was left of the Iraqi occupation. "Yes," he said, "I think we can make it."

"Okay," Franks agreed. "Go ahead."

The British attack in the south had turned into an expedition "rather like a grouse shoot," as one squadron commander put it. For two days Rupert Smith had shoved his division steadily eastward toward the Wadi al Batin, with the Desert Rats of 7th Armoured Brigade positioned north of 4th Brigade. Infantrymen crouched in the bays of their Warrior personnel carriers, listening to the pop of gunfire until the intercom crackled with the order: "De-bus! De-bus!" Then the armored doors swung open and soldiers scrambled out with their rifles locked and loaded, ready to fight an enemy who rarely fought back other than to fire a few desultory shots before surrendering.

One by one the objectives fell: Copper North, Zinc, and Platinum in the north, overrun by the Royal Scots Dragoon Guards, the Queen's Royal Irish Hussars, and 1st Battalion of the Staffordshire Regiment; in the south, Bronze, Brass, and Steel were captured by the 14th/20th King's Hussars, the 1st Battalion of the Royal Scots, and the 3rd Battalion of the Royal Regiment of Fusiliers. Everywhere the British found Iraqi dead to be buried — many shredded by artillery fire — or Iraqi prisoners to bundle off to detention camps.

In a few instances, the enemy resisted. At Objective Lead early in the evening of the 26th, Staffordshire C Company rolled up to a small cluster of buildings ringed with a sand berm and defended by an Iraqi battalion. When a long burst of machine gun fire failed to provoke any surrenders, the British poured 30mm cannon rounds into the berm and buildings. Dozens of Iraqis threw up their hands.

But when Private Carl Moult dismounted to round up the prisoners, an RPG round blew through his chest, killing the young soldier and setting his Warrior on fire. For nearly two hours a bloody firefight raged, garishly lit by flares pumped into the night sky from British mortar tubes. Warrior gunners lashed the buildings with cannon rounds, while infantry teams attacked four bunkers with grenades and rifle fire. When the shooting stopped just before 9 P.M., 285 Iraqis surrendered, abandoning a compound now carpeted with bodies. By early Wednesday morning, Objectives Lead and Tungsten had fallen — after a bizarre counterattack by several robed Iraqis mounted on camels — and the division stood poised on the western lip of the wadi. Smith had estimated that the fighting from breach to wadi might last as long as ten days; it had lasted two.

Here Franks's decision about what to do next with his British allies was influenced by a painful truth: most of the British dead and many of the wounded in the ground campaign had fallen victim not to the Iraqis but to the Americans. On Wednesday morning, while rounding up prisoners, two gunners from the Queen's Royal Irish Hussars were badly wounded by 1st Infantry Division tank rounds.

By far the worst episode, however, had occurred on Tuesday afternoon. Two A-10s flying over Objective Steel mistook a pair of Warriors from the Royal Regiment of Fusiliers for T-55 tanks. The British air liaison officer, known as Lobo 01, later insisted he had radioed target coordinates that were many miles from the point of attack; both pilots denied receiving any instructions from Lobo 01 other than the assurance that there were "no friendlies within ten kilometers." The flight leader made two reconnaissance passes with binoculars, the first at fifteen thousand feet, the second at eight thousand. Spotting neither the fluorescent panels nor inverted Vs displayed on the British vehicles — later determined to be invisible above five thousand feet — he launched a Maverick missile at 3:02 P.M. His wingman fired four minutes later. Both missiles ran true. The blasts killed nine and wounded eleven. Of the dead soldiers the eldest was twenty-one, the youngest seventeen.

Franks initially advised Rupert Smith to prepare to pursue the enemy north up the wadi on Wednesday toward the intersection of the Basrah highway and the northern Kuwaiti border. While perfectly willing to fight Republican Guardsmen instead of the blighters in the south, the British commander was less pleased at the prospect of parading his division past thousands of heavily armed Americans — a maneuver, he privately concluded, "likely to lead to the biggest blue-on-blue engagement of all time."

Franks soon shared those qualms. Less than six hours later he countermanded his own order and told 1st Armoured to continue east into Kuwait toward Objective Varsity. By midafternoon on Wednesday, against scant opposition, the British reached that goal.

VII Corps called with new instructions: the British now were to drive *south* to clear Ruqi Road, a four-lane blacktop that snaked through southwestern Kuwait past the triborder to a logistics base in Saudi Arabia. Again Franks quickly countermanded himself after concluding that Smith's forces should instead press eastward, cutting the Basrah highway at Objective Cobalt, roughly fifteen miles north of Mutlaa Ridge.

General Smith, accepting this change with good humor, issued a warning order to his troops: on final authorization from corps, the division would race twenty-five miles across the desert toward Cobalt. British soldiers, determined not to be again mistaken for Iraqis, broke out their Union Jacks and regimental colors. (A few crews even hoisted white pennants until directed to haul them down.) The order came. The brigades surged forward. Above the churning dust, thousands of flags snapped in the wind — a glorious flutter of reds and whites and royal blues. Like an avatar of the bygone empire, the British Army again charged east, proudly, at full pelt.

American intelligence now believed that of forty-two Iraqi divisions entrenched in the Kuwaiti Theater of Operations six weeks earlier, only the Madinah and Hammurabi possessed enough combat punch to threaten the allied attack. Pentagon staff officers joked that the Iraqi army had been reduced from the fourth largest in the world to the second largest in Iraq. The Hammurabi's whereabouts were uncertain, although the division appeared to have decamped northward. The Madinah was known, thanks in part to radio intercepts, to be forming a second screen behind the Tawalkana thirty miles southwest of Basrah.

As for the Tawalkana, it was no more. One brigade had been demolished at the 73 Easting; two others were obliterated by Franks's heavy divisions. The 3rd Armored Division, after rolling through a Tawalkana mechanized battalion and most of an armored brigade, continued east through fragments of the Iraqi 12th Armored and several other units. On more than one occasion the Americans fell on the enemy so suddenly that Iraqi armor crews were caught with their tank batteries removed to underground dugouts, where they were being used to power electric lights and heaters. By late afternoon on Wednesday, 3rd Armored began crossing into Kuwait.

On VII Corps's extreme northern flank, Ronald Griffith's 1st Armored Division pushed ahead with three tank brigades abreast. JSTARS had reported enemy forces — apparently a mechanized infantry unit from the Republican Guard's Adnan Division — moving south from the XVIII Corps sector. Twenty-three American soldiers were wounded by Adnan artillery fire on Tuesday night, provoking a devastating counterattack with MLRS rockets and Apache gunships. In a brief firefight at midmorning on Wednesday, 1st Armored's Black Lion battalion also destroyed several tanks and a pickup truck. (The latter reportedly was used to machine-gun Iraqi soldiers in the back as they tried to surrender.) The Adnan threat, feeble to start with, dissipated as quickly as it had appeared.

Shortly before noon on Wednesday, above the northwest corner of Kuwait, the Madinah's 2nd Brigade formed the last coherent defensive formation that Saddam's army would muster in the gulf war. Initially facing south, the enemy brigade recognized the American threat from the west in time to pivot around in a battle line seven miles long.

Bearing down on the Iraqis were 166 M1A1s from 1st Armored's 2nd Brigade. Led by Colonel Montgomery C. Meigs, the brigade edged forward with nine tank companies on line. At 12:17 P.M. American gunners reported enemy armor at three thousand meters, roughly two miles away. Meigs ordered his commanders to open fire.

The first sabot round punched a three-inch hole in an Iraqi hull. A tongue of flame spurted from the opening for a few seconds before the ammunition inside detonated, flipping the turret forty feet into the air. Within ten minutes the Iraqi line began to disintegrate in a caldron of black smoke and flaming hulks. Meigs, whose thermal sights had stopped working, stood on his turret fifty yards behind the firing line, squinting through the haze. "Enemy is direct front," he radioed to the division command post. "I count eighteen smoke plumes." Several minutes later he added, "Now I've got thirty-six."

The Madinah crews, some of whom had been caught outside their tanks lunching on tomatoes and rice, fired back haplessly at the M1A1 muzzle flashes. Iraqi rounds skipped across the desert floor, often hundreds of yards short of the mark. Like his Tawalkana counterpart at 73 Easting the previous afternoon, the Madinah commander had hoped to ambush the Americans as they came over the ridge before driving them back into an artillery kill sack. But he had fatally miscalculated the difference in range between the T-72s, which were ineffective beyond eighteen hundred meters, and the M1A1s, which were now destroying targets at nearly twice that distance. Iraqi artillery rounds exploded harmlessly behind the advancing Americans.

The Abrams's crews fired, reloaded, and fired again, slower now, searching out new targets and edging forward at ten miles per hour. "Don't get bushwhacked," Meigs ordered. "I want you to move gently but deliberately and kill all those people."

Forty minutes after it began, the fighting stopped. Without suffering a scratch, 2nd Brigade had destroyed sixty T-72s and dozens of personnel carriers. Artillery and Apaches struck other targets deeper on the battlefield; A-10s and F-16s picked off stragglers and logistics convoys fleeing north and east. Farther south, the 1st and 3rd Brigades rolled over pockets of resistance — a tank company here, an infantry platoon there. Bombarded with MLRS rockets, many Iraqis broke and ran, leaving their engines running, their radios on, and rounds chambered in their guns. All told, the 1st Armored Division destroyed approximately three hundred Iraqi armored vehicles at a cost of one American killed.

Like the 73 Easting, the fight at Madinah Ridge was waged with tactical acumen and devastating firepower, reducing the enemy to a pathetic rabble. Again the Americans displayed overwhelming superiority in weaponry, intelligence, gunnery, combined arms tactics, and leadership. It was, in short, like the war itself: a brilliant slaughter.

At 3:30 P.M. Franks flew in his Blackhawk to see Ron Griffith at the 1st Armored Division's command post. The two generals watched as artillery rounds screamed overhead and Meigs's tanks pushed forward toward Phase Line Italy. Jagged lightning split the sky and the rumble of thunder added to the cacophony of air strikes, Hellfire missiles, and booming tank fire. After several minutes the corps commander turned his back on the battlefield. "Okay," Franks said, bending over a map, "we've got some things to work out."

Two issues wanted resolution. The most immediate concern was fuel. Since February 23 the corps had burned nearly eight million gallons. Despite herculean efforts by logisticians to keep pace, combat units had outrun most of their trains. Having traveled farthest on the outside flank of the Left Hook, 1st Armored was in particularly desperate straits. Griffith's last stocks had been sucked dry by the 75th Artillery Brigade, which joined the division Tuesday night carrying many tons of ammunition but no gas. Now Meigs had only two hours of fuel remaining in his tanks, with less than five thousand gallons in the brigade reserve; each M1A1 guzzled five hundred gallons every eight to ten hours. Without replenishment, further advance was impossible.

Relief came from two sources. Paul Funk, much to the chagrin of his logisticians, dispatched twenty tankers from 3rd Armored's dwindling stocks. Then an emergency convoy of forty-six fuel trucks — driven by a press-ganged platoon of cooks and clerks — left Log Base Echo in Saudi Arabia early Wednesday. Guided north by one of Griffith's helicopters, the convoy arrived in time to begin refueling 1st Armored early Wednesday evening. Combined with Funk's bequest, the trucks provided 130,000 gallons — enough to permit the division to press the fight for at least another morning.

With the fuel crisis at least temporarily resolved, Franks concentrated on the grander question of where to steer his corps. Earlier in the afternoon he had sketched his battle plan on a piece of flimsy acetate, which he now showed Griffith. On the corps's left flank, 1st Armored would swing slightly south to make room for the 1st Cavalry Division, which would drive toward Basrah along the lower edge of the Rumaylah oil field.

On the right flank, having decided not to send the British north, Franks would push the 1st Infantry across the Basrah–Kuwait City highway, where the division would then angle toward Basrah. The 2nd Cavalry would wheel to the western edge of the highway before also pressing toward Basrah on a path parallel to the Big Red One's. Within a day, whatever remnants of the Iraqi army that had not fled across the Euphrates or into Basrah would be encircled.

The plan was, in Franks's description, "a classic, out-of-the-book double envelopment." Rare was the battle captain who had not dreamed of such an opportunity. In 216 B.C., Hannibal had similarly encircled the Roman army at Cannae, cutting the enemy to pieces in one of warfare's most celebrated victories. More than two thousand years later, the Carthaginian tactic had mesmerized von Schlieffen, whose scheme for defeating the French in 1914 stressed envelopment and "extermination by attack upon [the enemy's] rear." A generation later, the Germans had again used *Kesselschlachten* — battles of encirclement and annihilation — to conquer Poland.

With the 1st Infantry already angling toward the highway, Franks instructed Griffith to squeeze his force down into Kuwait, making room for the 1st Cavalry. The encirclement began.

18

# Closing the Gates

## Washington, D.C.

Despite Fred Franks's best-laid plans, there would be no double envelopment of the retreating Iraqi army, no *Kesselschlachten.* At midafternoon on Wednesday in Washington — approximately 10:30 P.M. in the theater — President Bush gathered his senior advisers in the Oval Office for what was billed as another routine war briefing. Powell, brandishing charts and maps as usual, used a small penlight to indicate the allied positions, from the Marines and Arabs in the south to VII and XVIII Corps in the west and north.

"Both Norm and I feel that we're within the window of opportunity to end this," the chairman said. "It's clear that the Iraqi army is broken. If anything, they're just trying to get out. That was our mission: to get them out. And I can report to you that they are well on their way to being out. In fact, we're crucifying large numbers of them."

Powell's conviction that the war must end had only strengthened during the day. He told Schwarzkopf in a morning phone call that the White House was nervous about the images of "wanton killing" in the theater; certainly the anxiety was Powell's, too. For months he had struggled to maintain a clear correspondence between political and military objectives. Now he was certain that those political goals had been achieved. The Iraqis had been so thoroughly dismembered that allied intelligence "can't find divisions, can't find brigades, can't find battalions," the chairman continued. "It's all just shattered."

"Do you want another day?" Bush asked.

"By tonight there really won't be an enemy there. If you go another day, you're basically just fighting stragglers," Powell replied. "You're no longer fighting against a serious, organized opponent. Nothing that is worthy of three corps, four corps really — the Marines, VII, XVIII, and the Arab coalition. The vaunted Republican Guard formations are no longer."

Others in the office agreed. Cheney felt that the foremost issue was not how many Iraqi tanks and gun tubes had been destroyed, but whether the allies' strategic aims had been satisfied. Like Powell, he believed they had: Kuwait was liberated, Saddam's war-making capability smashed. Baker inclined to accept the military's recommendations, particularly after hearing Powell's description of an unmitigated slaughter. Scowcroft concurred; another day of carnage like that along the Highway of Death served no one's interests. His deputy for Middle Eastern affairs, Richard Haass, observed, "We don't want to be seen as piling on."

If the rout of the Iraqi army was clear, the extent of the enemy's entrapment was not. Earlier in the day, Cheney had been told by the Joint Staff that the Republican Guard was already cut off from further retreat into Basrah. CENTCOM's daily intelligence summary for February 27 also reported that "the Republican Guards are encircled . . . They have few options other than surrender or destruction." At least one service chief believed that the only enemy soldiers escaping into Basrah were those carrying small arms; repeated bombing of bridges across the canal that angled east and south of the city had ostensibly cut escape routes for most heavy equipment. In the White House, several of Bush's advisers assumed that withdrawing Iraqis would have to pass through an allied checkpoint.

Only the fabled fog of war can explain this misapprehension. The remaining enemy forces were not yet encircled. No allied troops occupied the corridor leading into Basrah from the south or the roads stretching north from the city. Nor had American soldiers reached the Hawr al Hammar causeway to the northwest. ARCENT had hoped to have XVIII and VII Corps attack across a common front, but that proved difficult for Yeosock to coordinate from Eskan Village, hundreds of miles from the fighting. Consequently, a substantial gap had developed between Franks and Luck, with VII Corps nearly thirty miles farther east. Most of the three Republican Guard infantry divisions — the Adnan, Nebuchadnezzar, and Al Faw — exploited the seam to escape across the Euphrates or into Basrah.

Approximately two divisions' worth of Iraqi armor was stacked up waiting to cross makeshift pontoon bridges across the Shatt al Basrah, but the canal — not much wider than a four-lane highway in some places — was hardly impassable. More than twenty bridges and causeways led out of the Kuwaiti theater. Although bombers and Apaches braved the foul weather to whittle away at all of these escape avenues, few had been severed completely.

Schwarzkopf contributed to the confusion. During an hour-long televised press briefing shortly before Bush convened his meeting, the CINC had declared, "We've accomplished our mission, and when the decision makers come to the decision that there should be a cease-fire, nobody will be happier than me.

"The gates are closed," he said. "There is no way out of here." All Republican Guard divisions in the theater had been destroyed except for "a couple that we're in the process of fighting right now." When asked, though, whether ground forces were blocking the roads to Basrah, Schwarzkopf replied, "No."

"Well," a reporter persisted, "is there any way that they can get out that way?"

"No," the CINC answered cryptically. "That's why the gate is closed."

But before returning to the war room, Schwarzkopf amended his remarks. "When I say the gate is closed, I don't want to give the impression that absolutely nothing is escaping. Quite the contrary. What isn't escaping is heavy tanks, what isn't escaping is artillery pieces . . . I'm talking about the gate that is closed on the war machine that is out there."

This nuance was not pursued during the meeting in the Oval Office. For Powell, who could read a map as well as any soldier, the issue was subordinate to the larger one of military objectives. The chairman noted that some pilots had begun returning from their sorties with qualms about strafing and bombing a prostrate enemy. "There's almost a psychic cost to be borne if we ask our troops to continue military operations when it's clear we've won," he warned. Bludgeoning a defeated foe was not only distasteful but also "not American," Powell believed. "There is," he added, "chivalry in war."

Bush, much taken with his chairman's sentiments, seemed to concur. "What's Norm think?" the president asked.

Powell took the receiver from the secure satellite phone on the president's desk and quickly apprised Schwarzkopf of developments at the White House. "I've presented our views and the thinking is that we should end it today," the chairman said in a formulation that hardly invited demurral. Bush was considering calling a halt that evening on national television. "Would you have any problem with that?" Powell asked.

"I don't have any problem with it," Schwarzkopf replied. Yeosock had asked for one more day to complete the destruction of the Republican Guard, but the CINC did not press the point on Powell; already he had directed his staff to begin drafting plans for the return of troops to the United States. "Our objective was the destruction of the enemy forces," he told the chairman, "and for all intents and purposes we've accomplished that objective."

The final issue before the president and his advisers was how and when to stop the war. Rather than declare a cease-fire, which implied negotiation, Bush would impose a "cessation of hostilities," a formula seen as unilateral, keeping the initiative in the allied camp. As to timing, Powell initially favored 8 or 9 P.M., Washington time, but concluded that midnight — 8 A.M. Thursday in Riyadh — made more sense. That would allow several hours after the president's announcement on national television for the word to filter across the theater, both to allied and Iraqi forces. Moreover, it gave Schwarzkopf's troops two hours of morning light "to take a last look at the battlefield and make sure we haven't done anything that puts us in a compromising position."

Later in the afternoon Powell again called Riyadh from the White House to inform Schwarzkopf that the shooting would stop at midnight Eastern Standard Time, precisely one hundred hours after the ground war had begun. Bush and Cheney each took a turn on the phone to offer congratulations. "Norm," the secretary said, "you've done a helluva job."

## Riyadh

In Schwarzkopf's war room, the demands of a fast-moving ground offensive had left little time for discussion about whether the war had run its course. Robert Johnston, the CINC's chief of staff, believed that even without guidance from Washington, CENTCOM was running out of enemies to fight. Capturing Basrah, Johnston concluded, made sense only if the allies intended to push north toward Baghdad. Many of Schwarzkopf's officers shared the belief that Iraqi survivors would likely join a coup; sufficient combat power existed in the Basrah pocket, according to one intelligence assessment, "to ensure an orderly transfer of power when Saddam falls."

Up King Abdul Aziz Boulevard, in the basement of the Royal Saudi Air Force headquarters, a warning of the war's imminent end caught Buster Glosson and Dave Deptula by surprise. Recent intelligence had helped to pinpoint another of Saddam's subterranean command posts in Baghdad; Deptula drafted justification documents for Glosson to give Schwarzkopf in asking permission to bomb the target if the fighting lasted any longer. The two pilots joked sourly that the Army seemed determined to outshine the Israelis' triumphant Six Day War by a day or two.

Glosson's bombers already had made one final effort to decapitate the Iraqi leadership. Early Wednesday afternoon a C-141 cargo jet landed at Taif carrying a pair of 4700-pound bombs, still warm to the touch from the molten explosives poured into the casings by Air Force ordnance technicians in Florida. The immense munitions — known as GBU-28s — had been fashioned from surplus artillery barrels in a crash program to develop a "penetrator" capable of puncturing the deepest enemy bunkers. The day before, a prototype fired horizontally on a high-speed sled in New Mexico had sliced through twenty-two feet of reinforced concrete, skipped once across the desert floor, and continued down range for another half mile before finally coming to rest.

The primary target was a bunker more than forty feet below ground at Taji, northwest of Baghdad. In repeated strikes, beginning on January 17, F-117s had failed to penetrate the complex even with 2000-pound LGBs. Senior Iraqi commanders were known to be using the bunker; Air Force intelligence hoped, without solid evidence, that Saddam also had taken refuge there.

Five hours after the new bombs arrived, a pair of F-111s — Cardinal 7–1 and Cardinal 7–2 — took off. Each carried a GBU-28 under the left wing, with an ordinary I-2000 beneath the right wing for ballast. At 8 P.M. Cardinal 7–1 swooped over Taji from the east with afterburners roaring to lend the bomb as much velocity as possible. The huge cylinder broke clean from the wing, sliced through the night sky — and missed, hitting short and slightly south. Cardinal 7–2, flown by Lieutenant Colonel Dave White and Captain Tommy Himes, diverted from the secondary target — another bunker, three miles south — found the primary on the second pass, and let fly. The bomb burrowed through the roof and detonated far under ground. Smoke poured from ventilation shafts and doorways. The target was destroyed. But Saddam, as the Americans soon discovered, was not inside.

Schwarzkopf had fielded Powell's second call from the White House in his bedroom. After hanging up he walked into the war room to find Cal Waller. "I talked to Colin," the CINC said. "We're going to stop."

Waller's eyebrows shot up. "We're going to stop?"

"The more technical term is that they want a cessation of hostilities."

"What the hell does that mean?" Waller asked. "Is this a cease-fire?"

"No, it's not a cease-fire," Schwarzkopf said. "We're going to take up a protective posture where we'll only fire if fired on. We'll still be on a wartime footing."

"Why are we stopping at this point?"

"I'm told that we're stopping because there's some concern about our attacks and the carnage on the highway from Kuwait to Basrah."

The DCINC pondered this for a moment. Schwarzkopf seemed neither upset nor surprised. Waller wondered why the plan to seal off the theater with a double envelopment was not given time to unfold. "Why not go ahead and complete what we were going to do?" he asked, framing the debate that would continue long after the war ended.

"Because that's what the commander-in-chief wants," Schwarzkopf replied evenly. "The president has said we've accomplished enough."

Schwarzkopf made another round of calls to his field commanders — Yeosock, Boomer, Wayne Downing, and Stan Arthur. All seemed relieved that the end had come. On *Blue Ridge,* Arthur had monitored the relentless sorties by his carrier pilots with growing unease. He too believed the onslaught had reached a point that exposed the United States to charges of wanton killing. "Boy," he told his staff after learning that the attacks would stop, "I wish I'd heard that twelve hours ago."

## Washington, D.C.

News of the decision spread quickly from the Oval Office. A few officials in Washington and abroad questioned the move, but none with sufficient conviction to urge reconsideration. In London, Charles Powell, foreign affairs adviser to John Major, voiced misgivings, since British commanders recognized that the Iraqis had yet to be encircled. But Major offered no objections during a phone conversation with Bush. Foreign Minister Douglas Hurd, in a meeting with the president Wednesday afternoon at the White House, had gingerly inquired whether the land campaign had indeed run its course. Although Hurd retreated in the face of Bush's unqualified assurances, he subsequently told reporters in the White House driveway, "There will come a time when the military objectives of the allied forces have been achieved, but we're not at that point today."

At Foggy Bottom, Under Secretary of State Robert Kimmitt, while applauding Bush's decision as "very courageous," suggested to Baker that the war could be prosecuted for another day before Washington felt a "public and political backlash"; Baker shrugged off the advice.

At the Pentagon, Paul Wolfowitz concurred with Bush's timing. If the fighting continued and Iraqi resistance suddenly stiffened, resulting in the death of a few dozen more American soldiers, the government would be hard pressed to explain why they had died. But he favored "keeping the Iraqis off balance" by maintaining some ambiguity about whether the allies might yet push toward Baghdad. The under secretary abandoned that notion on learning that Schwarzkopf in his press conference had precluded just such a possibility. ("We were a hundred and fifty miles from Baghdad and there was nobody between us and Baghdad," the CINC had declared, referring to the 101st Airborne's assault into the Euphrates Valley. "If it had been our intention to overrun the country, we could have done it unopposed.")

For the Joint Chiefs, the decision arrived as a fait accompli. They had gathered Wednesday morning in Powell's office, taking time to call Schwarzkopf with congratulations. Although they were not polled on their sentiments regarding a cease-fire, the chairman had no reason to suspect opposition. Nor, when the deed was done, was any dissension voiced. "Look, we achieved our objectives," Carl Vuono reasoned. "We got what we wanted, so there's nothing to be gained by pressing on."

Vuono's view, though it reflected that of the majority, was not shared by all the chiefs. Tony McPeak harbored deep — if unspoken — misgivings. The war, he believed, was ending two or even three days too soon. A priceless opportunity had been forfeited. The Air Force chief had disagreed with Powell on other issues, notably the need for a second Army corps and the decision to stop bombing Baghdad bridges. Usually, on reflection, he had concluded that the chairman was correct in his judgment.

But not this time. McPeak's initial doubts were spun from emotion and a warrior's instinct for the jugular; over time they would harden, annealed by the belief that a truly decisive victory had slipped away. For many months after the war he would review the matter, chastising himself for a lack of courage in not confronting Powell while recognizing that a confrontation would have changed nothing. His admiration for the chairman's intelligence and sagacity grew steadily. History, he suspected, would treat Colin Powell generously on this matter. And yet McPeak could not shake the conviction that the chairman was wrong.

No American military decision since the Vietnam War provoked more controversy, more debate, more caustic commentary, than the choice to offer Iraq — in McPeak's later phrase — "a merciful clemency." The issue lay dormant for months, until the euphoria of victory had faded and the magnitude of Saddam's surviving force became apparent. But within a year two thirds of Americans polled said that the war had ended too soon. Some military analysts would liken the decision to Hitler's failure to wipe out the British at Dunkirk and Meade's reluctance to pursue Lee from Gettysburg. It would stain Bush's triumph, darkening the president's most conspicuous achievement when he ran for re-election in 1992.

Without question, men who were physically exhausted and emotionally spent made certain miscalculations. The enemy was not encircled. The gates were not closed. The Defense Intelligence Agency later concluded that seventy to eighty thousand Iraqi troops had fled into Basrah. An Army analysis estimated that "as many as one third of the [Republican] Guard's T-72s made it out of the KTO." Other intelligence analysts calculated that roughly eight hundred tanks and fourteen hundred armored personnel carriers escaped destruction. Within days of the cease-fire, Iraqis would use dozens of Hind and Hip helicopters — part of a surviving fleet of several hundred — to terrorize rebels in the Shi'ite south and Kurdish north. Although the Republican Guard was mauled and some units obliterated, others got away nearly intact.

Schwarzkopf, as has been noted, would blame VII Corps for failing to tighten the noose. Yet the CINC himself had personally halted the 24th Division in its drive to the Euphrates and delayed dispatching the 1st Cavalry to the fight. He also had directed against Iraqi conscript divisions air attacks that would otherwise have hit the Republican Guard. In all cases, he had legitimate tactical reasons, although not necessarily any more legitimate than Fred Franks's battlefield judgments. Moreover, when given the opportunity to urge that the fighting continue — to Powell, to Cheney, and to Bush — Schwarzkopf demurred. No thought appears to have been given to halting deep air attacks while letting the Army — with close support from the Air Force — either complete Franks's double envelopment or encircle all of Basrah.

The much-feared public uproar over the slaughter of retreating Iraqi troops never materialized, and there is no proof that it would have had the war persisted another day or so. Even the Soviets, who earlier had decried the pummeling administered by allied air power to their former client, said little during the ground war. And the wan hope that Iraqi commanders would topple Saddam, in retribution for the holocaust to which he had subjected his nation, was obviously a pipe dream. Indeed, his army remained distressingly loyal.

Nevertheless, the decision wears well with time. No substantive evidence suggests that the surviving Iraqi forces proved decisive in suppressing the postwar rebellions. Had not a single tank or personnel carrier escaped allied destruction south of the Euphrates, Saddam would still have possessed an immense force north of the river, which totaled more than three thousand armored vehicles, including hundreds of tanks. By one estimate, nearly 90 percent of Iraqi artillery in the theater was destroyed by the allies — almost three thousand tubes — yet there were at least a thousand additional heavy guns elsewhere in the country.

Certainly the destruction visited on Iraq's military during six weeks of war was prodigious, even without the annihilation of every battalion in the Kuwaiti theater. Two years after the invasion of Kuwait, American intelligence calculated that Saddam's armed forces had reached only 40 percent of their prewar manpower and heavy equipment strength; moreover, Iraq suffered disproportionate losses of its most modern weapons. The Iraqi air force became a shadow of its former self, particularly after Iran refused to return the 132 aircraft that had sought sanctuary in the east. The integrated air defense network was wholly smashed.

The number of Iraqi battle deaths would be debated by American military and intelligence analysts without resolution for more than two years after the war. Estimates of Iraqi killed and wounded ranged from the low thousands to 100,000. Perhaps the best guess, predicated mostly on enemy prisoner of war reports, would emerge from the Air Force's postwar air power study: ten to twelve thousand Iraqis were killed by allied bombs prior to the ground offensive, the study calculated, and perhaps another ten thousand during the four-day ground war.

But the issue hangs on more than hard tallies. Had the war persisted, more American and allied soldiers would have died — either at the hands of the enemy or at the hands of their comrades. They would have died for the dubious glory of slaughtering a few thousand more Iraqi teenagers, surely a cause unequal to their sacrifice and unworthy of their nation. Every slain soldier tells the same story, the author Michael Herr once observed: "Put yourself in my place."

Errors would be made in establishing the conditions of cease-fire and in the dithering neutrality effected during the Iraqi insurrections. But stopping the war was no mistake. Rather, it was a rare triumph for the better angels of our nature.

On a word processor in Scowcroft's outer office, Richard Haass finished the rough draft of Bush's televised announcement. Shortly after sunset on Wednesday, the president's advisers gathered in the West Wing to polish the address.

The terms for a formal cease-fire — beyond the provisional halt at midnight — were drawn in part from stipulations drafted earlier in the war by Bush's advisers. Iraq must release all prisoners of war and Kuwaiti hostages, disclose the location of all mines, and halt all Scud launches. Baghdad must comply with all Security Council resolutions, agree to pay reparations, and rescind the annexation of Kuwait.

Bush's remarks, the president's men agreed, would open with three declarative sentences: "Kuwait is liberated. Iraq's army is defeated. Our military objectives are met."

## Phase Line Victory, Kuwait

Even before the cease-fire order bounced from Washington to Riyadh to commanders in the field, VII Corps's double envelopment had become snarled in its own complexity. With the 1st Infantry Division now angling northeast and three other divisions above them pushing east, Fred Franks became concerned about a potentially disastrous convergence of American forces. Darkness had fallen. Bombers and helicopter gunships trolled the skies south of Basrah. Soldiers and commanders alike were exhausted, anxious, and jumpy.

From his command post near the northwest corner of Kuwait, Franks radioed the Big Red One shortly before 7 P.M. on Wednesday and advised the division not to advance beyond Phase Line Kiwi, approximately fifteen miles from the highway to Basrah. "We're already very much east of there," Rhame's deputy told the corps commander. The lead brigade had closed almost within sight of the roadway. Franks ordered the division to "stop in place." By 7:30, the Big Red One had come to a halt.

Farther north, the effort of 1st Armored to make room for the 1st Cavalry was stalled by a firefight in an Iraqi logistics camp. An enemy RPG round struck one Bradley; another shell — either from the Iraqis or an Abrams tank — hit a second personnel carrier, decapitating the driver and wounding a platoon sergeant, who subsequently lost a leg. Heeding Ron Griffith's advice, Franks prudently postponed the passage of lines, keeping 1st Cavalry out of the fight. Across the battlefield, the mounting toll from friendly fire had dampened the Army's audacity.

At 11:30 P.M. Franks's deputy, John Landry, radioed from the corps main headquarters with news of a possible cease-fire at five o'clock on Thursday morning. Franks called Yeosock, who confirmed the report. The error in this advisory remains a mystery. Some believe the confusion reflected debate in Washington over when to halt the war; another explanation is that the original order specified 5 A.M. Zulu — Greenwich Mean Time, which meant 8 A.M. in the theater — and was misunderstood along the chain of command when someone neglected to make the adjustment for local time. Similar errors had occurred during exercises and even in combat.

Whatever the reason, the U.S. Army operated for nearly four hours under the impression that the war would stop an hour before dawn. Instead of preparing "to take a last look at the battlefield," as Powell had hoped, most of the troops assumed the fighting was over. At 3:15 A.M. Yeosock woke Stan Cherrie with the truth. "Tell General Franks the cease-fire will be at 0800," Yeosock said. "We want you to drive. No more business as usual. Turn the heat up. Go, go, go." Franks, who had snatched a two-hour nap, radioed his divisions with new instructions in an effort to seize as much Iraqi territory as possible before 8 A.M.

Once again commanders lashed their troops forward. "I want a forty-five-minute artillery prep," Griffith told his subordinates in 1st Armored. "I want every gun and every rocket launcher in the division firing. I want it to be the most violent artillery prep in history. Five minutes after the fires lift, I want the Apache battalion flying fifty kilometers deep. Go through the zone and kill everything you see. Then we'll attack with three brigades abreast."

Tom Rhame was awakened with news that Franks now wanted the 1st Infantry moving by 4 A.M. to seize the highway. "You crazy sons of bitches!" he barked. "Tell him I can't do it by four; I'll be moving by five." Rhame then radioed his two easternmost brigades. "The cease-fire's at 0800. It is absolutely imperative that you cut that goddam highway as soon as possible. You must begin to move by 0500."

In XVIII Corps, the war's imminent end provoked jubilation, confusion, and regret. An early morning Apache attack on Engagement Area Thomas, north of Basrah, was canceled. Binnie Peay also scuttled his plan to insert a brigade from the 101st Airborne above the city. Peay chided himself for not having been more aggressive in urging Gary Luck to leapfrog over the Euphrates to cut the road leading north from Basrah to Al Amarah. "Damn," Peay told his staff, "I wish we'd just jumped the river and gone to the east. We could have encircled the whole thing."

The bogus report of a 5 A.M. cease-fire also caused Barry McCaffrey to cancel his attack toward Basrah with the 24th Division and 3rd

Armored Cavalry Regiment. On learning that the war would in fact last another three hours, McCaffrey was furious. "How could we blow the goddam time?" he demanded. "How could this happen?"

Corps headquarters suggested the division press ahead with a helicopter attack, but the Apaches could not be readied on such short notice. Instead, McCaffrey intensified the artillery barrage that had begun early Wednesday evening. His armored brigades edged forward before halting at Phase Line Victory, less than thirty miles from downtown Basrah.

At 7 A.M., in a chilly fog, Rhame's vanguard reached the highway at Objective Denver. A dozen burning oil wells blazed to the east, backlighting the lesser fires from shattered Iraqi trucks and armored vehicles. The rising sun hung like a pale medallion behind a greasy pall of smoke. Dazed Iraqi soldiers wandered through the gloom, flinging down their weapons and stepping around the dead, who lay strewn along the blacktop. Roaring up behind his lead battalions, Rhame surveyed the scene with immense satisfaction: the swirling mist, the dancing flames, the enemy wraiths shambling forward with hands held up in surrender. Not in twenty-seven years as an Army officer had he experienced a more exhilarating moment.

Yeosock had drawn Franks's attention specifically to the road junction at Safwan, several miles north of the Kuwaiti border. Troops from the Big Red One turned north toward the town, but at 7:20 A.M. an Army artillery unit issued a frantic distress call by FM radio, claiming to be under fire from 3rd Armored. Franks ordered a cease-fire throughout the corps. Within fifteen minutes the crisis was resolved, but Rhame, whose brigades had dutifully halted, never heard the subsequent order to resume the attack. Safwan thus remained in Iraqi hands, a detail that seemed of little consequence until Schwarzkopf discovered that the proposed site of his armistice conference was still in enemy control.

At 8 A.M. the guns fell silent. On the corps's southern flank, three British battle groups rolled onto the highway above Al Jahra amid the flapping of Union Jacks. Rupert Smith followed in close order, having paused en route for a cup of tea while his overheated engine cooled down. Burial details shooed away packs of scavenging dogs and gathered up Iraqi corpses. Soon the roadside was pocked with the graves of would-be conquerors.

On the northern flank, 1st Armored halted along Phase Line Monaco after an artillery barrage of four thousand shells. Some soldiers felt the sag of anticlimax. Griffith fretted over enemies still unfought. "Sir, I hope to hell we didn't stop too soon," the division commander told Franks. "We still don't know where the Hammurabi Division is."

Most soldiers, though, were swept with relief. In Monte Meigs's 2nd Brigade, armored vehicle crews clambered from their hatches, shaking hands, slapping backs, and draping their whip antennae with American flags. A psyops team rooted through its cache of audio tapes for a suitable victory paean. Unable to find a recording of Lee Greenwood's "God Bless the U.S.A.," the soldiers settled for a 1967 soul hit that had been popular in Vietnam, both for its snappy syncopation and its sassy lyrics, which could be belted out with ironic gusto. But on this day in this war, there was no irony, only exultation, as James Brown's voice rose as if in a hymn of thanksgiving:

> Whoa! I feel good!
> I knew that I would now!

## Rumaylah Oil Field, Iraq

Across most of the theater, the fighting was finished. Allied soldiers continued to round up prisoners — now in excess of eighty thousand, including several teenage militiamen whose mothers remanded them to American custody. Engineers blew up mountains of Iraqi ammunition and fleets of abandoned vehicles. U.S. cavalrymen patrolling southwest of Basrah discovered a villa, ostensibly built by a former Iraqi defense minister, replete with two swimming pools, a manmade lake, and an encircling wall surmounted by blockhouses and searchlights. A storeroom contained stacks of rocket-propelled grenades, several powder-blue bidets still in their shipping crates, and three thousand cases of Perrier mineral water. After admiring the teardrop crystal chandeliers, the silver candelabra, the Limoges china, and the Italian bath towels, troopers fired two 165mm artillery rounds into the main house, splashed fifty gallons of kerosene across the complex, and set it ablaze.

At 1:30 P.M. on February 28, ninety soldiers from 1st Battalion of the 3rd Special Forces Group swooped by helicopter into the U.S. Embassy compound in Kuwait City. Other than a few bullet holes in the Marine barracks, the embassy was unscathed. Finding the chancery still locked, a Green Berets demolition team blew an immense hole through the door, inadvertently igniting the curtains inside and triggering a frantic scramble for fire extinguishers. The liberation of Kuwait was complete.

To the north, however, sporadic shooting continued. McCaffrey had pushed his reconnaissance forces to Phase Line Crush at the western edge of the Rumaylah oil field, whose disputed reserves lay at the heart of Baghdad's grievance with Kuwait. For two days following the cease-fire, occasional Iraqi artillery and mortar rounds fell among the Americans, who answered with massive counterbattery barrages. Shortly after midnight on March 1, two buses full of surrendering enemy troops pulled up to a roadblock on Highway 8. Iraqis filed from the first bus in good order, but on the second a pair of recalcitrant soldiers abruptly opened up with AK-47s. The Americans riddled the bus with gunfire, killing seven Iraqis and wounding six.

The fragile truce collapsed completely on the morning of March 2. At 4:45, scouts from McCaffrey's 1st Brigade reported mysterious headlights and heavy traffic several miles east of Phase Line Crush. "Hold your position," advised McCaffrey's operations officer, Lieutenant Colonel Pat Lamar. "Just continue to observe. If they're not shooting at you, leave them alone."

At first light, helicopter scouts discovered the Iraqis had bulldozed enough dirt into the Hawr al Hammar to build a ramp that bypassed bomb craters in the long causeway spanning the lake. Two hundred vehicles already had crossed the Hawr and were headed north toward Ash Shanin; perhaps a thousand others were advancing through the oil field toward the causeway. The elusive Hammurabi Division had at last been found.

From his command post on Highway 8, McCaffrey shoved forward the 2nd Battalion of the 7th Infantry Regiment, commanded by Lieutenant Colonel Charles C. Ware. After struggling through a boggy *sabkah* that temporarily entrapped six of his fourteen tanks, Ware positioned four companies along the Rumaylah's western perimeter. Forbidden to fire unless fired upon, the Americans watched in frustration as the exodus of trucks, howitzers, and armored vehicles surged toward the causeway.

Shortly after 8 A.M., Charlie Company on Ware's right flank reported RPG fire from dismounted infantry in the south. Charlie fired back with 25mm cannon fire. Attracted by the shooting, Iraqi BMPs maneuvered toward the Americans, and several enemy tank turrets pivoted westward. "Hey, look, I need to do something," Ware pleaded in a radio call to the brigade commander, Colonel John Le Moyne. "I need to put some return fire on these guys." Le Moyne reported the action to the division command post. After asking for assurances that Ware's battalion was in jeopardy, McCaffrey approved a counterattack. "Okay, that's it," he ordered. "Take 'em apart."

"Ready on the left," Ware commanded after receiving Le Moyne's authorization. "Ready on the right. Fire! Fire!" A volley of Abrams and Bradley fire swept across the desert. Iraqi gunners replied with T-72 rounds and at least one Sagger missile.

The Iraqis appear to have fired first in the clash at Rumaylah; whether the resulting retribution was proportionate to the offense would long be debated, *sotto voce,* within the 24th Division and elsewhere in the Army. As Ware's guns boomed in the south, five American artillery battalions unleashed hundreds of artillery rounds and MLRS rockets. Helicopter gunships ripped through a convoy on the causeway, blocking further escape. At 8:50 A.M., two Apache companies swung across the Hawr from the northeast while a third attacked southeast of the oil field. For more than an hour the Apaches battered the Iraqis with cannon fire, rockets, and 107 Hellfire missiles, of which all but five reportedly struck their targets.

Meanwhile, the 4th Battalion of the 64th Armored Regiment looped past Ware's men to attack into the oil field beginning at 10:30 A.M. Three parallel roads crossed the Rumaylah from the south, all elevated on berms above the swampy ground and all converging on the causeway. The 4/64 battalion commander, Lieutenant Colonel John Craddock, sent one company up the westernmost road to the Hawr; on the eastern road, gunners destroyed two ammunition convoys. Craddock's main force rolled up the spine of the Iraqi procession on the heavily congested center road, slashing a seven-mile path of utter destruction. Another company veered south to demolish an Iraqi garrison, where enemy infantrymen had fired on Craddock's logistics train. Shortly before noon, McCaffrey landed by helicopter next to an oil pipeline, jumped into Ware's Bradley, and charged into the action. The division commander eventually captured more than a dozen frightened Iraqis who gave themselves up beneath a dirty white rag.

Shooting persisted until 3 P.M., when the Americans ran out of targets to destroy. Iraqi troops had offered little if any organized resistance. Many peeled off their boots and even their uniforms before fleeing into the marshes, which were now crisscrossed with thousands of footprints. Within a box measuring eight by twelve miles, countless smoke columns billowed from the raging pyres. Fuel tanks exploded, gun barrels melted, and ammunition cooked off in spectacular flaming pinwheels.

The final tally included thirty Iraqi tanks, 147 additional armored vehicles, and more than four hundred trucks and other wheeled vehicles. McCaffrey estimated that several hundred Iraqis had been killed. Among the casualties were a number of civilians — some of them children — riding in a bus struck during the air attack. Iraqi forces north of the Hawr and east of Rumaylah had been spared. American losses included one M1A1 damaged and another destroyed, both set ablaze by secondary explosions from burning Iraqi vehicles. A single U.S. soldier was wounded.

With the media focused on Kuwait City and Schwarzkopf's impending armistice talks at Safwan, the Rumaylah fight received scant attention. American field commanders considered the carnage just retribution, meted out after persistent enemy artillery fire and the RPG attack on Ware's flank. Few officials in Washington or Riyadh regretted tacking the pelt of a Hammurabi brigade next to trophies of the Madinah and the Tawalkana. Even those weary of bloodshed could muster little sympathy for the Republican Guard as the extent of Iraqi atrocities inflicted on Kuwait became apparent. Given the idiosyncrasies of the gulf war, it hardly seemed odd that one of its most destructive battles occurred two days after the war ostensibly had ended.

## Baghdad

For David Eberly and other allied prisoners, there was no perceptible demarcation between war and peace, no announcement of a cease-fire, no certainty that deliverance was at hand. Only a slight improvement in their lot during the final week of captivity — an increase in rations, an occasional gesture of civility — suggested that perhaps the conflict had run its course, that perhaps one could dare to dream of freedom.

After the bombing of the Biltmore on February 23 and the evacuation of POWs from the rubble, Eberly had spent four days in Abu Ghuraib Prison. Following that first night of cramped fellowship, the allied pilots were forced to sit for hours in an open courtyard before being dispersed to separate cells. The interrogations ceased; otherwise, prison life resumed its grim monotony. Initially denied a blanket, Eberly alternately paced in a tight circle and curled himself into a tight ball at night in a futile effort to keep warm. During the day he parted with a few precious bread crumbs to feed the sparrows that twittered on the walkway outside his cell. "I will come to America someday," a guard confided nonchalantly, "to kill your children."

The Arab madman resumed his nightly howling. On the afternoon of the 25th, four Iraqis dragged him into the narrow courtyard, where he was chained to a tree, savagely beaten with plastic rods, doused with water, and subjected to a game of Russian roulette. Peering through the steel bars of his door, Eberly realized that his neighbors in the adjacent cell block included a family with a baby. Mingled with his outrage at those who would imprison an infant was a nagging worry — a prisoner's instinct for self-preservation — that the child's crying would rob him of sleep's refuge.

The day passed, then another. On the morning of February 28, Eberly heard the distant sound of shooting: a long, puzzling fusillade of automatic weapons fire. At noon the POWs were hustled from their cells without handcuffs or blindfolds and herded onto a bus. "Keep your heads down. No talking," a guard ordered, covering the windows with yellow curtains. Eberly considered the possibilities. He had often imagined being bused to freedom, but surely the war hadn't ended yet; the allied ground offensive, he knew from his interrogators at the Biltmore, had not even begun as of February 18. It's too early, he told himself; don't get your hopes up. The bus pulled from the prison compound and picked up speed. Despite his caution, Eberly fancied that they were headed west, perhaps toward Jordan.

Forty-five minutes later the bus stopped. "Come," the guard commanded. "Hurry." The men filed into a single-story brick building, where they were segregated and herded into separate cells. Fighting back his despair, Eberly surveyed his new abode, a cell darker and damper than the one at Abu Ghuraib. The solid door was broken only by a small barred hole through which the guards badgered him with orders to sit down. He sat, wrapped in two wool blankets, until night fell and he was again at liberty, if only to pace.

Come morning he peeled a shard of rusty metal from the bottom of the door and scratched another calendar on the wall amid the hundreds of hatchmarks left by an earlier tenant. In the previous three prisons he had refused to carve out the month of March, hoping by force of will to keep it at bay. Yet here it was: thirty-one days hath March.

Despite the disheartening ritual of adding his daily strokes to the scarred wall — March 1, then 2, then 3 — certain signs lifted his spirits. The food improved, with feedings that now contained chicken and rice, extra bread, even an orange or a tangerine. He received ointment for his scaling skin and three more blankets, enough to fashion a cozy cocoon. For the first time since being shot down he could sleep without waking to numb toes and chattering teeth. At midmorning on Sunday, March 3 — as Schwarzkopf was negotiating his release at Safwan — Eberly was led outside the cell for a shave with warm water and a new razor. Later, after a quick dunking in a barrel — his first bath in six weeks — a guard delivered a clean POW uniform of yellow duck.

Another day passed. The weather turned foul. A chill draft swirled through the cell. For several days he had heard no air raid sirens. Was the war over? Or had the president declared a bombing moratorium?

Maybe the sirens were beyond earshot. An eerie silence descended over the cell block. He wondered whether the others had been freed, whether the Iraqis had decided to keep him as a pawn in their negotiations. The brief contact with his fellow pilots at Abu Ghuraib had awakened his yen for companionship. The solitude, once comforting, now was oppressive. When the guard fastened a cardboard patch over the small opening in his cell door, Eberly felt entombed. On returning from a trip to the toilet he plucked the cardboard from the door and laid it on the floor. "Please," he asked, "no." The guard nodded and left the small window uncovered.

On March 5, his forty-third day of captivity, he awoke to a veritable feast: cheese, a hard-boiled egg, an orange, and bread with marmalade. He ate slowly, then pulled a blanket around his shoulders, planning to rest for a few minutes before his daily hike around the cell. Suddenly the door opened. Four Iraqis stood in the corridor. Never certain whether to stand or remain seated in the enemy's presence, he hesitated before getting to his feet, wondering whether he was facing an execution squad. An older man in a clean military uniform stepped into the cell.

"David William Eberly?" he asked softly.

"Yes. Why?"

The man smiled. "I've come to take you home."

Eberly stood motionless for a moment, searching the men's faces for signs of a hoax. "Yes?" he asked. The Iraqi nodded. Eberly's emotions, so long suppressed, now cracked. Slowly he wrapped his arms around the stranger, his ostensible enemy. Tears rolled down his cheeks. The Iraqi, pulling him closer, whispered, "Remember: you're a man."

A long line of prisoners in yellow uniforms stood silently in the corridor. Eberly joined the file, searching for familiar faces. He was surprised to see two women among the men. How many times had he tried to imagine this moment? He had even fashioned a short speech to be delivered, if the chance arose, on returning to America. In the coming days he would polish those words, which he would deliver on behalf of his fellow prisoners at Andrews Air Force Base outside Washington, where Cheney, Powell, and a jubilant crowd came to welcome the POWs home on March 10: "God saved us, our families' love and your prayers sustained us," he would tell them. "The camaraderie of our squadrons brought us home to fly again. I'm proud to report that the conduct during captivity of the ladies and gentlemen beside me has been without question. Their sense of honor and duty to country is beyond reproach." Referring to the prisoners' families, he would add a phrase that could have applied to the POWs themselves: "You need to know that those who waited also served."

The line inched forward. An Iraqi wielding an atomizer sprayed Eberly's shoulders with a mist of cheap cologne. Another Iraqi tied a surgical mask over his eyes, one final blindfold. The voyage home would begin in a Baghdad hotel lobby, where the International Committee of the Red Cross would take the former prisoners into custody and escort them from Iraq — first to Riyadh, from there to the hospital ship U.S.S. *Mercy,* then to Washington, and finally, for David Eberly, back to Barbara and Timm in North Carolina.

He tugged the mask up high enough to see the floor as they shuffled from the dank corridor. The procession wended its way from the prison building toward a waiting bus in the parking lot. The stormy weather of the past two days had cleared. He felt the spring sun warm the nape of his neck. He saw his shadow, stretching forward.

# Epilogue

## Washington, D.C.

Parade Saturday — June 8, 1991 — dawned warm and clear. Morning light seeped across the city, gilding the dark Potomac, glinting from the Capitol dome, unmasking the gargoyles crouched atop the buttresses of the National Cathedral. As the rising sun inched higher, the long shadow of the Washington Monument drew back, like a sword slipping into its scabbard. Happy battalions of spectators — the vanguard of a crowd that would swell to 800,000 people — swept toward the Mall, staking out the choicest vantage points along Constitution Avenue to await the marching heroes.

Washington loved a military parade. Six weeks after Appomattox, Grant's Union Army had tramped down the city's boulevards to the strains of "Marching Through Georgia" in a two-day Grand Review. In 1919, John Pershing had ridden on horseback at the head of his victorious doughboys; a generation later, jubilant throngs cheered Dwight Eisenhower as he waved from a jeep, his conquering legions marching behind him.

Now the city gathered itself once again for a pageant of power and catharsis. Bunting stirred in the morning breeze. Children capered around the arsenal on display in front of the Smithsonian Institution. Soldiers, sailors, and airmen mustered in the lee of Capitol Hill, where the procession began to mass. For a few sunny hours, one could almost believe again in glory.

In the three months since the war ended, peace had taken an ugly turn. George Bush, in mid-February, had urged Iraqis "to take matters into their own hands, to force Saddam Hussein, the dictator, to step aside." But the anticipated coup by Saddam's vanquished army failed to materialize. Instead, bloody rebellions erupted in the Kurdish north and

Shi'ite south, where people were foolish enough to take Bush's counsel. The strength of the insurrections imperiled not only Saddam but also the Iraqi officer corps — predominantly Sunni Muslims aligned with the ruling Ba'athist Party — who rallied to Saddam, though more for self-preservation than through loyalty.

Here the Americans and their allies made several miscalculations more significant than the question of whether the cease-fire should have been delayed another day or two. Fearful of a Shi'ite victory that would strengthen pro-Iranian Muslim fundamentalists in the Persian Gulf, Washington failed to recognize that most Iraqi Shi'ites were a different faction from those in Tehran. Neither beholden to an Iranian ayatollah nor inclined to political separatism, they aspired chiefly to rightful representation in Baghdad, which had long favored the country's Sunni minority.

Saudi moderates like Prince Bandar, the ambassador in Washington, recognized this distinction but failed to convince the White House that the Shi'ites were worthy of support. Bandar also soon regretted not passing a sharper warning to Tehran — through the Syrians — to keep a discreet distance as the insurrection unfolded; consequently, Iran's overt support further galvanized the Iraqi army to unite around Saddam and reinforced the impression that Shi'ite rebels were Iranian stooges fighting to create another Islamic republic. The simplest course for Washington was to do nothing.

To the north, Saddam moved up several divisions to crush the rebellion by the Kurds. American inaction persisted until mid-April, by which time Kurdish refugees had fled by the hundreds of thousands through snowy mountain passes to Turkey and Iran. As European governments prepared to intercede — with or without American leadership — Bush reluctantly agreed to send ground troops into northern Iraq to protect a Kurdish enclave. American, French, and British warplanes flew combat patrols to block further Iraqi attacks above the 36th Parallel — at least temporarily.

In the south, no such protection obtained. During his armistice talks at Safwan, Schwarzkopf, in an unfortunate act of largesse, had consented to enemy helicopter flights by Iraqi officials who needed an expeditious means of transportation because of bomb damage to roads and bridges. Instead, helicopter gunships and loyalist ground troops slaughtered hundreds, perhaps thousands, of Shi'ites, including women and children.

Once the marauding nature of the flights was recognized, the U.S. government dithered for two weeks over how to respond. In Riyadh, CENTCOM drafted a communiqué warning Baghdad that armed helicopters would be destroyed. Paul Wolfowitz vigorously argued that American passivity was akin "to idly watching a mugging." Regardless of Schwarzkopf's Safwan pledge, Wolfowitz insisted, any Iraqi helicopter flown in combat against either Kurds or Shi'ites should be shot down. (Several Iraqi jets were destroyed for violating Schwarzkopf's prohibition against flights by fixed-wing aircraft.)

But Colin Powell countered that the helicopters were relatively unimportant in Saddam's suppression of the rebellions; moreover, shooting down helicopters could embroil other American troops in an Iraqi civil war on behalf of the rebels. "If you want us to go in and stop the killing of the Shi'ites, that's a mission I understand," the chairman argued in one Pentagon strategy session. "But to what purpose am I going to stop the killing? If the Shi'ites continue to rise up, do we then support them for the overthrow of Baghdad and the partition of the country? That's exactly the objective we said we weren't interested in." Washington would find itself "trying to sort out two thousand years of Mesopotamian history." At the White House, Richard Haass and others agreed. "What do we do if shooting down helicopters doesn't work?" Haass asked. "What if the Iraqis just use artillery instead? Do you take out the artillery?"

Caution prevailed. Bush limply suggested that helicopters "should not be used for combat purposes inside Iraq." Otherwise, in a feckless abdication of a victor's power and responsibility, the administration turned a blind eye.

An argument also erupted over the terms of peace. The administration's Deputies Committee — senior officials from State, Defense, and other government agencies — agreed in early March on the need for a demilitarized zone in Iraq. Although U.S. officials at the United Nations believed it might be possible to establish a DMZ across southern Iraq almost as far as Jordan, the deputies proposed a more modest zone that extended in an arc from south of Basrah to Nisab on the Iraqi-Saudi border. Patrolled by United Nations troops, the DMZ would provide both a trip wire to warn of future Iraqi aggression and a lever to vouchsafe Saddam's compliance with allied dictates.

This proposal found no favor in Schwarzkopf's headquarters. "What the hell is it for?" Cal Waller fumed. "It serves no useful purpose. How are you going to man the thing? We don't want another Korea, do we?" When James Baker arrived in Riyadh in early March to discuss postwar issues, Schwarzkopf immediately denounced the idea as "militarily irrelevant." Arab leaders, the CINC added, would see the DMZ as an unwarranted intrusion on Iraqi sovereignty. Others in Riyadh feared that Saddam might use the zone as a rallying point for Iraqi nationalism, as Hitler had harnessed German resentment over the demilitarized Rhineland after World War I.

Faced with such stiff opposition from the military, Baker let the proposal die. In truth the American military wanted nothing more than to leave Iraq and come home before its sterling victory became tarnished. The White House, ever fearful of a Middle Eastern quagmire, shared that sentiment. "We are not going to permit this to drag on in terms of [a] significant U.S. presence à la Korea," Bush decreed.

Considered separately, neither the refusal to shoot down helicopters nor the scuttling of the DMZ was a critical error. Yet both exemplified postwar American passivity, a policy of drift and inaction in the weeks following Saddam's capitulation. Three days after the UN Security Council adopted cease-fire terms in early April, Iraq accepted. American forces bolted from occupied Iraq as fast as their tanks and trucks would carry them.

Baghdad again controlled all of its territory except for Kurdistan. Saddam remained in power and gradually became strengthened and emboldened by American reluctance to pressure him militarily. Bush's most attractive characteristics as a wartime commander-in-chief — steely resolve, foresight, adherence to principle — now stood in sharp contrast to his waffling uncertainty. By the time American forces marched in triumph through Washington, the victory had begun to fade.

Before the parade, they honored the dead. On the leafy slopes of Arlington National Cemetery, Bush gathered with the families of the 390 American men and women who died during Desert Shield and Desert Storm, of whom 148 had been killed in action.

"We confront mysteries here," the president declared, standing in the warm June sunshine near the Tomb of the Unknowns. "We celebrate the fact that each person we commemorate today gave up life for principles larger than each of us, principles that at the same time form the muscle and strength of our national heart."

Bush scanned a sea of tear-stained faces turned toward him. His voice carried above the muffled sobs, seeking to console the inconsolable. "Through their sacrifice, as they caused brutal aggression to fail, they renewed our faith in ourselves." A sergeant sang "The Last Full Measure of Devotion," and then the doleful blare of Taps swept down the hill toward the river, echoing and re-echoing among the endless ranks of tombstones.

After the service, buses carried the mourners back across the Potomac to join the throng gathering on the Mall. Shortly before noon

Schwarzkopf and his staff wheeled onto Constitution Avenue, followed in close order by nine thousand troops representing the nearly six hundred thousand who had waged the war. Down the boulevard they marched at 116 steps per minute, a grand panoply of flags and battle streamers, crashing brass bands and lumbering tanks that scarred the soft asphalt. Overhead the first of eighty-three aircraft streaked past, a solitary F-117 that abruptly appeared from nowhere with a throaty roar and a flash of black wings. For an instant many of the startled spectators fell silent, as though sensing the fate of a hundred targets in Tikrit and Taji and Baghdad.

The parade pressed on. The crowd regained its voice, whooping in adulation. Past the reviewing stand the warriors marched, eyes right, exchanging salutes with Bush, Powell, and the service chiefs. Schwarzkopf peeled off to stand with the president. Then the procession snaked westward, past the Ellipse, before veering south around the graceful black walls of the Vietnam Veterans Memorial, which seemed oddly diminished, as if finally whittled into proper proportion.

For what had they died, those 390 Americans? What cause was worthy of their sacrifice, or that of the 458 Americans wounded in action, or the 510 allied casualties?

For months after the invasion of Kuwait, the president had struggled to articulate his motives for going to war. With the shooting now stopped, the reasons at last seemed clear: the conflict had been waged on behalf of cheap oil, friendly monarchies, and Washington's strategic goal of preventing the emergence of a hegemonic power inimical to American interests in the Middle East. When measured against those objectives, the war succeeded in making the world safe for the status quo ante bellum.

Saudi Arabia, soon cleansed of foreign troops and Western journalists, again became a closed kingdom, where women were forbidden to drive and democracy was the alien concept of infidels. Kuwait began rebuilding both its battered infrastructure and autocratic government. The emir, who tarried in exile for weeks after the liberation of Kuwait City, stalled the restoration of civil liberties and the return of parliament. In one particularly shameful episode, a young Palestinian who had worn a Saddam Hussein T-shirt was sentenced to fifteen years in prison.

The last of several hundred blazing oil wells would be extinguished in November — then reignited so that the emir could formally douse the blaze with suitable ceremony. The Kuwait Oil Company estimated that 3 percent of the country's total reserves had been lost, or nearly three billion barrels. The long-term environmental impact of the spillage remained uncertain. Grim predictions of "nuclear winter" from airborne soot or "greenhouse warming" aggravated by carbon dioxide from the flaming wells proved to be exaggerated. Nevertheless, a miasma of oil smoke spread across an area more than twice the size of Alaska. Black snow was reported in the Himalayas. Millions of barrels sloshed around the Persian Gulf, fouling the shorelines and killing at least thirty thousand sea birds. Saddam's legacy of ecological terrorism would endure for years.

The U.S. Embassy in Kuwait City issued a press release listing achievements in reconstructing the shattered nation, like the reopening of Kentucky Fried Chicken, Arby's, and Pizza Hut restaurants. The financial cost of the war to Arab countries was eventually calculated at $620 billion, with Iraq and Kuwait each bearing roughly a third of the total burden.

The war's achievements, of course, went well beyond the resurrection of a few fast-food franchises in liberated Kuwait. "By God," Bush declared after the armistice, "we've licked the Vietnam syndrome once and for all." This sentiment was neither fatuous nor insignificant. For twenty years the debacle of Vietnam had bred self-reproach, mistrust, and an abiding doubt in the efficacy of military power. The competence and potency of the American military was now beyond question.

"It is not big armies that win battles," Maurice de Saxe noted in 1732. "It is the good ones." The United States had built a military that was both big *and* good. The nation demonstrated that superpower status was calculated not simply in nuclear megatonnage but also in more prosaic capabilities: only America could have amassed more than nine million tons of matériel, hauled it six thousand miles to the Middle East, fought a war, then carted the stuff home again. (Nearly 90 percent of the cost of the campaign had been underwritten by the Japanese, Germans, and Arabs, thanks to aggressive American officials, who collected $54 billion in fund-raising expeditions dubbed Tincup I and Tincup II.) All in all, weapons and tactics worked well, troops performed with admirable skill, commanders showed themselves equal to the challenge.

In Schwarzkopf, the nation rediscovered the pleasure of adoring a military hero. For many months after the war the CINC seemed ubiquitous, appearing at the Kentucky Derby, at the Indianapolis 500, on Capitol Hill, in parades, on bubblegum cards. "What next to conquer?" asked an electronic message on the scoreboard at Tampa Stadium during one giddy celebration. Mentioned as a possible Army chief of staff or even for five-star rank, he would choose instead to retire after thirty-five years of service, becoming rich and ever more famous.

Yet history must draw back a bit. Schwarzkopf's generalship, not unlike the war itself, was hardly unblemished. He had, at first, grossly misjudged the time and forces needed to expel the Iraqi invaders from Kuwait. His overestimation of the enemy's size and capability persisted to the last shot and beyond. His imprint on the allied air campaign was virtually nil, other than to insist on an early thumping of Iraqi ground forces. After the catastrophe of the Al Firdos bunker, much of the strategic targeting authority was tugged from his hands by his superiors in Washington. He had stubbornly resisted turning his attention to the only Iraqi gambit that could have threatened him strategically — the Scud missile attacks against Israel. In the end, allied success was less a reflection of any particular brilliance in the war plan than of the stellar competence of Schwarzkopf's lieutenants in executing it, a contribution at times acknowledged with stingy reluctance.

Nevertheless, the man had risen to the task. He had unified the coalition forces and kept them unified. He had generally encouraged initiative and intrepidity among his subordinate commanders. His tactical assessments — in declining, for example, to launch an amphibious attack — proved sound. He had projected an image of strength and resolve. He brought home alive far more of his soldiers than even the most optimistic strategists in Washington had dared hope. He had won. Norman Schwarzkopf had earned his due.

The larger military lessons of the war were harder to parse. Few could doubt the full emergence of air power as the pre-eminent factor in modern combat, although sensible leaders recognized that the optimal conditions that had rendered air attacks so lethal in this war would by no means be assured in the next. Those who considered air power the linchpin of a new Pax Americana, as land power had characterized Pax Romana and sea power created Pax Britannica, needed only to recall Vietnam to remember the limitations of bomber fleets against a determined foe sheltered by mountains and thick foliage.

Even against Iraq, the impact of the air campaign was mixed and bitterly debated. A two-year independent study commissioned by the Air Force would find little evidence that the strategic attacks against Baghdad and other targets north of the Euphrates had been critical in the allies' ultimate success. Despite destroying nearly all of Iraq's petroleum refining capacity, for example, such attacks "bore no significant military results" — in part because the war did not last long enough for fuel shortages to severely hamper enemy forces. Also, despite 840 attacks against leadership and other command-and-control targets, Saddam was still alive and his regime still in power. On the other hand, the 22,000 strikes against the Iraqi army — although failing to reach Schwarzkopf's goal of fifty percent destruction before the ground offensive began — had clearly battered the enemy to near senselessness. The wrangling over air power's efficacy, begun nearly a century before, seemed likely to continue for at least another generation.

Sharp reductions in the U.S. military, begun even before the gulf war, also raised doubts about whether America could again mount the sort of expedition that had embarked to the Middle East. As Bush himself would warn in a speech at West Point in January 1993, the nation could not serve as "the world's policeman." War as an instrument of national policy, the president concluded, was justified only "where and when force can be effective" and "where the potential benefits justify the potential costs and sacrifice." Given such prudent constraints, relatively few of the world's conflicts — including the carnage in disintegrating Yugoslavia — seemed to lend themselves to resolution by American force of arms.

Even so, the gulf campaign reaffirmed the bond between those in uniform and the larger republic, a delicate relationship that has waxed and waned for more than two centuries. Although difficult to measure, that reaffirmation was important. As the author James Fallows once wrote, "A nation's military, especially in a democracy, can endure the hardships of war only if it feels tied to a nation by a sense of common purpose and respect."

The war also demonstrated Washington's capacity to assemble an extraordinarily diverse alliance, some of its members former blood enemies from the Cold War. It resuscitated and empowered the United Nations. It provided the United States with sufficient cachet to bring Israelis and Arabs together for comprehensive peace talks — offering at least faint hope for resolving the world's most dangerous and intractable conflict.

Finally, and perhaps most significantly, the war defanged Saddam Hussein. In testimony before Congress a few days after his triumphal march through Washington, Schwarzkopf claimed that Iraq's forty-two divisions in the Kuwaiti theater "were all destroyed." The CINC exaggerated. By subsequent Army calculations, about one third of all Iraqi forces got away; fully half of the Republican Guard tanks and other heavy equipment escaped. Nevertheless, the enemy host *had* been substantially dismembered. An avaricious despot had been stopped in his tracks, and the barbarity of his occupation avenged.

With a victor's gift for self-delusion, the Americans initially assumed that Iraq's nuclear, biological, and chemical weapons programs had been destroyed by six weeks of bombing. This, as United Nations teams soon discovered, was hardly so. More than a hundred Scud missiles survived, as well as missile-production equipment, at least nineteen mobile launchers, and components for a new, two-stage missile. Seventy tons of nerve agents and four hundred tons of mustard gas also escaped destruction.

Allied bombers had concentrated on three nuclear facilities, the last of which, Al Athir, was not discovered until late February, and was struck in the last F-117 sortie of the war. "Our pinpoint attacks have put Saddam out of the nuclear bomb-building business for a long time to come," Bush had declared. In fact, Iraq had more than twenty facilities linked to a mammoth bomb-making effort resembling the Manhattan Project. Air strikes had only "inconvenienced" Iraqi plans to build a bomb, United Nations analysts later concluded.

But that which the bombers failed to destroy fell prey to UN inspectors, who roamed through Iraq under provisions of the April cease-fire agreement. These search-and-destroy missions would be tedious, protracted, and periodically thwarted by the Iraqis. Yet two years after the invasion of Kuwait, there seemed little doubt that Saddam's programs for weapons of mass destruction had been gutted. The wholesale demolition of chemical stocks began in the fall of 1992 in a UN-built incinerator and hydrolysis plant. In September, the leader of the United Nations' nuclear inspection team declared Iraq's bomb-building effort to be "at zero." Such optimism may have been premature, given the additional caches of bomb-making equipment discovered in subsequent months. Saddam's most sinister aspirations remained intact, and he manifestly possessed sufficient conventional firepower to terrorize his own citizenry. But as an immediate threat to its neighbors, Iraq was finished.

Why, then, given this catalogue of accomplishments, did the sweet savor of victory so quickly turn to the taste of ashes? At the time of the parade in Washington, three quarters of all Americans polled believed the war had been worth fighting; nine in ten said that the United States should be proud of its accomplishments. Bush's approval rating leaped to an astounding 88 percent, comparable to Harry Truman's at the end of World War II.

Yet these numbers steadily plummeted to the point that the president failed to muster the majority needed for re-election. Scarcely two years after the invasion of Kuwait, Bush, Thatcher, Gorbachev, and Yitzhak Shamir were all swept from power. Saddam, still entrenched, celebrated

Bush's defeat at the polls by unholstering his pistol and firing a few defiant shots into the air.

Before the war began, Bush had established limited objectives for a limited campaign. Through six weeks of combat he stuck to those goals with fixed determination — a quality Clausewitz held among the highest attributes of a successful commander. He resisted the temptation to march on Baghdad, certainly a decision that spared countless lives and incalculable political complications.

But at the same time Bush had encouraged the nation to consider the war a great moral crusade — a struggle of good versus evil, right against wrong. "Nothing of this moral importance," he had proclaimed in absurd overstatement, had occurred "since World War II." By demonizing Saddam, Bush aroused passions that would remain unsated. "I pledge to you: there will not be any murky ending," the president had vowed in November 1990. "I will never, ever agree to a halfway effort."

But, by definition, limited wars achieve limited results. "Murky endings" have been the rule rather than the exception in modern warfare. In the months following Schwarzkopf's triumph, the ending grew ever murkier. By mid-1992 the CIA concluded that Saddam was tightening his grip on Iraq — partly through the prodigal employment of firing squads — and had become more secure than at any time since the cease-fire. Despite the allied naval blockade and hyperinflation estimated somewhere between 300 and 4000 percent, Saddam turned the reconstruction of Iraq into his own national crusade.

By late 1992, the Justice, Defense, and Local Government ministries had been rebuilt. Only one bridge remained down over the Tigris. The so-called AT&T building — the first structure destroyed in downtown Baghdad, on January 17, 1991 — was refurbished, as were highways and oil refineries. At least half a dozen secret police agencies again terrorized Iraqi society in a relentless hunt for subversives, real or imagined. The army rank-and-file was purged of Shi'ites. Countless new portraits of Saddam — smiling, omnipresent, and rarely defaced — appeared on billboards and street corners. The man who had made innumerable tactical and strategic miscalculations nevertheless correctly reckoned that he could lose the war and still remain in power — thus demonstrating that he knew his country better than the Americans did.

Some American policymakers argued that Saddam's survival provided a strategic benefit by keeping Saudi Arabia dependent on the United States for protection. But such arguments soon seemed as fatuous as they were clever. Emboldened by Bush's reluctance to risk further entanglement, Saddam reasserted his claim to Kuwait and refused to recognize the border drawn by the United Nations between Iraq and the emirate. Washington responded by selling $20 billion worth of arms to Iraq's neighbors, hardly an innovative means to regional security. The CIA also undertook a covert $40 million program to destabilize the regime, which, among other projects, reportedly called for flooding Iraq with counterfeit dinar bank notes to further stoke inflation. In July 1992 Iraqi troops intensified attacks on Shi'ite rebels through air strikes from Talil Air Base. Such behavior violated United Nations resolution 688, which forbade the repression of Iraqi citizens. This time the United States and its allies intervened by imposing a "no-fly" zone that prohibited flights south of the 32nd Parallel.

Tensions heightened in December 1992 when Baghdad harassed United Nations inspection teams and humanitarian relief convoys headed for Kurdistan. On December 27, U.S. jets shot down an Iraqi MiG in the south. Ten days later, Washington ordered Iraq to remove surface-to-air missiles from the southern no-fly zone. On January 10, 1993, two hundred Iraqi soldiers raided a United Nations bunker not far from Safwan and snatched a weapons cache that included fifteen Silkworm missiles. Three days later, 110 U.S. and British warplanes attacked four SA-3 SAM sites and four air defense centers, hitting about two thirds of thirty-three specific aim points. Bush also ordered a U.S. armored battalion back to Kuwait.

More raids followed against the southern air defense network. On Sunday, January 17, 1993, U.S. warships fired more than forty Tomahawk missiles at Zaafaraniya, a research complex southeast of Baghdad allegedly capable of making machinery used to enrich plutonium for nuclear warheads. One TLAM went awry, hitting the Al Rashid Hotel and killing three civilians.

The attack would be the last major event of George Bush's presidency. On January 20, he gracefully handed over the office to Bill Clinton of Arkansas. For two more days the skirmishing continued in Iraq, with U.S. fighters striking antiaircraft and missile sites in the north before a tacit cease-fire again took hold.

After more than two months of study, the Clinton administration affirmed most of the tenets of Bush's postwar policy toward Iraq. Although Saddam's removal was no longer explicitly required to ease international economic sanctions, other conditions — such as an end to attacks on Kurds and Shi'ites, the acceptance of Kuwaiti sovereignty, and the continuation of UN inspections — remained intact.

Few Western analysts believed Saddam likely to accept such strictures willingly. Sooner or later, he would again test the resolve of his

Western adversaries. Something close to the Korean precedent, which Bush and his brain trust had so ardently sought to avoid, had taken shape in the Persian Gulf. The U.S. military faced a hostile foe across not one, but two lines of demarcation — the 36th Parallel in the Kurdish north and the 32nd Parallel in the Shi'ite south. American forces in the gulf region comprised twenty-six ships, more than 150 aircraft, and 24,000 troops — compared with 37,000 in Korea.

George Bush had always drawn a distinction between wicked Saddam and the Iraqi people, "who have never been our enemy." Yet this appeared ever more simplistic. Iraqis had supported Saddam through two catastrophic wars of aggression. Few showed remorse for the brutal sack of Kuwait. Schoolchildren in the spring of 1993 again lined up before class to chant, "America, the Zionists, France, and Britain are the enemy of Iraq!" A thwarted plot to assassinate Bush during a visit to Kuwait in April 1993 appeared to have links to the Iraqi intelligence service. Clinton retaliated by ordering another Tomahawk salvo on June 26. Twenty missiles destroyed the Iraqi Intelligence Service headquarters in Baghdad; three others landed outside the compound in a residential neighborhood, killing eight people and wounding twelve.

The Iraqis answered with more vitriol and more harassment of UN inspectors. Iraq might require allied vigilance for years or even decades, regardless of Saddam's fate. A policy akin to the NATO containment strategy of the Cold War would be necessary to restrain Iraq and its equally volatile neighbor, Iran. (It is worth noting that after World War I the Inter-Allied Control Commission had roamed Germany for years in search of arsenals and munition factories; the commission withdrew in 1926, and a decade later the German military was the most powerful in the world.)

These developments helped eclipse the war's genuine accomplishments, as did the gradual revelation of the U.S. government's unsavory support for Saddam through the 1980s. The "new world order," so grandly trumpeted by Bush during and after the war, proved an empty slogan. The world's sole remaining superpower still felt compelled to be the world's armory, peddling $63 billion worth of weaponry and military products to 142 nations in 1991 alone. Restoring the status quo in the Persian Gulf also permitted Americans to indulge their addiction to foreign petroleum with hardly a second thought. A nation with 5 percent of the world's population continued to devour 25 percent of its energy, a gluttony made possible only by importing half the country's oil.

Moreover, the war did nothing to address crippling social and economic problems at home. Having earned his laurels, Bush would rest on them with a complacency that cost him his job. The president showed neither the inclination nor political wherewithal to harness his popularity in the cause of domestic achievement. His prowess as commander-in-chief underscored a lackluster domestic record, giving rise to cynicism. SADDAM HUSSEIN STILL HAS HIS JOB, a popular bumper sticker observed. HOW ABOUT YOU?

"They set out to confront an enemy abroad," Bush had said of his legions in early March of 1991, "and in the process they transformed a nation at home." Sadly, no such transformation took place. The sense of the war as a watershed proved ephemeral. Americans soon recognized that expeditionary warfare offered no panacea for the nation's most profound challenges.

Someday, perhaps, the tide would change and revisionist sentiment would bring the war into proper perspective. A jaundiced discontent was common after other American wars. Franklin Roosevelt and his subordinates were long accused of winning World War II but losing the peace. Harry Truman left office badly stained by Korea; not for years was that conflict seen as a plausible achievement on behalf of American interests and principles.

The Persian Gulf War was neither the greatest moral challenge facing America since 1945, as Bush had declared, nor the pointless exercise in gunboat diplomacy portrayed by his severest critics. The truth lay somewhere on the high middle ground, awaiting discovery.

The parade surged around the Lincoln Memorial, gleaming in the midday sun. The immense figure of Abraham Lincoln sat as though watching in review, a faint smile chiseled on his marble lips. The battle captains trooped by: John Yeosock and Gary Luck, helmeted and somber; Barry McCaffrey and Binnie Peay; Walt Boomer and Ron Griffith and Tom Rhame; Chuck Horner in his flight suit and pilot scarf, followed by Dave Deptula. At the head of the VII Corps contingent, Fred Franks hobbled slightly, the stump of his left leg rubbing uncomfortably against his artificial limb.

Onto Memorial Bridge they marched, past the equestrian statues of Valor and Sacrifice. Police had cleared spectators from the bridge. Cheering faded to a distant murmur. Ahead lay the greensward of Arlington. A tranquil silence settled over the ranks, broken only by the tramp of boots and the sharp rap of a drum.

AUTHOR'S NOTE

WITH APPRECIATION

CHRONOLOGY

BATTLE MAPS

NOTES

BIBLIOGRAPHY

INDEX

# AUTHOR'S NOTE

History, the author Simon Schama has written, deals with "the teasing gap separating a lived event and its subsequent narration." Rarely is that gap wider or more vexing than when one writes about war. The search for an answer to the simplest question — what really happened? — is constantly challenged by conflicting memories, secrecy, and vainglory. The emotional pitch of battle ensures that no two participants recall events precisely alike, even if they shared a cockpit, tank, or tent. In reconstructing the history of the Persian Gulf War, where recollections and opinions diverge I have attempted to sift through the evidence and draw conclusions to the best of my ability.

To research this book, I traveled to Israel, Germany, Great Britain, and more than a dozen American military bases. I made two visits to the Persian Gulf, including a six-week stint in March and April 1991 with allied forces in Saudi Arabia, Iraq, and Kuwait. More than five hundred participants consented to interviews, often on condition that their remarks remain unattributed. A number of key figures were interviewed repeatedly, in as many as a dozen separate sessions. I'm deeply grateful for their time and assistance. Although the heart of this narrative is drawn from such interviews, I also examined thousands of pages of documents, including afteraction reports, investigative files, and personal notes. Dialogue is recounted only when at least one direct witness helped in its reconstruction. Several participants in the gulf campaign were kind enough to review portions of the manuscript for accuracy. Nevertheless, any shortcomings are my own responsibility.

I am indebted to many officers and civilians who helped in locating documents or in arranging interviews. Thanks to Pete Williams, Lt. Col. Rick Oborn, and Lt. Cmdr. Ken Satterfield from the Defense Department Office of Public Affairs, as well as Col. Bill Smullen and Cmdr. Dave Barron of the Joint Staff.

From the U.S. Army I appreciate the aid of Sgt. Robert Austin, Maj. Craig Barta, Maj. Kevin Bergner, Capt. Lewis M. Boone, Capt. Bob Dotson, Col. James L. Fetig, Capt. Steve Hart, Lt. Col. John Head, Maj. Peter A. Hnatiuk, Maj. Pete Keating, Col. Rick Kiernan, Col. Don Kirchoffner, Brig. Gen. Charles W. McClain, Jr., Lt. Col. Don McGrath, Steve Moore, Col. Bill Mulvey, Col. George Norton, Maj. George Rhynedance, Col. George Stinnett, and Maj. Mark Tolbert III. Special thanks to Col. George T. Raach.

My thanks in the U.S. Navy to Rear Adm. Brent Baker, Senior Chief J. M. Burke, Lt. Cdr. Steve Chesser, Lt. Fred Henney, Cmdr. Steve Honda, Lt. Taylor Kiland, Capt. Jim Mitchell, Capt. Mark Neuhart, Lt. Rob D. Raine, Cmdr. John Woodhouse, and especially Lt. Mark Walker.

In the U.S. Air Force, particular thanks to Lt. Col. Mike Gannon and Brig. Gen. Ed Robertson, as well as Jay Barber, Capt. Tom Barth, Sgt. Gary Boyd, Capt. Dave Cannon, Chief Master Sgt. James Chumley, Capt. Becky Colaw, Lt. Col. Darrell C. Hayes, Staff Sgt. Dee Ann Heidercheit, Sgt. Linda Mitchell, Staff Sgt. Alvin J. Nall, Tech. Sgt. Mike Otis, Master Sgt. Bobby Shelton, and Tech. Sgt. Rick Shick.

From the U.S. Marine Corps, I am indebted to Capt. Scott R. Campbell, Capt. Dan Carpenter, Maj. Jay C. Farrar, CWO Randy Gaddo, Capt. Steve Manuel, Col. Fred Peck, Sgt. René Reyna, Col. John Shotwell, Capt. Bill Taylor, the Marine Corps Historical Center, and especially Maj. Nancy Laluntas.

I also appreciate the efforts of Col. John W. Dye III and George B. Grimes of the U.S. Special Operations Command; Maj. Olin Saunders and Maj. James L. Royster of Central Command; Joseph DeTrani and Peter Earnest of the Central Intelligence Agency; Judith C. O'Neil of the State Department; Steve Wilmot of the Royal Air Force; and Tim Downes and Andrew Silverman of the British Ministry of Defence.

*The Washington Post,* my employer, again granted me the time and freedom to complete this book. Thanks to Dan Balz, Benjamin C. Bradlee, Ann Devroy, Karen DeYoung, Jackson Diehl, Leonard Downie, Jr., Barton Gellman, Michael Getler, Brad Graham, Don Graham, David Hoffman, David Ignatius, Robert G. Kaiser, John Lancaster, Steve Luxenberg, Molly Moore, Caryle Murphy, Don Oberdorfer, R. Jeffrey Smith, Ben Weiser, and Bob Woodward. The contributions of hundreds of correspondents from various newspapers, magazines, wire services, and television networks were invaluable.

My researcher, Lucy Shackelford, displayed tenacity, ingenuity, and good humor, for which I'm deeply grateful. In Paris, Kelly Couturier was helpful in researching the role of the French military during the war. Brad Wye was an admirably conscientious and competent cartographer.

My agent and good friend, Rafe Sagalyn, was invariably supportive and enthusiastic. John Sterling, editor-in-chief of Houghton Mifflin, was, once again, simply brilliant. Frances Apt, the manuscript editor, again improved every page with her sharp eye and devotion to the English language. I am in her debt. Thanks also to Rebecca Saikia-Wilson and Dan Maurer.

Finally, I wish to thank my family for their love and patience.

# WITH APPRECIATION

Lt. Col. Larry Adair; Brig. Gen. John Admire; Brig. Gen. Frank H. Akers, Jr.; Col. Roy Alcala; Margaret Aldred; Lt. Col. Keith Alexander; Seaman Jeselito Alino; Charles Allen; Capt. James B. Andersen; Capt. Gregg Andreachi; Moshe Arens; William M. Arkin; Brig. Gen. David Armstrong; Lt. Bill Armstrong; Lt. Joseph D. Armstrong; Richard L. Armitage; Maj. Gen. Steven L. Arnold; Adm. Stanley R. Arthur; Lt. Cdr. Greg Atchison; Col. Richard Atchison; James A. Baker III; Capt. James Baldwin; WO R. F. Balwanz; Prince Bandar bin Sultan; LCpl. (U.K.) Johan Bangsund; Col. Marv Bass; Maj. John Batiste; Maj. Pat Becker; Lt. Col. Tom Belisle; Maj. Austin Bell; Rear Adm. (Isr.) Avraham Ben-Shoshon; Maj. Kevin Bergner; Sgt. (U.K.) Clive Bergstran; Maj. Dale Berry; Lt. Col. Randy Bigum; Capt. Lyle Bien; Capt. David S. Bill; Capt. John Bley; Sgt. Reginald Blossom; Lt. Col. Jon Boettcher; Maj. Jerry R. Bolzak, Jr.; Gen. Walter E. Boomer; Col. Garrett D. Bourne; Capt. Larry Bowers; Capt. Mike Bowers; Christopher J. Bowie; Lt. Gen. Martin L. Brandtner; Col. Emil Brominski; Lt. Col. Daniel P. Brownlee; Capt. Bill Bruner; Lt. Col. Bruce Brunn; Capt. Charles Brunson; William Burns; Sgt. Maj. (U.K.) Keith But; Lt. Paul Calvert; Maj. Jason Camia; Capt. Jay Campbell; Lt. Cdr. Kevin Campbell; Maj. Steve Campbell; Maj. Bill Carrothers; Maj. Paul Casinelli; Maj. Jim Chambers; Richard B. Cheney; Brig. Gen. Stanley F. Cherrie; Lt. Cdr. Steve Chesser; Capt. Ernest E. Christensen, Jr.; Brig. Gen. Dan Christman; Col. Bob Clark; Maj. Gen. Wesley K. Clark; Lt. Rivers Cleveland; HT3 Ron Clift; Lt. Col. Dick Cody; SFC. Bob Cole; Lt. Col. Ray Cole; Lt. Col. Gary Collenborne; Lt. John Cooper; Lt. (U.K.) Alex Cormack; L. Neal Cosby; Lt. Col. John Craddock; Col. James C. Crigger, Jr.; Lt. Col. Bruce Crimin; WO Jim Crisafulli; Lt. Col. Marty Crumrine; Maj. Mark Curry; Capt. Pete Curry; Maj. Kent Cuthbertson; Col. John Davidson; Capt. Gerald S. Davie, Jr.; Maj. Dan Davis; Col. Daniel O. Davis, Jr.; Lt. Col. William J. Davis; Capt. Cutler Dawson; Lt. Col. Mike Decuir; Col. Ray W. Dehncke; Lt. Col. David A. Deptula; Lt. Tony Despirito; Maj. Bob Desrosiers; SK2 Richard Devlin; Corp. (U.K.) Ian Dewsnap; Capt. Tom Dietz; Capt. Bob Dobson; Lt. Gen. Wayne Downing; Brig. Gen. Thomas V. Draude; Lt. Col. James Dubik; Capt. Lee Duckworth; Gen. Michael Dugan; Col. Layton G. Dunbar; Capt. David Dykes

Lawrence Eagleburger; Barbara Eberly; Col. David W. Eberly; Timm Eberly; Col. Bob L. Efferson; Maj. Doug Eliason; Col. Bob Ellis; Sgt. Donald Eveland; Capt. Rich Faulkner; Capt. (U.K.) Nick Fenton; Sgt. Donald Ferra; Col. James L. Fetig; Capt. John Fleming; Col. Robert B. Flowers; Rear Adm. William Fogarty; Col. Greg Fontenot; Capt. Charlie Forshee; Col. Jerry Foust; Lt. Col. Chuck Fox; Capt. David Francavilla; Gen. Frederick Franks, Jr.; Maj. Ben Freakley; Lt. Jon Fredas; Capt. Robert Freehill; Maj. Ken Fugett;

Col. James A. Fulks; Lt. Gen. Paul E. Funk; Col. Tony Gain; Lt. Keith Garwick; Robert M. Gates; Lt. Col. Norm Gebhard; Lt. Col. Ralph Getchell; Lt. Mike Gilday; Maj. Tom Gill; WO Gregory M. Gilman; Lt. Greg Glaros; Lt. Col. Jim Gleisberg; Maj. Russell W. Glenn; Lt. Gen. Buster C. Glosson; Lt. Col. Thomas R. Goedkoop; Lt. Col. Lewis J. Goldberg; Lt. Col. (U.K.) Robert Gordon; Lt. Col. Tom Greco; Maj. Gen. Jerome H. Granrud; Sgt. Gordon Gregory; Capt. David J. Grieve; Capt. Ken Griffin; Sgt. (U.K.) Kevin Griffin; Lt. Gen. Ronald H. Griffith; Maj. Dan Grigson; Sgt. Christopher Groninger; Lt. Col. Clay Grubb

Richard Haass; Capt. Dan Hacker; Capt. Mark Hackler; Brig. Gen. F. N. Halley; Lt. Col. Chuck Hallman; Capt. Terry Halton; Spec. Anthony Ham; Dave Hamlin; LCpl. (U.K.) Ian Hammond; Capt. Randy Hammond; Lt. Clint Hancock; Capt. (U.K.) Piers Hankinson; Dr. John C. Harper; Capt. Jim Hart; Capt. Steve Hart; Lt. Col. Ben Harvey; Richard L. Haver; Col. Mike Hawk; Lt. Col. Stephen Hawkins; Lt. Col. Tom Hayden; Lt. John Hayes; Group Capt. (U.K.) Bill Hedges; Spec. Glenn Heil; Capt. Dean M. Hendrickson; Col. Fred Hepler; Capt. Steve Herczeg; Spec. Michael Herzog; Lt. Andrew Hewitt; Lt. Col. Sam Higdon; Col. Tom Hill; Lt. Col. James Hillman; Sgt. Michael Himes; Col. Johnnie B. Hitt; Col. Richard W. Hodory; Lt. Col. Phil Hoffman, Jr.; Lt. Col. Keith T. Holcomb; Capt. Gordon Holder; Lt. Col. Ken Holder; Brig. Gen. L. Don Holder; Capt. Edward B. Hontz; Lt. Gen. Charles A. Horner; Maj. Ed Howard; Capt. Antonio Huggar; Corp. (U.K.) David Hughes; Lt. Mary Hughes; Lt. Col. Jerry Humble; Lt. Pete Hunt; Capt. J. R. Hutchison; Maj. John Imhoff; David Ivri; Maj. Mike Ivy; Capt. Al Jackson; LCpl. (U.K.) Sion James; Col. Cash Jasczak; Lt. Col. Richard Jemiola; Maj. Gen. Harry W. Jenkins, Jr.; Maj. Wilbur C. Jenkins; Adm. David Jeremiah; Capt. Carlos Johnson; Capt. Greg Johnson; Lt. Gen. James Johnson; Col. Jesse Johnson; Capt. Robert L. Johnson, Jr.; Sgt. Maj. William Johnson; Lt. Gen. Robert Johnston; Capt. Dan Jones; WO David A. Jones; Lt. Col. Taylor Jones; Sgt. George Jons; DC3 Michael Jurkowski; Lt. Cdr. Jim Kear; Lt. Col. Al Keith; Lt. Gen. Thomas W. Kelly; Capt. Newton Kendrick; Capt. Marcel E. Kerdavid, Jr.; Lt. Col. Charles Kershaw; Maj. Gen. William M. Keys; Maj. Roger King; Robert M. Kimmitt; Lt. Col. Hank L. Kinnison IV; Maj. John Klemencic; Capt. Gary Klett; Lt. Col. Mike Kobbe; Col. James W. Kraus; Capt. Merrick E. Krause; Lt. Col. Bruce Kriedler; Brig. Gen. Charles C. Krulak; Capt. Chris Kupko; Capt. Ken Lacy; Lt. Col. Pat Lamar; Brig. Gen. John R. Landry; Lt. Rodney LaPearl; Maj. Jerry Leatherman; Maj. Willes Lee; Maj. Gen. John Leide; Col. John Le Moyne; Lt. Col. Thomas J. Leney; Col. Thomas Lennon; Sgt. Lawrence Lentz; BM Onzie Level; Col. George E. Lewis, Jr.; I. Lewis Libby; Col. Larry Livingston; Brig. Gen. (Isr.) Ze'ev Livne; Col. Andy Lloyd; Lt. Col. Ned Longsworth; Rear Adm. Joe Lopez; Lt. Gen. Gary Luck; Lt. Col. Doug Lute; Maj. (U.K.) Julian Lyne-Pirkis

Maj. (U.K.) Hamish MacDonald; Lt. Col. (U.K.) Christian D. Mackenzie-Beevor; Capt. (U.K.) David Madden; Maj. Mike Mahar; Col. Lon Maggart; DC Terry Marable; Sgt. Stewart Marin; Lt. Col. Steve Marshman; Lt. Col. Toby Martinez; SSgt. Jeffrey Mauer; Maj. James P. Mault; Maj. Mike Mauro; Vice Adm. Henry H. Mauz; Lt. Gen. Barry R. McCaffrey; Lt. Col. Jeffrey D. McCausland; Rear Adm. Mike McConnell; Maj. William B. Mc-

Cormick; Maj. Gen. Billy McCoy; Sgt. David McDonald; Col. James McDonough; Lt. Col. Rick McDow; Capt. G. Bruce McEwen; Brig. Gen. Robert P. McFarlin; Lt. Col. Don McGrath; Capt. Edward J. McHale; Lt. Rich McKinley; Sgt. Tony McKinney; Capt. H. R. McMaster; Lt. Charles L. McMurtrey; Gen. Merrill A. McPeak; Cmdr. Donald McSwain; Brig. Gen. Montgomery C. Meigs; Cmdr. Michael J. Menth; Ens. Dale Meyers; Capt. Alan Miller; Lt. Col. Denny Miller; Maj. Dan Miltenberger; Rear Adm. Riley Mixson; Sgt. David Moody; Capt. Mickey Moore; Capt. Franklin J. Moreno; Col. Tony Moreno; Sgt. Maj. (U.K.) Charles Morton; BM Bill Morvey; Lt. Col. Steve Moses; SFC. Ron Mullinix; SSgt. Robert Muto; Maj. Gen. Mike Myatt; Lt. Col. Cliff Myers; Maj. (U.K.) James N.C. Myles; Brig. Gen. Richard I. Neal; Benjamin Netanyahu; Capt. Scot Newport; Spec. Terrence Newton; Patrick Nixon; Maj. Dan Nolan; Capt. Joe Nuti; Lt. Bill Nutter; Maj. William J. O'Connell; Pvt. (U.K.) David O'Connor; Lt. Col. Mike O'Connor; Col. David H. Ohle; Capt. Van Oler; Lt. Col. Donald C. Olson; WO Thomas R. O'Neal, Jr.; Lt. Tony Onorati; Maj. Don Osterberg; Maj. (U.K.) Mark O'Reilly; Seaman Michael Padilla; Col. Leslie M. Palm; Lt. Gen. Gus Pagonis; Spec. Kenneth Parbel; Maj. Bill Parry; Sgt. Douglas Patterson; Maj. Richard Pauly; Capt. Steve Payne; Corp. (U.K.) Mark Pearcey; Lt. Gen. J. H. Binford Peay III; Maj. (U.K.) Fred Perry; Col. David Peterson; Cdr. J. W. Phillips; Capt. Allen Pierson; Lt. Col. Steve Pingel; Roman Popaduik; Gen. Colin L. Powell; Lt. Col. Robert Purple; Col. Joseph H. Purvis; MSgt. Bruce Quickell

Lt. Col. Mike Rapp; Capt. Eddie Stephen Ray; Maj. Robert S. Rea; Gen. Dennis J. Reimer; Sgt. Alex Remington; Maj. Gen. Thomas Rhame; Sgt. Mary Rhoads; Col. Ron Richard; Col. William M. Rider; CSgt. Maj. Robert S. Rivera; WO James E. Roberts, Jr.; Capt. James H. Robinette; Col. Steve Robinette; Sgt. (U.K.) Terry Robinson; Maj. (U.K.) John Rochelle; Lt. Erich Roeder; Maj. Horst Roehler; Maj. Buck Rogers; Lt. Col. (U.K.) Charles Rogers; Maj. Jack Rogers; Col. Ronald F. Rokosz; SSgt. Luis A. Roman; Maj. Jim Rose; Harry S. Rowen; Col. Bill Rutherford; Lt. Justin Ryan; Capt. Norbert R. Ryan; Cmdr. Rodney Sams; Lt. Col. Rick Sanchez; Capt. Joseph Sartiano; Col. David A. Sawyer; Brig. Gen. Robert H. Scales, Jr.; Maj. Gen. Charles F. Scanlon; Ze'ev Schiff; Col. L. S. Schmidt; Brig. Gen. Edison E. Scholes; Capt. Liz Schwab; Gen. H. Norman Schwarzkopf; Lt. Cdr. Brian Scott; Maj. Douglas R. Scott; Lt. Col. R. E. Scott; Lt. Gen. Brent Scowcroft; Capt. Larry Seaquist; Capt. Bradley A. Seipel; Lt. Cdr. Steve Senk; Lt. Col. Rick Shatzel; Corp. (U.K.) Mark Shaw; Lt. Col. Michael D. Shaw; Maj. Tom Shaw; Capt. Todd K. Sheehy; Master Sgt. Jeffrey Sims; Capt. Bill Simmons; Master Chief PO Steve Skelley; Maj. Doug Slater; Sgt. Clay Slaton; Lt. Cdr. Jeff Smallwood; Lt. (U.K.) Tim Smart; Sgt. Jonathan Smith; Sgt. Robert Smith; Maj. Gen. (U.K.) Rupert Smith; Col. William Smullen; Lt. Joe Smyder; Lt. Cdr. John Snedeker; Cmnd. Sgt. Maj. Ronald Sneed; Sgt. Rick Sommerfelt; Lt. Gen. Harry E. Soyster; Spec. Michael Stapleton; Senior Chief PO William Stapp III; Col. Billy Steed; Maj. Gen. John F. Stewart, Jr.; Gen. Carl W. Stiner; PFC. Jeremy Storm; Capt. Joe Strabala; Lt. Col. Tom Strauss; Brig. (U.K.) Tim Sulivan; Gen. Gordon Sullivan; Col. George Summers; Col. Bryan Sutherland; Capt. Linda Suttlehan; Capt. Don Swartout; Lt. Col. John Sweeney; Maj. Mike Sweeney; Lt. Robert

Sweet; Capt. Ernie Szabo; Maj. (U.K.) Jim Tanner; Lt. Steve Taylor; PFC Todd Taylor; Lt. (U.K.) Rod Tevaskus; Maj. Bob Tezza; Col. Charles Thomas; Lt. Col. Rod Thomas; BM Harry Thompson; Lt. John Tieu; Lt. Col. Frank Toney; Lt. Col. Terry Lee Tucker; Cmdr. R. J. Turner; Robert C. Turrell; Margaret Tutwiler; Ed Valentine; Col. John Van Alstyne; Gen. Henry Viccellio, Jr.; SSgt. John Villanueva; Capt. Eric von Tersch; Gen. Carl E. Vuono

Sgt. Kevin Walden; Capt. Rick Walker; Lt. Gen. Calvin A.H. Waller; Maj. Gen. William F. Ward; Col. John A. Warden III; Lt. Col. Charles C. Ware; Lt. Laura Warn; Capt. Chris Warren; Maj. Dave Waterstreet; Lt. Glenn G. Watson; William H. Webster; Col. David S. Weisman; Lt. Robert Wetzel; Lt. Col. Roy S. Whitcomb; Lt. Col. Dave White; Capt. Frank White; Sgt. (U.K.) Paul Whiteley; Col. Alton C. Whitley; Maj. (U.K.) Jerry Whittingham; Lt. Col. Jerry Wiedewitsch; Lt. Col. Chuck Wiley; Winston Wiley; PFC Arah Williams; Lt. Col. Bennie Williams; Maj. Dick Williams; Pete Williams; Col. Eugene Wilson; Lt. Col. G. I. Wilson; Maj. Gary Wines; Paul Wolfowitz; Col. James D. Woodall; Sgt. Joseph Woytko; Sgt. (U.K.) Stephen Wright; Lt. Buck Wyndham; Lt. Frank Wynne; Ruth Yaron; Lt. Gen. John J. Yeosock; Lt. Guy W. Zanti; Brig. Gen. John Zierdt; MSgt. Richard Zimmerman; Vice Adm. R. J. Zlatoper; Maj. Chuck Zvarich

# CHRONOLOGY

| | |
|---|---|
| July 1990* | *Internal Look*, a U.S. war game, shows Saudi Arabia could be defended against Iraqi invaders, but at terrible cost. |
| August 2 | Iraq invades Kuwait. |
| August 5 | President Bush declares invasion "will not stand." |
| August 6 | King Fahd meets with Richard Cheney, requests U.S. military assistance. |
| August 8 | Initial U.S. Air Force fighter planes arrive in Saudi Arabia. |
| August 10 | John Warden first meets with Schwarzkopf in Tampa to outline proposed air campaign. |
| August 28 | Secret Israeli delegation flies to Washington to stress likelihood of Iraqi attack on Israel if war begins. |
| September 18 | Schwarzkopf asks four Army planners to begin work on ground offensive. |
| October 10 | CENTCOM's One Corps Concept unveiled at White House. |
| October 21 | Colin Powell flies to Riyadh to discuss offensive plans. |
| October 31 | Bush decides to double U.S. forces in Saudi Arabia; decision kept secret until November 8. |
| November 29 | UN Security Council authorizes use of "all means necessary" to eject Iraq from Kuwait. |
| December 6 | First ship carrying VII Corps equipment arrives in Saudi Arabia from Germany. |
| January 9, 1991 | James Baker meets Tariq Aziz in Geneva in unsuccessful effort to find a peaceful solution. |
| January 12 | Congress authorizes use of force. |
| January 15 | UN deadline for Iraqi withdrawal.<br>Schwarzkopf accuses Air Force of ignoring orders by not including Republican Guard in initial bombing sorties. |
| January 17 | Allied attack begins with Apache strike at 2:38 A.M. |

*Dates and hours are in Riyadh time.

| | |
|---|---|
| January 18 | First Scuds hit Israel.<br>Navy aircraft losses during attack on Scud sites leads to recriminations about low-altitude bombing tactics.<br>First American air attacks are launched from Turkey. |
| January 19 | David Eberly and Thomas Griffith are shot down. |
| January 20 | Lawrence Eagleburger and Paul Wolfowitz arrive in Tel Aviv. |
| January 22 | Navy attacks Iraqi oil tanker, triggering Schwarzkopf's threat of court-martial.<br>British high command, alarmed at aircraft losses, abandons low-altitude attacks against airfields. |
| January 23 | Intense attack against Iraqi aircraft shelters begins. |
| January 26 | U.S. Marines in Oman participate in Sea Soldier IV, rehearsal for amphibious landing on Kuwaiti coast.<br>F-111s attack oil manifolds at Al Ahmadi in effort to counter Iraqi sabotage. |
| January 29 | Iraqis attack Khafji and other border positions.<br>Allied pilots begin flying combat air patrols to thwart Iraqi flights to Iran. |
| January 30 | Richard Cheney dispatches Delta Force to Saudi Arabia to hunt for Scuds. |
| January 31 | Khafji is recaptured.<br>David Eberly is moved to new prison, the "Biltmore." |
| February 1 | Last Tomahawk missiles are launched in attack on Baghdad airfield. |
| February 2 | Schwarzkopf formally decides against amphibious landing in Kuwait. |
| February 3 | First battleship gunfire against targets in Kuwait. |
| February 5 | First "tank-plinking" mission flown. |
| February 6 | VII Corps finishes closing in theater with arrival of final 3rd Armored Division equipment. |
| February 7 | CIA, in daily intelligence brief, notes large discrepancy between Washington and Riyadh regarding destruction of Iraqi armor in air attacks. |
| February 8 | Cheney and Colin Powell fly to Riyadh for final review of ground war plans. |
| February 11 | Al Firdos bunker in Baghdad suburb added to master attack plan.<br>Moshe Arens in Washington complains about ineffectiveness of Patriot missile against Scuds.<br>Yevgeni Primakov arrives in Baghdad to urge Iraqi withdrawal. |

| | |
|---|---|
| February 13 | Strike on Al Firdos bunker kills more than two hundred civilians and leads to restrictions on strategic bombing campaign. |
| February 15 | Radio Baghdad suggests Iraq willing to withdraw. Bush rejects proposal as a "cruel hoax." |
| February 16 | VII Corps moves into final attack positions. |
| February 18 | U.S.S. *Tripoli* and U.S.S. *Princeton* strike mines.<br>Army complaints about insufficient air support lead to confrontation with Air Force. |
| February 20 | 1st Cavalry Division feints up the Wadi al Batin; pulls back with three dead and nine wounded. |
| February 21 | CIA and Pentagon officials meet at White House to air differences over battle-damage assessment.<br>Bush sets deadline of noon, February 23, for Iraqi withdrawal. |
| February 22 | Marines begin infiltrating into Kuwaiti bootheel. |
| February 23 | Stealth fighters attack Iraqi intelligence headquarters, unaware that allied POWs are inside.<br>Army Special Forces teams inserted deep into Iraq. |
| February 24 | Ground attack begins.<br>Schwarzkopf decides to accelerate main attack of VII Corps by fifteen hours. |
| February 25 | Schwarzkopf explodes at slow pace of VII Corps.<br>101st Airborne Division cuts Highway 8 in Euphrates Valley.<br>Iraqis counterattack 1st Marine Division.<br>Scud destroys barracks in Al Khobar, killing twenty-eight Americans and wounding ninety-eight. |
| February 26 | Iraqis flee Kuwait City.<br>VII Corps hits Republican Guard in Battle of 73 Easting. |
| February 27 | 24th Infantry Division attacks toward Basrah in Euphrates Valley.<br>1st Armored Division fights Madinah Division.<br>Kuwait City is liberated.<br>Bush and advisers agree to stop the war. |
| February 28 | Cease-fire takes effect at 8 A.M. |
| March 2 | 24th Infantry Division fights Hammurabi Division as it flees; destroys six hundred vehicles. |
| March 3 | Schwarzkopf meets Iraqi generals at Safwan. |
| March 5 | David Eberly and most other POWs are released. |
| June 8 | Victory parade in Washington. |

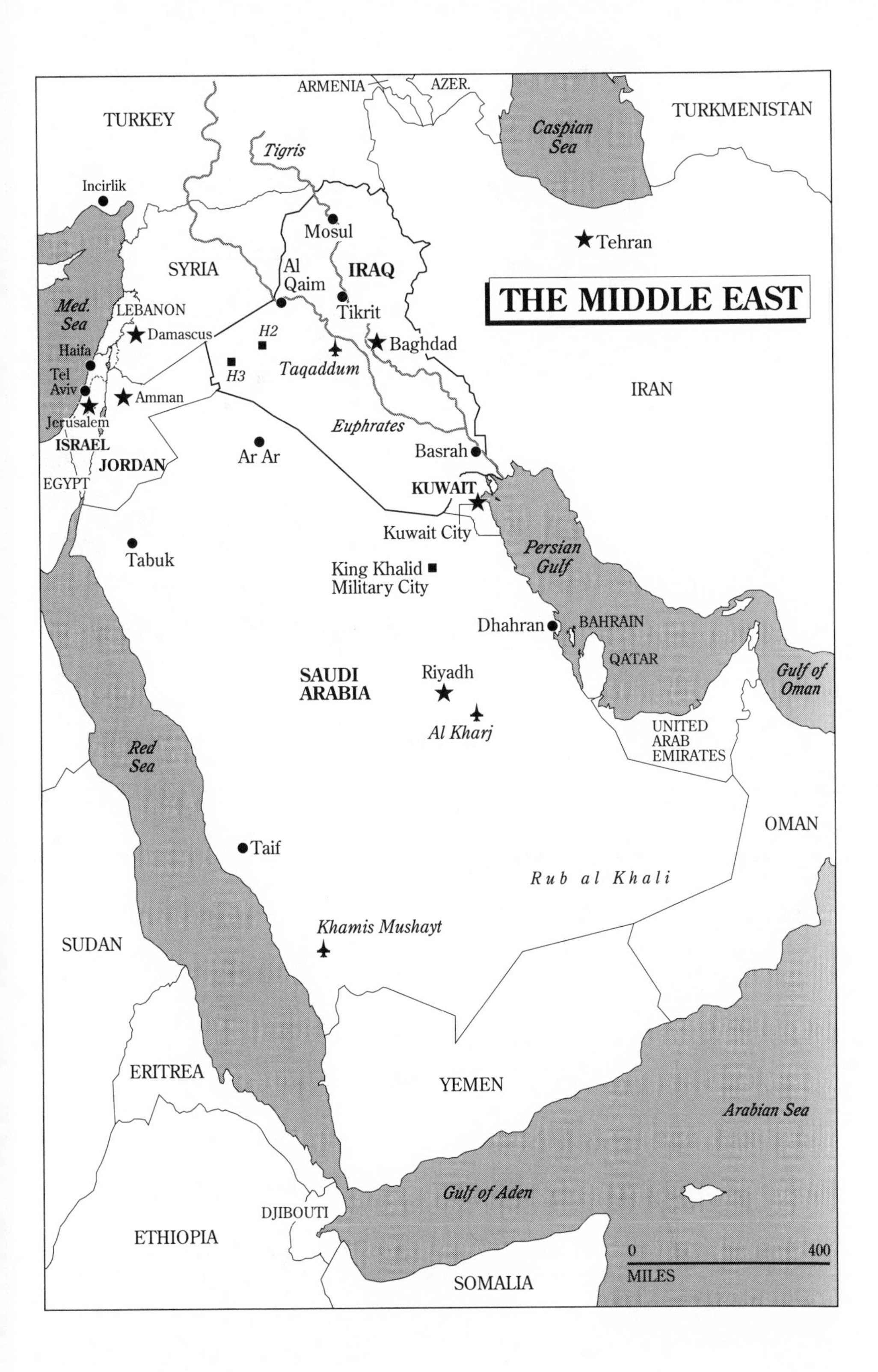
THE MIDDLE EAST
TURKEY
ARMENIA
AZER.
TURKMENISTAN
Caspian Sea
Tigris
Incirlik
Mosul
Tehran
SYRIA
Al Qaim
IRAQ
Med. Sea
LEBANON
Tikrit
Damascus
H2
Baghdad
Haifa
H3
Taqaddum
Tel Aviv
Amman
IRAN
Jerusalem
Euphrates
ISRAEL
Ar Ar
Basrah
JORDAN
EGYPT
KUWAIT
Kuwait City
Tabuk
Persian Gulf
King Khalid Military City
Dhahran
BAHRAIN
QATAR
SAUDI ARABIA
Riyadh
Gulf of Oman
Al Kharj
UNITED ARAB EMIRATES
Red Sea
OMAN
Taif
Rub al Khali
Khamis Mushayt
SUDAN
ERITREA
YEMEN
Arabian Sea
Gulf of Aden
DJIBOUTI
ETHIOPIA
0
400
MILES
SOMALIA

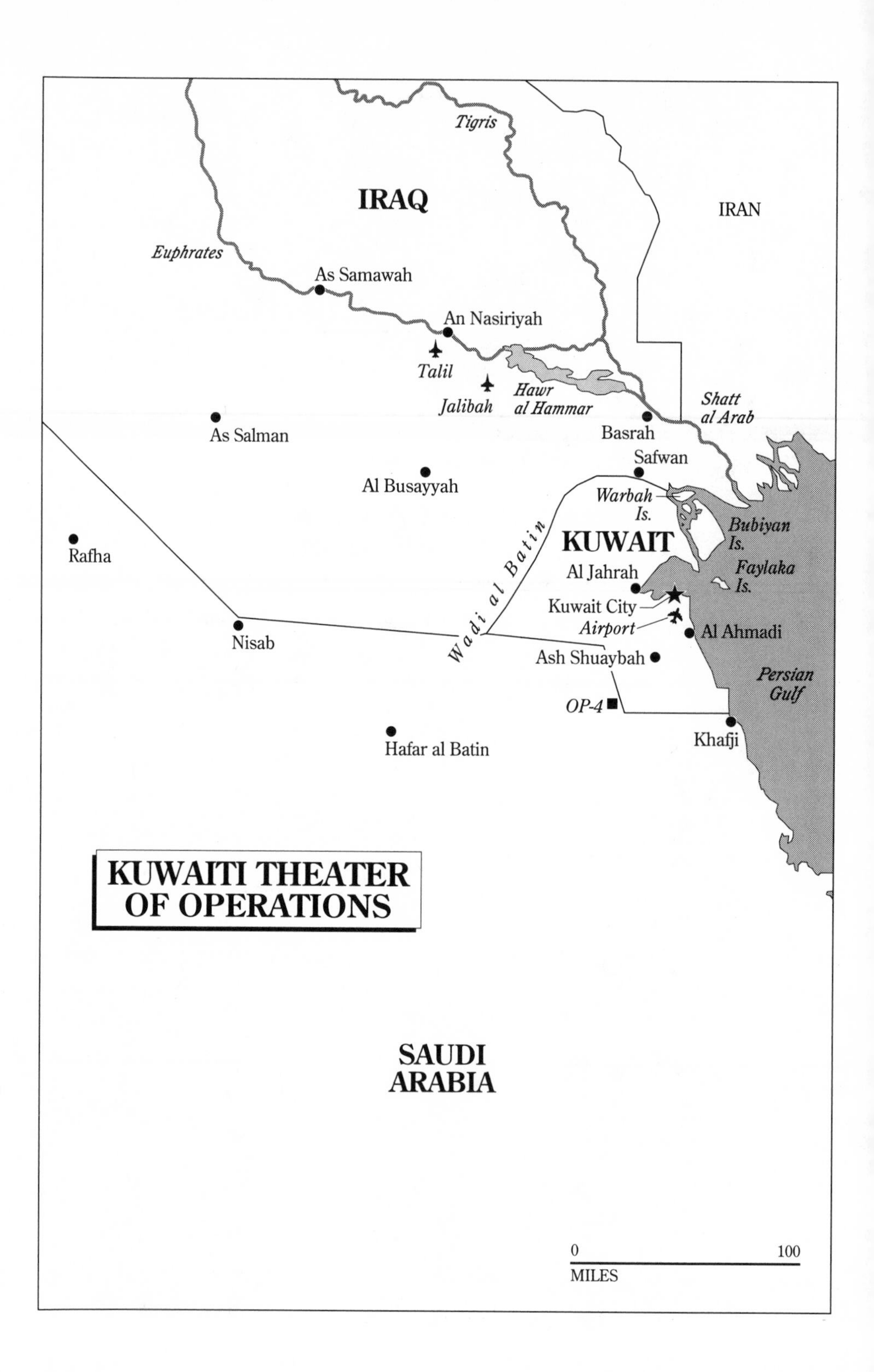

Tigris
IRAQ
IRAN
Euphrates
As Samawah
An Nasiriyah
Talil
Jalibah
Hawr
al Hammar
Shatt
al Arab
Basrah
As Salman
Safwan
Al Busayyah
Warbah
Is.
Bubiyan
Is.
KUWAIT
Wadi al Batin
Rafha
Al Jahrah
Faylaka
Is.
Kuwait City
Airport
Al Ahmadi
Nisab
Ash Shuaybah
Persian
Gulf
OP-4
Khafji
Hafar al Batin
KUWAITI THEATER
OF OPERATIONS
SAUDI
ARABIA
0
100
MILES

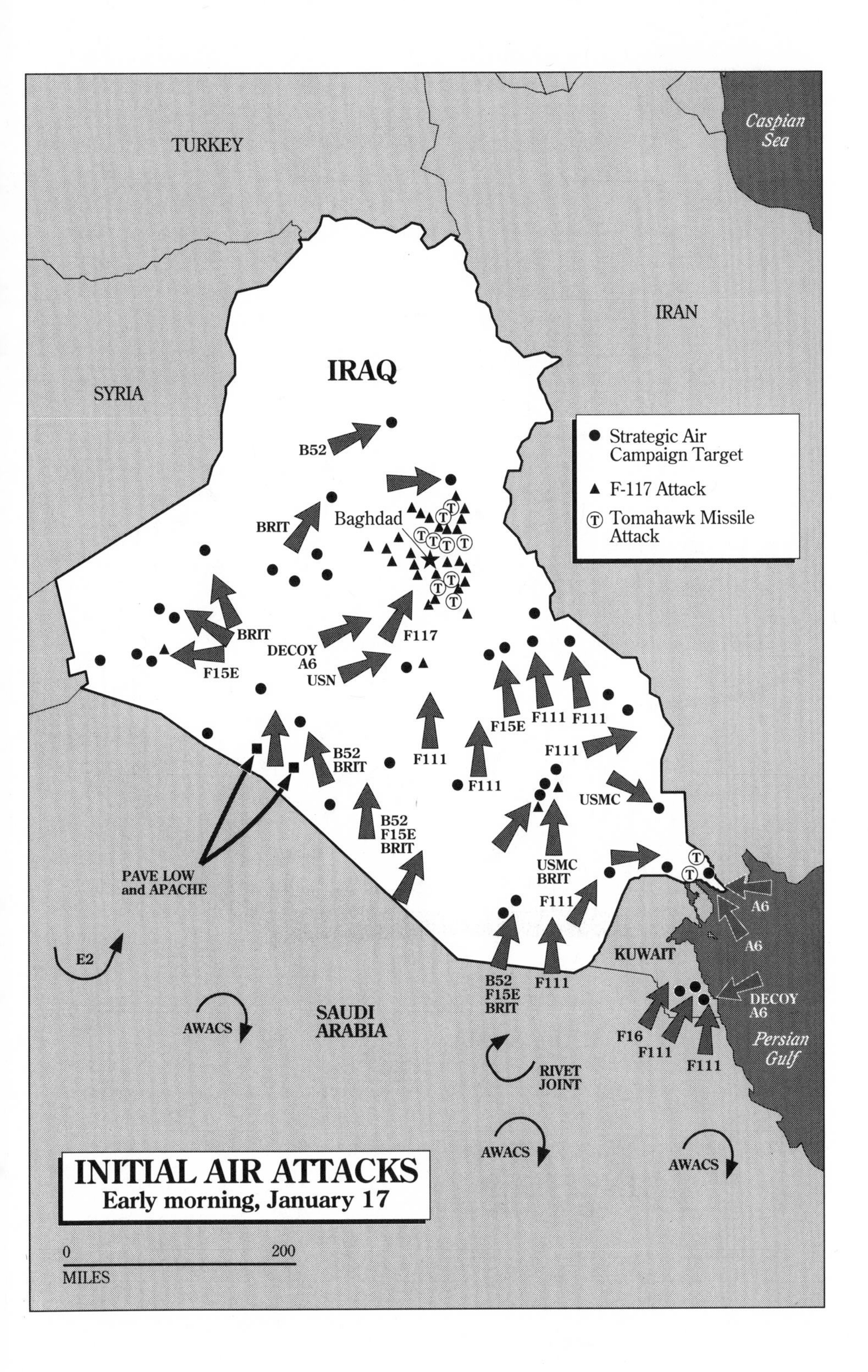

TURKEY
Caspian Sea
IRAN
IRAQ
SYRIA
Strategic Air Campaign Target
F-117 Attack
Tomahawk Missile Attack
B52
BRIT
Baghdad
BRIT
DECOY A6
USN
F117
F15E
F15E
F111
F111
F111
B52 BRIT
F111
F111
USMC
B52 F15E BRIT
USMC BRIT
F111
PAVE LOW and APACHE
A6
A6
E2
KUWAIT
B52 F15E BRIT
F111
DECOY A6
SAUDI ARABIA
AWACS
F16
F111
F111
Persian Gulf
RIVET JOINT
AWACS
AWACS
INITIAL AIR ATTACKS
Early morning, January 17
0
200
MILES

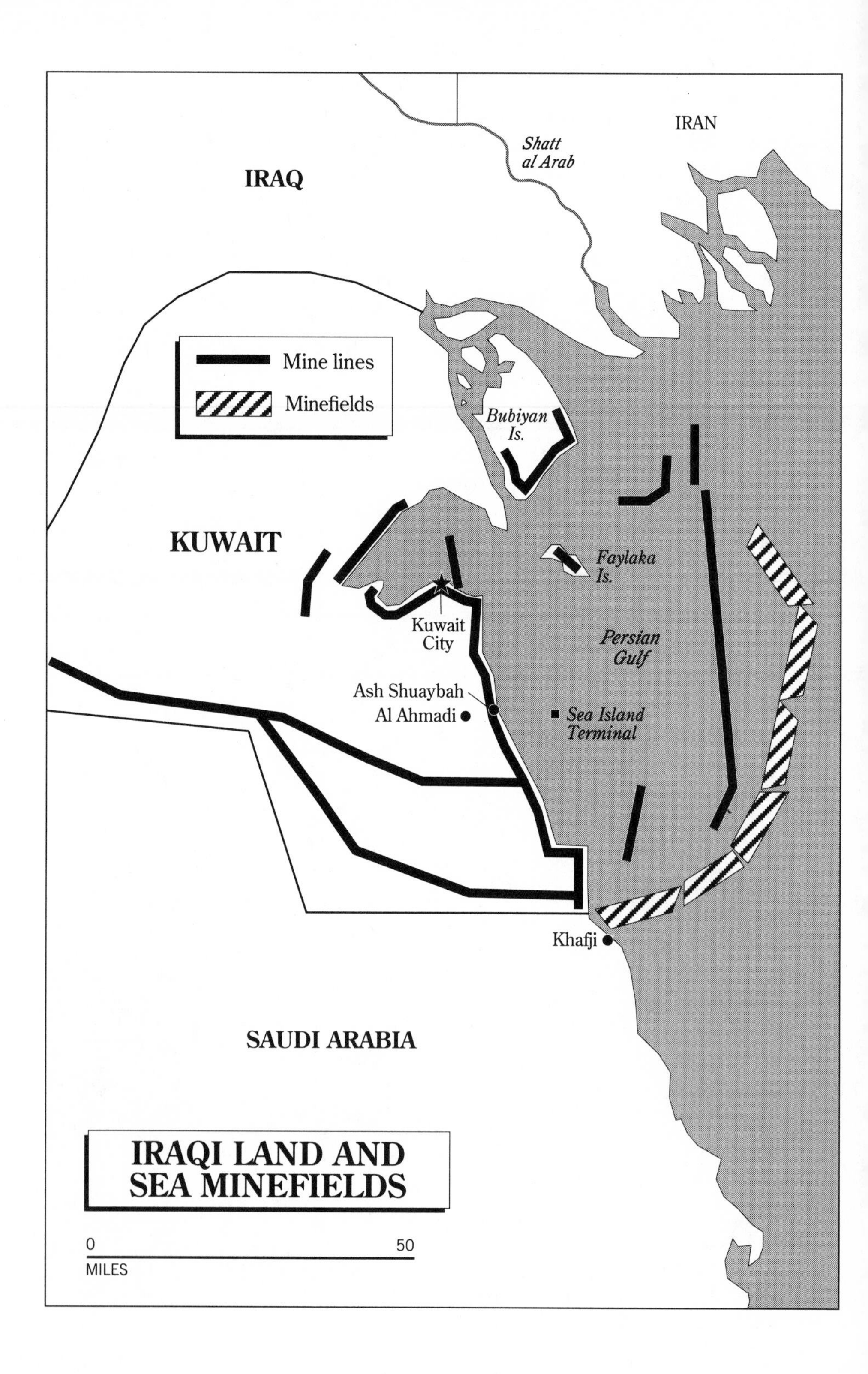
IRAN
Shatt
al Arab
IRAQ
Mine lines
Minefields
Bubiyan
Is.
KUWAIT
Faylaka
Is.
Kuwait
City
Persian
Gulf
Ash Shuaybah
Al Ahmadi
Sea Island
Terminal
Khafji
SAUDI ARABIA
IRAQI LAND AND
SEA MINEFIELDS
0
50
MILES

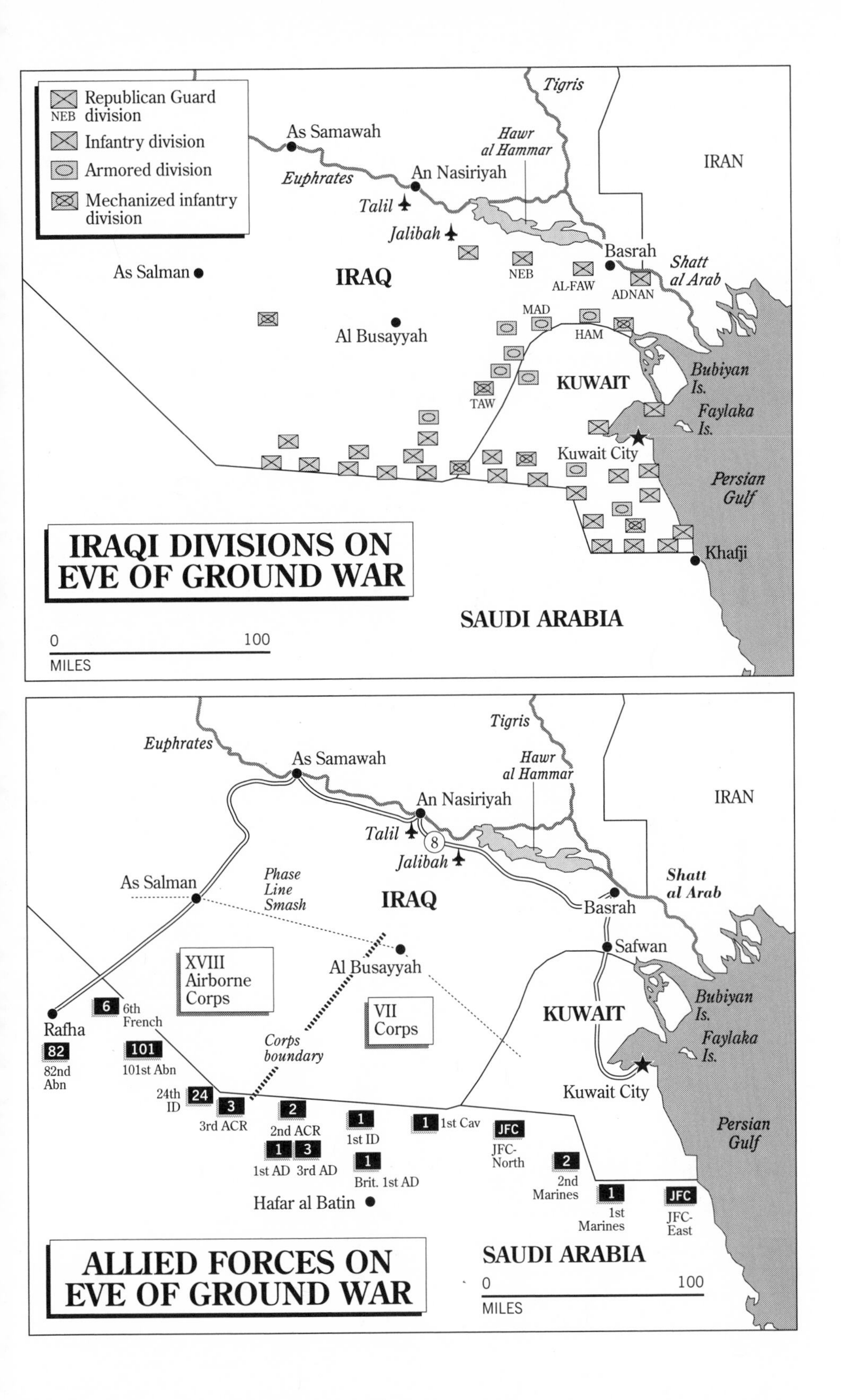

Republican Guard division
NEB
Infantry division
Armored division
Mechanized infantry division
Tigris
Hawr al Hammar
IRAN
As Samawah
Euphrates
An Nasiriyah
Talil
Jalibah
Basrah
Shatt al Arab
As Salman
IRAQ
NEB
AL-FAW
ADNAN
MAD
HAM
Al Busayyah
TAW
KUWAIT
Bubiyan Is.
Faylaka Is.
Kuwait City
Persian Gulf
Khafji
IRAQI DIVISIONS ON EVE OF GROUND WAR
SAUDI ARABIA
0
100
MILES
Tigris
Euphrates
As Samawah
Hawr al Hammar
IRAN
An Nasiriyah
Talil
8
Jalibah
As Salman
Phase Line Smash
IRAQ
Shatt al Arab
Basrah
Safwan
XVIII Airborne Corps
Al Busayyah
6
6th French
Rafha
VII Corps
KUWAIT
Bubiyan Is.
Faylaka Is.
82
82nd Abn
101
101st Abn
Corps boundary
24th ID
24
3
3rd ACR
2
2nd ACR
1
1st ID
1
1st Cav
JFC
JFC-North
Kuwait City
Persian Gulf
1
3
1st AD
3rd AD
1
Brit. 1st AD
2
2nd Marines
Hafar al Batin
1
1st Marines
JFC
JFC-East
ALLIED FORCES ON EVE OF GROUND WAR
SAUDI ARABIA
0
100
MILES

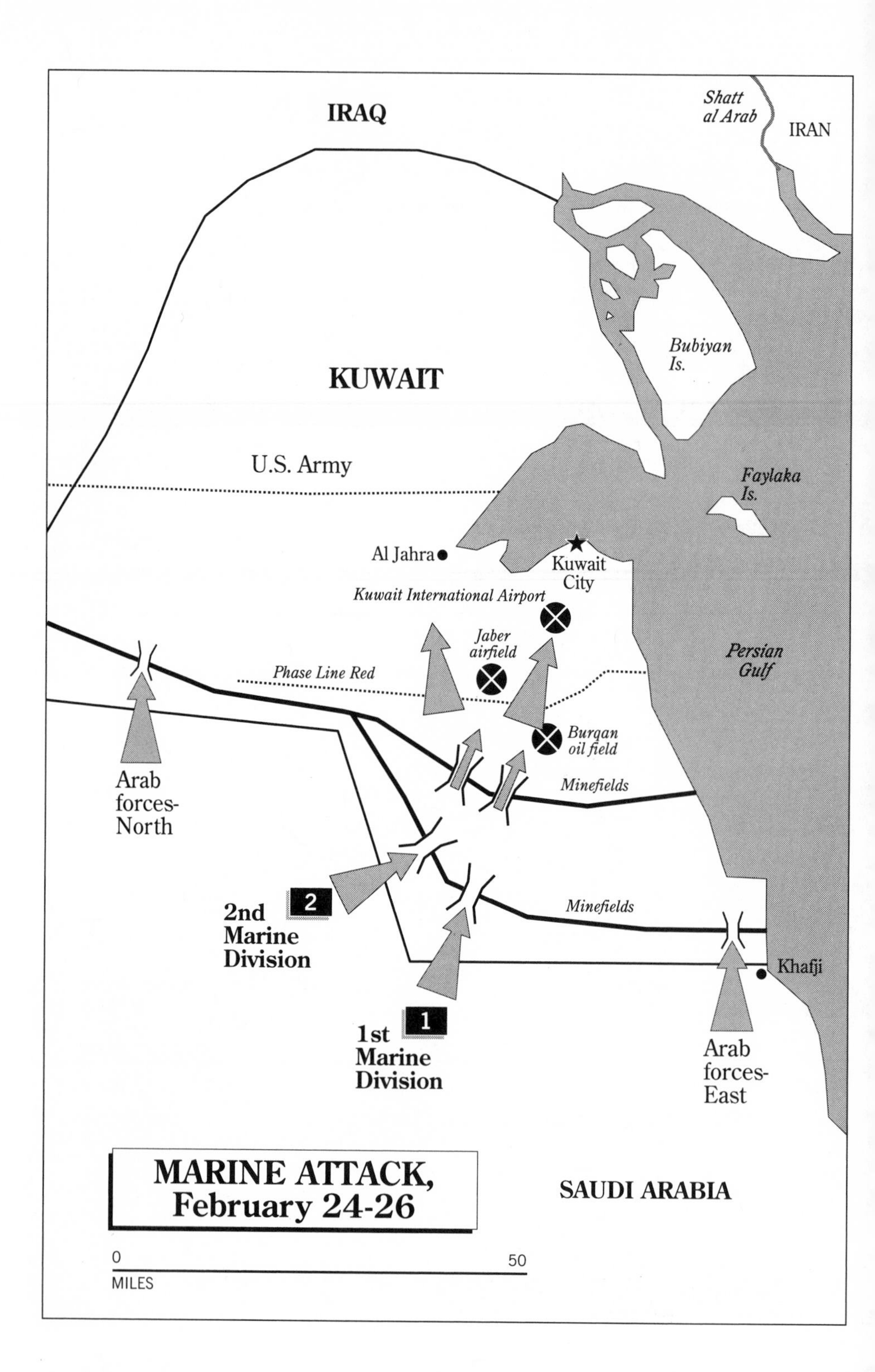


**MARINE ATTACK, February 24-26**

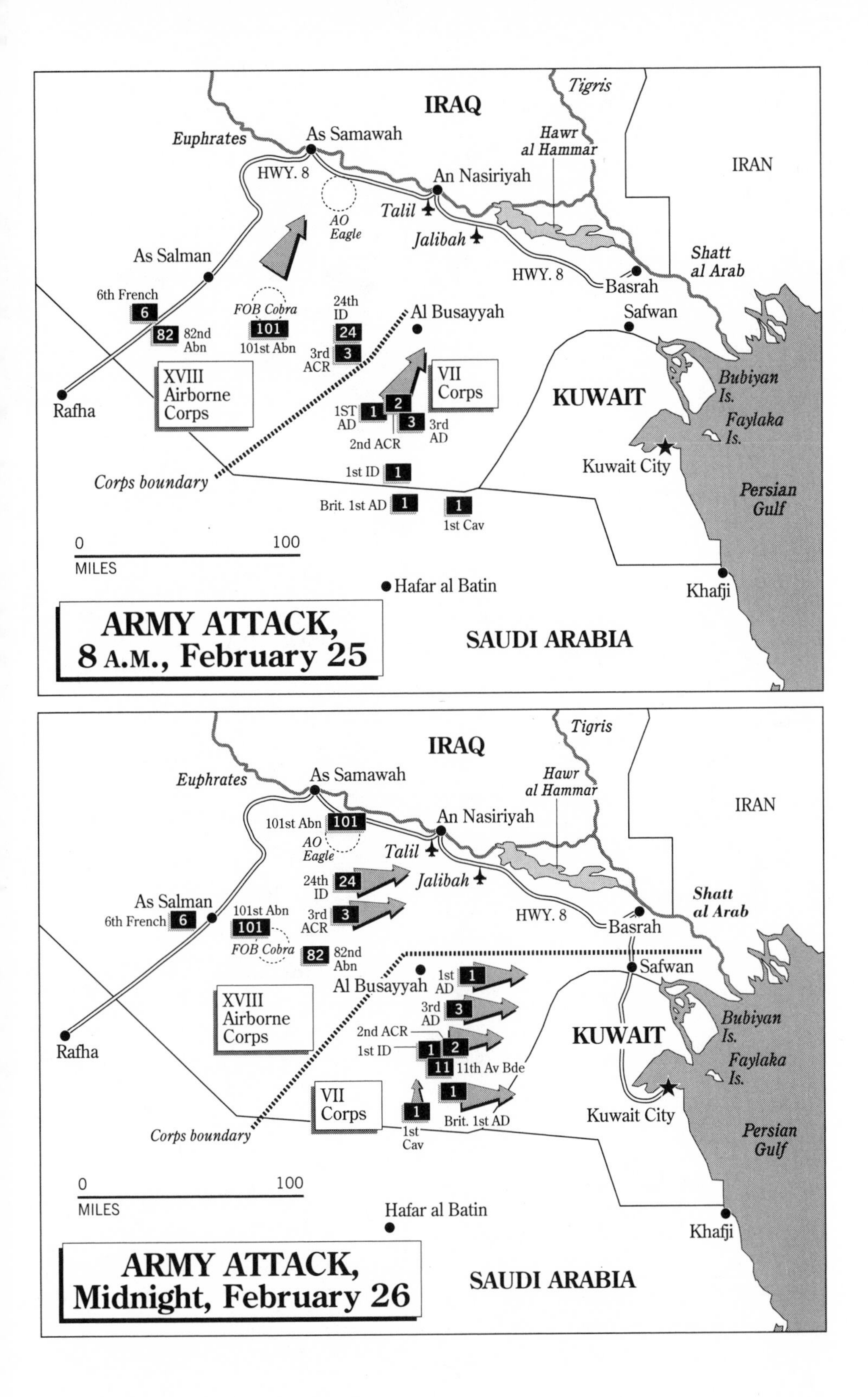
IRAQ
Tigris
Euphrates
As Samawah
Hawr al Hammar
HWY. 8
An Nasiriyah
IRAN
AO Eagle
Talil
Jalibah
As Salman
Shatt al Arab
HWY. 8
Basrah
6th French
FOB Cobra
24th ID
Al Busayyah
Safwan
82nd Abn
101st Abn
3rd ACR
VII Corps
XVIII Airborne Corps
Rafha
1ST AD
3rd AD
KUWAIT
Bubiyan Is.
Faylaka Is.
2nd ACR
Kuwait City
1st ID
Corps boundary
Persian Gulf
Brit. 1st AD
1st Cav
0
100
MILES
Hafar al Batin
Khafji
ARMY ATTACK, 8 A.M., February 25
SAUDI ARABIA
IRAQ
Tigris
Euphrates
As Samawah
Hawr al Hammar
An Nasiriyah
IRAN
101st Abn
AO Eagle
Talil
24th ID
Jalibah
3rd ACR
As Salman
6th French
101st Abn
Shatt al Arab
HWY. 8
Basrah
FOB Cobra
82nd Abn
Safwan
1st AD
Al Busayyah
3rd AD
XVIII Airborne Corps
2nd ACR
1st ID
KUWAIT
Bubiyan Is.
Faylaka Is.
Rafha
11th Av Bde
VII Corps
Kuwait City
Brit. 1st AD
Corps boundary
1st Cav
Persian Gulf
0
100
MILES
Hafar al Batin
Khafji
ARMY ATTACK, Midnight, February 26
SAUDI ARABIA

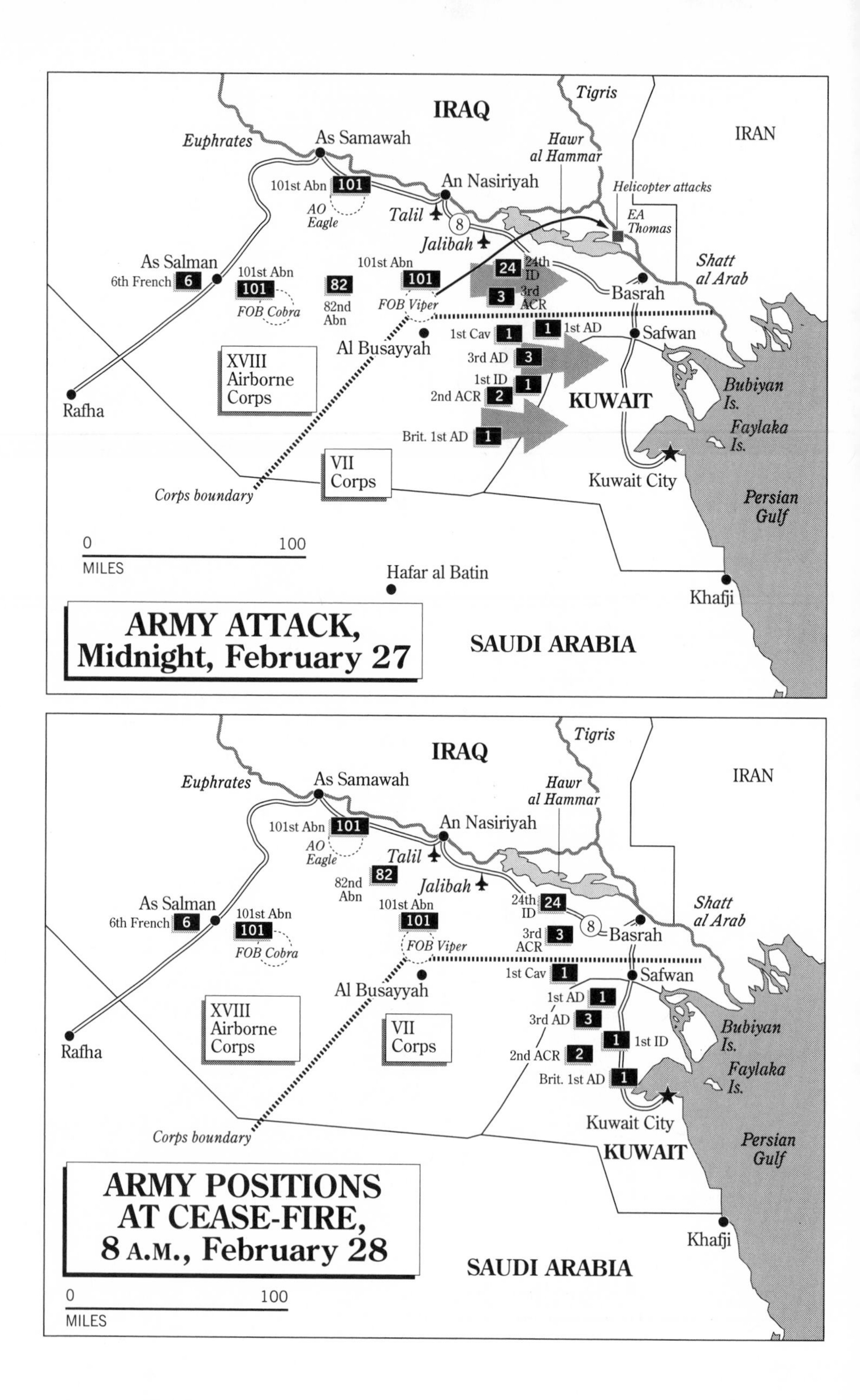
IRAQ
IRAN
Tigris
Euphrates
As Samawah
Hawr al Hammar
An Nasiriyah
Helicopter attacks
101st Abn 101
AO Eagle
Talil
8
EA Thomas
Jalibah
Shatt al Arab
As Salman
101st Abn
6th French 6
101
101st Abn
101
82
24
24th ID
3
3rd ACR
Basrah
FOB Cobra
82nd Abn
FOB Viper
1st Cav 1
1 1st AD
Safwan
Al Busayyah
3rd AD 3
XVIII Airborne Corps
1st ID 1
2nd ACR 2
Bubiyan Is.
KUWAIT
Rafha
Faylaka Is.
Brit. 1st AD 1
VII Corps
Kuwait City
Corps boundary
Persian Gulf
0
100
MILES
Hafar al Batin
Khafji
ARMY ATTACK, Midnight, February 27
SAUDI ARABIA
IRAQ
Tigris
IRAN
Euphrates
As Samawah
Hawr al Hammar
An Nasiriyah
101st Abn 101
AO Eagle
Talil
82nd Abn 82
Jalibah
As Salman
101st Abn
24th ID 24
Shatt al Arab
6th French 6
101st Abn
101
101
8
Basrah
3rd ACR 3
FOB Cobra
FOB Viper
1st Cav 1
Safwan
Al Busayyah
1st AD 1
XVIII Airborne Corps
3rd AD 3
VII Corps
1 1st ID
Bubiyan Is.
Rafha
2nd ACR 2
Faylaka Is.
Brit. 1st AD 1
Kuwait City
Corps boundary
Persian Gulf
KUWAIT
ARMY POSITIONS AT CEASE-FIRE, 8 A.M., February 28
Khafji
SAUDI ARABIA
0
100
MILES

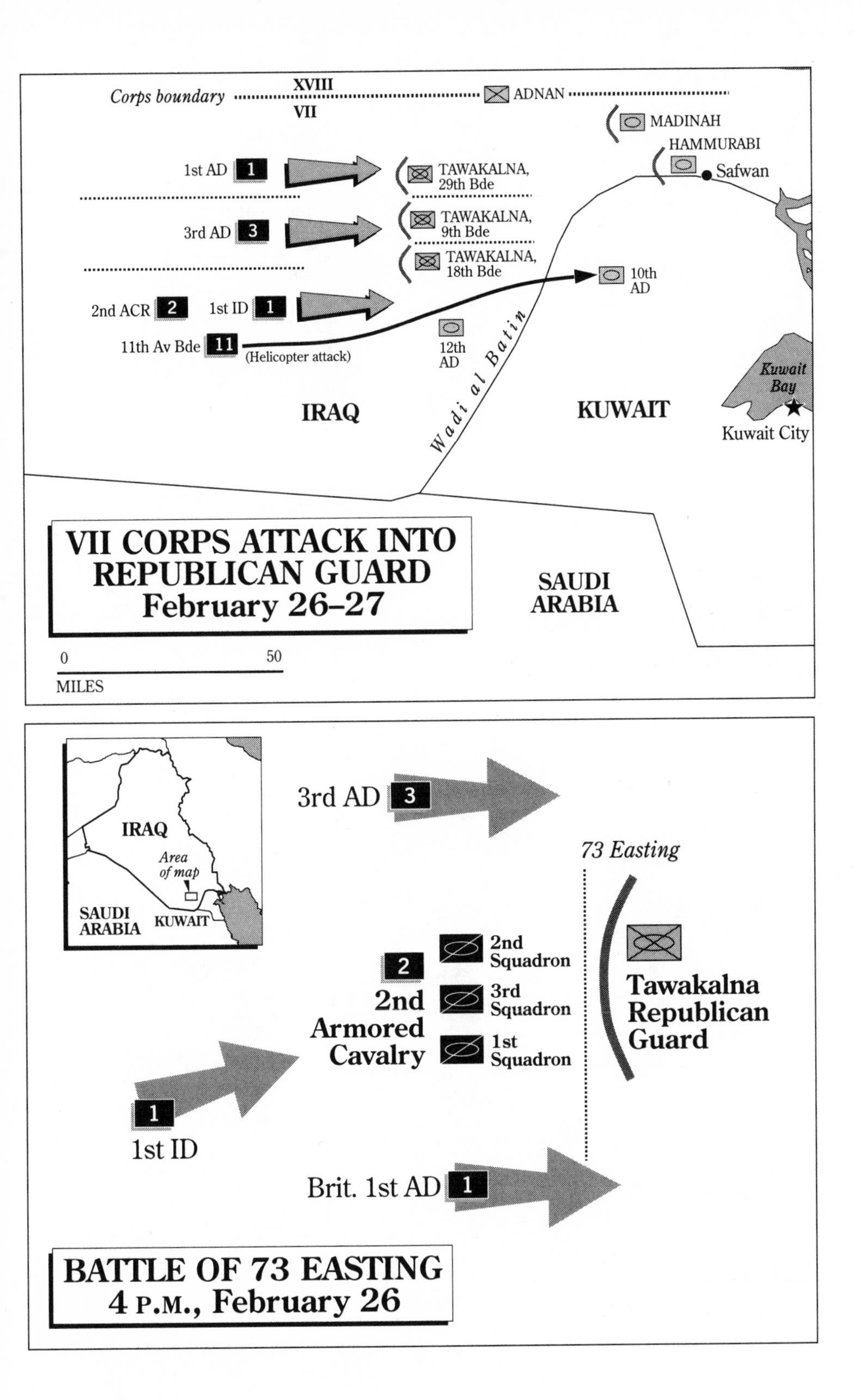
Corps boundary
XVIII
VII
ADNAN
MADINAH
HAMMURABI
Safwan
1st AD 1
TAWAKALNA, 29th Bde
3rd AD 3
TAWAKALNA, 9th Bde
TAWAKALNA, 18th Bde
10th AD
2nd ACR 2
1st ID 1
11th Av Bde 11
(Helicopter attack)
12th AD
Wadi al Batin
IRAQ
KUWAIT
Kuwait Bay
Kuwait City
VII CORPS ATTACK INTO REPUBLICAN GUARD February 26–27
SAUDI ARABIA
0
50
MILES
IRAQ
Area of map
SAUDI ARABIA
KUWAIT
3rd AD 3
73 Easting
2
2nd Armored Cavalry
2nd Squadron
3rd Squadron
1st Squadron
Tawakalna Republican Guard
1
1st ID
Brit. 1st AD 1
BATTLE OF 73 EASTING 4 P.M., February 26

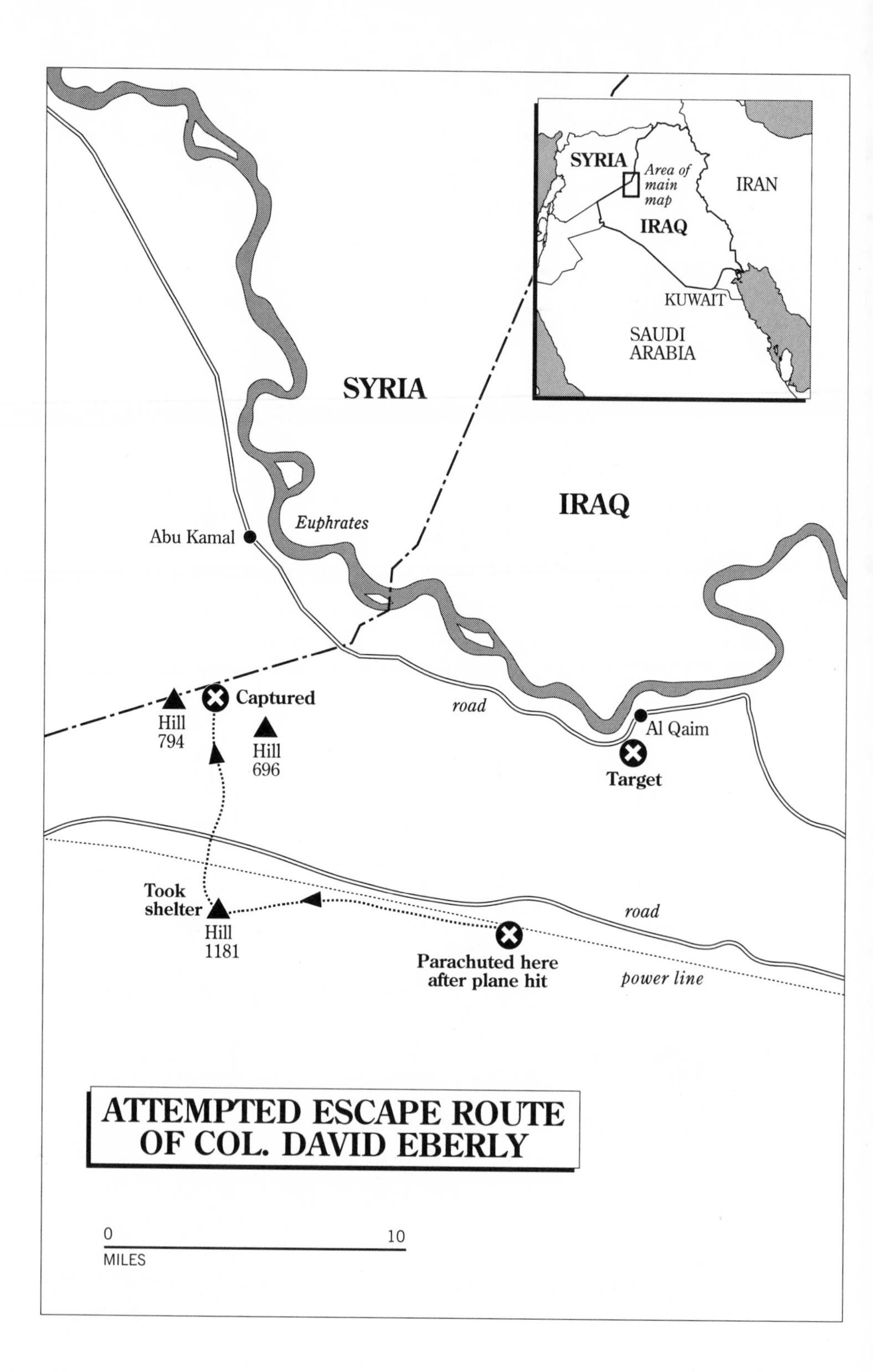
SYRIA
Area of
main
map
IRAN
IRAQ
KUWAIT
SAUDI
ARABIA
SYRIA
IRAQ
Euphrates
Abu Kamal
Captured
Hill
794
Hill
696
road
Al Qaim
Target
Took
shelter
Hill
1181
road
Parachuted here
after plane hit
power line
ATTEMPTED ESCAPE ROUTE
OF COL. DAVID EBERLY
0
10
MILES

# NOTES

**Prologue**

*page* ***Safwan***

2 British marched up the Tigris: John Simpson, *From the House of War,* p. 218; Perrett, *Desert Warfare,* pp. 51, 54.

3 a flight into hell: *Life,* March 18, 1991, p. 32.

"roaring flux of forces": quoted in Eric Larrabee, *Commander in Chief,* p. 5.

"conspicuous in the matter of temperament": quoted in Ibid., p. 312.

mailed home Christmas presents: Brenda Schwarzkopf in *People,* "Heroes of the Gulf," Spring-Summer 1991.

4 "a just war . . . far better for a man's soul": quoted in Edmund Morris, *The Rise of Theodore Roosevelt.*

7 "kill all his forces": quoted in Jim Tice, *Army Times,* Aug. 26, 1991, p. 12.

"never any intention to disobey orders": quoted in Tom Donnelly, *Army Times,* March 2, 1991, p. 8.

8–9 Dialogue of peace talks drawn from transcript, "Cease-Fire Discussions with Iraqis at Safwan Airfield," obtained under Freedom of Information Act (hereafter FOIA). Some details drawn from Steve Coll, *Washington Post,* March 4, 1991, and Norman Schwarzkopf's account in *It Doesn't Take a Hero* (hereafter *Hero*), pp. 473–491.

8 Iraqis appeared stunned: Department of Defense, *Conduct of the Persian Gulf War* (hereafter *Conduct*), p. 215.

10 "splendid little war": quoted in Larry H. Addington, *The Patterns of War,* p. 117.

**1. First Night**

***U.S.S.* Wisconsin**

13 they had scrubbed the teak deck: war preparations described in pool reports by George Rodrigue of *Dallas Morning News* and Robert Ruby of *Baltimore Sun.*

15 The missile's navigation . . . by terrain-contour matching: "Special Report: Gulf Legacy," IEEE *Spectrum,* Sept. 1991, pp. 25–50; "The Mind of a Missile," *Newsweek,* Feb. 18, 1991, p. 40.

16 "Alpha, alpha": Official Navy account says U.S.S. *Bunker Hill* was the first shooter from Persian Gulf, but *Wisconsin*'s log, the captain, and her crew — responsible for gulf missile shots — believe it was *Paul F. Foster.* First TLAM of war fired from Red Sea by U.S.S. *San Jacinto.* Office of

*page*

the Chief of Naval Operations, "The United States Navy in Desert Shield/ Desert Storm," May 15, 1991.

***Riyadh***

20 Allied objectives in the coming war: operations order 91–001 obtained from Central Command under FOIA.

22 Flamboyance . . . element in the general's art: Morris Janowitz, *The Professional Soldier*, p. 47.

***Washington***

24 A top secret warning order . . . sent to Schwarzkopf: Bob Woodward, *The Commanders*, p. 367.

***Baghdad***

26 Baghdad's last day of peace: description drawn from Milton Viorst, *The New Yorker*, June 24, 1991, pp. 55–74; Simpson, *From the House of War*, pp. 283–286; "Witness to War," *Los Angeles Times*, special section, March 12, 1991, p. H4; Marc Fisher, "Underground a Luxurious Lair," *Washington Post*, Jan 23, 1991, p. A23; Merle Severy, "Iraq: Crucible of Civilization," *National Geographic*, May 1991, p. 103.

28 a plunge in oil prices: Janice Gross Stein, "Deterrence and Compellence in the Gulf, 1990–1991," *International Security*, Fall 1992, p. 147.

cheating on oil production quotas: Milton Viorst, *Washington Post*, Oct. 25, 1992, p. C1.

(Kuwait's historical autonomy): Stein, *International Security*, "Deterrence and Compellence in the Gulf, 1990–1991," Fall 1992, p. 159.

Saddam felt particularly aggrieved: *Conduct*, p. 6.

(Per capita income in Kuwait): Jean Edward Smith, *George Bush's War*, p. 24.

a feudal monarchy: Caryle Murphy, *Washington Post*, Oct. 3, 1992, p. A13.

Saddam would reap a windfall: Woodward, *The Commanders*, p. 226; *Conduct*, p. 22.

***U.S.S.* Wisconsin**

31 Poobah's Party: David A. Fulghum, *Aviation Week*, April 27, 1992, p. 18.

***Objective Oklahoma***

32 Hellfire leaped from its rail: details of how the missile works from *Aviation Week*, "U.S. Army Foresees Key Role for Hellfire in Ground War," Feb. 25, 1991; also, Richard MacKenzie, "Apache Attack," *Air Force Magazine, Oct.* 1991, p. 54.

***Baghdad (stealth attacks)***

34 Stealth had been developed in deepest secrecy: Atkinson, "Project Senior C.J.," *Washington Post*, Oct. 8, 1989; Peter Grier, *Christian Science Monitor*, Dec. 2, 1991, p. 9.

smart munitions: details drawn from Richard Hallion, *Storm Over Iraq*, pp. 21, 303–307; Norman Friedman, *Desert Victory*, pp. 326–329; Malcolm Browne, *New York Times*, Feb. 26, 1991, p. C1; *Conduct*, Appendix T.

37 Of seventeen bombs dropped: Office of History, 37th Fighter Wing, "Nighthawks Over Iraq: A Chronology of the F117A Stealth Fighter in Operations Desert Shield and Desert Storm."

*page*

37 The first wave struck: precise number of Tomahawks hitting Baghdad the first night depends in some measure on how the city limits are defined. Navy sources put the number around sixty.

38 Iraq's electrical power was supplied: Harvard Study Team Report, "Public Health in Iraq After the Gulf War," May 1991, pp. 19–25.

***Riyadh (Royal Saudi Air Force headquarters)***

39 In a single raid . . . at Schweinfurt: Addington, *Patterns of War Since the Eighteenth Century,* p. 210.

Over Normandy: John A. Warden III, *The Air Campaign,* p. 91.

Forty Israeli jets: Eliot A. Cohen and John Gooch, *Military Misfortunes,* p. 110.

40 Horner anticipated coalition losses: Hallion, *Storm Over Iraq,* p. 195.

died in an auto crash: Associated Press, Jan. 18, 1991.

"fights like a savage": quoted in Warden, *The Air Campaign,* p. 15.

41 sweeps had been flown to good effect: William W. Momyer, *Airpower in Three Wars,* pp. 113–117.

***Al Taqqadum***

42 the potent amphetamine Dexedrine: some details drawn from Mark Sauter, Tacoma, Washington, *Morning News Tribune,* April 9, 1992, p. 1; Janet D'Agostino, *Air Force Times,* April 27, 1992, p. 3. Of 464 pilots questioned in a survey after the war, 65 percent said they had used amphetamines during Desert Shield, 57 percent during Desert Storm.

44 The aircraft were not bombers: "Special Report: Gulf Legacy," IEEE *Spectrum,* Sept. 1991, pp. 25–50.

45 more than two hundred HARM missiles: John D. Morrocco, *Aviation Week,* July 2, 1991.

the ALARM popped open a parachute: Stan Morse, ed., *Gulf Air War Debrief,* p. 155.

***Riyadh***

47 Scott Speicher, flying from the U.S.S. *Saratoga:* Mark Crispin Miller, *New York Times,* Sept. 15, 1992, op-ed page.

48 B-52s . . . loosed hundreds of bombs on four airfields: *Conduct,* Appendix T, p. 26.

only one sizable war had been decided: Warden, *The Air Campaign,* p. 165.

the nation had been chronically unprepared to fight: Charles E. Heller and William A. Stoft, ed., *America's First Battles, 1776–1965.*

**2. First Day**

50 paving the way for Saddam: Dean Baquet, *New York Times,* April 27, 1992, p. 1.

the CIA covertly began sharing intelligence: George Lardner, Jr., and R. Jeffrey Smith, *Washington Post,* April 28, 1992, p. A6.

several Arab nations also shipped: Murray Waas and Craig Unger, *The New Yorker,* Nov. 2, 1992, p. 64.

State Department report noted: Elaine Sciolino, *New York Times,* Sept. 22, 1992, p. A15; R. Jeffrey Smith, *Washington Post,* April 21, 1992, p. A15.

*page*

50 National Security Decision Directive 26: R. Jeffrey Smith and John Goshko, *Washington Post*, June 27, 1992, p. A7; Elaine Sciolino, *New York Times*, May 29, 1992, p. A4. Also, some of the most illuminating reporting on the preinvasion period was done by Doug Frantz and Murray Waas for the *Los Angeles Times*.

51 the Atlanta branch of Italy's Banca Nazionale del Lavoro: Elaine Sciolino, *New York Times*, July 19, 1992, p. 12.

Baker . . . pressed for White House approval: R. Jeffrey Smith, *Washington Post*, Oct. 25, 1992, p. A4.

"human rights record [remains] abysmal": quoted in R. Jeffrey Smith and John Goshko, *Washington Post*, June 27, 1992, p. A7.

(U.S. taxpayers would be stuck): Elaine Sciolino, *New York Times*, Oct. 18, 1992, p. D4; R. Jeffrey Smith, *Washington Post*, Oct. 20, 1992, p. A4.

nearly eight hundred export licenses: R. Jeffrey Smith, *Washington Post*, July 13, 1992, p. A6; July 22, 1992, p. 1; Oct. 20, 1992, p. A21.

the British . . . permitted: Dean Baquet, *New York Times*, Oct. 27, 1992.

52 Glaspie . . . informed the Iraqis: Elaine Sciolino, *New York Times*, Sept. 24, 1992, p. A6; R. Jeffrey Smith, *Washington Post*, Oct. 21, 1992, p. A17; U.S. News & World Report, *Triumph Without Victory*, p. 25.

Schwarzkopf believed — that at most: Schwarzkopf, *Hero*, p. 295.

At 1 A.M. on August 2: *Conduct*, p. 1; Woodward, *The Commanders*, p. 222.

53 military contributions from thirty-eight nations: *Conduct*, p. 23.

King Fahd recognized: Jean Edward Smith, *George Bush's War*, p. 92, quoting Eliot Cohen.

54 White House telephone logs: Michael Duffy and Dan Goodgame, *Marching in Place*, p. 153.

offered sundry compensations: David Hoffman, *Washington Post*, Aug. 14, 1992, p. A19.

not beyond the use of deception: Duffy and Goodgame, *Marching in Place*, pp. 134, 150; Smith, *George Bush's War*, p. 98; Woodward, *The Commanders*, p. 279.

55 his greatest ally . . . was Saddam Hussein: Stein, "Deterrence and Compellence," *International Security*, pp. 166–167, 175, 178.

Omar Bradley had warned: David McCullough, *Truman*, p. 854.

56 "it will just come down to fate": quoted in U.S. News, *Triumph Without Victory*, p. 203.

"For whatever purpose God has": quoted in Schwarzkopf, *Hero*, p. 412.

***Washington***

57 the first bomb had fallen: Larrabee, *Commander in Chief*, p. 584.

Claire Chennault's boast: Ibid., p. 545.

Arthur (Bomber) Harris had predicted: Addington, *Patterns of War*, p. 208.

the national telephone system worked: Cornelius Ryan, *The Last Battle*, p. 458.

Strategic bombing in 1945 was discredited: John Keegan, *Daily Telegraph*, Feb. 10, 1991.

*page*

58 "the point where the enemy is most vulnerable": quoted in Warden, *The Air Campaign,* p. 9.
*points sensibles:* Lee Kennett, *A History of Strategic Bombing,* p. 28.
a modern, centralized state: Ibid., p. 43.
Giulio Douhet: Ibid., p. 179.
"The potential . . . was greater than its achievement": quoted in Ibid., p. 178.
("can hit a town"): quoted in Ibid., p. 49.

59 (ninety-four strategic targets in North Vietnam): Momyer, *Airpower in Three Wars,* pp. 15, 32.

60 "Dead battles, like dead generals": quoted in Barbara Tuchman, *The Guns of August,* p. 38.
He had been leery of Warden: Schwarzkopf, *Hero,* p. 318.
"I think the Iraqis will withdraw": quoted in James P. Coyne, *Airpower in the Gulf,* p. 45.

64 could eliminate half of the Republican Guard in just five days: U.S. Air Force, *Gulf War Air Power Survey,* summary report (hereafter *GWAPS*), p. II–25.

***Riyadh***

66 a log of such events: Schwarzkopf, *Hero,* pp. 414–416.
"There does not appear to be a coordinated Iraqi plan": CENTCOM daily intelligence summary, obtained under FOIA.

67 further cause for good cheer: Schwarzkopf, *Hero,* p. 416.
Tolstoy's famous observation: Leo Tolstoy, *Anna Karenina,* p. 1.
("he is always in a rage"): quoted in Larrabee, *Commander in Chief,* p. 155.

70 the prima donna George Patton: James Charlton, ed., *The Military Quotation Book,* p. 68.

71 He often slept badly: "A Soldier's Conscience," *Newsweek,* March 11, 1991, p. 32; "Sayings of Stormin' Norman," *Time,* March 11, 1991, p. 27; Schwarzkopf testimony to House Armed Services Committee, June 12, 1991.
"nobody to turn to but God": quoted in Molly Moore, *Washington Post,* Feb. 5, 1991.
imagining the bloody bag: Schwarzkopf, *Hero,* p. 194.
lions in action: Barbara Tuchman, *The Guns of August,* p. 237.
"to keep people sharpened up": quoted in de la Billière, *Storm Command,* pp. 42, 149. In his dealings with the Saudis and other allies, Schwarzkopf was greatly assisted by Maj. Gen. Paul Schwartz.

73 "an essentially thespian general": quoted in Larrabee, *Commander in Chief,* p. 350.
George Marshall's . . . "white fury to cold and quiet contempt": Ibid., p. 100.
"they send for the sonsabitches": quoted in Ibid., p. 153.

74 "For what art can surpass that of the general": A. von Bogulawski, quoted in C. D. B. Bryan, "Operation Desert Norm," *The New Republic,* March 11, 1991, p. 20.

***Al Kharj***

An hour's drive southeast of Riyadh: details drawn from pool reports by Edith M. Lederer and David Evans; also, author's visit in March 1991.

*page*

75 toward Tikrit, Saddam's summer home: F-111 and B-52 attacks against, respectively, the Tikrit palace and Republican Guard units took place after dark on January 17.

***Dhahran***

77 the U.S. Army had positioned 132 Patriot missile launchers: some Patriot details drawn from Robert M. Stein, "Patriot ATBM Experience in the Gulf War," article published by Raytheon Co.; Robert M. Stein, "Patriot Experience in the Gulf," *International Security,* Summer 1992, p. 199; Theodore A. Postol, "Lessons of the Gulf War Experience With Patriot," *International Security,* Winter 1991–2, p. 119; Ben Sherwood, "The Blip Seen 'Round the World," Sept. 20, 1992, p. C2.

"hitting a bullet with another bullet": Hallion, *Storm Over Iraq,* pp. 300–302.

78 A pair of Air Force Space Command satellites: Craig Covault, *Aviation Week,* Feb. 4, 1991, p. 25.

**3. An Event in Israel**

***Washington***

83 ("This is no time to go wobbly"): quoted in American Enterprise Institute and Discovery Channel, "The Gulf Crisis: The Road to War," Jan. 1992 three-part television series.

Baker placed quick calls: Duffy and Goodgame, *Marching in Place,* pp. 162–163. Participants disagree on the sequence of Baker's calls that night.

84 Baker called Arens: some details drawn from official memorandum of conversation, U.S. State Department. Call to Shamir recounted in *New York Times,* March 3, 1991, p. 1.

85 That Iraq possessed immense stocks: Thomas Whiteside, "The Yellow Rain Complex," *The New Yorker,* Feb. 11, 1991, p. 38.

86 By Central Intelligence Agency estimates: William H. Webster, comments at *Washington Post* luncheon, Dec. 14, 1990.

Chuck Horner had laid out seventeen targets: U.S. Air Force, *GWAPS,* p. II–8.

88 (United Nations inspectors would discover): UN Department of Public Information, press releases of Aug. 11, Oct. 31, Dec. 11, 1991.

Botulinum is three million times more potent: *Conduct,* p. 18.

biological combat was rare: Jeanne McDermott, *The Killing Winds: The Menace of Biological Warfare,* pp. 21–22.

89 The British . . . had leased a fleet of freezer wagons: de la Billière, *Storm Command,* p. 119.

kill up to 2 percent of the spores per minute: Rick R. Smith, *Raleigh News and Observer,* Jan. 13, 1992, p. 1.

***Tel Aviv***

90 Israel endured the first salvo: Yossi Klein Halevi, "The Longest Week," *The Jerusalem Report,* Jan. 31, 1991, pp. 12–15.

91 A group of Hassidic Jews: Ann Husarska, "Holed Up," *The New Republic,* Feb. 11, 1991, pp. 13–14.

*page*

91 sheltered in the concrete stalls: *The Jerusalem Report,* Jan. 31, 1991, p. 20.
Phone calls from the United States to Israel: *The Jerusalem Report,* Feb. 7, 1991, p. 5.
Isaiah 26:20: *The Jerusalem Report,* Jan. 31, 1991, p. 15.
Nuclear missile crews stood at full alert: Seymour M. Hersh, *The Sampson Option: Israel's Nuclear Option and American Foreign Policy,* quoted by Joel Brinkley, *New York Times,* Oct. 19, 1991, p. 1.

93 Arens . . . retained a deep emotional attachment: Yossi Klein Halevi, "The Iceman," *The Jerusalem Report,* Feb. 14, 1991, pp. 8–11; Dennis Hevesi, *New York Times,* Dec. 21, 1988, p. 8.
Shamir talked to Bush: Jackson Diehl, *Washington Post,* March 19, 1991, p. A21.

***Washington***

94 he dismissed . . . Woerner: Woodward, *The Commanders,* pp. 96–99. Dugan's firing came after the Air Force chief talked too much on a trip to Saudi Arabia with three reporters, including the author.

95 he subjected himself to fifteen tutorials: Woodward, *The Commanders,* p. 330.

***U.S.S.* Saratoga**

98 a giant aircraft hatchery: Larrabee, *Commander in Chief,* p. 373.
The first *Saratoga:* Ibid., p. 165.

99 the critical Nazi oil refinery at Ploesti: Ibid., pp. 241–251.
70 percent of . . . American aircraft fell: Momyer, *Airpower in Three Wars.*

**4. The Left Hook**

***Riyadh***

107 His heroes: interview with *U.S. News & World Report,* Feb. 3, 1991.
at a cost of nearly half the Americans' fighting strength: U.S. Army, *Certain Victory: The American Army in the Gulf War,* draft version, p. II–44.
the Saudis . . . favored a line of defense: *Conduct,* pp. 42–43.
Schwarzkopf . . . would need eight to ten months: *Conduct,* p. 84.

108 conquering Kuwait . . . "can't be done": Schwarzkopf, *Hero,* p. 315.
45,000 German casualties: Eliot A. Cohen, "The Unsheltering Sky," *The New Republic,* Feb. 11, 1991, p. 23.
"Be more than you appear to be": quoted in U.S. Army, *Certain Victory: The American Army in the Gulf War,* p. I–28.
Joe Purvis and his team: Other officers included Majors Gregory Eckert, Bill Pennypacker, and Daniel Roh.

109 Guderian's Panzer attack: U.S. Army, *Certain Victory,* p. III–129.
drive toward the high ground above Mutlaa Pass: *Conduct,* pp. 85–88.

110 the CINC was wary of underestimating: C. D. B. Bryan, "Operation Desert Norm," *The New Republic,* March 11, 1991, p. 20.
Any officer offering a personal opinion: Schwarzkopf's account of Left Hook evolution in *Hero,* pp. 354–359.

*page*

111 "CINC's Assessment": from CENTCOM After Action Report, Vol. I, obtained under FOIA.

113 (Several proposals drafted by the Joint Chiefs' staff): *Triumph Without Victory,* p. 170.
"Tell me what you need": CENTCOM After Action Report, Vol. I.
"the don't-screw-around school": quoted in Woodward, *The Commanders,* p. 307.

114 The Marines struck him as "mustard keen": de la Billière, *Storm Command,* p. 94.

***Incirlik***

115 Many Turks resisted: Clyde Haberman, *New York Times,* Jan. 19, 1991.

116 Buster Glosson wrote a memo: Tony Capaccio, *Defense Week,* Jan. 13, 1992, p. 1; also, *Conduct,* p. 200.
the country's electrical output in 1920: Barton Gellman, *Washington Post,* June 23, 1991, p. 1.

117 The widespread loss of power was ruinous: William M. Arkin, Damian Durrant, Marianne Cherni, "Modern Warfare and the Environment: A Case Study of the Gulf War," Greenpeace, Washington, D.C., May 1991; Harvard University Study Team, "Public Health in Iraq After the Gulf War," 1991; William M. Arkin, briefing on postwar Iraq, National Press Club, Jan. 8, 1991; Michael Gordon, *New York Times,* Feb. 23, 1992, p. 1; Barton Gellman, *Washington Post,* March 6, 1992.

***Washington***

118 two large waves of warplanes: some details of Israeli plan in Schwarzkopf, *Hero,* p. 417.

119 The CINC was particularly incensed: Schwarzkopf, *Hero,* p. 418.

120 Born in Harlem: profile of Powell drawn from Howard Means, *Colin Powell;* Rudy Abramson and John Broder, *Los Angeles Times Magazine,* April 7, 1991; Rick Hampson, Associated Press, Feb. 10, 1991.

122 Powell had resisted war with Iraq: Woodward, *The Commanders,* pp. 299–300, 307.

***Al Qaim***

125 as these Ravens began their second orbit: interview with F-111 pilot.

***Tel Aviv***

130 80 percent of Israeli citizens: poll by Guttman Institute of Applied Social Research, *Jerusalem Report,* Jan. 13, 1991, p. 6.
*Kippy of Rechov Sumsum:* Yossi Klein Halevi, "The Longest Week," *Jerusalem Report,* Jan. 31, 1991, pp. 12–15.
ten senior rabbis: William Claiborne, *Washington Post,* Jan. 22, 1991.

133 Four hundred newborn males: *Wall Street Journal,* Feb. 6, 1991, p. 1; *Jerusalem Report,* Feb. 21, 1991, pp. 28–29.

***Al Qaim***

136 folklore persisted from Vietnam: Harry G. Summers, Jr., *Vietnam Almanac,* p. 308.
only seven rescue missions would be launched: briefing paper by Gen. Carl W. Stiner, U.S. Special Operations Command; also, *Conduct* (interim report to Congress), p. 5–5.

*page*

**5. Delta**

***Washington***

142 Slim . . . once declared: Larrabee, *Commander in Chief,* p. 567.

143 SAS . . . formed in North Africa in 1941: Bryan Perrett, *Desert Warfare.*

144 "the darndest search-and-destroy effort": Bush in White House briefing, Jan. 18, 1991.

145 a close relative of Nazi *Vergeltungswaffen:* Addington, *The Patterns of War,* pp. 211–215; Warden, *The Air Campaign,* p. 176; also, Dilip Hiro, *The Longest War.*

a "circular error probable" of more than three thousand meters: *Conduct,* pp. 15–16.

147 Iraqi decoys were so cleverly constructed: U.S. Air Force, *GWAPS,* p. III–29.

***U.S.S.* Blue Ridge**

148 a wave of A-6 bombers: Department of the Navy, "The United States Navy in Desert Shield/Desert Storm," May 1991, p. A-18.

150 "I simply have not got enough Navy": quoted in Larrabee, *Commander in Chief,* p. 66.

Schwarzkopf . . . had more ships than water: de la Billière, *Storm Command,* p. 115.

151 "The violence of interservice rivalry": quoted in Larrabee, *Commander in Chief,* p. 105.

***Tobuk***

153 London's contribution had been briefly overshadowed: *Aerospace Daily,* July 2, 1991, p. 12; *Times* of London, June 28, 1991; *Independent,* June 29, 1991.

"Brits look stupid": de la Billière, *Storm Command,* p. 181.

British training stressed very low altitude flying: testimony of Air Vice Marshal William J. Wratten, House of Commons Defence Committee, "Preliminary Lessons of Operation Granby," July 1991.

154 only two of the five Tornados: Ibid.; and Peter Almond, *Daily Telegraph,* May 3, 1991.

"bad luck doesn't last forever": Wratten briefing, Riyadh, Jan. 22, 1991.

the pilots sometimes wept openly: Reuter pool report, Feb. 4, 1991.

155 The fleet included: Lt. Gen. Walter E. Boomer, "MARCENT Operations in the Campaign to Liberate Kuwait," 1991.

But by the third day of combat the number was down to sixty: Gen. M. A. McPeak, briefing on Desert Storm, Pentagon, March 15, 1991.

Fifteen planes had been shot down or lost: Norman Friedman, *Desert Victory,* pp. 357–358.

(Seventeen allied planes were lost): Department of the Navy, "The United States Navy in Desert Shield/Desert Storm."

when they destroyed four thousand Soviet aircraft in a week: Warden, *The Air Campaign,* p. 42.

***Washington (Powell-Cheney briefing)***

159 Polls showed that four of every five: *New York Times,* Jan. 18, 1991, p. A11, and Jan. 22, 1991, p. A12.

*page*

159 the press corps's own disgruntlement: *Conduct,* Appendix S; John J. Fialka, *Hotel Warriors: Covering the Gulf War.*

160 seventy-six missed their targets: Office of History, Headquarters 37th Fighter Wing, "Nighthawks Over Iraq," pp. 8–12.

161 "you can win the battle [but] lose the war": quoted in Woodward, *The Commanders,* p. 155.

162 the chairman used a sentence he had carefully rehearsed: Jason DeParle, *New York Times,* May 5, 1991, p. 1; May 6, 1991, p. A9.

**6. Mesopotamia**

***Suqrah Bay***

169 Iraq had shifted several divisions: *Conduct,* pp. 293–299.

170 Marines had been storming beaches since 1776: Larrabee, *Commander in Chief,* p. 276.

At Gallipoli in 1915: Edward L. Beach, *The United States Navy,* p. 465.

At Dieppe in August 1942: Merrill L. Bartlett, ed., *Assault From the Sea: Essays on the History of Amphibious Warfare,* p. 249.

known as Terrible Tarawa: Ibid., pp. 210–216.

"the greatest tactical innovation of World War II": quoted in Ibid., p. xiii.

Omar Bradley said in 1949: Ibid., p. xvii.

MacArthur's grim estimate: Ibid., p. 352.

171 the Clausewitz dictim: Ibid., p. 210.

Marines preferred to attack: Marine Corps Combat Development Command, *Joint Doctrine for Landing Force Operations,* Quantico, Va., Dec. 1989, p. IV–11.

***Washington***

174 known informally as Starship Enterprise: Woodward, *The Commanders,* p. 327.

175 Allied aircraft now averaged more than a hundred counter-Scud sorties: *Conduct,* pp. 224–226.

178 (the SAS would suffer four dead and five captured): Reuters, *Chicago Tribune,* March 1, 1992, p. 18; Michael Fleet, *Daily Telegraph,* Feb. 29, 1992, p. 3; Michael Evans, *Times* of London, Feb. 29, 1992; *Times* of London, May 15, 1991, p. 1. Also, de la Billière, *Storm Comand,* pp. 192, 221–267.

179 nine Iraqi armored vehicles drew so close: Rick R. Smith, *Raleigh News and Observer,* Jan. 14, 1992, p. 7A.

***Andover***

182 initiated the launching of thirty-one Patriots: Central Command, "Desert Storm Chronology of Significant Events," obtained under FOIA.

183 body bag production: John Kifner, *New York Times,* Jan. 16, 1991, p. 15.

***Al Ahmadi***

184 an estimated three dozen Iraqi engineers and a thousand troops: Lee Hockstader, *Washington Post,* April 1, 1991, p. D1. Also, Arkin, Durrant, and Cherni, "Modern Warfare and the Environment: A Case Study of the Gulf War."

*page*

187 The effect of this attack would remain in dispute: T. M. Hawley, *Against the Fires of Hell*, pp. 45–48.

**7. Khafji**

***Washington***

190 Bessmertnykh had expressed dismay: Michael R. Beschloss and Strobe Talbott, *At the Highest Levels: The Inside Story of the End of the Cold War*, pp. 326–333.

192 He was . . . thoughtful, gracious, and remarkably well versed: For a fine, insightful analysis of Bush's character and presidency, see Michael Duffy and Dan Goodgame, *Marching in Place*.

194 "a successful backbone transplant": quoted in Smith, *George Bush's War*, p. 68.

extemporaneous comments to the Republican National Committee: Ann Devroy, *Washington Post*, Jan. 26, 1992, p. 1.

"that lying son of a bitch": quoted in Jack Nelson, *Los Angeles Times*, Feb. 17, 1991, p. 1.

196 none eclipsed Bandar: Roxanne Roberts, *Washington Post*, Aug. 16, 1990, p. D1.

The prince, *Newsweek* magazine once observed: Smith, *George Bush's War*, p. 78.

Bandar had served as a secret middleman: Woodward, *The Commanders*, p. 203.

***Observation Post 4***

199 Delta Company, commanded by Captain Roger Pollard: some details of battle drawn from accounts compiled by U.S. Marine Corps Historical Center and Marine friendly fire investigative reports obtained under FOIA. Also, *Conduct*, pp. 174–175; Appendix I, pp. 36–37.

***R'as al Khafji***

203 Barry ordered his men to fall back: Barry's account in taped oral history, USMC Historical Center.

***OP-4***

205 Krulak marshaled clerks, typists, and other troops: U.S. Naval Institute, *Proceedings*, Nov. 1991, pp. 47–81.

***Khafji***

210 three AC-130 Spectre gunships: some details drawn from Air Force Special Operations Wing investigative documents obtained under FOIA.

Boomer . . . vented his exasperation: transcripts of I MEF staff meetings, USMC Historical Center.

212 (Among the CINC's contributions was to dissuade the Saudis): Schwarzkopf, *Hero*, p. 424.

the Arab forces were determined: John H. Admire, *Marine Corps Gazette*, "The 3rd Marines in Desert Storm," Sept. 1991; also, Admire speech to Honolulu Chamber of Commerce, 1991.

***Dover***

213 this first clutch of native sons to fall in ground combat: *New York Times*, Jan. 31, 1991, p. A9; *Los Angeles Times*, Feb. 24, 1991, p. 11A; *Deseret*

*page*

News (Salt Lake City), Feb. 12, 1991, p. 1; *Salt Lake City Tribune,* Feb. 7, Feb. 12, and March 17, 1991; *Dallas Morning News,* Feb. 1, Feb. 6, and Feb. 10, 1991.

214 Daniel B. Walker came home to Whitehouse: David Maraniss, *Washington Post,* Feb. 10, 1991, p. A1, a brilliant account.

**8. The Riyadh War**

***Riyadh***

216 Billy Mitchell insisted: Hallion, *Storm Over Iraq,* p. 7.
George Patton complained bitterly: Martin Blumenson, *Patton,* p. 185, and Momyer, *Air Power in Three Wars,* p. 49.
should continue hammering strategic targets in Germany: Warden, *The Air Campaign,* p. 112.

217 four of every ten sorties . . . had been canceled: *Conduct* (interim version), pp. 4–6.

218 an Army directive in the early days of manned flight: Morris Janowitz, *The Professional Soldier,* p. 25.
target locations . . . to be pinpointed on the map within a hundred meters: U.S. Army, *Certain Victory,* pp. IV–15–17.

219 the Marines also possessed their own air force: House Armed Services Committee, "Defense for a New Era: Lessons of the Persian Gulf War," 1992.

220 "like trying to stuff spaghetti": quoted in de la Billière, *Storm Command,* p. 43.

222 of 3067 targets nominated: U.S. Army, *Certain Victory,* p. IV–23.

223 Army commanders concluded that the air planners: Ibid., p. IV–13.
a RAND Corporation study: C. Kenneth Allard, *Command, Control and the Common Defense,* p. 12.
(drove Eisenhower to a four-pack-a-day cigarette habit): McCullough, *Truman,* p. 738.

224 Slightly more than half had struck their intended targets: Barton Gellman, "Gulf Weapons' Accuracy Downgraded," *Washington Post,* April 10, 1992, p. 1.
a strategy presented to Winston Churchill: Larrabee, *Commander in Chief,* p. 589.

225 "We have been doing everything we can": Tom Kelly, quoted in *Washington Post,* Feb. 2, 1991, p. 1.
air strikes killed nearly 2300: William M. Arkin, Greenpeace, press briefing at National Press Club, Washington, Jan. 8, 1992.
During the Normandy invasion in 1944: calculations of Normandy deaths in 1944 and Vietnamese in 1972 by Robert Pape, Jr., a historian at the University of Michigan, quoted by R. Jeffrey Smith, *Washington Post,* Feb. 3, 1991. Ratio of bombs to civilian deaths in the gulf war is based on total of 88,500 tons dropped, but does not include Kuwaiti civilian deaths, which are unknown. The calculation also excludes indirect deaths from disease, etc.

226 93 percent were dumb bombs: Coyne, *Airpower in the Gulf,* p. 95.

*page*

226 his first bomb would hit sixty feet short: James P. Coyne, "Bombology," *Air Force Magazine*, June 1990, p. 64.

B-52s . . . would drop 72,000 bombs: Coyne, *Airpower in the Gulf*, p. 79.

"One lost control of bodily functions": quoted in Hallion, *Storm Over Iraq*, p. 58.

Dumb bombs . . . hit their targets only 25 percent of the time: Barton Gellman, *Washington Post*, March 16, 1991.

(only one British bomber in five): Kennett, *A History of Strategic Bombing*, p. 129.

("comical oxymoron"): Paul Fussell quoted in *Philadelphia Inquirer*, April 25, 1992, p. 8.

bad weather, smoke, and haze: *Conduct*, Appendix T, p. 6.

The three waves of stealth fighters: Office of History, Headquarters, 37th Fighter Wing, "Nighthawks Over Iraq," pp. 17–18.

227 The consequences of inaccuracy could be horrific: details drawn from report and briefing by William M. Arkin, Greenpeace, Jan. 8, 1992.

struck a hospital near Kuwait City: William Branigin, *Washington Post*, March 25, 1991, p. A10.

"Its mystery . . . is half its power": quoted in Kennett, *A History of Strategic Bombing*, p. 178.

***Al Kharj***

229 fewer than 3 percent of all American fighter pilots were aces: Donald B. Rice, secretary of the Air Force, speech, Feb. 23, 1991.

Over North Korea in September 1952: Momyer, *Airpower in Three Wars*, p. 150.

***Riyadh***

233 A half dozen satellites wheeled overhead: Kathy Sawyer, *Washington Post*, Feb. 19, 1991; *Space News*, Feb. 3, 1991, p. 1; Craig Covault, *Aviation Week*, Feb. 4, 1991, p. 25; Jeffrey T. Richelson, *The U.S. Intelligence Community*.

236 A green sticker marked a division: In his memoirs, Schwarzkopf refers to the stickers representing Iraqi divisions as changing from red to green as the units were progressively battered. Leide's recollection is that on CENTCOM's "cartoon," the colors were reversed. Schwarzkopf, *Hero*, p. 439.

***U.S.S.* Blue Ridge**

237 sunk by mines in the Korean War: J. M. Martin, *Proceedings*, "We Still Haven't Learned," July 1991, p. 64.

238 "tantamount to suicide": quoted in de la Billière, *Storm Command*, p. 258.

Iraq possessed one to two thousand mines: *Conduct*, p. 274.

Schwarzkopf . . . had restricted surveillance: Ibid., p. 282.

The American minesweeping force in Desert Storm: Ibid., p. 277.

none of which could effectively detect mines in less than thirty feet of water: House Armed Services Committee, "Defense for a New Era," p. 28.

*page*

238 In addition to five British mine hunters: *Conduct,* p. 280; U.S. Navy, "Mine Warfare Plan," Jan. 29, 1992.
During the U.S. campaign against the Japanese: John Keegan, *The Price of Admiralty,* p. 318.

239 The CINC had slept badly: Molly Moore, *Washington Post,* Feb. 5, 1991, p. 1. Schwarzkopf: "I get enough sleep, but I don't get enough rest because I wake up fifteen, twenty, twenty-five times in the middle of the night and my brain is just in turmoil over these agonizingly difficult decisions that I have to make."

240 the new focus . . . would be . . . Faylaka Island: *Conduct,* p. 299.

**9. The Desert Sea**
***Riyadh***

244 Boomer wanted to strike: some details of Marine deliberations over their attack plans from U.S. Naval Institute, *Proceedings,* Nov. 1991, pp. 47–81. Also, postwar briefing by Boomer in archives of U.S. Marine Corps Historical Center.

247 Allied forces at Arnhem bridge: Louis L. Snyder, *Historical Guide to World War II,* p. 33.

248 The Egyptians, whom the CINC considered "indispensable": Schwarzkopf, *Hero,* p. 388.

249 Since the age of Napoleon: Keegan, *The Price of Admiralty,* p. 43.
Schwarzkopf once described donning Arab robes: Bryan, "Operation Desert Norm," *The New Republic,* March 11, 1991, p. 20
the primeval confusion of Vietnam's triple-canopy jungle: Neil Sheehan, *A Bright Shining Lie,* p. 639.
"The desert suits the British": William Joseph Slim, *Defeat Into Victory,* p. 4.
"was fought like a polo game": quoted in Correlli Barnett, *The Desert Generals,* p. 1.
"agoraphobic vastness": quoted in Ibid., pp. 84–85.
"in their mobility, ubiquity, their independence": quoted in Bryan Perrett, *Desert Warfare,* p. 60.

250 Kitchener, the British hero of Khartoum: Ibid., p. 26.
the battle for North Africa became a struggle between supply officers: Ibid., p. 162.
The British smartly kept their men and machines stoked: Ibid., p. 133.
included a motorized brothel: Barnett, *The Desert Generals,* p. 28.
stricken fleets, could not rely on topography: Keegan, *The Price of Admiralty,* p. 124.
Crassus . . . led nearly forty thousand soldiers: Perrett, *Desert Warfare,* pp. 13–14.
Operation Compass in December 1940: Ibid., p. 104.
At El Alamein, Rommel lost 55,000 dead: Ibid., p. 166.

251 "battles fought therein result in total victory or total defeat": quoted in Ibid., p. 11.
"pell mell battle": quoted in Keegan, *The Price of Admiralty,* p. 53.

*page*

251 "cash transaction": quoted in Ibid., p. 4.
the tank was so named: Addington, *The Patterns of War*, pp. 142–143.
Fuller imagined armored contraptions: Ibid., p. 163.
Guderian effectively massed his tank fleets: John Keegan, *Daily Telegraph*, Feb. 10, 1991.
Israel used Guderian's blitzkrieg tactic: Addington, *Patterns of War*, pp. 285–286.
The M1A1 Abrams . . . was a sixty-seven-ton behemoth: *Conduct*, Appendix T, pp. 143–145.
In World War II, a stationary American tank: U.S. Army, *Certain Victory*, p. I–10.
"the force of a race car striking a brick wall": quoted in U.S. Army, *Certain Victory*, p. I–10.
approximately ten thousand armored vehicles: Department of Defense briefing on fratricide, Col. Roger Brown, Aug. 13, 1991.

252 Depuy's effort bore fruit: John J. Romjue, *From Active Defense to AirLand Battle: The Evolution of Army Doctrine, 1973–1982*. Also, Rick Atkinson, *The Long Gray Line*, pp. 492–494.
"meeting the strength of the Soviet attack head-on": Allard, *Command, Control and the Common Defense*, p. 174.

253 Matthew Arnold's haunting image: G. B. Harrison, ed., *A Book of English Poetry*, "Dover Beach," p. 403.
Marshall . . . warned that mobile combat: Larrabee, *Commander in Chief*, p. 111.

***Al Qaysumah***

the army corps was a legacy of Napoleon Bonaparte: Addington, *The Patterns of War*, p. 19.

257 VII Corps had fewer than two hundred HETs: *Conduct*, Appendix F, p. 45.

258 Franks . . . planned to have the 1st Infantry Division: Tom Donnelly, "The Generals' War," *Army Times*, March 2, 1992, p. 8.

**U.S.S. Wisconsin**

259 naval theorists had periodically consigned the battleship to obsolescence: Addington, *Patterns of War*, p. 103.

260 "a shattering, blasting, overbearing force": quoted in Robert K. Massie, *Dreadnought*, p. 783.
great shoots of yore: Naval Historical Center, *Operational Experience of Fast Battleships: World War II, Korea, Vietnam*.
a concussive shock that hammered the throat: George Rodrigue, pool report, *Dallas Morning News*, Feb. 8, 1991.

***Riyadh (Cheney and Powell)***

268 "slow, ponderous, pachyderm mentality": quoted in Schwarzkopf, *Hero*, pp. 433–434.

269 The Army's youngest division commander: James Blackwell, "Georgia Punch," *Army Times*, Dec. 2, 1991, p. 13.

270 "I think we should go with the ground attack now": Schwarzkopf's account of meeting in *Hero*, p. 435.
the CINC had hinted that he would . . . resign if pressed to strike: De la Billière mentions this twice in *Storm Command*, pp. 84, 196.

*page*

**10. Al Firdos**

***Riyadh***

272 "Command is a true center of gravity": Warden, *The Air Campaign*, p. 53.

273 Dugan. . . . had publicly disclosed: Rick Atkinson, "U.S. to Rely on Air Strikes if War Erupts," *Washington Post*, Sept. 16, 1990, p. 1.

five assassination attempts in 1981: Dilip Hiro, *The Longest War.*

Bush's secret order, in August 1990: Woodward, *The Commanders*, p. 282.

more than four hundred sorties: *Conduct*, p. 214.

274 an Air Force operations order predicted: U.S. Air Force, *GWAPS*, p. II–20.

"by daring to win all": quoted in Michael Howard, *Clausewitz*, p. 39.

276 the Mukhabarat . . . a successor to the Ba'ath Party's secret police: Judith Miller and Laurie Mylroie, *Saddam Hussein and the Crisis in the Gulf*, pp. 48–50.

***Washington***

278 (as few as 9 percent of engagements resulted in confirmed "warhead kills"): General Accounting Office report; Barton Gellman, *Washington Post*, Sept. 30, 1992, p. A4.

Some Patriots had apparently caused damage: Robert M. Stein, "Patriot Experience in the Gulf," *International Security*, Summer 1992, p. 233. After the war, Ehud Olmert, the Israeli health minister, reported that Scuds had caused roughly $50 million in damage. Of more than twelve thousand apartments damaged, 195 were destroyed. Two deaths were directly attributed to the attacks; sixty other citizens were wounded severely enough to require hospitalization. Eleven others died from heart attacks or suffocation from improperly adjusted gas masks. Another 765 Israelis injected themselves needlessly with atropine or suffered symptoms of hysteria and anxiety. Stein, *International Security*, Summer 1992, pp. 221, 232.

281 another Scud had struck Tel Aviv: *Jerusalem Report*, Feb. 14, 1991, pp. 16–18.

***Baghdad***

Primakov's journey had been singularly unpleasant: Moscow Domestic Service, trans. by Foreign Broadcast Information Service, Feb. 12, 1991; *Los Angeles Times* special section, "Witness to War," March 12, 1991; Yevgeni Primakov, "My Final Visit with Saddam Hussein," *Time*, March 11, 1991.

282 In October, he had urged that Kuwait present Saddam: Michael R. Beschloss and Strobe Talbott, *At the Highest Levels*, p. 275.

Four weeks of war . . . had reduced the Iraqi capital to nineteenth-century privation: Harvard Study Team report, "Public Health in Iraq After the Gulf War," May 1991; William M. Arkin, Damian Durrant, Marianne Cherni, Greenpeace, May 1991; "Modern Warfare and the Environment: A Case Study of the Gulf War"; John Simpson, *From the House of War*; Patrick Tyler, *New York Times*, July 5, 1991, p. A4; news accounts in *Washington Post*, *Wall Street Journal*, Reuters.

*page*

283 (more air combat sorties): William M. Arkin, "The Teddy Bear's Picnic," *The Nation,* Nov. 2, 1992, p. 510.

284 his strategy was similar to that of the Japanese: Larrabee, *Commander in Chief,* p. 12.
to liken himself to Nebuchadnezzar: Judith Miller and Laurie Mylroie, *Saddam Hussein and the Crisis in the Gulf,* p. 58.

***Amariyah***

285 Screams ripped through the darkness: scene at Shelter Number 25 drawn from *Daily Telegraph,* Feb. 14, 1991; John Simpson, *From the House of War,* pp. 329–331; *Washington Post,* Feb 14, 1991, p. 1; *New York Times,* July 5, 1991, p. A4; Laurie Garret, "The Dead," *Columbia Journalism Review,* May-June 1991, p. 32; "Needless Deaths in the Gulf War," Middle East Watch Report, pp. 129–137; televised reports by Peter Arnett, CNN; Mamoun Youssef, Reuters, Feb. 13, 1991.

286 At Yarmuk Hospital: Patrick E. Tyler, *New York Times,* July 5, 1991 p. A4.
Residents of Amariyah insisted that the shelter had recently been opened: "Needless Deaths in the Gulf War," Middle East Watch Report, p. 129.

***Washington (Powell)***

289 six million tons of explosives dropped: Harry G. Summers, Jr., *Vietnam Almanac,* p. 100.
A national poll taken immediately after Al Firdos: Richard Morin, *Washington Post,* Feb. 16, 1991, p. A19, Washington Post–ABC News Poll taken February 14.

***Riyadh***

291 an Italian pilot flipped an oversized grenade: Lee Kennett, *A History of Strategic Bombing,* p. 13.
"a depressing effect": quoted in Ibid., p. 7.
"It is particularly humiliating": quoted in Ibid., p. 24.
the psychological "yield" of RAF attacks: Ibid., p. 51.
one French town endured: Ibid.
Marshal Foch warned: Ibid., p. 53.
Hitler in 1939 envisioned: Ibid., p. 118.
Bomber Command subsequently decreed: Ibid., p. 129.
J. M. Spaight's warning in 1930: Ibid., p. 54.
Harris . . . concluded that in a police state: Ibid., p. 130.

292 even the bellicose Winston Churchill: Ibid., p. 188.
after the Italian capital was bombed: Ibid., pp. 150–151.
Strategic Bombing survey concluded: Warden, *The Air Campaign,* p. 62.
the devastating Linebacker II campaign: Momyer, *Airpower in Three Wars,* p. 243.
"a war-winning weapon in its own right": quoted in Russell F. Weigley, *The American Way of War,* p. 240.
Dugan . . . had spoken almost mystically: Atkinson, *Washington Post,* "U.S. to Rely on Air Strikes," Sept. 16, 1990, p. 1.

293 the firebombing of Tokyo: Kennett, *A History of Strategic Bombing,* p. 171.

294 one message . . . dropped after an attack on Frankfurt: Ibid., p. 144.

295 only five would be hit: U.S. Air Force, *GWAPS,* p. VIII–14.

*page*

295 U.S. intelligence concluded that fiber-optic cables led . . . to Basrah: *Conduct,* p. 182.

**11. Misadventure**

***Washington***

297 Clausewitz had sorted wars into two types: Russell F. Weigley, *The American Way of War,* p. xx; Michael Howard, *Clausewitz,* p. 16.

298 as the political scientist Robert E. Osgood noted: Weigley, *The American Way of War,* p. 412.

the American objective of ejecting North Korean invaders: Clay Blair, *The Forgotten War,* pp. 325–328.

at a cost of 54,000 dead Americans: Ibid., p. 975.

300 MacArthur had warned: McCullough, *Truman,* p. 804.

characterized by a pernicious inconclusiveness: Weigley, *The American Way of War,* p. 468.

301 "getting an arm caught in the mangle": quoted in de la Billière, *Storm Command,* p. 17.

302 Gorbachev cabled Bush: *New York Times,* March 3, 1991, p. 1.

the Soviet president suggested: Beschloss and Talbott, *At the Highest Levels,* p. 334.

"No way, José!": Ibid.

Bush, alerted to the communiqué: *Washington Post,* Feb. 16, 1991, p. 1; *New York Times,* Feb. 17, 1991, p. 18.

***Eskan Village***

307 "the human heart is the starting point": attributed to Marshal de Saxe, "Reveries on the Art of War," 1732.

***King Fahd International Airport***

311 bridges . . . tumbling at a rate of seven to ten each week: *Conduct,* p. 210.

313 Sweet saw the fireball of his Warthog: Robert Sweet, "Oh, Man, I'm Hit!" *People,* Spring-Summer 1991, p. 91; also, Steven Phillis's Silver Star certificate, copy provided by U.S. Air Force.

***Phase Line Minnesota***

314 the Defense Advanced Research Projects Agency began to evaluate: *Conduct,* Appendix M, p. 1.

French casualties from friendly artillery: Charles R. Shrader, Combat Studies Institute, *Amicicide: The Problem of Friendly Fire in Modern War,* p. x.

315 Germans dubbed their 49th Field Artillery: Ibid., p. 2.

stopped using large metal canisters crammed with steel flechettes: Ibid., p. 31.

at Monte Cassino and Venafro: Ibid., p. 35.

decimated the 30th Infantry Division: Ibid., pp. 40–43.

At the battle of Hill 875: Atkinson, *The Long Gray Line,* pp. 247–248.

"something less than 2 percent": Shrader, *Amicicide,* p. 105. A more recent analysis by Army Col. David M. Sa'adah, a physician, estimates that friendly fire deaths in earlier wars were considerably higher than Shrader's numbers. Sa'adah put the figure at "something like 15%" of total battle deaths in World War II and Vietnam. See Robert Mackay, *Washington Post,* March 8, 1993, p. A4.

*page*

315 (In Southeast Asia . . . deaths were classified under "misadventure"): Shrader, *Amicicide,* p. xii.
Several other variables: *Conduct,* Appendix M, p. 1.

316 Tank gunnery . . . stressed the importance: Ibid., p. 1.
positive identification was difficult beyond seven hundred meters: U.S. Army, *Certain Victory,* p. VI–61.
"The impact of amicicide . . . is geometric": quoted in Shrader, *Amicicide,* p. 106.
(A young corporal): Michael E. Mullen, subject of *Friendly Fire,* C. D. B. Bryan.
a reluctance . . . to promptly and fully disclose the details: Barton Gellman, " 'Friendly Fire' Reports: A Pattern of Delay, Denial," *Washington Post,* Nov. 5, 1991, p. 1.

317 A few "DARPA lights" arrived: *Conduct,* Appendix M. Fifteen thousand simple infared beacons, nicknamed "Bud lights," also came into the theater in late February. Not having trained with the devices, many soldiers feared the lights would betray their position to the enemy and therefore turned them off.
the Big Red One slashed twenty holes: 1st Infantry Division, after-action review, March 14, 1991, signed by Maj. Gen. Thomas G. Rhame.
Hillman had anchored his right flank: The incident involving Ralph Hayles drawn from interviews with several participants; Army investigative documents obtained under FOIA; "Friendly Fire," *60 Minutes,* Nov. 10, 1991; Robert Johnson and Caleb Solomon, *Washington Post,* Sept. 10, 1991, p. 1.

**12. Echo to Foxtrot**

***U.S.S.* Tripoli**

321 Farragut . . . at Mobile Bay: Addington, *Patterns of War,* pp. 70–71.
Mines so bedeviled . . . in the Korean War: Tamara Moser Melia, *Damn the Torpedoes: A Short History of U.S. Naval Mine Countermeasures, 1777–1991,* p. 76.

322 Six fields lay *east: Conduct,* pp. 274–283.
in mid-February the CINC had administered another tongue lashing: Schwarzkopf, *Hero,* p. 437.
"Intelligence is perhaps a little like first love": House of Commons Defence Committee, "Preliminary Lessons of Operation Granby."
At 7 P.M. on February 17: The details from mine strikes are drawn largely from author's visits to *Tripoli* in San Diego and *Princeton* at shipyard in Long Beach, California.

***Base Weasel***

330 Stonewall Jackson once decreed: Warden, *The Air Campaign,* p. 162.
laws of modern war: *Conduct,* Appendix O, p. 21.
American commanders in the southwest Pacific: Warden, *The Air Campaign,* pp. 32–33.
Chennault periodically repainted: Ibid., p. 164.
British commanders showed a particular gift: Larrabee, *Commander in Chief,* p. 449.

*page*

331 the celebrated "man who never was": Michael Dewar, *The Art of Deception in Warfare,* p. 17.
"Force and fraud are in war": James Charlton, ed., *The Military Quotation Book,* p. 18.

332 a feint into the wadi: Some details are drawn from J. Paul Scicchitano, *Army Times,* Aug. 23, 1991, p. 8.

***King Fahd International Airport***

335 "the moral is to the material": quoted in Robert Debs Heinl, Jr., *Dictionary of Military and Naval Quotations,* p. 196.

336 Operation Tintinnabulation: Peter Watson, *War on the Mind,* pp. 408–410.
"killing of the enemy's courage": quoted in Howard, *Clausewitz,* p. 44.

337 audio cassettes . . . were smuggled: de la Billière, *Storm Command,* p. 126.

***Riyadh***

338 "Who's running the goddam war?": quoted in Molly Moore, *A Woman at War,* p. 155.

340 tens of thousands fewer: House Armed Services Committee, "Defense for a New Era," pp. 29–34.
Saddam's ground forces were believed to have grown: U.S. Army, *Certain Victory,* p. III–113.

341 U.S. analysts occasionally lost track of entire divisions: Ibid., pp. IV–4–5.
An intelligence system that for forty years had concentrated: Ibid., p. III–112, pp. IV–5–6.
the untimely retirement . . . of the SR-71: Ibid., pp. IV–4–5.
In mid-January the Pentagon estimated: *Conduct,* p. 353.
Another twenty-four divisions . . . remained north of the Euphrates: U.S. Army, *Certain Victory,* p. IV–3.
Two infantry divisions: Ibid., IV–40.
inadvertently confused the identities: Ibid., p. V–16.
Schwarzkopf's overestimation of the enemy's size: briefing in Riyadh, Feb. 27, 1991.

342 Schwarzkopf would continue to assert: testimony to House Armed Services Committee, June 12, 1991.
some divisions may have had as few as four thousand soldiers: Bob Woodward, *Washington Post,* March 17, 1991, p. A24.
would prove to be overestimated by 20 to 25 percent: U.S. Air Force, *GWAPS,* pp. III–43 and IV–7.
the Americans shipped nearly 220,000 rounds: *Conduct,* p. 188.
A study by the House Armed Services Committee: House Armed Services Committee, "Defense For a New Era," pp. 29–34. A former DIA analyst subsequently estimated that desertions reduced the Iraqi numbers from "less than 400,000" in mid-January to "perhaps even below 200,000" by late February. He also estimated Iraq's military casualties at no more than 9500 dead and 26,500 wounded. See John O. Heidenrich,

page

"The Gulf War: How Many Iraqis Died?," *Foreign Policy*, Spring 1993, p. 108.

***Washington***

344 Schwarzkopf . . . feared that the coalition could lose five thousand: Schwarzkopf, *Hero*, pp. 442, 445.
the CINC accused Powell of political expediency: For Schwarzkopf's account of this episode see Ibid., pp. 441–443.

345 CENTCOM's estimates of Iraqi tanks destroyed: *Conduct*, p. 188.
"a beagle chasing a rabbit": quoted in David Lamb, *Los Angeles Times*, Feb. 20, 1991, p. 1.
Assessments compiled by the Army's John Stewart: *Certain Victory*, pp. IV–37–39.

***Moscow***

347 "The timing is crucial": quoted in Yevgeni Primakov, "My Final Visit with Saddam Hussein," *Time*, March 11, 1991.

348 "The clearer this gets, the worse it gets": Beschloss and Talbott, *At the Highest Levels*, p. 336.
Aziz . . . balked . . . at Soviet attempts to impose the withdrawal timetable: Ibid.

***Washington (Bush's meeting)***

349 "We've got to get this thing back under control": quoted in Beschloss and Talbott, *At the Highest Levels*, p. 338.

350 Bush . . . strode upstairs to his study: some details drawn from "The Night Bush Decided," *Time*, March 4, 1991, p. 20.

351 allied pilots would suspend their attacks for two hours: de la Billière, *Storm Command*, p. 280.

352 Bush took a phone call from France's Mitterrand: American Enterprise Institute and Discovery Channel broadcast, "The Gulf Crisis: The Road to War," Jan. 1992.
Satellites had detected many new pillars of smoke: *Defense News*, March 18, 1991.

**13. The Biltmore**

***Riyadh***

361 The survival rate of soldiers and airmen in captivity: Robert E. Mitchell, *Foundation*, "The Vietnam Prisoners of War," Fall 1991, p. 28.
raid on the Son Tay prison camp: Daniel P. Bolger, *Americans at War*, pp. 114, 144.

362 the sleek structure was known as the White Ship: Bob Simon, *Forty Days*, p. 91.

***Baghdad***

364 guards savagely pummeled the crew's correspondent: Ibid., pp. 180–185.

***Task Force Grizzly***

365 The CINC's timetable: *Conduct*, p. 338.

366 the jellied gasoline was formally known as Incendergel: *New York Times*, Feb. 23, 1991, p. 8.

367 "The effect of an FAE explosion": *Defense Week*, Jan. 22, 1991, p. 12.

*page*

***King Khalid Military City***

369 SEALs slipped into Zodiac rubber raiding boats: *Newsweek*, "Secret Warriors," June 17, 1991.

***Phase Line Becks***

372 (Other map demarcations named after beers): Steve Vogel, "A Swift Kick," *Army Times*, Aug. 5, 1991, p. 10.

**14. G-Day**

***Washington***

375 caught the president and James Baker in the Camp David gymnasium: American Enterprise Institute and Discovery Channel broadcast, "The Gulf Crisis: The Road to War," Jan. 1992.

Gorbachev halfheartedly argued: Beschloss and Talbott, *At the Highest Levels*, p. 340.

***Umm Gudair***

376 CENTCOM meteorologists: weather forecasts obtained from CENTCOM under FOIA.

377 "We probably didn't get seventy-two minutes": quoted in U.S. Naval Institute, *Proceedings*, Nov. 1991, pp. 47–81.

378 half the line charges would not "command detonate": *Conduct*, Appendix T, p. 134.

380 Ripper had fired fifty-five tank rounds: Molly Moore, *A Woman at War*, p. 210.

381 "Progressing smoothly — timely manner": quoted in Ibid., p. 215.

"None of our fears materialized": quoted in Ibid., p. 231.

***Objective Rochambeau***

a battle order of succinct élan: Details of French attack are drawn from "Guerre Éclair dans le Golfe," Éditions Jean-Claude Lattes et Addim, 1991 (official version of French campaign, written in collaboration with French participants); Erwin Bergot with Alain Gandy, "Opération Daguet," Presses de la Cité, 1991; Jane Kramer, "Letter from Europe," *The New Yorker*, March 18, 1991; "The Gulf Crisis," *France Magazine*, Spring 1991; Frederic Prater, "France's Role in the Gulf Crisis," *Foreign Policy*, Summer 1991.

383 the Legion brought to the desert a reputation: Bryan Perrett, *Desert Warfare*, pp. 165–173.

385 Wave upon wave of helicopters swept into Cobra: "Flight of Eagles," *Army Times*, July 22, 1991, p. 8.

***Riyadh***

391 his elation tempered only by frustration: Schwarzkopf, *Hero*, p. 452.

392 Allied forces driving across Europe: Larrabee, *Commander in Chief*, p. 473.

393 radio intercepts from the Iraqi 3rd and 4th corps: U.S. Army, *Sudden Victory*, p. VI–8.

their Challenger tanks would have to move . . . under their own power: U.S. Army, *Sudden Victory*, p. VI–8.

394 Iraqi forces had blown up the desalination plant: Schwarzkopf, *Hero*, pp. 453–454.

*page*

***Phase Line Colorado***

394 Six hundred thousand MLRS bomblets: U.S. Army, *Certain Victory*, pp. VI–10–11.

(Napoleon's famed gunners): John Keegan, *The Face of Battle*, p. 217.

the British cannonade at the Somme: Ibid., p. 235; J. B. A. Bailey, *Field Artillery and Firepower.*

397 No psyops surrender appeals: Barton Gellman, *Washington Post*, Sept. 13, 1991, p. A21.

the bloody fight for Tulagi: Larrabee, *Commander in Chief*, p. 268.

a United Nations conference on conventional weaponry: *Conduct*, Appendix O, p. 33.

***Washington (Saint John's)***

398 Saint John's had long offered refuge: Constance McLaughlin Green, *The Church on Lafayette Square, 1815–1970.*

"Ever since the war in the Persian Gulf began": transcript of Dr. John C. Harper's sermon on February 24, provided courtesy of Saint John's.

***Phase Line Dixie***

400 "Britain's last great cavalry charge": William Manchester, *The Last Lion: Winston Spencer Churchill, Visions of Glory*, p. 277.

the Royal Scots . . . vied with France's Picardy Regiment: Robert Fox, pool report, *Daily Telegraph*, Feb. 12, 1991.

402 Schwarzkopf thus far had refused to cut loose: Peter S. Kindsvatter, "VII Corps in the Gulf War: Ground Offensive," *Military Review*, Feb. 1992.

Franks radioed his commanders for advice: some details drawn from VII Corps documents, including "Combat Assessments, Orders, Results — Decision Points."

**15. On the Euphrates**

***Riyadh***

405 Yeosock fielded one irate call: Schwarzkopf's account of his conversations with Yeosock varies in minor details from the recollections of officers in both the CENTCOM and ARCENT war rooms; see Schwarzkopf, *Hero*, pp. 455, 461.

406 after a call from Colin Powell: Tom Donnelly, "The Generals' War," *Army Times*, March 2, 1992, p. 8; Schwarzkopf, *Hero*, p. 463.

Schwarzkopf would further suggest that VII Corps's dawdling: Schwarzkopf, *Hero*, p. 482.

McCaffrey's . . . 2nd Brigade was about to be stalled: *A History of the 24th Infantry Division Combat Team During Operation Desert Storm* (division chronicle), p. 23.

a major precept of American military thought since the 1840s: C. Kenneth Allard, *Command, Control and the Common Defense*, p. 49.

"There is no higher and simpler law": Ibid., p. 55.

***Al Khidr***

410 a sergeant emptied two full clips . . . and killed thirteen Iraqis: Sean D. Naylor, "Flight of Eagles," *Army Times*, July 22, 1991, p. 8.

*page*

***Haql Naft Al Burqan***

411 Task Force Papa Bear held the 1st Division's right flank: some details drawn from Papa Bear chronicle, dated March 18, 1991, and 1st Division documents.

412 One corporal courageously grabbed an AT-4: *Conduct,* p. 386.

***Al Khobar***

416 At 8:32 an early-warning satellite spotted: Department of the Army Review Team (DART), "Analysis of the 25 February 1991 Dhahran Scud Incident," memorandum for the record, June 14, 1991, and other Army investigation documents obtained under FOIA (hereafter DART memorandum). Also, U.S. General Accounting Office, "Patriot Missile Defense: Software Problems Led to System Failure at Dhahran, Saudi Arabia," Feb. 1992, GAO/IMTEC 92–26.

417 an infinitesimal "timing error" of one microsecond: "Special Report: Gulf Legacy," *IEEE Spectrum,* Sept. 1991, p. 52.
caused the Patriot radar to miscalculate: DART memorandum, p. 6.
The new software, Version 35: Ibid.

418 Saudi parents herded their children together: Richard H. P. Sia, *Baltimore Sun,* Feb. 26, 1991, p. 1.
Of the 245,000 reservists: *Conduct,* Appendix H, p. 1.
In civilian life the soldiers were clerks, secretaries, college students: *New York Times,* March 4, 1991, p. A12.

419 the warning . . . came less than thirty seconds: DART memorandum, p 2.
one soldier . . . scurried away after spotting a mouse: Associated Press report, *Philadelphia Inquirer,* March 7, 1991, p. A11.
possibly the only Iraqi missile that failed to break apart: Eric Schmitt, *New York Times,* June 6, 1991, p. 9; also May 20, 1991, p. 6.
gouging . . . a crater four feet deep: DART memorandum.
Anthony Drees had been fumbling for his helmet: "Heroes of the Gulf War," *People,* Spring-Summer 1991; also Associated Press account, March 2, 1991, contained in Army records.
firemen and military policemen swarmed: Sue Ann Pressley, *Washington Post,* March 6, 1991, p. A21; Bob Drogin and Patt Morrison, *Los Angeles Times,* Feb. 26, 1991, p. 1; Donatella Lorch, *New York Times,* p. 18.

420 Marlene Wolverton had called the Greensburg armory: details drawn from Paul Ciotti, "When the Scud Hit Home," *Philadelphia Inquirer Magazine,* June 30, 1991, p. 19; also, interviews conducted by Lucy Shackelford with Lt. Mary Hughes, Sgt. Donald Ferra, and Sgt. Mary Rhoads.

***Phase Line Smash***

422 (the British in North Africa captured 130,000 Italians): Perrett, *Desert Warfare,* pp. 94, 104.

423 Another armored brigade, the 50th: U.S. Army, *Certain Victory,* p. VI–15.
the Tawalkana Division intended to form a blocking force: postwar memorandum from Col. L. Don Holder to Lt. Gen. Fred Franks, April 1, 1991.

page

424 Frag Plan 7 . . . was predicated on the Republican Guard's: Peter S. Kindsvatter, "VII Corps in the Gulf War: Ground Offensive," *Military Review,* Feb. 1992.

For Schwarzkopf . . . the enemy was all but beaten: *Hero,* p. 456.

425 these units were repositioning: U.S. Army, *Certain Victory,* p. VI–26.

## 16. Upcountry March

### *King Khalid Military City*

427 Waller began to protest: Schwarzkopf's account of Waller stalking from the war room differs significantly from Waller's recollection. The CINC portrays the DCINC as frustrated largely by the slow pace of VII Corps. See Schwarzkopf, *Hero,* p. 460.

### *Phase Line Bullet*

428 No sight is dearer to a soldier: Warden, *The Air Campaign,* p. 92.

429 Hornburg rousted his F-15E squadron commanders: James P. Coyne, *Airpower in the Gulf,* p. 169.

430 The 7th Cavalry had . . . been decimated: Neil Sheehan, *A Bright Shining Lie,* p. 579.

431 an M1A1 sabot round sliced into the Bradley: Some details of 7th Cavalry fratricide are drawn from Army investigative documents obtained under FOIA.

433 General Funk . . . ordered his division to halt until daylight: U.S. Army, *Certain Victory,* VI–41.

### *As Salman*

434 Half-eaten meals of rice and onions: pool reports filed by Joseph Albright of Cox Newspapers and Bob Davis of the *Wall Street Journal,* Feb. 27, 1991.

two soldiers were killed and several wounded: *Le Monde,* March 1, 1991.

436 loudly chanting a jody call: pool report, Carol Morello, *Philadelphia Inquirer,* and Laurence Jolidon, *USA Today,* Feb. 11, 1991.

### *Riyadh*

437 Perhaps the most celebrated retreat was that of Xenophon: Xenophon, *The Persian Expedition,* Rex Warner, trans., pp. 1–4.

("It is always good to let a broken army return home"): John Toland, *Adolf Hitler,* pp. 609–611.

Chinese soldiers poured into North Korea: Blair, *The Forgotten War,* pp. 496, 520; McCullough, *Truman,* p. 818.

"miles long and a hundred yards wide": Bruce Catton, *This Hallowed Ground,* pp. 57–58.

438 Tiger Brigade . . . had rolled over Mutlaa Ridge: *Conduct,* p. 397; J. Paul Scicchitano, "Eye of the Tiger," *Army Times,* June 10, 1991, p. 12.

(Among the U.S. jamming systems . . . was the so-called Sandcrab): U.S. Army, *Certain Victory,* IV–20.

440 The Defense Intelligence Agency now assessed: *Conduct,* p. 387.

The Egyptians had finally reached: Ibid., p. 395.

Franks called the war room: some details of conversation drawn from VII Corps and CENTCOM logs. Also, Schwarzkopf, *Hero,* pp. 463–464.

*page*

441 "The way home is through the Republican Guard": Tom Donnelly, "The Generals' War," *Army Times,* March 2, 1991, p. 8.

***73 Easting***

443 McMaster dropped down inside the hatch: some battle details drawn from "Eagle Troops Summary of Action, 26 February 1991"; U.S. Army, *Certain Victory,* I–1–4; Michael D. Krause, "The Battle of 73 Easting," U.S. Army Center of Military History and Defense Advanced Research Projects Agency, Aug. 1991; and particularly a reconstruction assembled by the Institute for Defense Analysis Simulation Center (special thanks to L. Neale Cosby and Robert C. Turrell).

447 two thousand howitzer rounds and a dozen MLRS rockets: U.S. Army, *Certain Victory,* V–33.

**17. Liberation**

***Washington***

450 (Franklin Roosevelt during World War II): Larrabee, *Commander in Chief,* p. 15.

Tuesday had been the biggest day of the air war: Office of the Chief of Naval Operations, "The United States Navy in Desert Shield/Desert Storm."

451 Zlatoper reminded his pilots: Rowan Scarborough, *Washington Times,* Feb. 26, 1991.

approximately fifteen hundred destroyed vehicles: William Branigin and William Claiborne, *Washington Post,* March 11, 1991, p. A14.

"no significant Iraqi movement to the north": Steve Coll and William Branigin, *Washington Post,* March 11, 1991, p. A1.

452 the law of war . . . permitted an attack: *Conduct,* Appendix O, p. 35; John H. Cushman, Jr., *New York Times,* Feb. 27, 1991, p. A21.

"lots of little Jeffersonian democrats": *Christian Science Monitor,* Sept. 11, 1991, p. 9.

a "strategy of annihilation": Russell F. Weigley, *The American Way of War,* pp. xxii, 418, 465.

453 Schwarzkopf's avowed intention was to "drive to the sea": Schwarzkopf, *Hero,* p. 469.

no pictures of the Highway of Death had yet appeared: U.S. Air Force, *GWAPS,* p. IX–17.

"an old woman trying to shoo her geese": Allard, *Command, Control and the Common Defense,* p. 58.

the Falaise pocket west of the Seine: Larrabee, *Commander in Chief,* p. 467; John Keegan, *Daily Telegraph,* Feb. 27, 1991.

"The will to conquer": Tuchman, *The Guns of August,* p. 49.

454 Sherman's description of Shiloh: James M. McPherson, *Battle Cry of Freedom,* p. 413.

***Jalibah Air Base***

the French had considered the Ardennes: Weigley, *The American Way of War,* pp. 350–1.

1st Brigade reached Battle Position 102: Jason K. Kamiya, "A History of the 24th Mechanized Infantry Division Combat Team During Operation Desert Storm," official history, p. 26.

*page*

455 Contingency Plan Ridgway: U.S. Army, *Certain Victory*, p. V–32.

456 Oil smoke cut visibility: *Conduct*, p. 401.

457 Blackhawk 214 . . . flew headlong: Rhonda Cornum, *She Went to War: The Rhonda Cornum Story*, pp. 9–10.

Two accompanying Apaches also were hit: These gunships evidently were struck by friendly fire. U.S. Army, *Certain Victory*, p. IV–29.

Peay . . . suspended further raids: Ibid., p. V–64.

***Kuwait City***

458 one enemy battalion encircled themselves: Bill Gannon, pool report, *Newark Star-Ledger*, Feb. 26, 1991.

459 he retained his conviction: Schwarzkopf, *Hero*, p. 355.

Butch Neal called Boomer's headquarters: Molly Moore, *A Woman at War*, p. 291.

the Kuwaiti crown prince . . . warned: Milton Viorst, "After the Liberation," *The New Yorker*, Sept. 30, 1991, p. 37.

460 Palestinians were beaten severely: Ibid.

The Iraqis had caused more than a thousand Kuwaiti civilian deaths: John Lancaster, *Washington Post*, March 20, 1993, p. A18; Michael Gordon, *New York Times*, March 20, 1993, p. 3.

The once pristine capital had been sacked: details of looting drawn from United Nations Department of Public Information, *Kuwait: Report to the Secretary-General*, Sept. 1991; Milton Viorst, "After the Liberation," *The New Yorker*, Sept. 30, 1991, p. 37; John Simpson, *From the House of War*, pp. 167–168; *Times* of London, June 28, 1991.

Most of Kuwait's 1330 oil wells: T. M. Hawley, *Against the Fires of Hell*, p. 96.

461 Boomer led a small Marine convoy: George Rodrigue, *Dallas Morning News*, Molly Moore, *Washington Post*, Denis Gray, Associated Press, all pool reports, Feb. 27, 1991.

***Objective Norfolk***

462 Army pilots had been prohibited: U.S. Army, *Certain Victory*, pp. V–51, 68–69, VII–15.

confused and exhausted 1st Division troops: U.S. Army investigative records obtained under FOIA; Steve Vogel, "Hell Night," *Army Times*, Oct. 7, 1991, p. 8.

463 Rhame radioed Franks late Wednesday morning: memo from Rhame to Franks, April 5, 1991, VII Corps archives.

464 Smith had estimated that the fighting: de la Billière, *Storm Command*, p. 297.

two gunners from the Queen's Royal Irish Hussars: Ministry of Defence report on friendly fire, London, July 1991; eyewitness accounts.

Two A-10s flying over Objective Steel: U.S. Department of Defense incident report sent to coroner of Oxfordshire, 1992.

465 Twenty-three American soldiers were wounded by Adnan artillery fire: U.S. Army, *Certain Victory*, p. V–39.

466 the Madinah commander had hoped to ambush: U.S. Army, *Certain Victory*, pp. 54–57; Steve Vogel, "Metal Rain," *Army Times*, Sept. 16, 1991, p. 8.

*page*

468 von Schlieffen, whose scheme for defeating the French: Tuchman, *The Guns of August,* p. 41.
*Kesselschlachten:* Addington, *The Patterns of War Since the Eighteenth Century,* p. 184.

## 18. Closing the Gates

***Washington***

469 Bush gathered his senior advisers: some details drawn from *Newsweek,* March 11, 1991, p. 28; "The Day We Stopped the War," *Newsweek,* Jan. 20, 1992, pp. 16–25; Ann Devroy and R. Jeffrey Smith, *Washington Post,* March 28, 1991, p. 1; Elizabeth Drew, *The New Yorker,* May 6, 1991, p. 101.

470 a substantial gap had developed between Franks and Luck: U.S. Army, *Certain Victory,* pp. V–68–70.
More than twenty bridges and causeways: Ibid.

471 Powell . . . quickly apprised Schwarzkopf: CINC's account of phone calls, Schwarzkopf, *Hero,* pp. 469–470.

***Riyadh***

473 a C-141 cargo jet landed at Taif: some details drawn from interviews with pilots and aircrews; also, John Morrocco and David A. Fulghum, *Aviation Week,* May 6, 1991; *Air Force Times,* June 3, 1991; *Aerospace Daily,* May 14, 1991, p. 253; Gregg Jones, *Dallas Morning News,* July 6, 1991, p. 1.

***Washington***

474 Charles Powell . . . voiced misgivings: interview with BBC, quoted in Sunday *Times* of London, Aug. 18, 1991; *New York Times,* Aug. 20, 1991, p. A7.

475 McPeak harbored deep — if unspoken — misgivings: McPeak first hinted at his doubts in a briefing at the Pentagon on March 15, 1991: "I personally wasn't so sure that we were making the right move when our ground forces . . . stopped and offered a really merciful clemency."

476 two thirds of Americans polled said that the war had ended too soon: Peter Applebome, *New York Times,* Jan. 16, 1992, p. 1.
Defense Intelligence Agency later concluded that seventy to eighty thousand Iraqi troops had fled into Basrah: *Conduct,* p. 399.
"as many as one third of the [Republican] Guard's T-72s made it out of the KTO": U.S. Army, *Certain Victory,* p. V–68.

477 nearly 90 percent of Iraqi artillery: Michael R. Gordon and Eric Schmitt, *New York Times,* March 25, 1991, p. 1.
Saddam's armed forces had reached only 40 percent: Barton Gellman, "Buildup Reported Slow in Iraq," *Washington Post,* Aug. 6, 1992.
ten to twelve thousand Iraqis were killed: U.S. Air Force, *GWAPS,* p. IX–16.
Michael Herr once observed: Michael Herr, *Dispatches,* p. 31.

***Phase Line Victory***

478 Franks radioed the Big Red One: memo from Rhame to Franks, April 5, 1991, VII Corps archives.
friendly fire had dampened the Army's audacity: U.S. Army, *Certain Victory,* p. V–62.

*page*

480 at 7:20 A.M. an Army artillery unit issued a frantic distress call: The Army's postwar history puts the time of this event at 6:45 A.M., but VII Corps documents suggest it was thirty-five minutes later.

***Rumaylah Oil Field***

481 U.S. cavalrymen patrolling southwest of Basrah: The author was with 2nd Squadron of the 4th Cavalry Regiment at the defense minister's villa.

482 The Americans riddled the bus with gunfire: Maj. Jason Kamia, 24th Infantry Division afteraction report.

***Baghdad***

486 He was surprised to see two women: The female prisoners were Maj. Rhonda Cornum, shot down in Blackhawk 214 on Feb. 27, and Spec. Melissa A. Coleman, captured during the battle of Khafji.

## Epilogue

***Washington***

488 Grant's Union Army had tramped: William S. McFeely, *Grant,* p. 230.

489 Bush reluctantly agreed to send ground troops: Michael Duffy and Dan Goodgame, *Marching in Place,* p. 168.

491 of the 390 American men and women who died: *Conduct,* Appendix A.

492 Kuwait began rebuilding: Milton Viorst, "After the Liberation," *The New Yorker,* Sept. 30, 1991, p. 37.

The last of several hundred blazing oil wells: Matthew L. Wald, *New York Times,* Nov. 7, 1991, p. A3.

493 3 percent of the country's total reserves: Hawley, *Against the Fires of Hell,* pp. 39, 96.

The financial cost of the war to Arab countries: Youssef M. Ibrahim, *New York Times,* Sept. 8, 1992, p. A8.

"we've licked the Vietnam syndrome": quoted by Harry Summers, Jr., *On Strategy II: A Critical Analysis of the Gulf War,* p. 19.

"It is not big armies that win battles": Hallion, *Storm Over Iraq,* p. 83.

(Nearly 90 percent of the cost): *Conduct,* Appendix P.

494 "bore no significant military results": U.S. Air Force, *GWAPS,* p. III–21.

495 As Bush himself would warn: Michael Wines, *New York Times,* Jan. 6, 1993, p. 1.

James Fallows once wrote: James Fallows, *National Defense,* p. 108.

Schwarzkopf claimed that Iraq's forty-two divisions: testimony before House Armed Services Committee, June 12, 1991.

half of the Republican Guard tanks and other heavy equipment escaped: U.S. Air Force, *GWAPS,* p. III–43.

496 More than a hundred Scud missiles survived: *Conduct,* p. 208; Rolf Ekeus, *Washington Post,* Aug. 6, 1992, op-ed article.

Seventy tons of nerve agents and four hundred tons of mustard gas: Associated Press report, *Washington Post,* Sept. 13, 1992, p. A34.

Al Athir . . . was struck in the last F-117 sortie: *Conduct,* p. 128.

Air strikes had only "inconvenienced": cited in U.S. Air Force, *GWAPS,* p. III–26.

Iraq's bomb-building effort to be "at zero": Reuters, *Washington Post,* Sept. 3, 1992, p. A39.

*page*

496 optimism may have been premature: Diana Edensword and Gary Hilhollin, *New York Times,* "Iraq's Bomb — an Update," April 26, 1993, p. 17.
three quarters of all Americans polled believed: *New York Times,* June 11, 1991, p. 1.

497 a quality Clausewitz held among the highest attributes: Michael Howard, *Clausewitz,* p. 27.
"Nothing of this moral importance": Smith, *George Bush's War,* p. 238.
the CIA concluded that Saddam was tightening his grip: Caryle Murphy and R. Jeffrey Smith, *Washington Post,* June 20, 1992, p. 1.
the reconstruction of Iraq: Trevor Rowe, *Washington Post,* Nov. 11, 1992, p. 1; Trevor Rowe, *Washington Post,* Nov. 8, 1992, p. A41; Marie Colvin, *Sunday Times of London,* Oct. 4, 1992, p. 16; Trevor Rowe, *Washington Post,* Nov. 19, 1992, p. A31.

498 Washington responded by selling $20 billion worth of arms: Paul Lewis, *New York Times,* July 31, 1992, p. A6.
The CIA also undertook a covert $40 million program: Paul Lewis, *New York Times,* Aug. 5, 1992, p. A8; Elaine Sciolino with Michael Wines, *New York Times,* June 27, 1992, p. 1.
U.S. and British warplanes attacked: Barton Gellman and Julia Preston, *Washington Post,* Jan. 15, 1993, p. 1; Ann Devroy and Julia Preston, *Washington Post,* Jan. 16, 1993, p. 1.
the Clinton administration affirmed: R. Jeffrey Smith and Julia Preston, *Washington Post,* March 27, 1993, p. 1.

499 Bush had always drawn a distinction: address to Joint Session of Congress, March 6, 1991.
Few showed remorse for the brutal sack: Youssef M. Ibrahim, *New York Times,* Aug. 2, 1992, p. 12.
Schoolchildren . . . again lined up before class: Michael Gregory, *New York Times,* Feb. 27, 1993, p. 4.
plot to assassinate Bush: Douglas Jehl, *New York Times,* May 20, 1993, p. 1.
the Inter-Allied Control Commission had roamed Germany: Kennett, *A History of Strategic Bombing,* p. 70.
peddling $63 billion worth of weaponry: Center for Defense Information, *The Defense Monitor,* vol. XXI, no. 5, 1992.
A nation with 5 percent of the world's population: Nick Kotz and Rick Young, *Washington Post,* Oct 19, 1992, p. A21.

# BIBLIOGRAPHY


Abramson, Rudy, and John Broder. "Four Star Power." *Los Angeles Times Magazine*, April 7, 1991.

Addington, Larry H. *The Patterns of War Since the Eighteenth Century.* Bloomington: Indiana University Press, 1984.

Allard, C. Kenneth. *Command, Control and the Common Defense.* New Haven: Yale University Press, 1990.

Arkin, William M., Damian Durrant, and Marianne Cherni. "Modern Warfare and the Environment: A Case Study of the Gulf War." Washington: Greenpeace, 1991.

Bailey, J. B. A. *Field Artillery and Firepower.* Oxford: Military Press, 1989.

Bamford, James. *The Puzzle Palace.* New York: Penguin Books, 1983.

Barnet, Richard J. "Reflections: The Disorders of Peace." *The New Yorker*, January 20, 1992.

Barnett, Correlli. *The Desert Generals.* Bloomington: Indiana University Press, 1960.

Bartlett, Merrill L., ed. *Assault From the Sea: Essays on the History of Amphibious Warfare.* Annapolis: Naval Institute Press, 1983.

Beach, Edward L. *The United States Navy.* New York: Henry Holt, 1986.

Bergot, Erwin, with Alain Gandy. *Opération Daguet.* Paris: Presses de la Cité, 1991.

Beschloss, Michael R., and Strobe Talbott. *At the Highest Levels: The Inside Story of the End of the Cold War.* Boston: Little, Brown, 1993.

Billière, General Sir Peter de la. *Storm Command: A Personal Account of the Gulf War.* London: HarperCollins, 1992.

Blair, Clay. *The Forgotten War.* New York: Times Books, 1987.

Blumenson, Martin. *Patton: The Man Behind the Legend, 1885–1945.* New York: William Morrow, 1985.

Boettcher, Thomas D. *First Call: The Making of the Modern U.S. Military, 1945–1953.* Boston: Little, Brown, 1992.

Bolger, Daniel P. *Americans at War: 1975–1986, An Era of Violent Peace.* Novato, California: Presidio, 1988.

Brownlee, Romie L., and William J. Mullen III. *Changing an Army: An Oral History of General William E. DePuy.* Carlisle, Pennsylvania: U.S. Military History Institute, 1979.

Bryan, C. D. B. *Friendly Fire.* New York: Bantam, 1976.

Catton, Bruce. *This Hallowed Ground.* New York: Pocket Books, 1955.

*Certain Victory: The American Army in the Gulf War.* U.S. Army Special Study Group. Washington, February 1993 (draft version).

Charlton, James, ed. *The Military Quotation Book.* New York: St. Martin's, 1990.

Cohen, Eliot A., and John Gooch. *Military Misfortunes: The Anatomy of Failure in War.* New York: Free Press, 1990.

*Conduct of the Persian Gulf Conflict: An Interim Report to Congress.* U.S. Department of Defense. Washington, July 1991.

*Conduct of the Persian Gulf War.* U.S. Department of Defense. Washington, April 1992.

Cornum, Rhonda, told to Peter Copeland. *She Went to War: The Rhonda Cornum Story.* Novato, California: Presidio, 1992.

Coyne, James P. *Airpower in the Gulf.* Washington: Air Force Association, 1992.

"Defense for a New Era: Lessons of the Persian Gulf War." House Armed Services Committee. Washington: U.S. Government Printing Office, 1992.

Dewar, Michael. *The Art of Deception in Warfare.* Devon (U.K.): David Charles, 1989.

Doughty, Robert A. *The Evolution of U.S. Army Tactical, 1946–1976.* Fort Leavenworth, Kansas: Combat Studies Institute, 1979.

Douhet, Giulio. *The Command of the Air.* Dino Ferrari, trans. Washington: Office of Air Force History, 1983.

Duffy, Michael, and Dan Goodgame. *Marching in Place: The Status Quo Presidency of George Bush.* New York: Simon & Schuster, 1992.

Dyer, Gwynne. *War.* New York: Crown, 1985.

Fallows, James. *National Defense.* New York: Random House, 1981.

Fialka, John. *Hotel Warriors.* Washington: Woodrow Wilson Center Press, 1991.

Friedman, Norman. *Desert Victory: The War for Kuwait.* Annapolis: Naval Institute Press, 1991.

Green, Constance McLaughlin. *The Church on Lafayette Square.* Washington: Potomac Books, 1970.

*Guerre Éclair dans le Golfe,* editions Jean-Claude Lattes et Addim. 1991 (official French account).

*Gulf War Air Power Survey,* U.S. Air Force summary report (draft), May 1993.

Halevi, Yossi Klein. "The Iceman." *The Jerusalem Report.* February 14, 1991.

Hallion, Richard P. *Storm Over Iraq: Air Power and the Gulf War.* Washington: Smithsonian Institution Press, 1992.

Harrison, G. B., ed. *A Book of English Poetry.* New York: Penguin, 1968.

Hartmann, Gregory K. *Weapons That Wait.* Annapolis: Naval Institute Press, 1991.

Hawley, T. M. *Against the Fires of Hell: The Environmental Disaster of the Gulf War.* New York: Harcourt Brace, 1992.

Heinl, Robert Debs, Jr. *Dictionary of Military and Naval Quotations.* Annapolis: Naval Institute Press, 1966.

Herr, Michael. *Dispatches.* New York: Avon, 1978.

Herzog, Chaim. *The Arab-Israeli Wars.* New York: Vintage Books, 1984.

Hiro, Dilip. *The Longest War: The Iran-Iraq Military Conflict.* New York: Routledge, 1991.

Hopple, Gerald W., and Bruce W. Watson, eds. *The Military Intelligence Community.* Boulder: Westview Press, 1986.

Howard, Michael. *Clausewitz.* Oxford: Oxford University Press, 1983.
Insight Team of the Sunday Times, *The Yom Kippur War.* London: André Deutsch Ltd., 1975.
Janowitz, Morris. *The Professional Soldier.* New York: Free Press, 1971.
Jones, R. V. *Reflections on Intelligence.* London: William Heinemann, 1989.
Keegan, John. *The Mask of Command.* New York: Viking, 1987.
———. *The Price of Admiralty.* New York: Viking Penguin, 1989.
Kelly, Michael. *Martyrs' Day.* New York: Random House, 1993.
Kelly, Orr. *Brave Men — Dark Waters: The Untold Story of the Navy Seals.* Novato, California: Presidio, 1992.
Kennett, Lee. *A History of Strategic Bombing.* New York: Scribner's, 1982.
*Kuwait: Report to the Secretary-General on the Scope and Nature of Damage Inflicted on the Kuwaiti Infrastructure During the Iraqi Occupation.* New York: United Nations Department of Public Information, September 1991.
Larrabee, Eric. *Commander in Chief: Franklin Delano Roosevelt, His Lieutenants, and Their War.* New York: Simon & Schuster, 1988.
Liddell Hart, B. H., ed. *The Rommel Papers.* New York: Harcourt Brace, 1953.
Macy, Melinda. *Destination Baghdad.* Las Vegas: M&M Graphics, 1991.
Massie, Robert K. *Dreadnought: Britain, Germany and the Coming of the Great War.* New York: Random House, 1991.
McCullough, David. *Truman.* New York: Simon & Schuster, 1992.
McDermott, Jeanne. *The Killing Winds: The Menace of Biological Warfare.* New York: Arbor House, 1987.
McFeely, William S. *Grant.* New York: W. W. Norton, 1981.
McPherson, James M. *Battle Cry of Freedom.* New York: Oxford University Press, 1988.
Means, Howard. *Colin Powell.* New York: Donald I. Fine, 1992.
Melia, Tamara Moser. *"Damn the Torpedoes": A Short History of U.S. Naval Mine Countermeasures, 1777–1991.* Washington: Naval Historical Center, 1991.
Miller, Judith, and Laurie Mylroie. *Saddam Hussein and the Crisis in the Gulf.* New York: Times Books, 1990.
Momyer, William W. *Airpower in Three Wars.* Washington: Department of the Air Force, 1978.
Moore, Molly. *A Woman at War: Storming Kuwait with the U.S. Marines.* New York: Scribner's, 1993.
Morris, Edmund. *The Rise of Theodore Roosevelt.* New York: Ballantine, 1979.
Morrocco, John. *Rain of Fire: Air War, 1969–1973.* Boston: Boston Publishing Co., 1985.
———. *Thunder From Above: Air War, 1941–1968.* Boston: Boston Publishing Co., 1984.
Morse, Stan, ed. *Gulf Air War Debrief.* Westport, Connecticut: Airtime Publishing, 1991.
"Needless Deaths in the Gulf War." Middle East Watch Report. Human Rights Watch, 1991.

O'Ballance, Edgar. *No Victor, No Vanquished.* San Rafael, California: Presidio, 1978.
*Operational Experience of Fast Battleships: World War II, Korea, Vietnam.* Washington: Naval Historical Center, 1989.
"Patriot Missile Defense: Software Problem Led to System Failure at Dhahran, Saudi Arabia," U.S. General Accounting Office, February 1992.
Perrett, Bryan. *Desert Warfare.* Wellingsborough: Patrick Stephens, 1988.
Postol, Theodore A. "Lessons of the Gulf War Experience With Patriot." *International Security,* Winter 1991–2.
*Preliminary Lessons of Operation Granby.* House of Commons Defence Committee. London, July 1991.
Primakov, Yevgeni. "The Inside Story of Moscow's Quest for a Deal." *Time,* April 4, 1991.
———. "My Final Visit with Saddam Hussein." *Time,* April 11, 1991.
"Public Health in Iraq After the Gulf War." Harvard Study Team, May 1991.
Richelson, Jeffrey T. *The U.S. Intelligence Community.* New York: Ballinger, 1989.
Ridgway, Matthew B. *The Korean War.* New York: Doubleday, 1967.
Romjue, John L. *From Active Defense to AirLand Battle: The Development of Army Doctrine, 1973–1982.* Fort Monroe, Virginia: U.S. Army Training and Doctrine Command, 1984.
Ryan, Cornelius. *The Longest Day.* New York: Simon & Schuster, 1959.
Schwarzkopf, H. Norman, with Peter Petre. *It Doesn't Take a Hero.* New York: Bantam, 1992.
Sheehan, Neil. *A Bright Shining Lie.* New York: Random House, 1988.
Shrader, Charles R. *Amicicide: The Problem of Friendly Fire in Modern War.* Fort Leavenworth, Kansas: Combat Studies Institute, 1982.
Simon, Bob. *Forty Days.* New York: Putnam, 1992.
Simpson, John. *From the House of War.* London: Arrow Books, 1991.
Slim, William Joseph. *Defeat Into Victory.* London: Macmillan, 1956.
Smith, Jean Edward. *George Bush's War.* New York: Henry Holt, 1992.
Smith, Perry M. *How CNN Fought the War: A View from the Inside.* New York: Birch Lane Press, 1991.
Snyder, Louis L. *Historical Guide to World War II.* Westport, Connecticut: Greenwood Press, 1982.
Stein, Janice Gross. "Deterrence and Compellence in the Gulf." *International Security,* Fall 1992.
Stein, Robert M. "Patriot Experience in the Gulf." *International Security,* Summer 1992.
Stewart, John F., Jr. "Operation Desert Storm: The Military Intelligence Story." 3rd U.S. Army document, April 1992.
Stokesbury, James L. *A Short History of Air Power.* New York: William Morrow, 1986.
Summers, Harry G., Jr. *On Strategy II: A Critical Analysis of the Gulf War.* New York: Dell, 1992.
Toland, John. *Adolf Hitler.* New York: Ballantine, 1976.
Tuchman, Barbara. *The Guns of August.* New York: Dell, 1962.

"The United States Navy in Desert Shield/Desert Storm." Washington: Department of the Navy, May 1991.

U.S. News & World Report. *Triumph Without Victory: The Unreported History of the Persian Gulf War.* Times Books, New York, 1992.

Viorst, Milton. "After the Liberation." *The New Yorker,* September 30, 1991.

Waas, Murray, and Craig Unger. "In the Loop: Bush's Secret Mission." *The New Yorker,* November 2, 1992.

Warden, John A., III. *The Air Campaign.* Washington: National Defense University Press, 1988.

Watson, Peter. *War on the Mind.* New York: Basic Books, 1977.

Weigley, Russell F. *The American Way of War.* New York: Macmillan, 1973.

Whiteside, Thomas. "The Yellow Rain Complex." *The New Yorker,* February 11, 1991.

Witherow, John, and Aidan Sullivan. *The Sunday Times War in the Gulf.* New York: St. Martin's, 1991.

Woodward, Bob. *The Commanders.* New York: Simon & Schuster, 1991.

Xenophon. *The Persian Expedition.* Rex Warner, trans. New York: Penguin, 1981.

# INDEX